Magna Verada

Magna
TE, TF, TH, TJ, TL and TW Series

Verada
KE, KF, KH, KJ, KL and KW Series

1996 – 2005

Service and Repair Manual No. 272

Gregory's

Published by Haynes Manuals, Inc.
ABN 74 079 052 485

Trademark pending, used under license from Universal Press Pty Ltd
ABN 23 000 087 132 to Haynes Manuals, Inc. ABN 74 079 052 485

Haynes

Haynes Manuals, Inc
ABN 74 079 052 485
63 Fairford Road
Padstow, NSW 2211 Australia

Haynes Publishing Group
Sparkford Nr Yeovil
Somerset BA22 7JJ England

Haynes North America, Inc
861 Lawrence Drive
Newbury Park
California 91320 USA

ABCDE
FGHIJ
KLMNO
P

(272-10P2-3)

GREGORY'S AUTOMOTIVE PUBLICATIONS
Service and Repair Manual No. 272

Published by
Haynes Manuals, Inc.
ABN 74 079 052 485

Printed in the U.S.A.

ISBN-13: 978-1-56392-939-7
ISBN-10: 1-56392-939-2

11-352

Introduction

This Gregory's Service and Repair Manual covers the Australian built V6 Mitsubishi Magna TE, TF, TH, TJ, TL and TW series released in April 1996 in sedan and wagon variants, and Verada KE, KF, KH, KJ, KL and KW series released in September 1996 in sedan only.

The V6 fuel injected engine features single overhead camshafts operating four valves per cylinder on each bank. The camshafts operate the valves via rocker arms equipped with hydraulic tappets and rollers on the camshaft end to reduce friction. Two versions of the engine are available with capacities of 3.0 and 3.5 litres. The camshafts are driven by a toothed rubber timing belt that must be renewed at 100 000 km intervals.

The power of the engine is transmitted to the front wheels via a five speed manual or a four or five speed, electronically controlled automatic transmission. On Magna Sports models a tiptronic sequential shift style automatic transaxle is available.

Rack and pinion power assisted steering is standard on all models.

The independent front suspension comprises MacPherson strut assemblies with coil springs, A type control arms and a stabiliser bar.

On sedan models, the independent rear suspension is equipped with multi links, upper and lower control arms, coil spring struts and on Sports models, a stabiliser bar.

On station wagon models, the rear suspension comprises a tubular rear axle assembly supported by coil springs and located by four trailing arms and a panhard rod.

The dual circuit braking system has ventilated disc brakes installed to the front and solid disc brakes incorporating the handbrake drum fitted to the rear.

An antilock braking system (ABS) is available on some models. The system electronically controls the brake fluid pressure to prevent wheel lock, improving directional and steering stability.

On some models electronic driver and passenger airbags or supplemental restraint system (SRS) is installed within the steering wheel horn pad and the dashboard crash pad. If the vehicle is involved in a heavy impact collision, the airbag deploys to form a cushion to the driver/passenger. Refer to Body (chapter 17) for safety precautions when working on an airbag equipped vehicle.

Chapter 1
General Information

Contents

Specifications

Length:
- Magna —
  - Sedan ... 4 787 mm
  - Station wagon ... 4 812 mm
- Verada —
  - Sedan ... 4 929 mm
  - Station wagon ... 4 878 mm

Width ... 1 785 mm

Wheelbase:
- Sedan ... 2 720 mm
- Station wagon ... 2 722 mm

Front track ... 1 545 mm

Rear track:
- Sedan ... 1 536 mm
- Station wagon ... 1 500 mm

Overall height:
- Sedan ... 1 435 mm
- Station wagon ... 1 482 mm

Turning circle ... 11.2 m

Fuel tank capacity:
- Sedan ... 71 litres
- Wagon ... 68 litres

Towing capacity:
- Without trailer brakes ... 500 kg
- With trailer brakes ... 1 500 kg
- Maximum tow ball vertical load —
  - Without a load distributing hitch ... 75 kg
  - With a load distributing hitch ... 150 kg

* Fuel consumption:

	L/100 km (City)	L/100 km (Highway)
Magna man transaxle	10.0	6.6
Magna auto transaxle	10.5	6.6
Verada auto transaxle	12.0	7.4

* The fuel consumption information is based on tests made according to Australian Standard 2877. The actual fuel consumption obtained will depend on many factors including prevailing conditions, driving habits and the vehicle's equipment and condition.

About this Manual

The purpose of this manual is to help you learn more about the working components of your vehicle while saving money on its upkeep and operation. It will guide you in knowing what work must be done to keep the vehicle running as it should.

To produce this manual, a vehicle of the type covered was taken into our workshop, where it was progressively dismantled, pictures taken and procedures checked in practice.

The manual is designed primarily for the well equipped home mechanic but the comprehensive specifications, detailed wiring diagrams and many procedures would also be valuable for the professional motor mechanic.

The manual concentrates on routine maintenance to help prevent problems occurring and also gives comprehensive trouble shooting procedures with a logical course of action should random problems occur. Substantial savings in time and money can be made by performing your own service and repair work.

Most of the procedures in this manual are straightforward and can be performed by anyone reasonably handy with tools. However do not attempt a service or repair operation before completely reading the appropriate chapter (or other chapters that may be referred to) in this manual.

Reading the chapter will help you prepare a list of things that will be needed for the job, such as replacement seals, O rings or gaskets as well as any sealer or thread locking compound. Also, certain repairs in this manual may require the use of a special tool not normally found in the home tool kit. When special tools are required, the equipment and its function will be outlined at the start of the relevant procedure.

If when reading a procedure, doubts occur as to whether you have the knowledge or tools to complete the job, it may be better to refer the problem to a reliable mechanic or an authorised dealer. You can then use the knowledge gained from the manual to check that the job has been performed satisfactorily.

Some repairs, such as an automatic transmission overhaul, should be left to an authorised dealer or specialist who has the extensive knowledge and equipment required. In these cases, the removal and installation procedures are fully covered enabling the unit to be removed for repair, or a reconditioned unit to be installed.

Reference in the manual to the left and right hand sides of the vehicle are from the point of view of someone standing at the back of the vehicle and looking forward.

Unless otherwise stated, nuts and bolts are removed by turning anticlockwise, and tightened by turning clockwise.

USING THE MANUAL

This manual is divided into chapters, each covering particular systems and their related components. Each chapter is divided into main headings for easy reference, and has a contents listing at the start.

There are two ways to find a particular component:

1 Refer to the index at the back of the manual.

2 Refer to the relevant chapter and use the contents listing.

For servicing the vehicle, refer to the Maintenance Chart in Routine Maintenance (chapter 2) and then to the Routine Maintenance heading in the appropriate chapters.

Comprehensive trouble shooting procedures are included at the start of most chapters to help pinpoint problems and ensure that the vehicle is back in service as quickly as possible.

Illustrations (photographs and line drawings) are used throughout to help you accomplish the tasks described in the accompanying text. The illustrations also show the order of the components.

Icons are used to draw the readers attention to specific details in the text.

Information — additional information expanding on a procedure

Caution — warning of danger of injury to yourself or damage to the component

TIP *Tip — tips and tricks, alternative methods, things to look out for*

Special equipment — special tools required to perform a particular job

A conversion table is included at the end of the manual if converting metric to imperial or vice versa.

Tools and Equipment

To successfully perform any maintenance or repair work on a motor vehicle, suitable hand tools are essential. The use of tools for other than their intended purpose or the use of incorrectly fitting tools can cause damage to the component and/or injury to the operator.

BASIC TOOL KIT

It is recommended that every motor vehicle be permanently equipped with a basic tool kit. If travelling in isolated areas, it is recommended that the tool kit should be more comprehensive. Even if the driver does not have the skill to use the tool kit in the event of a breakdown, a friendly motorist who has the necessary skill may be prepared to help.

Following is a suggested list of basic tools to carry in the vehicle. If the kit is carefully packed, it would take up very little room.

Set of ring spanners
Set of open ended spanners
Set of sockets
Assorted bladed screwdrivers
Assorted pliers — combination, long nose, multigrips, vice grip
Assorted adjustable spanners
Cold chisel
Hammer
Scraper
Assorted files, flat, half moon
Feeler gauges
Hacksaw/blade
Pump or siphon
Trouble lamp or torch
Jumper leads with a surge protection device
An aerosol of WD40 or similar

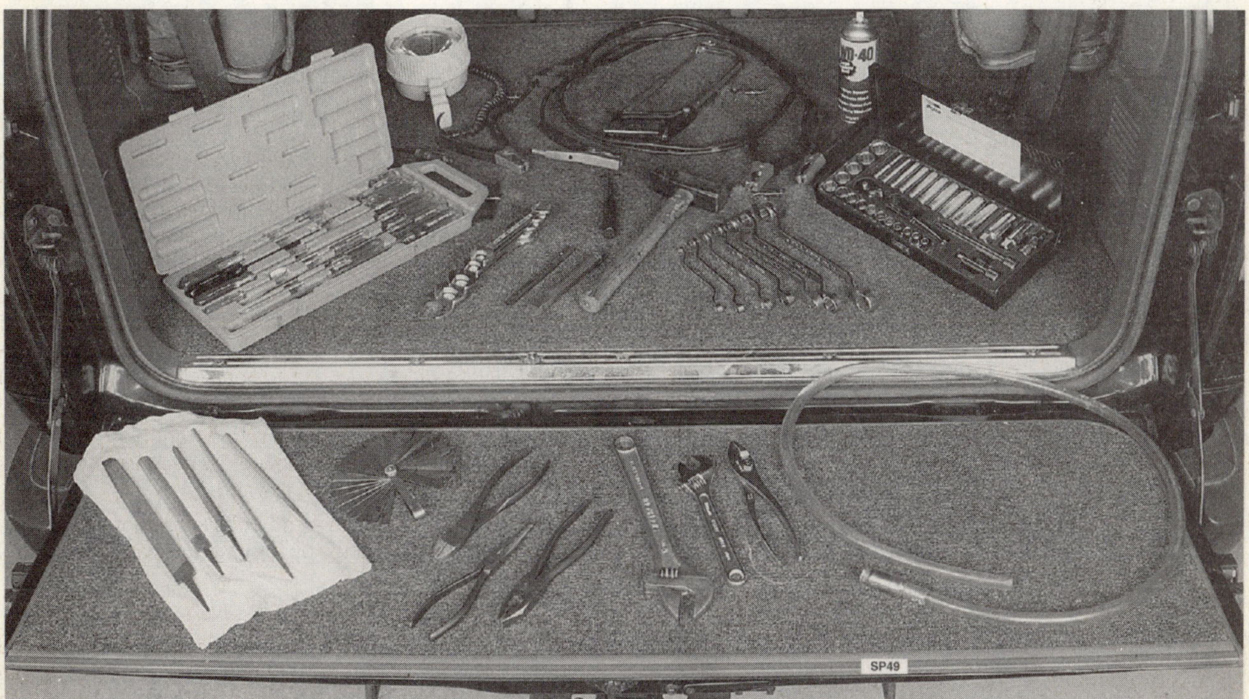

A good basic set of tools and equipment is handy for emergency repairs.

HOME TOOL KIT

Besides the basic tool kit, the following is a suggested list of tools and equipment that should be kept at home for the majority of the maintenance and repair procedures described in this manual. Of course, not all of the tools are required for all the jobs, so purchasing tools on an as needed basis is wise.

Snap ring pliers (internal and external)
Pin punches and centre punch
Torque wrench
Hydraulic jack
Chassis stands
Car ramps
Grease gun
Oil can
Oil gun
Oil filter removal tool
Funnel
Containers for draining oil and washing components
Inspection lamp
Test lamp
Tachometer
Digital multimeter
Piston ring compressor
Valve spring compressor
Micrometer — inside, outside, multi range
Vernier calipers
Assorted pullers
Magnet
Mirror
Safety goggles
Allen keys

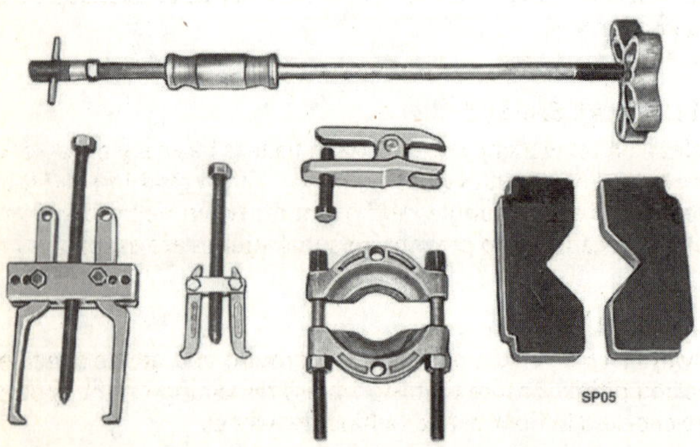

Slide hammer, ball joint puller, assorted pullers and press plates.

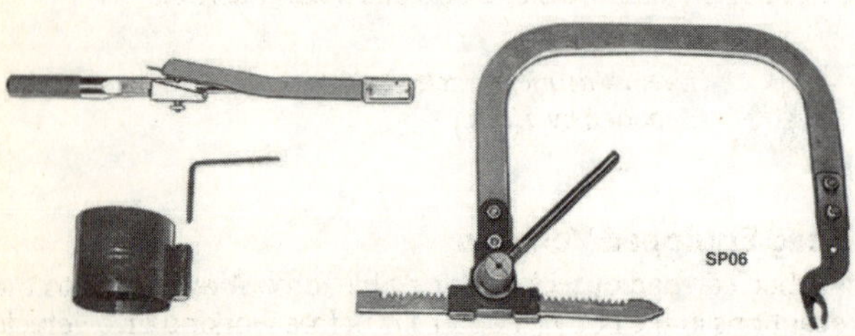

Torque wrench, piston ring compressor and valve spring compressor.

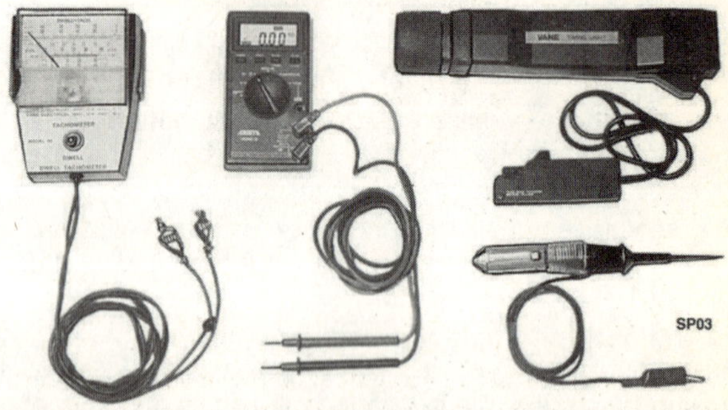

Assorted test equipment, tachometer, digital multimeter, inductive timing light and test lamp.

When purchasing tools, it is good policy to select the highest quality that can be afforded, as the working life of cheaper tools is often very short and their accuracy cannot be relied upon. Ensure that the tools are suitable for the system of bolt and nut sizing on the vehicle. The range of vehicles covered by this manual use metric sizes.

TOOL CARE

To ensure the longest possible life for hand tools, it is important that time be spent maintaining them.

At the conclusion of each job, all tools used should be washed thoroughly in kerosene or similar cleaning solvent. Ensure that all dirt and grease is removed, particularly from tools with moving parts such as pliers and adjustable spanners. The tools should then be wiped dry with a clean cloth.

Measuring devices should be given particular attention as their accuracy can be affected if not properly maintained. Feeler gauges should be kept clean at all times and the blades should be wiped with an oily cloth after use to prevent rusting.

If the tools are to be stored for any length of time, it is good policy to wipe them with an oily cloth.

Bladed screwdrivers should be checked for damage to the tip. If necessary, the tip can be returned to its original profile by careful grinding. Do not grind screwdriver tips to a sharp point.

Hammer heads should be secure on the handles and should be regularly checked for cracking or other damage.

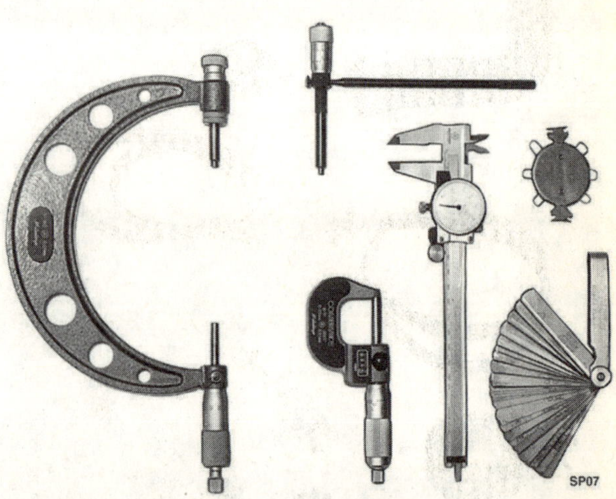

Assorted measuring devices – inside and outside micrometers, vernier calipers, wire gauges and feeler gauges.

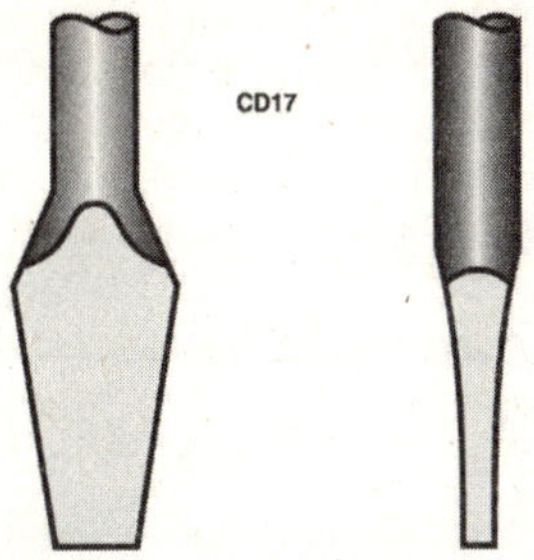

Views showing the correct tip profiles for a bladed screwdriver. Bladed screwdrivers should not be ground to a sharp point.

Chisels and punches should be checked for damage or "mushrooming" of the head. Any faults should be rectified by grinding.

Hydraulic jacks should be regularly checked for fluid leaks. Chassis stands and car ramps should be checked for damage and cracks. Any equipment that is suspect should not be used.

SPARES

It is recommended that every motor vehicle be equipped with some basic spare parts.

If travelling in isolated areas, it is recommended that the spares carried should be more comprehensive.

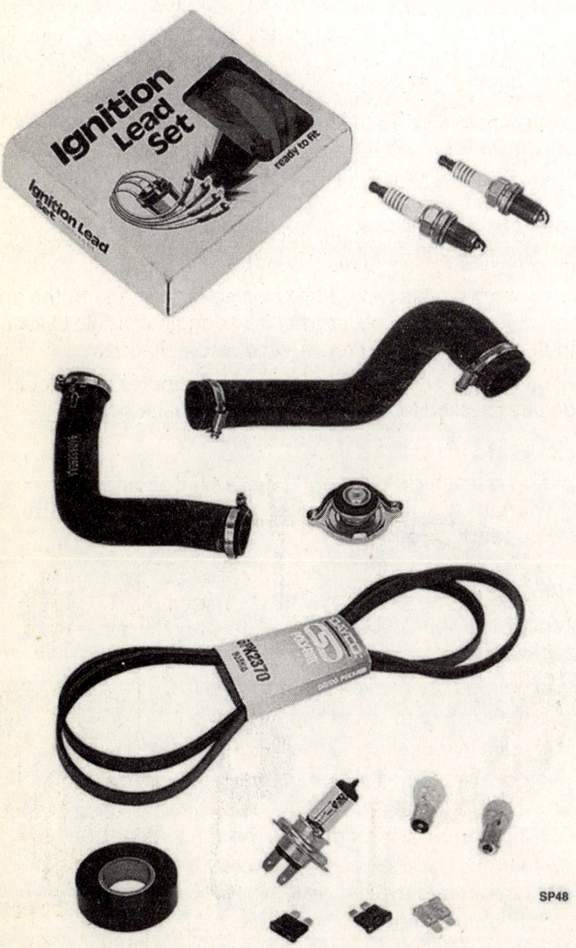

A selection of spares should be carried on long trips.

Set of radiator hoses
Radiator cap
Drive belts(s)
Bulbs
Fuses — different amperages
Roll of electrical insulation tape
A metre of 3 mm electrical wire
Assortment of screws, nuts, split pins etc.
Spark plugs
High tension leads
Distributor cap and rotor
12 volt air compressor (cigarette lighter attachment)

It is also a good idea to carry a sheet of plastic to lie on and a few pieces of cloth to wipe greasy hands on, and of course, a Gregory's manual.

STORES

For routine maintenance, stores of automotive oils, greases and additives should be kept on hand. The following is a suggested list.

Engine oil
Brake fluid
Manual or automatic transmission oil
Power steering fluid
Cooling system inhibitor or mixed coolant
Chassis grease
High melting point grease for hub bearings etc.
WD 40 or similar
Kerosene or similar cleaning solvent
Methylated spirits

Safety

PERSONAL SAFETY

Safety, when working on a motor vehicle is basically a matter of commonsense. Always wear the correct safety clothing and footwear. Wear safety goggles and ear muffs when necessary. Some safety precautions to prevent personal injuries are as follows.

Raising a Vehicle

Always jack a vehicle on firm, level ground and at the specified jacking points. Ensure that the wheels remaining on the ground are chocked in front of and behind the wheel.

After raising the vehicle, place chassis stands underneath and allow the weight of the vehicle to rest on them.

Do not use bricks, blocks of wood or similar material.

⚠ *Never work under a vehicle that is only supported by a jack.*

Airbag Equipped Vehicles

An airbag can deploy with considerable force. Always refer to the precautions in the Body (chapter 17) before working on a vehicle equipped with an airbag.

Electrical System

Always disconnect the negative battery terminal when working on any electrical components. Avoid wearing metal watches, rings and chains that may short across live terminals.

As battery gases are explosive, keep naked flames and sparks clear of the work area. When connecting and disconnecting jumper leads, use extreme caution to avoid sparking.

Do not reverse the connection of the battery terminal leads.

Work Area

Do not run the engine in a confined space. Ensure that the work area is adequately ventilated.

Spilt oil or water should be cleaned immediately to avoid the possibility of slipping.

Fuel System

Always disconnect the negative battery terminal when working on any fuel components. Do not smoke. Keep naked flames and sparks clear of the work area.

Do not siphon fuel using the mouth. Use a hand pump or suitable siphon.

Do not attempt to repair a fuel tank by welding it. This is an extremely hazardous procedure as a fuel tank contains explosive vapour long after fuel has been drained. Repairs should be entrusted to a specialist.

Cooling System

To avoid scalding, use caution when releasing the radiator cap on an engine that is at normal operating temperature. The best procedure is to allow the engine to cool for a period until the upper radiator hose can be easily squeezed together. If this is not possible, using a thick cloth turn the cap anticlockwise to the first stop. Allow any pressure in the system to release before removing the cap from the radiator.

Brakes

As asbestos and other hazardous substances may be used in brake lining material, avoid inhaling brake dust. Do not use compressed air to remove the dust. Gentle brushing with a small brush or using a vacuum cleaner with a hose attachment are the safest methods of cleaning the brakes.

The above precautions also apply to the clutch plate lining material.

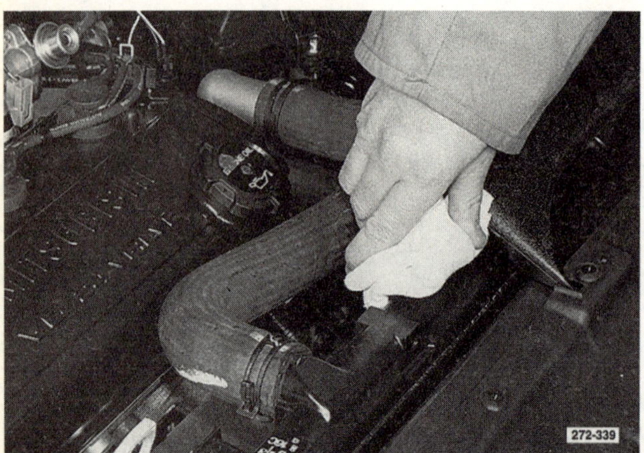

To avoid scalding when the engine is hot, cover the radiator cap with a thick cloth and turn the cap anticlockwise to the first stop.

Lifting Equipment

When using lifting equipment to lift heavy components such as the engine and/or transmission, use metal slings or chain in preference to rope. If rope must be used, ensure that it is not placed against sharp edges on the component.

Automotive Lubricants and Solvents

Avoid prolonged skin contact with oils, greases and solvents as some can cause skin irritations and dermatitis.

Exercise caution when using cleaning solvents as many are inflammable. Do not smoke. Keep naked flames and sparks clear of the work area.

Compressed Air

Never point an air hose at another person or allow compressed air to blow onto your skin. High pressure air forced against the skin can enter the bloodstream and prove fatal.

Suspension and Steering Components

Damaged suspension and steering components should not be welded. Many of these components are fabricated from toughened metals. If welded, they may lose their strength or become brittle. Damaged components should be renewed.

Air Conditioning

The air conditioning system contains refrigerant which is highly flammable and produces a poisonous gas when burnt. Escaping refrigerant can cause frostbite.

If disconnecting any of the refrigerant lines is necessary, the system must first be evacuated by an authorised workshop. Refrigerant must not be discharged to the atmosphere.

Welding

Extreme caution should be exercised when welding on the vehicle.

Consult the manufacturers instructions before using the welding plant and become familiar with its operation and associated precautions.

VEHICLE SAFETY

To prevent damage to the vehicle during servicing or repair work, note the following precautions.

Brake Fluid

If spilt on the vehicle paintwork, brake fluid should be immediately washed away with clean water and allowed to dry naturally, not wiped with a cloth.

Catalytic Converter

The following should be observed to prevent damage to the catalytic converter.

- Do not operate the vehicle on leaded fuel
- Do not push or tow start the vehicle
- Do not allow the engine to idle for prolonged periods
- Do not switch the ignition off while the vehicle is in motion and the transmission is in gear
- Do not prime the engine by pouring fuel into the inlet manifold
- Do not operate the vehicle if the engine is misfiring
- Avoid running the vehicle out of fuel
- Ensure that the engine oil is formulated to contain low phosphorus levels

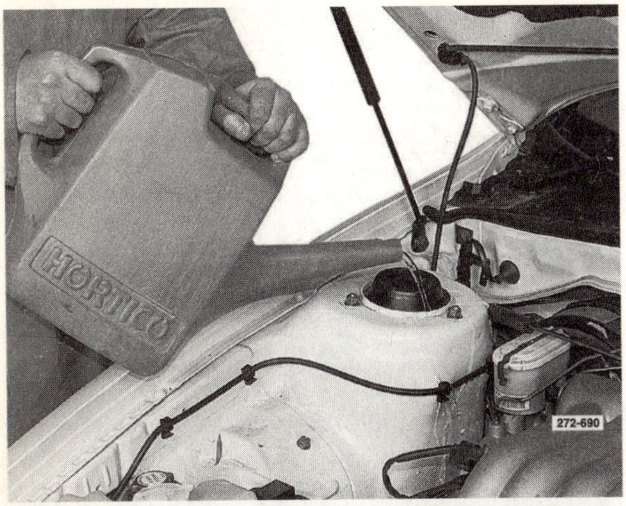

If brake fluid is spilt on the paintwork, immediately wash the area with plenty of clean water. Do not wipe the paintwork with a cloth. The area should be allowed to dry naturally.

Electronic Components

Electronic components can be damaged by the use of incorrect testing equipment.

It is essential in all tests where voltage or resistance is to be measured that a digital display multimeter with a minimum 10 megohm input impedance is used.

If using older style equipment, it is recommended that the manufacturer of the test equipment be consulted before using the equipment for any test procedures described in this manual.

Jump starting, or being jump started by another vehicle can cause damage to the electronic components of the vehicle. It is important that reference be made to Roadside Trouble Shooting (chapter 3) and the jump starting procedure described there is closely followed.

Welding

Before performing any electric welding operations on the vehicle, disconnect the battery leads, the alternator wiring and the wiring connectors from all electronic control units (ECUs) to prevent damage to the electronic circuitry.

When using oxy acetylene welding equipment use care not to damage fuel lines, brake lines, floor pans, carpets or insulation with heat or flame.

Always keep a fire extinguisher handy and keep the flame well away from the fuel tank and evaporative system.

Test Equipment and Procedures

When working on the electrical system, a test lamp and jumper leads can be very useful to check circuits.

TO MAKE A JUMPER LEAD

The minimum materials required to make a jumper lead are one length of 4 mm wire and two small alligator clips.

1. Bare both ends of a suitable length of 4 mm wire.
2. Connect a small alligator clip to each end of the wire. Solder and tape the connections.
3. Test the jumper lead for continuity by connecting one end of the jumper lead to the alligator clip on a test lamp and

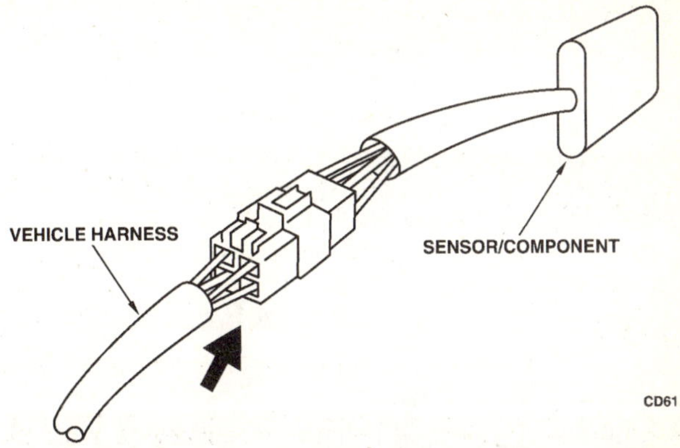

Large arrow indicates how insets used in this manual are viewed, unless otherwise stated.

the other end of the jumper lead to the negative battery terminal. Touch the probe of the test lamp to the positive battery terminal. The test lamp should operate indicating a completed circuit through the jumper lead.

TO MAKE A TEST LAMP

The electronic components on the range of vehicles covered by this manual can be damaged by excessive current draw. Excessive current draw can be caused by the use of conventional (filament type) test lamps. For this reason, the use of an LED (light emitting diode) test lamp is recommended.

A simple LED test lamp can be made from a suitable length of 3 mm wire, an alligator clip, an LED, a 1/4 watt 560 ohm resistor, a suitable case such as an old ball point pen case and a length of rod or wire sharpened to form a probe.

> *When using an LED test lamp, it should be remembered that, as the LED will only allow current flow in one direction, the lamp will only operate with the lead connected to negative and the probe to positive.*

1. Solder the length of wire to the K (cathode) lead of the LED. The K lead is the shorter of the two leads and is adjacent to the flat on the LED.
2. Solder the 560 ohm resistor to the A (anode) lead of the LED.
3. Solder a short length of wire to the other lead of the resistor and to the probe.

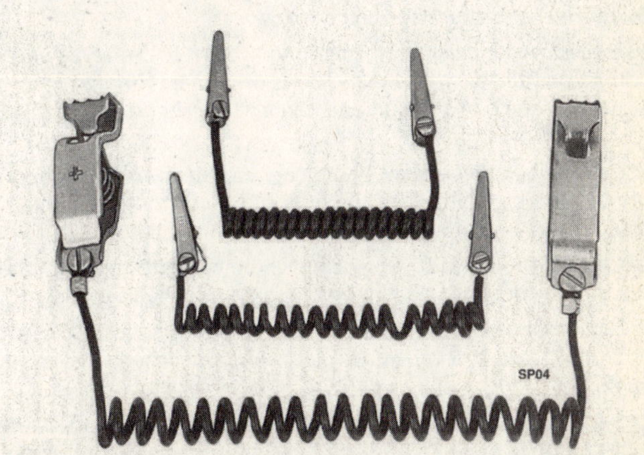

Selection of various sized jumper leads.

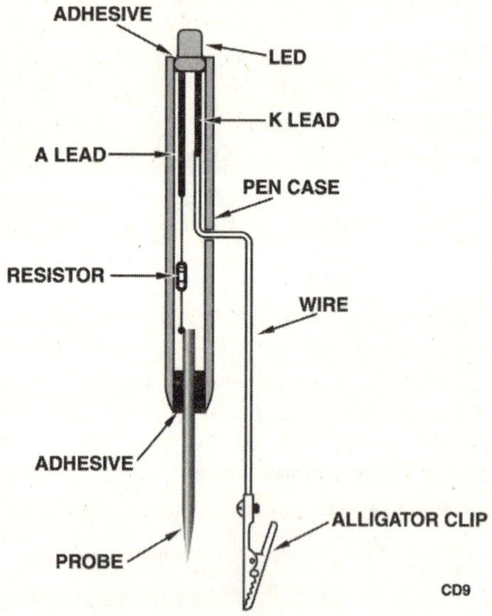

Schematic of an LED test lamp. Ensure that the connections to the K (cathode) lead and A (anode) lead of the LED are as illustrated.

4 Drill a hole in the side of the case to accept the wire attached to the K lead of the LED.
5 Mount the probe in the case. Push the probe through the lower end of the case after removing the refill.
6 Push the lead attached to the K lead of the LED through the hole made in the side of the case and solder the alligator clip to the lead.
7 Mount the LED to the case. The LED can be mounted in the plug at the top of the case after drilling the plug to accept the LED.
8 Suitably attach the probe to the case using adhesive or tape.

General Repair Procedures

MARKING OR NOTING POSITIONS OF COMPONENTS

Good motor vehicle repairers take particular care in the marking or noting of components as they are being removed from a vehicle or vehicle subassembly.

Besides ensuring correct positioning of these components, when assembling, it also makes for easier assembly. A few very simple methods that can be used are as follows:

Use a centre punch, scriber, quick drying paint or even typewriter correction fluid to mark mating parts or to mark the positioning of parts. Make one mark on one part and another mark adjacent to it on the other part.

Use containers (ice cream containers are ideal) for the storing of nuts and bolts etc. Keep the nuts and bolts from each unit or assembly in a separate container.

It is also a good idea to write from where the parts came on the side of the container, eg. timing cover, cylinder head, etc.

If dismantling something that looks to be complicated, use a notebook to draw or write down each component as it is removed.

Tagging is also very useful for parts identification or parts mating. One application is for the identification and mating of wiring, where tags can be made from masking tape and the applicable component or mating wire colour written on the masking tape with a pen. Another application is for the identification of vacuum hoses.

An impact driver is handy for loosening seized fasteners.

SEIZED FASTENERS

Seized bolts, nuts or screws should first have a liberal amount of penetrating oil applied. The fastener should be left for a period of time to allow the oil to penetrate and soften the corrosion that is causing the binding.

Often a sharp hammer blow to the head of the fastener can dislodge the corrosion and permit it to be loosened.

An impact driver, which can be fitted with a socket or screwdriver bit, can be used to loosen a seized fastener.

Another method is to heat the component in which the fastener is seized. However, extreme caution should be exercised when heating aluminium alloy components as the melting point is much lower than that of steel.

If the above methods fail to free a seized nut, carefully hacksaw through one side of the nut until it can be split. Care should be taken that the threads of the bolt or stud are not damaged.

Should a bolt or stud break below the surface of the component, it will be necessary to use a screw extractor to remove the remaining part. Follow the screw extractor manufacturers instructions.

Damaged threads can be repaired using a die nut on studs and bolts and a tap on nuts and threaded holes in castings. If the

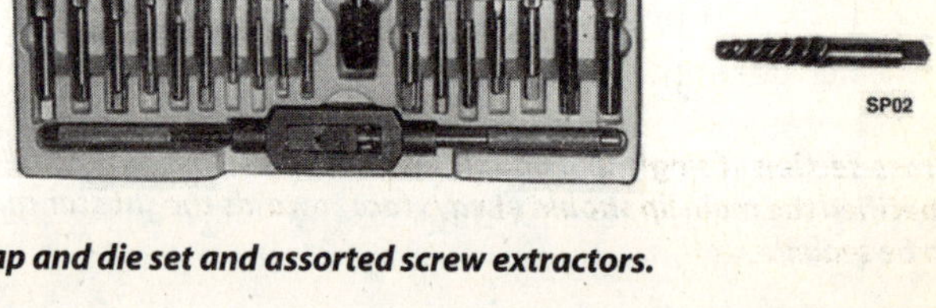

Tap and die set and assorted screw extractors.

threads of a threaded hole are damaged beyond repair, it will be necessary to drill and tap the hole to a larger size. Alternatively, a Helicoil insert can be used to restore the hole to the original thread size.

STUDS

The simplest method for removing studs is to lock two nuts together on the threaded section. The stud should then be able to be removed by applying an unscrewing action to the lower nut.

Alternatively, there are various makes of stud extracting tools available.

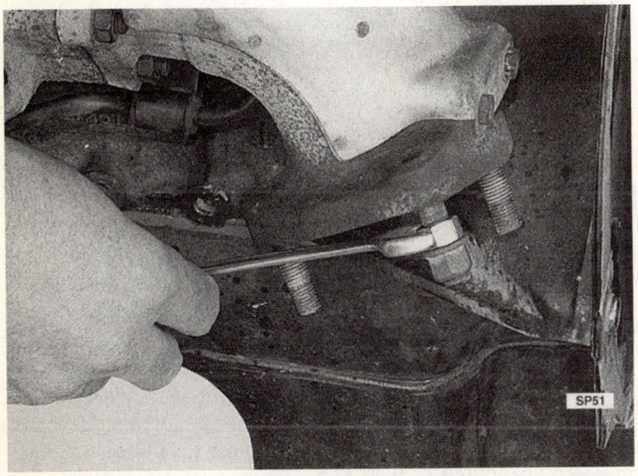

Using two nuts locked together to remove a stud.

OIL SEALS

Oil seals can usually be removed by levering out with a flat screwdriver or other suitable lever. Care should be taken not damage the surface of the component which the seal lip runs on.

Seals can also be removed by inserting a number of self tapping screws into the seal body. The seal can then be withdrawn using pliers gripping the self tapping screws.

Always apply a smear of grease or oil to the seal lip prior to installation to provide initial lubrication.

Unless otherwise stated, oil seals should always be installed with the lip facing inwards or towards the substance to be sealed. During installation, the seal lip should be protected from damage from sharp components such as shaft splines by wrapping tape around the sharp edges.

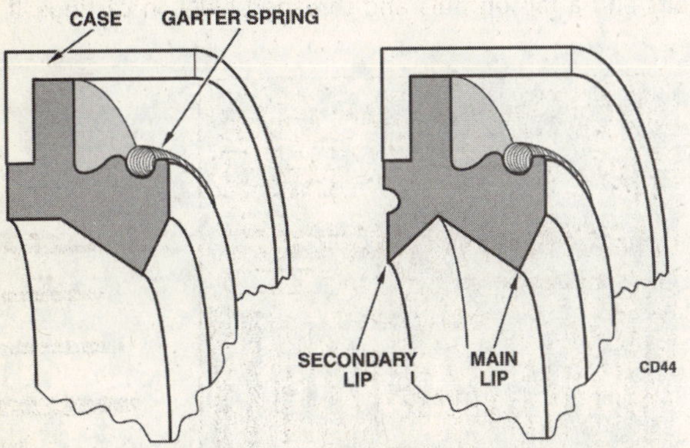

Cross-section of single and double lip oil seals. Unless otherwise specified the main lip should always face towards the substance to be sealed.

Install the new seal using a wooden block or a socket or length of tube of the appropriate diameter. Ensure that the seal is installed squarely or distortion and subsequent leakage may occur. If an installation depth is not specified, the seal should be installed flush with the component surface.

GASKETS

When separating mating components (i.e. a cylinder head and cylinder block), do not insert screwdrivers or similar levers between the components in an attempt to lever them apart. This can cause severe damage to the sealing surfaces, particularly if the components are made of alloy compounds.

The components can be separated by tapping along the joint with a soft faced hammer or piece of wood. Before installing a new gasket, the mating surfaces should be cleaned of all traces of old gasket material and sealant.

Check that the new gasket is correct by comparing the bolt holes and passages on the component face with the openings in the gasket. Cork and paper gaskets stored for some time may suffer from shrinkage. This can be rectified by soaking the gasket in water.

BEARINGS AND BUSHES

If the correct equipment is not available when removing and installing bearings and bushes, it is often possible to improvise.

Bearings can often be removed from shafts by tapping alternately on opposite sides with a hammer and drift.

If a press is unavailable, bushes can be installed by placing the bush and component between the jaws of a vice and screwing the jaws together until the bush is fully inserted. A vice can also be used to remove bushes by using suitably sized spacers against either vice jaw, one bearing on the bush and the other on the component. This method can also be used with a G clamp.

A simple removing and installing tool can be made using a long bolt, large and small washers, a nut and a tubular spacer.

Refer to the illustrations for the method and applications.

Rubber bushes and bushes in blind holes can be removed using an expanding type masonry bolt (Loxin, Rawlbolt). Install a neat fitting masonry bolt to the bush. Install and tighten the bolt until it grips the bush. The bolt and bush can then be removed using a slide hammer or levers.

Bushes in blind holes can also be removed using hydraulic force. With the hole packed with grease, insert a neat fitting drift and tap the drift into the hole. The pressure of the grease on the rear of the bush should force it from the hole.

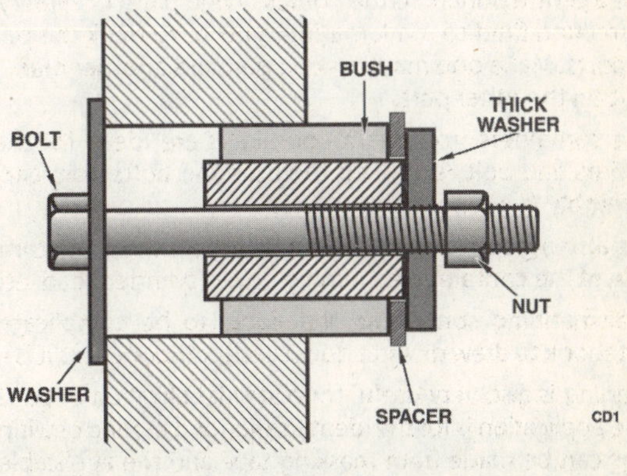

A simple bush installing tool using a bolt, nut and two washers.

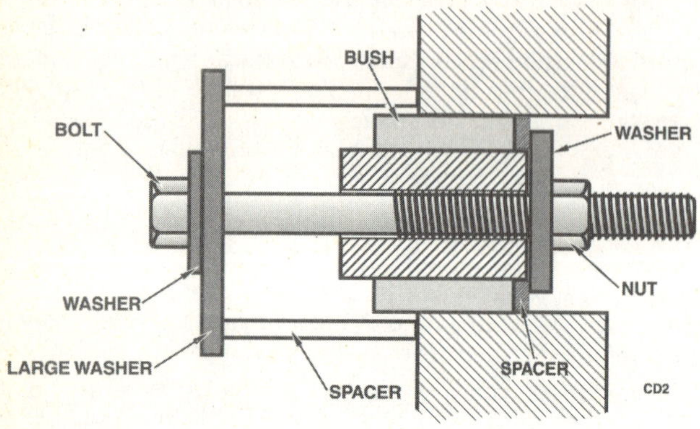

A simple bush removing tool using a bolt, nut, suitable spacer and large and small washers.

ROTATING THE CRANKSHAFT

There are many occasions when rotating the crankshaft to a certain position is necessary. These include timing and valve adjustment and checking valve timing.

To make the engine easier to rotate, remove the spark plugs. This will relieve compression pressure.

The simplest method is to use a socket or ring spanner on the crankshaft pulley nut and turn the engine in the normal direction of rotation.

On manual transaxle models, the engine can be turned by selecting top gear and rolling the vehicle forward. A variation of this method is to raise one of the driving wheels and, with fourth gear selected, rotate the driving wheel in a forward direction to turn the engine.

Alternatively, the engine can be brought to the desired position by briefly turning the ignition switch to the Start position and actuating the starter motor.

Vehicle Identification

When purchasing spare parts or when registering or insuring a vehicle, quoting various vehicle identification codes may be necessary. The locations of these codes are as follows:

The engine number is stamped approximately half way up on the front of the cylinder block adjacent to the transaxle mounting.

The vehicle identifications number (VIN) is stamped on firewall towards the drivers side of centre.

The compliance plate contains information on the approval number, seating capacity, VIN and date of manufacture. The plate is mounted close to the centre of the firewall.

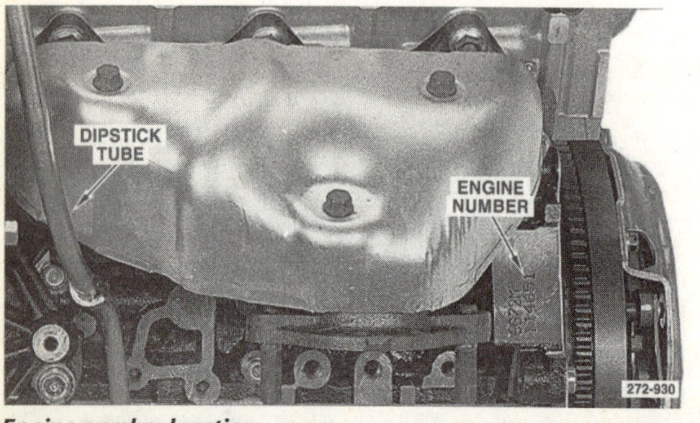

Engine number location.

The vehicle data plate is mounted beside the compliance plate. The plate contains codes relating to body style, engine, type of transaxle, paint and trim colours and the build date.

The tyre placard, which is located on the drivers door B pillar, contains information on tyre size, rim size and tyre pressures.

Buying Spare Parts

Spare parts are available from many sources, including authorised dealers, parts/accessory shops, wreckers and exchange component specialists. To be sure of obtaining the correct parts, it is important to note the engine number, chassis number, vehicle type and the build date. If possible, take the old part for positive identification.

Before commencing work on the vehicle, checking the availability and price of the parts that may be required to repair the problem is advisable. Checking the availability of these parts will help to determine the length of time the vehicle will be off the road and whether some parts should be ordered in advance. Ten minutes on the telephone comparing prices and working out the costing of a particular job may save many dollars.

Authorised dealers are usually the best source of parts peculiar to just one model of vehicle. Genuine parts purchased from a dealer are covered by a warranty and comply with ADR regulations. Genuine parts nearly must always be used if the vehicle is under warranty.

Parts/accessory shops are often the best places to buy oils, greases, filters, bulbs, drive belts etc. because of convenience, trading hours and competitive prices.

Wreckers are a good source of complete assemblies such as doors, lamps, bumpers etc. where the serviceability of the component can be easily gauged. Some wreckers specialise in just one make of vehicle. Parts purchased from wreckers are usually half the price of the same component bought new from an authorised dealer but often carry no warranty.

Exchange component specialists rebuild complete assemblies such as drive shafts, transmissions, brake components, power steering pumps, starter motors, alternators, engines etc. Most exchange components have a warranty and because the specialist rebuilds many similar components, the prices are often comparable to or even cheaper than the cost of doing the job yourself, especially when labour is taken into account.

Ultimately, the selection of parts must be weighed up by the individual. Large savings can be made by shopping around for parts. On the other hand, a cheap sub standard part that fails in service far from home can be costly in the long run.

Vehicle identification information.

Towing and Jacking

The vehicle should never be towed with the four wheels on the ground if the transaxle assembly is suspected of being faulty. In this situation, employing a professional towing service is advisable.

If towing with four wheels on the ground is unavoidable, the following limitations must be followed:

The vehicle must be towed in a forward direction only.

When towing the vehicle by its towing eyes, never tow on the eyes sideways, always apply a straight pull (in line with the front or rear towing eyes of the towed vehicle.

Automatic transmission vehicles should at most be towed 25 km at a maximum speed of 50 km/h to avoid damage through a lack of lubrication of internal parts.

JACKING POINTS

When jacking a vehicle always have the vehicle on firm level ground, preferably concrete or asphalt. Ensure that the vehicle is clear of passing traffic and is empty of passengers.

Trolley Jack and Chassis Stands

When the vehicle has to be raised using a trolley jack, locate the lifting pad of the jack at the front or rear positions indicated in the illustration.

Place wheel chocks under the rear wheels when the front end of the vehicle is raised and under the front wheels when the rear end of the vehicle is raised.

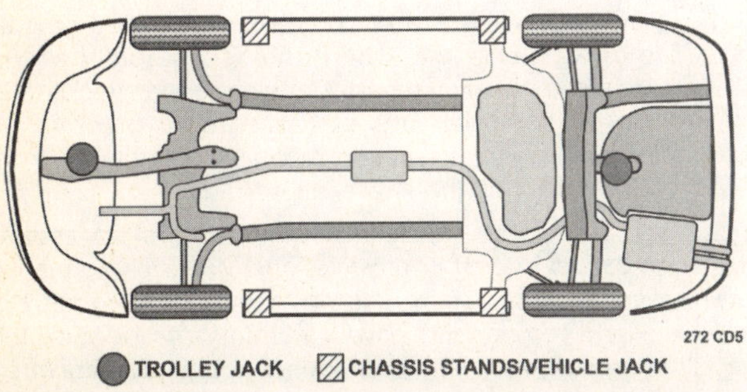

Vehicle jack, trolley jack and chassis stand locations. Sedan.

Never get under a vehicle that is only supported by the jack. Always support the vehicle with chassis stands at the positions shown in the illustration.

Insert a wooden or rubber block between the chassis stand and the vehicle underbody when the supported surface is flat.

Under no circumstances use a trolley jack, chassis stands or car ramps on sloping or unstable ground. Always use a flat solid surface, preferably concrete or ashphelt.

Car Ramps

When using car ramps, ensure that the wheels are centrally located on the ramps.

If someone is guiding you onto the ramps, have them stand well away from the front of the vehicle.

When the vehicle is on the car ramps, firmly apply the hand-brake. Also, chock the wheels that are on the ramps and the wheels that are on the ground.

Vehicle Jack

When using the vehicle jack, ensure that the edge of the body sill panel fits into the jack head slot. Use care as the vehicle jack can be unstable.

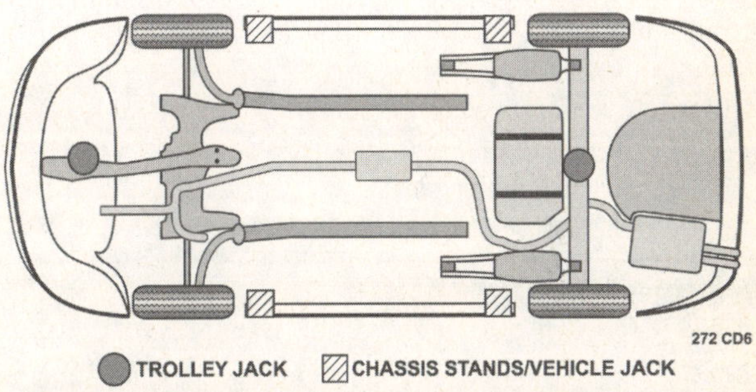

Vehicle jack, trolley jack and chassis stand locations. Station wagon.

Chapter 2
Routine Maintenance

Contents

Specifications

LUBRICANTS AND FLUIDS

Note: Listed here are manufactures recommendations at the time this manual was written. Manufacturers occasionally upgrade their fluid and lubricant specifications, so check with your local auto parts store for current recommendations.

Engine:
- Lubricant .... Refer to the chart
- Grade
  - TE/TF/TH/TJ and KE/KF/KH/KJ models .... API grade SA to SH
  - TL/TW and KL/KW models .... API grade SJ

Sump capacity (with filter):
- 3.0 litre .... 4.3 litres
- 3.5 litre .... 4.7 litres

Cooling system:
- Capacity, including heater .... 9.5 litres
- Coolant type .... Clean water and 30 – 50% inhibitor containing ethylene glycol

Clutch fluid type .... Dot 4

Manual transaxle:
- Lubricant .... SAE 75W/85 GL 4 gear oil
- Capacity .... 2.8 litres

Automatic transaxle:
- Lubricant
  - TE/TF/TH/TJ and KE/KF/KH/KJ models .... Mitsubishi ELC 4-SP2
  - TL/TW and KL/KW models .... Mitsubishi ELC 4-SP3
- Capacity —
  - Dry refill .... 8.5 litres
  - [1] Service refill .... 5.5 litres

Power steering lubricant .... Mitsubishi Power Steering Fluid

Brake fluid type .... Dot 4 brake fluid

Body:
- Hinges, door straps, etc .... Engine oil or spray grease
- Lock cylinders .... Graphite powder
- Door strikers .... Dry lube

[1] Twice this amount is required for flushing purposes

Description

This chapter contains the lubricant specifications for all items which would normally be part of regular servicing of the vehicle.

Also included is a Maintenance Chart which lists all service items and the intervals at which they should be checked.

Procedures and specifications relating to engine tune-up are covered at the start of Engine (chapter 4).

When performing a scheduled service, refer to the chart and note the items which are listed under the appropriate distance or time period. The information for servicing the various items is listed in the relevant chapter under the Routine Maintenance heading.

Lubricant capacities shown are approximate only. The correct lubricant level should be checked at the filler plug or dipstick.

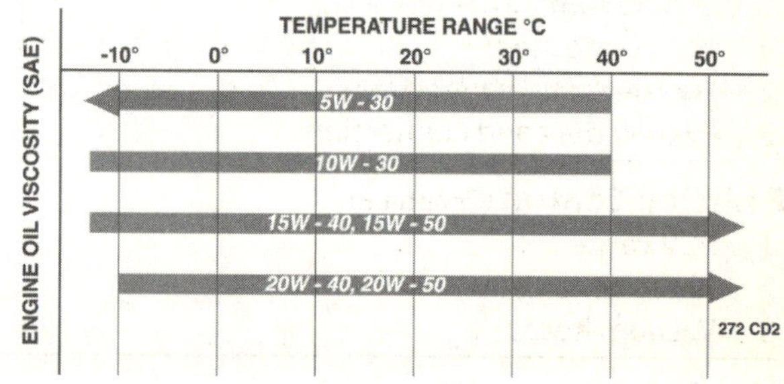

Engine oil chart. 5W-30 grade oil is factory fill. Thicker grade oils may effect fuel consumption slightly.

Maintenance Chart

The following maintenance chart applies to vehicles operating in normal conditions and should be performed at the specified intervals of distance or time, whichever occurs first. The service period must be reduced if the vehicle operates primarily in one or more of the following conditions: (a) Heavily laden or towing a trailer; (b) Repeatedly travelling short distances; (c) Repeatedly travelling for long periods at low speed or with extensive periods of idling. Refer to the relevant chapter in the manual for the routine maintenance procedures.

Service Items	KM X 1000	WEEKLY	15	30	45	60	75	90	105	120	135	150	165	180	195	210	225	240
	Years		1	2	3	4	5	6	7	8	9	10	11	12	13	14	15	16
Engine (Chapter 4)																		
1	Oil [1]	I	R	R	R	R	R	R	R	R	R	R	R	R	R	R	R	R
2	Oil filter [1]		R	R	R	R	R	R	R	R	R	R	R	R	R	R	R	R
3	Air cleaner element [2]		I	I	R	I	I	R	I	I	R	I	I	R	I	I	I	I
4	Fuel filter								R							R		
5	Spark plugs (conventional)				R			R			R			R			R	
6	Spark plugs (platinum)								R							R		
7	Distributor cap, rotor and leads		I		I		I		I		I		I		I		I	
8	Ignition timing and idle speed				I			I			I			I			I	
9	Engine drive belts		A	A	A	A	A	A	A	A	A	A	A	A	A	A	A	A
10	Timing belt		Renew every 100 000 km or 5 years															
11	Exhaust system		I	I	I	I	I	I	I	I	I	I	I	I	I	I	I	I
12	Oil leakage		I	I	I	I	I	I	I	I	I	I	I	I	I	I	I	I
Cooling System (Chapter 5)																		
1	Coolant	I	Change every 2 years (maximum 45 000 km)															
2	Hoses and connections		I		I		I		I		I		I		I		I	
3	Radiator exterior surfaces		I		I		I		I		I		I		I		I	
Fuel and Engine Management (Chapter 7)																		
1	Engine compartment wiring and connections		I		I		I		I		I		I		I		I	
2	Fuel lines and connections		I		I		I		I		I		I		I		I	
3	Inlet air duct and connections		I		I		I		I		I		I		I		I	
Emission Control (Chapter 8)																		
1	PCV valve			I		I		I		I		I		I		I		I
2	Charcoal canister				I		I		I		I		I		I		I	
3	Vacuum hoses				I		I		I		I		I		I		I	
Clutch and Manual Transaxle (Chapters 9 & 10)																		
1	Clutch pedal travel		I	I	I	I	I	I	I	I	I	I	I	I	I	I	I	I
2	Transaxle oil [3]		I	I	I	I	I	I	R	I	I	I	I	I	I	R	I	I
3	Oil leakage		I	I	I	I	I	I	I	I	I	I	I	I	I	I	I	I
Automatic Transaxle (Chapter 11)																		
1	Fluid level [3]		I	I	I	I	I	I	R	I	I	I	I	I	I	R	I	I
2	Filter								R							R		
3	Fluid leakage		I	I	I	I	I	I	I	I	I	I	I	I	I	I	I	I
4	Shift cable, pipes and hoses			I		I		I		I		I		I		I		I

A – Check and Adjust I – Inspect L – Lubricate R – Renew T – Tighten

[1] Halve the specificed interval if more than 50% of driving is in city traffic, if driving at high speeds at temperatures above 32° C, if driving in dusty or salty conditions, heavily loaded or towing.
[2] Halve the specificed intervals in sandy or dusty conditions.
[3] Renew the transaxle fluid every 45 000 km if used for heavy towing, taxi or courier operation.

Service Items	KM X 1000 / Years	WEEKLY	15 / 1	30 / 2	45 / 3	60 / 4	75 / 5	90 / 6	105 / 7	120 / 8	135 / 9	150 / 10	165 / 11	180 / 12	195 / 13	210 / 14	225 / 15	240 / 16
Steering (Chapter 12)																		
1 Power steering fluid			I	I	I	I	I	I	I	I	I	I	I	I	I	I	I	I
2 Hydraulic lines and hoses			I	I	I	I	I	I	I	I	I	I	I	I	I	I	I	I
3 Steering gear and tie rod ends			I	I	I	I	I	I	I	I	I	I	I	I	I	I	I	I
4 Front wheel alignment			As required															
Front Axle and Suspension (Chapter 13)																		
1 Suspension components			I	I	I	I	I	I	I	I	I	I	I	I	I	I	I	I
2 Front wheel bearings					I			I			I			I			I	
3 Drive shafts			I		I		I		I		I		I		I		I	
Rear Axle and Suspension (Chapter 14)																		
1 Suspension components			I	I	I	I	I	I	I	I	I	I	I	I	I	I	I	I
2 Rear wheel bearings					I/L			I/L			I/L			I/L			I/L	
Brakes (Chapter 15)																		
1 Brake fluid		I	Change every 2 year (maximum 45 000 km)															
2 Brake pads, discs and calipers			I	I	I	I	I	I	I	I	I	I	I	I	I	I	I	I
3 Brake hoses and lines including vacuum system			I	I	I	I	I	I	I	I	I	I	I	I	I	I	I	I
4 Handbrake operation				A		A		A		A		A		A		A		A
5 Brake pedal freeplay			A	A		A		A		A		A		A		A		A
Electrical system (Chapters 6 & 16)																		
1 Battery electrolyte, terminals and mounting			I	I	I	I	I	I	I	I	I	I	I	I	I	I	I	I
2 Lamps, horn, wipers and washers		I	I	I	I	I	I	I	I	I	I	I	I	I	I	I	I	I
3 Headlamp aim				I		I		I		I		I		I		I		I
Body (Chapter 17)																		
1 Mounting bolts and nuts				T		T		T		T		T		T		T		T
2 Bonnet, doors, locks and hinges			L	L	L	L	L	L	L	L	L	L	L	L	L	L	L	L
3 Seat belts, retractors and anchors			I	I	I	I	I	I	I	I	I	I	I	I	I	I	I	I
4 Underbody condition			I	I	I	I	I	I	I	I	I	I	I	I	I	I	I	I
Ventilation (Chapter 18)																		
1 Operate air conditioning system for minimum 5 minutes		I																
2 Heating system hoses and clamps				I		I		I		I		I		I		I		I
3 A/C condenser exterior surfaces, lines and refrigerant level			I	I	I	I	I	I	I	I	I	I	I	I	I	I	I	I
4 A/C evaporator drain tube			I	I	I	I	I	I	I	I	I	I	I	I	I	I	I	I
Wheels and Tyres (Chapter 19)																		
1 Tyre pressure		A	A	A	A	A	A	A	A	A	A	A	A	A	A	A	A	A
2 Tyre and wheel condition			I	I	I	I	I	I	I	I	I	I	I	I	I	I	I	I

A – Check and Adjust I – Inspect L – Lubricate R – Renew T – Tighten

[1] Halve the specificed interval if more than 50% of driving is in city traffic, if driving at high speeds at temperatures above 32° C, if driving in dusty or salty conditions, heavily loaded or towing.
[2] Halve the specificed intervals in sandy or dusty conditions.
[3] Renew the transaxle fluid every 45 000 km if used for heavy towing, taxi or courier operation.

Chapter 3
Roadside Trouble Shooting

Contents

Description

The ignition system in this vehicle is capable of producing dangerous voltages in the high tension circuits. Extreme care must be used when working on or near the ignition system to prevent severe electrical shock.

This chapter deals with the common causes of failure of an engine to start. Roadside breakdowns other than engine failure can be identified by reference to the Trouble Shooting heading in the chapter on the particular component affected.

For an internal combustion engine to run, three basic requirements are ignition, fuel and compression. Failure to start can usually be attributed to a fault in one of these three systems, with ignition or electrical faults the most common, followed by fuel system faults and lastly, mechanical or compression failure.

Jump Starting a Vehicle

Jump starting a vehicle can be dangerous if the procedure described below is not performed correctly. If any doubt exists, it is recommended that the services of a competent mechanic be obtained.

The range of vehicles covered by this manual are equipped with complex electronic circuitry which can be damaged by voltage surges. These voltage surges can be generated when jump starting or being jump started by another vehicle.

If available, use jumper leads equipped with a surge protection device and follow the lead manufacturers instructions carefully, particularly regarding the connection and disconnection of the leads.

1 Ensure that the booster battery is 12 volts and the negative terminal is earthed.

2 Ensure that the vehicles are not touching and that the ignition and all accessories on both vehicles are switched Off.

3 Ensure that the transmissions on both vehicles are in Park or Neutral and the handbrakes are firmly applied.

4 Open the cover on the battery box and remove the vent caps from the battery. Check the electrolyte level and replenish with distilled water as necessary.

5 Place the vent caps loosely over the cell apertures.

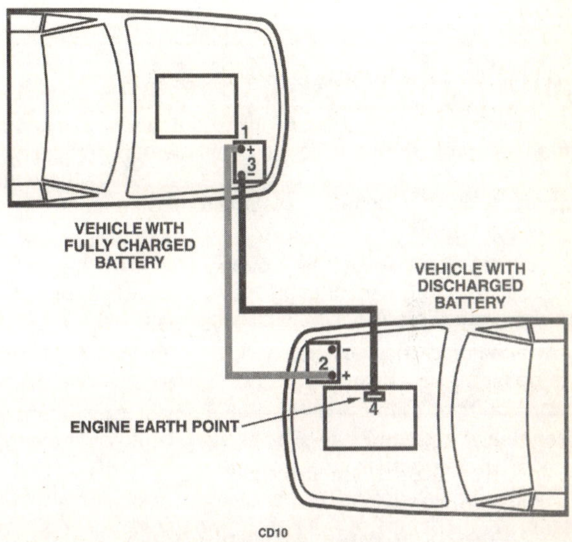

Illustration of jumper lead connections. Connect in the order shown and disconnect in the reverse sequence.

6 Connect one end of the red jumper lead to the positive (+) battery terminal of the booster battery and the other end of the red lead to the positive (+) battery terminal of the discharged battery.

The battery emits hydrogen gas which is explosive. Do not expose the battery to naked flames or sparks.

Do not lean over the battery when connecting the jumper leads.

Do not allow the ends of the jumper leads to touch one another or any part of the vehicle.

7 Connect one end of the black jumper lead to the negative (–) battery terminal of the booster battery and the other end of the black lead to a good earthing point on the engine of the vehicle with the discharged battery.

Do not connect the jumper lead directly to the negative (–) battery terminal of the discharged battery.

8 Start the engine on the vehicle with the booster battery and run the engine at a moderate speed for a few minutes.

9 Start the engine on the vehicle with the discharged battery.

10 Leave the engines of both vehicles running for at least 10 minutes. This will partially charge the discharged battery and reduce the risk of damage to electronic circuitry from voltage surges.

11 Switch the engines of both vehicles Off and disconnect the jumper leads in the reverse order of the connecting sequence.

12 Attempt to start the engine of the vehicle with the discharged battery.

13 If the battery has not charged sufficiently to start the engine, reconnect the jumper leads as previously described and start the engines of both vehicles.

14 Switch On the headlamps of the vehicle with the discharged battery.

15 Disconnect the jumper leads in the reverse order of the connecting sequence. Switch the headlamps Off.

Ignition and Electrical System Checks

All vehicles covered by this manual are equipped with an engine immobiliser which relies on coded signals sent from the ignition switch to the immobiliser ECU, which then signals the engine control unit (ECU) to permit the vehicle to be started.

Due to the inability to isolate a fault to the immobiliser, it is recommended that if the following checks fail to isolate the fault, refer to Fuel and Engine Management (chapter 7) for a more comprehensive immobilier system checking procedure.

1 Switch the ignition On and check for warning lamp illumination on the dashboard.

2 Operate the starter and check that the starter rotates the engine at a steady speed. If the starter rotates the engine normally, proceed to step 4.

3 Switch the headlamps On and check for good light intensity. Should the lamps not illuminate or the starter motor not rotate the engine, proceed as follows:

a Disconnect the battery terminals and clean both terminals and posts. Connect the terminals and where applicable, tighten firmly but not excessively.

When disconnecting or connecting battery terminals, ensure that the ignition is in the Off position to prevent damage to the electronic components and always disconnect the negative (earth) terminals first and connect it last.

b Check that the earth lead from the battery to the transaxle or body frame is not broken and that the connections are clean and secure.

Cleaning the battery terminals.

c Check that the lead from the battery to the starter motor solenoid is intact and has a clean and secure connection.

d Check the condition of the engine compartment fusible links.

e Repeat step 1. Should the starter motor still not operate or the lamps not illuminate, one or more of the following faults may be the cause:

No starter motor operation or lamps: Battery flat or defective.

Lamps illuminate but no starter operation: Starter motor drive jammed in mesh with flywheel ring gear. Starter motor or solenoid defective. Ignition switch faulty. On automatic transmission models, faulty inhibitor switch or the transmission is not in Park or Neutral.

Lamps dim and starter operation sluggish: Discharged battery or fault in starter motor. Battery flat due to broken drive belt or defective alternator. Faulty battery due to cell breakdown.

If the battery and starter motor operation proves satisfactory but the engine still fails to start, continue as follows:

Electronic ignition systems produce dangerous voltages in the high tension circuits. For this reason, extreme care must be taken when performing the following checks.

When disconnecting the wiring from any component, ensure that the ignition switch is Off to prevent possible damage to the electronic components.

4 Open the electrode gap of a serviceable spark plug to 6 mm. Securely earth the spark plug using a jumper lead or by tying the plug to an earthed engine component.

Checking for spark using a test spark plug.

5 Disconnect a high tension lead from a spark plug and connect it to the test spark plug.

6 Have an assistant operate the starter motor and check that a spark repeatedly jumps the gap on the test spark plug.

7 If the spark is strong and blue, remove the spark plugs and check their condition and the electrode gap as described in Engine (chapter 4).

8 If there is no spark, check the ECU relay located below the driver side of the centre console.

9 Access the immobiliser ECU as described in Fuel and Engine Management (chapter 7) and check as follows:

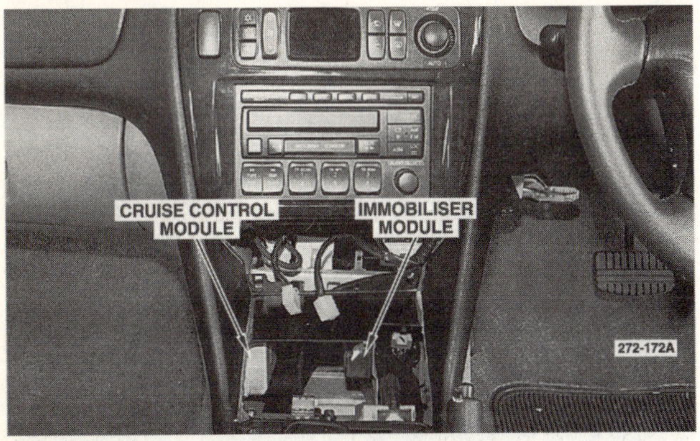

Immobiliser and cruise control module location.

- **a** Using a voltmeter or test lamp, check for battery power at the immobiliser red wire and black/white wire with the ignition On. If battery voltage is not available, use the wiring diagrams in Body Electrical (chapter 16) as a guide to trace the fault.
- **b** Using an ohmmeter, check for continuity to earth at the immobiliser connector black wires. Repair the wiring as necessary.

10 Check the high tension leads to ensure that they are dry and that the insulation is not cracked or perished. Check the ends of the leads for burning or corrosion. Remove the distributor cap and check for cracks or tracking. Also wipe the inside of the cap clean. If there is still no spark, refer to Fuel and Engine Management (chapter 7) for further ignition system checks.

11 Ensure that all electrical connections and earth leads are clean and secure.

Fuel System Checks

Due to the complex nature of the EFI system, it is recommended that should the following checks prove satisfactory but the engine fails to start and the mechanical system is operational, reference be made to Fuel and Engine Management (chapter 7) of this manual or an authorised workshop be consulted.

1 Check that the fuel tank contains a reasonable quantity of fuel.

2 Check that all fuel hose connections are secure.

3 Check for a damaged vacuum hose or loose connections between the brake booster and the inlet manifold.

Check the high tension leads for burnt or corroded terminals.

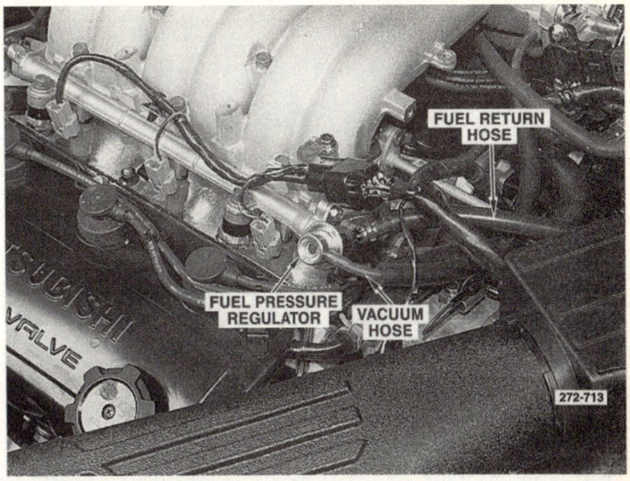

Fuel return hose location.

4 Grasp the fuel return hose and have an assistant switch the ignition On. A slight vibration should be felt through the hose for a few seconds, indicating fuel flowing through the hose. Switch the ignition Off.

If the pump is not operating, check the fuel pump relay located below the driver side of the console. Refer to Fuel and Engine Management (chapter 7).

5 If the engine still fails to start and no fault is found with the mechanical system, check the fuel pump and injector circuits as described in Fuel and Engine Management (chapter 7).

Mechanical System Checks

Faults in the mechanical system that will prevent engine operation can be caused by either loss of cylinder compression or component failure which prevents the crankshaft rotating.

1 Remove the front spark plugs and attempt to rotate the crankshaft pulley manually. If the crankshaft turns, proceed to step 4.

2 Remove the front timing belt cover and check the timing belt for breaks or missing teeth on the belt. Also check the belt tension. A loose belt can indicate a damaged belt tensioner or broken timing belt.

Check the distributor cap for cracks or tracking between the terminals. Also check the carbon brush in the centre.

Check the timing belt for breaks or missing teeth.

Attempt to rotate the crankshaft in either direction while observing the timing belt. If the crankshaft will rotate but the camshaft does not, a broken timing belt is indicated. Refer to the Timing Belt heading in Engine (chapter 4).

3 If the crankshaft pulley cannot be turned, remove the starter motor and again attempt to turn the crankshaft pulley. If the pulley now turns, it indicates that the fault is in the starter motor drive mechanism.

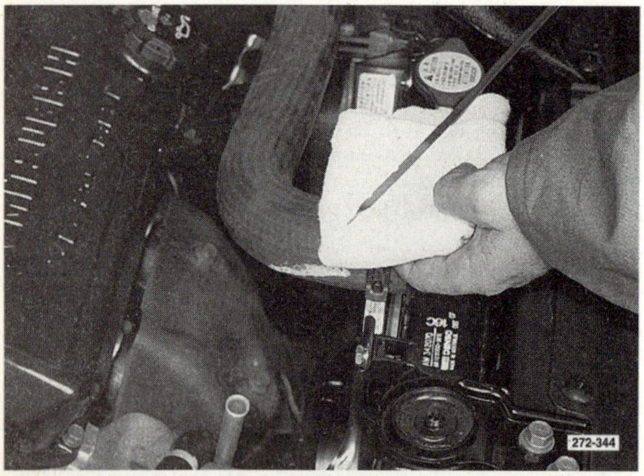

Check the engine oil dipstick for emulsified oil indicating internal coolant leakage.

On manual transmission models, depress the clutch pedal to disengage the transmission from the engine to eliminate the transmission as the problem source.

4 If the crankshaft pulley cannot be turned, it indicates engine seizure or broken components such as the timing belt, connecting rods, pistons or crankshaft. Refer to the Engine Mechanical Trouble Shooting heading in Engine (chapter 4).

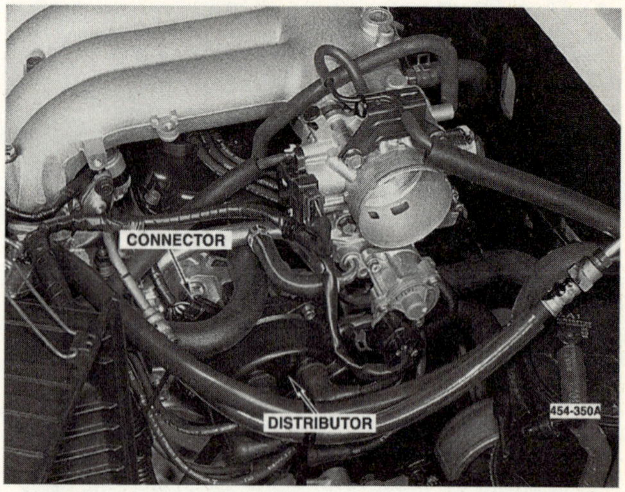

Distributor wiring connector viewed with the air inlet duct removed.

Engine seizure can be caused by overheating due to loss of coolant. If this has occurred, the engine may free up when it has cooled. Allow the engine to cool before filling the cooling system.

5 With the engine cool, check the radiator for loss of coolant. Do not remove the radiator cap from an engine at operating temperature. If coolant loss is evident, check carefully for any signs of external leakage. Remove the engine oil dipstick and check for emulsification of the oil. When oil mixes with water it turns creamy and the oil level may increase.

If the oil is emulsified, it indicates internal coolant leakage. Check the following possible causes:

Blown cylinder head gasket.

Distorted cylinder head or cylinder block faces.

Cracked cylinder head or block.

6 Disconnect the distributor wiring connector and do an engine compression test. Refer to Engine (chapter 4) for the procedure if necessary.

If coolant is discharged from any of the spark plug holes, internal coolant leakage is indicated. Check for a blown head gasket, cracked cylinder head or block, or distorted cylinder head or block sealing faces.

7 If any of the cylinders have low compressions, check the following possible causes:

Burnt or broken valve(s).

Broken piston(s).

Chapter 4
Engine

Contents

Specifications

ENGINE ASSEMBLY

Type ... V6 SOHC 24 valve
Capacity:
- 6G72 ... 2 972 cc
- 6G74 ... 3 497 cc

Maximum power:
- 3.0 litre ... 140 kW @ 5 500 rpm
- 3.5 litre ... 147 kW @ 5 000 rpm

Maximum torque:
- 3.0 litre ... 255 Nm @ 4 500 rpm
- 3.5 litre ... 300 Nm @ 4 000 rpm

Bore:
- 3.0 litre ... 91.1 mm
- 3.5 litre ... 93.0 mm

Stroke:
- 3.0 litre ... 76.0 mm
- 3.5 litre ... 85.8 mm

Compression ratio ... 9:1
Lubricant:
- Grade ... Refer to Routine Maintenance (chapter 2)
- Capacity (including filter) —
  - 3.0 litre ... 4.3 litres
  - 3.5 litre ... 4.7 litres

TUNE UP

Firing order ... 1 - 2 - 3 - 4 - 5 - 6
High tension lead resistance limit ... 22 k/ohms
Idle speed:
- Kerb ... 700 ± 100 rpm
- Base ... 700 ± 50 rpm

CO concentration at idle ... 0 – 0.5%
Compression pressure:
- Standard ... 1 177 kPa
- Minimum ... 875 kPa
- Maximum variation between cylinders ... 98 kPa

Spark plugs, 3.0 litre:
- Front bank —
  - Champion ... RC8YCC4
  - NGK ... BKR6E–11
- Rear bank (platinum) —
  - Champion ... RC8PYP4
  - NGK ... PFR6J–11
- [1] Gap ... 1.0 – 1.1 mm

Spark plugs, 3.5 litre:
- Front bank —
  - Champion ... RC10YCC4
  - NGK ... BKR5E–11
- Rear bank (platinum) —
  - Champion ... RC10PYP4
  - NGK ... PFR5J–11
- [1] Gap ... 1.0 – 1.1 mm

[2] Ignition timing:
- 3.0 litre ... 15° ± 3° BTDC
- 3.5 litre ... 10° ± 3° BTDC

[3] Drive belt deflection at 100 N (10 kg):
- Alternator/air conditioner —
  - New ... 6 – 7 mm
  - Used ... 8 – 10 mm
- Power steering pump —
  - New ... 8 – 9 mm
  - Used ... 11 – 14 mm

If the Specifications differ from the information on emission control information label located on the underside of the bonnet, use the information on the label.

TIMING BELT

Tensioner rod protrusion removed ... 11.7 – 12.3 mm

CYLINDER HEADS

Type ... One piece crossflow
Material ... Aluminium alloy
Cylinder head face distortion limit ... 0.20 mm
[4] Machining limit ... 0.20 mm

Height (new) .... 119.9 – 120.1 mm
Manifold face distortion limit:
- Inlet .... 0.10 mm
- Exhaust .... 0.15 mm

Valve seat:
- Angle .... 44°
- Valve contact width .... 0.9 – 1.3 mm

Valve seat service size .... 0.030, 0.060 mm oversize
Valve guide hole diameter:
- 0.05 mm oversize .... 11.05 – 11.07 mm
- 0.25 mm oversize .... 11.25 – 11.27 mm
- 0.50 mm oversize .... 11.50 – 11.52 mm

ROCKER ARMS, SHAFTS AND HYDRAULIC TAPPETS

Rocker arm:
- Inside diameter .... 20.02 – 20.04 mm
- Oil clearance —
  - Standard .... 0.02 – 0.05 mm
  - Limit .... 0.10 mm

Rocker shaft outside diameter .... 19.90 – 20.00 mm
Hydraulic tappet:
- Dry clearance .... 0.5 – 1.3 mm
- Leak down rate (in diesel fuel at 20°C) .... 1 mm in 4 – 20 seconds

CAMSHAFTS

Number of bearing journals 4
Journal diameter .... 44.93 – 44.94 mm
Journal oil clearance .... 0.06 – 0.10 mm
Cam lobe height:
- Standard —
  - Inlet .... 37.58 mm
  - Exhaust .... 36.95 mm
- Limit —
  - Inlet .... 37.08 mm
  - Exhaust .... 36.45 mm

Camshaft valve timing:
- Inlet —
  - Opens .... 9° BTDC
  - Closes .... 59° ABDC
- Exhaust —
  - Opens .... 47° BBDC
  - Closes .... 21° ATDC

VALVES AND SPRINGS

Valve face angle .... 45 – 45.5°
Valve head margin:
- Standard —
  - Inlet .... 1.0 mm
  - Exhaust .... 1.2 mm
- Limit —
  - Inlet .... 0.5 mm
  - Exhaust .... 0.7 mm

Valve length:
- Standard —
  - Inlet .... 112.30 mm
  - Exhaust .... 114.10 mm
- Limit —
  - Inlet .... 111.80 mm
  - Exhaust .... 113.61 mm

Valve protrusion from spring seat:
- Standard .... 49.3 mm
- Limit .... 49.8 mm

Valve stem diameter .... 6.0 mm
Valve stem to guide clearance:
- Standard —
  - Inlet .... 0.02 – 0.05 mm
  - Exhaust .... 0.04 – 0.07 mm
- Limit —
  - Inlet .... 0.10 mm
  - Exhaust .... 0.15 mm

Valve guide:
- Type .... Renewable
- Length —
  - Inlet .... 45.5 mm
  - Exhaust .... 50.5 mm
- Bore diameter .... 6.0 mm
- Bore in cylinder head .... 11.06 – 11.07 mm
- Protrusion from cylinder head .... 14 mm

Valve springs:
- Free length —
  - Standard .... 51.0 mm
  - Limit .... 50.0 mm
- Maximum out of square .... 4°
- Installed height .... 44.2 mm

CYLINDER BLOCK

Type .... 6 cyl 60° V configuration
Material .... Cast iron
Height .... 210.4 – 210.6 mm
Face distortion limit .... 0.10 mm
[4] Machining limit .... 0.20 mm
Cylinder bore:
- Diameter —
  - 3.0 litre .... 91.1 mm
  - 3.5 litre .... 93.0 mm
- Ovality .... 0.01 mm

CRANKSHAFT AND MAIN BEARINGS

Number of main bearings 4
Crankshaft end float:
- Standard .... 0.05 – 0.25 mm
- Limit .... 0.30 mm

Main bearing journal diameter .... 60 mm
Main bearing and crankpin journal:
- Runout and taper limit .... 0.005 mm

Main bearing journal ovality limit .... 0.003 mm
Crankpin journal diameter .... 50 mm
[5] Main bearing oil clearance:
- Standard .... 0.02 – 0.04 mm
- Limit .... 0.10 mm

Maximum flywheel runout .... 0.13 mm

CONNECTING RODS AND BEARINGS

Connecting rod:
- Maximum bend .... 0.05 mm
- Maximum twist .... 0.10 mm
- Side clearance —
  - Standard .... 0.10 – 0.25 mm
  - Limit .... 0.4 mm

[6] Connecting rod bearing oil clearance:
- Standard .... 0.02 – 0.05 mm
- Limit .... 0.10 mm

PISTONS AND RINGS

Piston diameter:	
3.0 litre	91.1 mm
3.5 litre	93.0 mm
Piston to cylinder bore clearance	0.02 – 0.04 mm
Piston oversizes	0.25, 0.50, 0.75 and 1.00 mm
Number of rings:	
Compression	2
Oil control	1
Piston ring side clearance:	
Standard —	
Upper ring	0.03 – 0.07 mm
Lower ring	0.02 – 0.06 mm
Oil ring	Not specified
Limit	0.1 mm
Piston ring end gap:	
Standard —	
Upper compression ring	0.30 – 0.45 mm
Lower compression ring	0.45 – 0.60 mm
Oil ring —	
3.0 litre	0.20 – 0.60 mm
3.5 litre	0.10 – 0.35 mm
Limit —	
Compression rings	0.80 mm
Oil ring	1.00 mm

LUBRICATION

Oil pump type	Trochoid rotor
Filter type	Full flow, disposable, spin on
Oil pressure at idle speed	More than 80 kPa
Oil pump clearances:	
Inner rotor tip to outer rotor	0.06 – 0.18 mm
Rotor assembly end clearance	0.4 – 0.10 mm
Outer rotor to housing —	
Standard	0.10 – 0.18 mm
Limit	0.35 mm

TORQUE WRENCH SETTINGS

Camshaft:	
Sprocket bolt	88 Nm
Thrust plate retaining bolts	12 Nm
Coolant drain plug in cylinder block	39 Nm
Connecting rod nuts	51 Nm
Crankshaft pulley bolt	181 Nm
Cylinder head bolts	108 Nm
Distributor retaining bolts	23 Nm
Driveplate to crankshaft bolts	74 Nm
Engine lifting hook bolts	17 Nm
Engine mountings (RHS):	
Bracket bolts	44 Nm
Damper bolts	12 Nm
[7] Through bolt	98 – 118 Nm
Retaining nuts and bolt	98 – 118 Nm
Engine mountings (LHS):	
Mounting nuts	69 – 93 Nm
[7] Through bolt	69 Nm
Engine mountings (front and rear):	
Bracket to transaxle bolts	69 Nm
[7] Through bolts	69 Nm
Retaining bolts	44 Nm
Exhaust:	
Catalytic converter retaining bolts	49 Nm
Engine pipe flange nuts	49 Nm
Exhaust manifold heat shields	14 Nm
Exhaust manifold to cylinder head nuts	49 Nm
Fuel rail:	
Pressure pipe bolts	9 Nm
Pressure regulator bolts	9 Nm
Retaining bolts	12 Nm
Transfer pipe bolts	9 Nm
Flywheel to crankshaft bolts	98 Nm
Inlet manifold to cylinder head nuts	Refer to text
Main bearing bolts:	
3.0 litre	93 Nm
3.5 litre	70 – 80 Nm
Oil pump:	
To cylinder block bolts	13 Nm
Cover screws	10 Nm
Pickup pipe bolts	18 Nm
Pressure relief valve plug	44 Nm
Oxygen sensor	44 Nm
Plenum chamber bolts	17 Nm
Rear crankshaft oil seal carrier bolts	11 Nm
Rocker cover retaining bolts	3 – 4 Nm
Rocker shaft retaining bolts	31 Nm
Spark plugs	25 Nm
Sump:	
Baffle plate bolts	11 Nm
Drain plug	39 Nm
Long bolts	6 Nm
Retaining bolts	11 Nm
Timing belt:	
Guide pulley bolt	44 Nm
Tensioner retaining bolts	23 Nm
Tensioner pulley lock bolt	48 Nm
Tensioner arm pivot bolt	44 Nm
Covers —	
6 mm bolts	11 Nm
8 mm bolts	13 Nm
Lower cover bolts	8 Nm
Thermostat cover and housing bolts	18 Nm
Throttle body bolts	11 Nm
Water outlet bolts	18 Nm
Water pump:	
8 mm bolts	24 Nm
10 mm bolts	41 Nm

[1] Platinum plugs are pre gapped and should not have the gaps adjusted.

[2] Non adjustable.

[3] A used belt is one which has been in use on a running engine for five minutes or more.

[4] Combined machining of cylinder block and cylinder head not to exceed 0.20 mm

[5] There are 3 standard grades of bearings available to achieve the correct clearance.

[6] On 3.5 litre models there are 5 standard grades of bearings available to achieve the correct clearance.

[7] Tighten with the weight of the engine on the mountings.

Description

Two types of engines are available in the range of vehicles covered by this manual; the 3.0 litre 6G72 and the 3.5 litre 6G74 engine. Both engines are of six cylinder, transversely mounted, water cooled, V configuration. The 3.5 litre engine has a larger bore and stroke than the 3.0 litre, thus giving the extra capacity.

The single overhead camshafts drive 12 valves in each cylinder head, and are located in crossflow alloy heads. Both camshafts run directly in the cylinder head. The camshafts are driven by a toothed rubber timing belt which must be renewed every 100 000 km. Belt tension is regulated by an automatic tensioner assembly.

The aluminium rocker arms are equipped with roller followers to reduce friction, and hydraulic tappets. The hydraulic tappets maintain zero clearance between the valves and the camshaft, thus the need for valve clearance adjustment is eliminated.

The oil pump, which is driven directly by the crankshaft, is a trochoid rotor type installed on the front of the engine. The relief valve is installed in the passage on the delivery side of the oil pump regulating the maximum oil pressure supplied to the engine.

The crankshaft is supported by four main bearings and the end float is taken at number three main bearing by renewable thrust plates. The crankshaft has a special surface treatment and should not be ground undersize. If the crankshaft is worn or damaged, it must be renewed. The crankshaft bearing caps are united by a cradle. This cradle reduces vibration in the cylinder block wall, and reinforces the crankshaft support. The connecting rods are of forged carbon steel.

The pistons are made of a special aluminium and are secured to the connecting rods by an alloy steel type gudgeon pin. The 3.0 litre engine has semi floating gudgeon pins, while the 3.5 litre model is equipped with fully floating gudgeon pins. The main and connecting rod bearings are of the renewable aluminium alloy type.

Both type of engines are fitted with a two piece sump. The upper sump is an alloy casting to provide additional stiffness to the engine and to reduce noise and vibration. The lower sump is a conventional pressed metal type sump.

The engine and transaxle assembly are supported on fluid filled, rubber bonded mountings.

Routine Maintenance

ENGINE TUNE-UP

In the past, an engine tune-up was an important part of regular maintenance. As mechanical and electrical components became worn, the engine would not perform as well as it should. It was necessary to have certain items checked, cleaned and adjusted or renewed.

On vehicles covered by this manual, the engine ECU (electronic control unit) controls many things that were previously manually adjustable. However, the term tune-up is still commonly used when referring to engine maintenance procedures performed in a set order. When trying to locate a fault such as a misfire or poor performance, it is good mechanical practice to perform a tune-up before starting other diagnosis. Despite the use of electronic engine management systems, a tune-up can be confidently performed at home without using specialised equipment.

The components which are renewed during a tune-up will depend on the age and condition of the vehicle. With a new, well maintained vehicle you may only need to renew the spark plugs and filters. With an older (high mileage) vehicle you may have to renew the above components as well as drive belts, high tension leads, distributor cap and rotor.

Due to restricted access to the rear spark plugs on this engine, platinum plugs are used on the rear bank to extend service life to 105 000 km.

Similarly, the adjustments required will vary with the age and condition of the vehicle. Ignition timing and the idle speed are set in the factory and controlled by the ECU, and should not normally require adjustment. Further more, the ignition timing adjustment requires the use of special equipment.

The following is a summary of the procedures required for an engine tune-up and the order in which they should be done. The major procedures are described individually in more detail later in this chapter.

Tune-up Procedures

1 Check the engine compartment for any obvious problems:

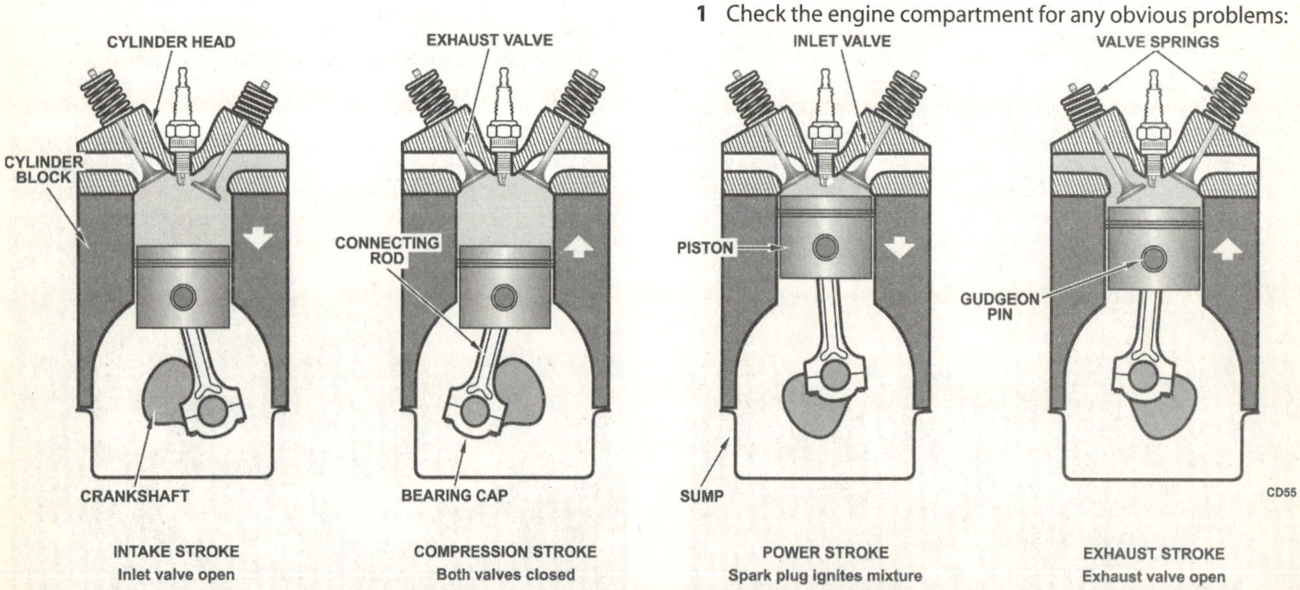

Major components and operating principle of the engine.

a Battery, ensuring that the case top is clean, the terminals are tight and not corroded. Top up the battery if necessary and ensure that it is firmly mounted.
b Engine compartment fluid levels, including coolant condition.
c Fuel, coolant and vacuum hoses for hardening, softening, leakage, etc.
d Electrical connections for looseness, corrosion, frayed or deteriorated wiring.
2 Air cleaner.
3 Fuel filter.
4 Engine drive belts.
5 Cylinder compression test. This need only be done if trying to rectify a fault, eg. misfire, rough idle, or if the mechanical condition of the engine is unknown, perhaps after the car has been purchased. It is impossible to tune an engine that is not in good mechanical condition. If the compression test indicates a problem, mechanical repairs must be made before continuing with the tune-up. It is also a good idea to write down the compression figures for future reference.
6 Spark plugs.
7 High tension leads.
8 Distributor.
9 Idle speed.

ENGINE OIL AND FILTER

To Check Oil Level

1 Park the vehicle on a level surface, turn off the engine and raise the bonnet.
2 Wait approximately one minute to allow the oil to drain into the sump.
3 Remove the dipstick and wipe it clean, using a lint free cloth. Insert the dipstick fully into the tube.
4 Withdraw the dipstick and check the oil level. The oil level should be maintained between the marks on the dipstick. The difference between the Low and Full marks on the dipstick represents approximately 1.0 litres of engine oil. If oil to be added to the engine, proceed as follows:
a Clean around the oil filler cap on the top of the engine and remove the cap.

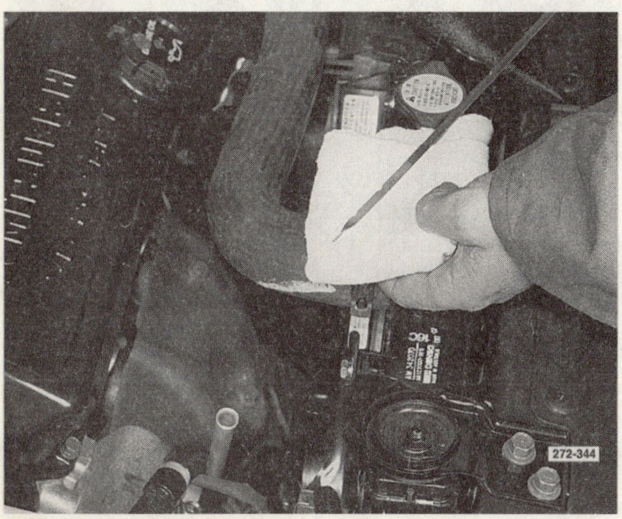

Checking the engine oil level.

Equipment necessary for an oil change. Note the spare drain plug ring in case the one on the vehicle is damaged.

b Add the required quantity of engine oil, in small quantities at a time. Never overfill with engine oil as this can result in oil leaks and sometimes an increase in oil consumption.
c Repeat steps 2 to 4 until the oil level reaches the Full mark.

To Renew Oil and Filter

TIP *It is best to drain the engine sump with the oil at operating temperature. However, when the oil is hot take care to avoid scalding.*

1 Drive the front of the vehicle onto car ramps and stop the engine. Chock the front wheels.
2 Raise the rear of the vehicle with a jack and support the vehicle with chassis stands under the recommended points. Refer to General Information (chapter 1) for the correct trolley jack and chassis stand positions to prevent damage or injury.

TIP *Keep the vehicle as level as possible to avoid false readings when checking the lubricant level.*

3 Using a wire brush and a cloth, clean around the engine sump drain plug.

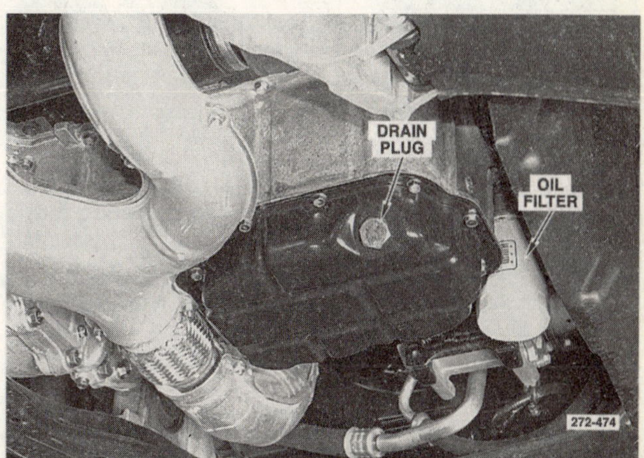

Engine sump drain plug and oil filter location.

Removing the oil filter, using a filter removal tool.

4 Place a container under the engine sump, remove the drain plug with the appropriate size ring spanner or socket and allow the engine sump to completely drain.

5 Remove the oil filter with a filter removal tool, turning it in an anticlockwise direction. Allow the residual engine oil to drain into the container.

6 Fill the oil filter with clean oil as much as possible, to help with initial oil pressure when the engine is started again. Smear oil onto the sealing gasket of the new filter and tighten the filter by hand as per the tightening instructions supplied with the new filter. If tightening instructions are not supplied, tighten the filter by hand three quarters of a turn after it contacts the filter housing.

TIP *Before installing the new filter ensure that the sealing gasket from the old filter has not adhered to the filter sealing seat on the engine.*

7 When the engine has completely drained, check the drain plug O ring to ensure that it is serviceable. Renew if necessary.

8 Install and tighten the sump drain plug to the torque specified. Wipe around the plug after installation.

9 Using the funnel, fill the engine with the specified amount and grade of engine oil. Check that oil is showing on the dipstick. Start the engine and check that the oil warning lamp extinguishes within a few seconds.

TIP *To prevent overfilling, do not to pour all the specified amount of oil into the engine. Retain at least half a litre and top up to the level mark on the dipstick after running the engine for a few minutes.*

10 Stop the engine, wait approximately one minute and check the oil level on the dipstick, as previously described. If necessary, add oil to bring the level to the full mark on the dipstick. Never overfill with engine oil as this can result in oil leaks and sometimes an increase in oil consumption.

11 Check thoroughly for oil leaks at the engine sump drain plug and engine oil filter.

12 Perform all other routine maintenance items that coincide with the oil change intervals.

Oil Leaks

Prior to checking for oil leaks, the engine should be completely degreased and cleaned. Excessive oil leaks can be easily detected

Run the engine over white paper to check for oil leaks.

by placing white paper on the floor directly beneath the engine while it is running at operating temperature. It may be necessary to remove the front splashguard, so the exact position of the leak can be pinpointed.

1 Check the front and rear crankshaft oil seals, rocker cover gaskets and around the perimeter of the sumps for excessive dampness. Check the oil filter for leakage.

2 If applicable, tighten the retaining nuts and bolts as required, or renew faulty oil seal(s) or the oil filter, referring to the appropriate heading.

3 Check the oil pressure switch for leakage and renew if necessary.

Check the oil pressure switch for leaks.

4 Refill the engine with the required quantity and grade of engine oil.

AIR CLEANER ELEMENT

The air cleaner is equipped with a paper type element which should be regularly inspected.

The element should be renewed at the intervals recommended in Routine Maintenance (chapter 2). This distance is only a guide for normal operating conditions and should be reduced accordingly if the vehicle is operating under dusty conditions.

Paper air cleaner elements should not be washed in petrol or any other type of cleaning solvent. If the element has been washed in solvent or has become oil soaked, it should be discarded and a new element installed.

Service the air cleaner element regularly.

To Inspect and Renew

The air cleaner element should be inspected and renewed at the specified intervals recommended in the Maintenance Chart.

1. Release the retaining clips, separate the air cleaner front cover and the rear housing and slide the air cleaner element from between the cover and the housing.
2. Inspect the front side of the element. If it is only lightly dusty, clean the element by blowing low pressure air through or by gently tapping it against a hard surface. Renew the element if it is excessively dirty.
3. Clean any dirt from the front cover.

Installation is a reversal of the removal procedure with attention to the following points:

1. Ensure that the element correctly seated in the rear housing.
2. Ensure that the front cover retaining clips are secure.
3. After installation, run the engine and check for air leaks.

FUEL FILTER

To Renew

1. Raise the rear of the vehicle and support it on chassis stands. Refer to General Information (chapter 1) for the correct trolley jack and chassis stand positions to prevent damage or injury.
2. Reduce the fuel system pressure as described in Fuel and Engine Management (chapter 7).
3. Place a container beneath the fuel filter.

TIP *To prevent the loss of fuel, open the fuel filler cap to release the pressure from the fuel tank. Leave the cap loose while the fuel hoses are disconnected.*

4. Using two spanners, one to hold the union on the filter and the other on the flare nut on the steel pipe, separate the steel fuel pipe from the union.
5. While lightly pulling the fuel supply hose, press the quick connector retaining tabs together and pull the hose away from the filter.
6. Remove the bolt retaining the filter to the bracket and withdraw the filter from the vehicle.

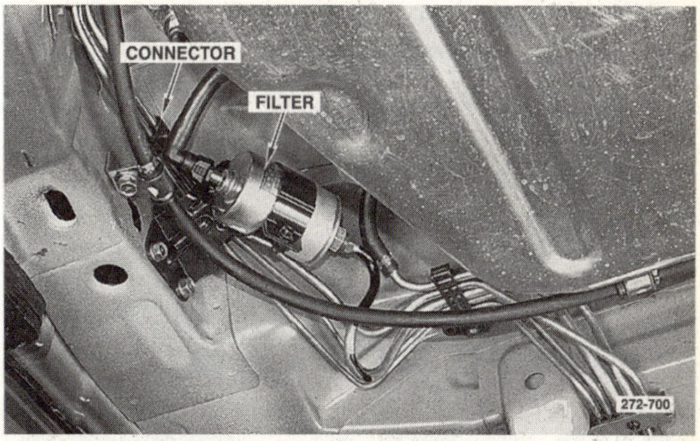

Location of the fuel filter on sedan models.

Installation is a reversal of the removal procedure with attention to the following points:

1. Install the new filter with the arrow on the filter pointing towards the steel pipe.
2. Attach the flare nut to the union and hand tighten the flare nut. Ensure that the nut correctly engaged the threads of the union. Using two spanners, tighten the fuel pipe securely.
3. Push the fuel supply hose onto the fuel filter, until the retaining tabs lock onto the filter. When the hose is fully seated on the filter, gently pull the hose to ensure that the connector is secure.
4. Tighten the fuel filler cap securely.
5. Test the fuel system for external leakage as described in Fuel and Engine Management (chapter 7).

DRIVE BELTS

To Inspect

Check the drive belts for cracks or separation between the rubber and the fibre cords by examining the belt. Cracks on the ribs are acceptable, however if adjacent ribs have missing sections exceeding 13 mm, the belt should be renewed. It is essential that all drive belts be adjusted to prevent slip, but without imposing excessively upon the component bearings.

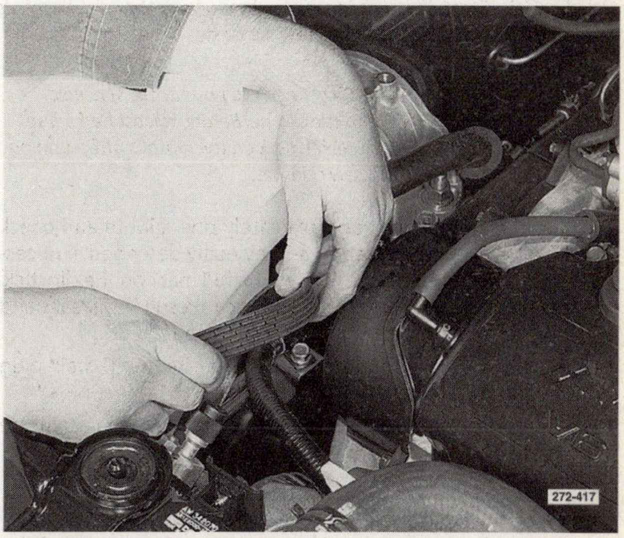

Checking the condition of the drive belt. A drive belt with this many cracks and missing sections should be renewed.

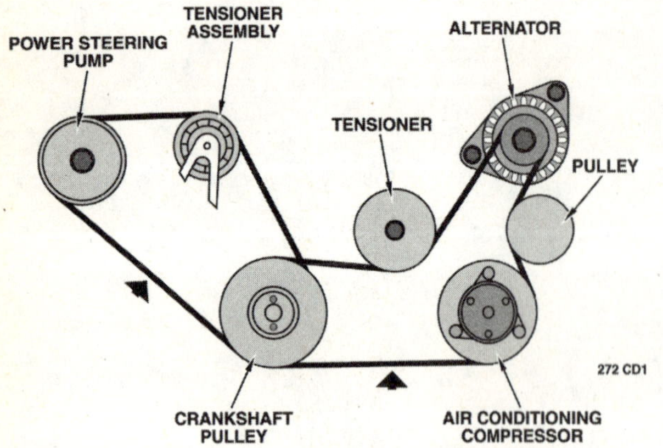

The drive belt deflection measuring points are indicated by the large arrows. Ensure that the drive belt is pushed in the direction shown. Models with air conditioning.

If a drive belt is slipping and the belt tension is correct, remove the drive belts and check that the pulleys operated by each drive belt rotate smoothly without binding. A belt is considered used after 5 minutes on a running engine.

To Renew or Adjust – Alternator and A/C Belt

On models with air conditioning, the alternator belt drives the air conditioning compressor. The procedure to renew or adjust the alternator drive belt is similar on all models.

1. Remove the retainers and withdraw the driver side splash-guard.
2. To check belt tension, apply a force of 100 Newtons (10 kg) to the point shown in the illustration. A new belt should deflect 6 – 7 mm, while the deflection of a used belt should be 8 – 10 mm.
3. If the belt is to be renewed, loosen the tensioner locknut and turn the adjusting bolt anticlockwise, until the belt can be removed.
4. If adjustment is required, loosen the tensioner locknut and turn the adjusting bolt clockwise to increase, or anticlock-wise to decrease the tension.

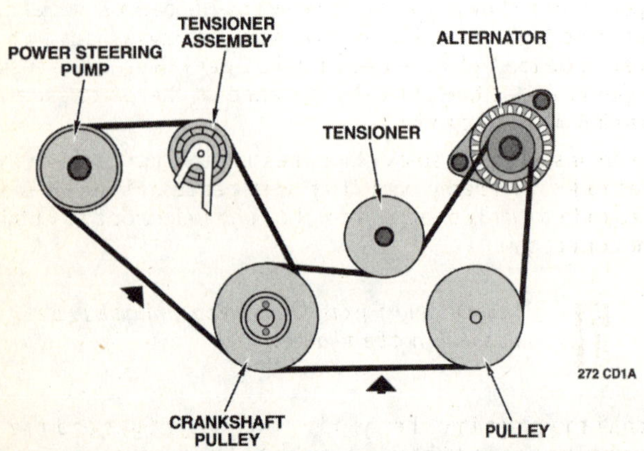

The drive belt deflection measuring points are indicated by the large arrows. Ensure that the drive belt is pushed in the direction shown. Models without air conditioning.

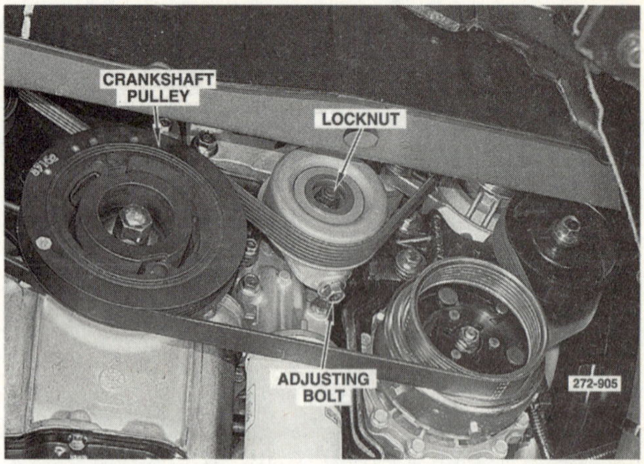

Alternator drive belt adjusting point. Models with air conditioning.

5. Securely tighten the locknut and the pivot bolt on the tensioner.
6. If a new belt has been installed, run the engine for approxi-mately five minutes and recheck belt tension. Adjust if necessary.

To Renew or Adjust – Power Steering Belt

1. Remove the retainers and withdraw the driver side splash-guard.
2. To check belt tension, apply a force of 100 Newtons (10 kg) to the point shown in the illustration. A new belt should deflect 8 – 9 mm, while the deflection of a used belt should be 11 – 14 mm.
3. If the power steering drive belt to be renewed, remove the alternator belt as previously described. To adjust the belt it is not necessary to remove the alternator belt.
4. Loosen the locknut and the pivot bolt on the power steering pump belt tensioner assembly.
5. Turn the adjusting bolt anticlockwise to increase, or clock-wise to decrease the tension.
6. Securely tighten the locknut on the tensioner assembly.
7. If a new belt has been installed, run the engine for approxi-mately five minutes and recheck belt tension. Adjust if necessary.

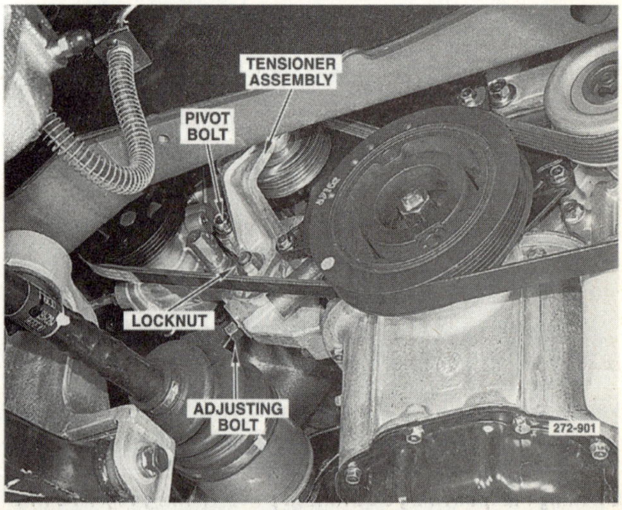

Power steering drive belt adjusting point.

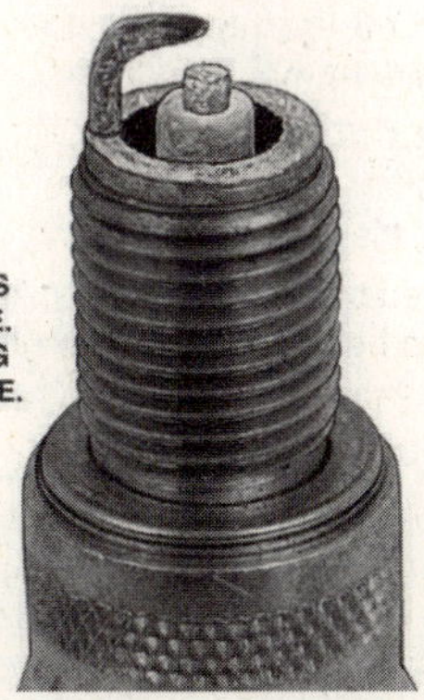

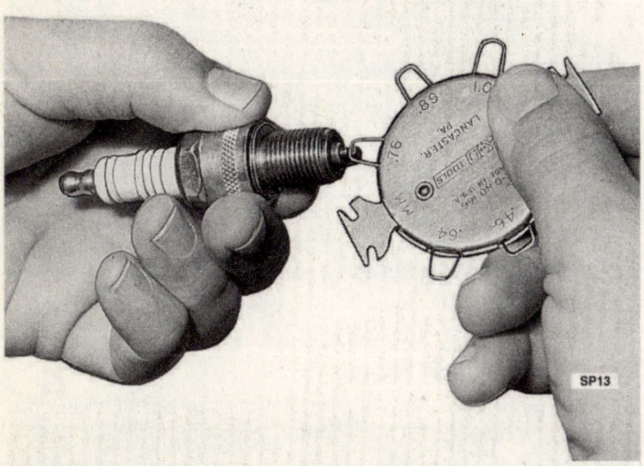

Checking the spark plug electrode gap with a wire gauge. if a wire gauge is not available, use a feeler gauge.

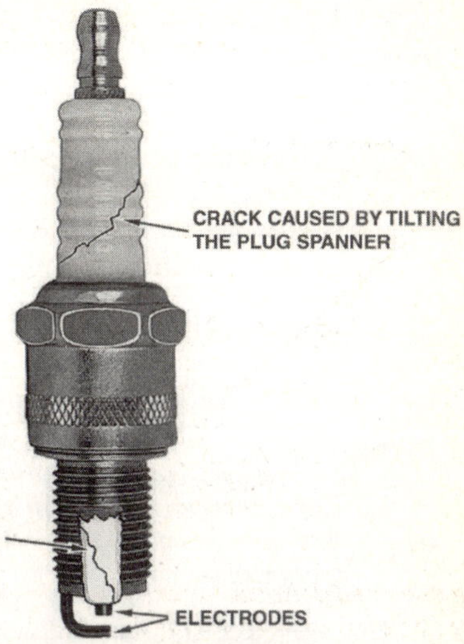

SPARK PLUGS

The spark plugs should be renewed at the intervals recommended in Routine Maintenance (chapter 2). The spark plugs in the rear cylinders are platinum tipped, long life plugs, which should be renewed at every 105 000 km. To remove the spark plugs from the rear cylinders, the plenum chamber must be removed, as described later in this chapter.

Before removing the spark plug leads, ensure that the area around the top of the rocker cover is clean to prevent foreign matter entering the cylinders when the plugs are removed.

Disconnect the high tension leads, pulling on the rubber boots only to prevent damage to the leads. Ease the leads out of the retaining clips, where necessary, noting the correct position of each lead.

Unscrew each spark plug in an anticlockwise direction, using a suitable thin spark plug socket and an extension bar.

Spark plugs removed from an engine in good mechanical condition should have a light powdery deposit ranging from light brown to greyish tan in colour. After considerable service the electrodes will show signs of wear or normal burning.

Spark plugs showing a thick black oily deposit indicate an engine in poor mechanical condition or possibly that spark plugs with a too low a heat range has been installed.

Spark plugs showing a white or yellowish deposit indicates sustained high speed driving or possibly that spark plugs with a heat range too high have been installed, particularly when these deposits are accompanied by blistering of the porcelain and burning of the electrodes.

Before installing the spark plugs, measure the gap between the electrodes, preferably with clean wire gauges. Move the side electrode towards or away from the centre electrode to obtain the correct gap.

The spark plug gap on platinum tipped plugs is preset, and should not be adjusted.

Screw the plugs into the cylinder head finger tight and use a torque wrench to tighten the plugs to the specified torque. A piece of rubber or plastic hose, that is a snug fit over the spark plug insulator may prove helpful on this engine to get the spark plugs started in the cylinder head threads.

If a torque wrench is unavailable, extreme care must be taken when tightening the spark plugs as overtightening may lead to stripping of the thread from the cylinder head.

Tighten the plugs until the seat of the plug just contacts the cylinder head and then tighten an additional quarter of a turn with a spark plug wrench.

DISTRIBUTOR CAP AND LEADS

When working on or near electronic ignition systems, care should be taken as dangerous high voltages are present.

The ignition switch should be turned Off before removing or installing any electrical connections, otherwise damage to the ignition system as well as severe electrical shock could result.

1 Remove the plenum chamber as described later in this chapter, to access the rear high tension leads.
2 Clean the rocker cover and distributor cap exterior surfaces with a brush or similar to prevent dirt entering the spark plug tubes or the distributor.
3 Disconnect the leads from the spark plugs and the rocker cover clips.

TIP *Always remove the lead from a spark plug by twisting and pulling on the rubber boot only. Use the same care when connecting the lead to the plug.*

The high tension leads have a carbon impregnated core. Care must be taken when removing the leads from the spark plugs to ensure that the leads are not damaged by stretching which will break the core and render the lead unserviceable. If a lead has a broken core it will cause misfiring.

4 Remove the distributor cap retaining screws and manoeuvre the cap and high tension leads from the distributor.
5 Check the inside and outside of the distributor cap for cracks and tracking. Renew the cap and rotor arm if cracks, excessive burning of the terminals or a worn carbon brush are apparent. Wipe the inside of the cap clean with a lint free cloth moistened with some methylated spirit or similar.

If the leads are serviceable, they may be carefully cleaned using a cloth moistened with kerosene. After they are clean, wipe completely dry.

When the cap and leads are removed it is recommended that an ohmmeter is used to check the leads as follows:

a Test one lead at a time by connecting the meter probes to the spark plug end of the lead and the corresponding terminal inside the cap. Resistance for each lead should be within Specifications.
b If the resistance is higher than the specified amount, remove the lead from the cap and check the resistance in the lead only. Renew the lead if necessary.
c High distributor cap resistance may be due to corrosion deposits on the cap terminals. These deposits can be removed with a small scraper or emery cloth.

6 When installing the distributor cap, ensure that the cap is correctly seated on the distributor housing.

Check inside the distributor cap for dirt, moisture, cracks and tracking between the terminals. Also check the carbon brush for wear and deterioration.

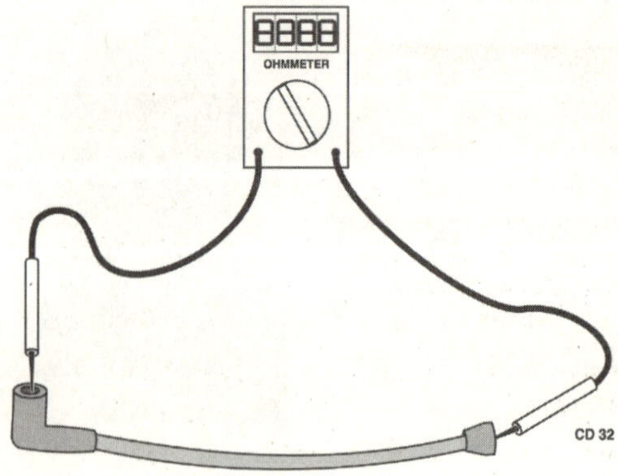

Checking the high tension lead resistance using an ohmmeter.

IGNITION TIMING AND IDLE SPEED

Connecting the timing light to the engine is not normally necessary, because the timing is set at the factory and is non adjustable. Timing is controlled by the engine ECU and can only be accurately checked using the manufacturers dedicated test equipment.

Connecting a Timing Light

1 Connect the timing light lead inductive pickup to the No 1 spark plug lead.
2 Where applicable, connect the power leads of the timing light to the battery.

Do not disconnect the timing light power leads with the engine running as voltage surges could damage the alternator.

Connecting a Tachometer

1 Connect the positive tachometer lead to terminal 3 (white wire) of the engine check connector.
2 Connect the negative tachometer lead to a good earthing point.

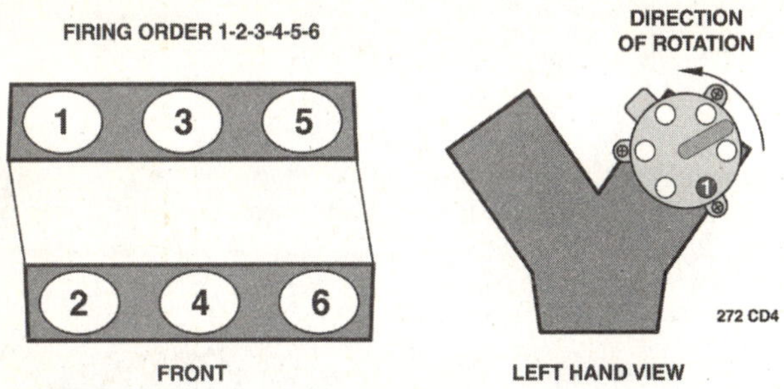

Cylinder numbering sequence showing the location of the No. 1 spark plug lead.

Do not allow the positive tachometer lead connector to short to earth as damage to the test equipment or ignition system may result.

To Check and Adjust Idle Speed

1 Run the engine until it reaches normal operating temperature.

2 Ensure that all accessories are turned off and, on automatic transaxle models, ensure that the transaxle is in Neutral or Park.

3 Stop the engine and connect the tachometer.

4 Start the engine and check the kerb idle speed. If the kerb idle speed is not to Specifications, proceed as follows:

a Stop the engine and bridge the diagnosis connector terminals 1 and 4 using a paper clip or split pin.

b Start the engine and check the base idle speed. Adjust the idle speed screw if the base idle speed is outside Specifications. If the idle speed is high and cannot be adjusted using the idle speed screw, check for vacuum leaks, an incorrectly adjusted throttle stop screw or a faulty fast idle air valve. The throttle stop screw adjustment is covered under the Throttle Body heading in Fuel and Engine Management (chapter 7).

c Remove the jumper lead from the diagnosis connector. After the engine has been idling for 10 minutes, check that the kerb idle speed is to Specifications.

Location of the engine check connector. Inset shows the terminal identification. Terminal 2 is the fuel pump test and terminal 3 is the tachometer terminal.

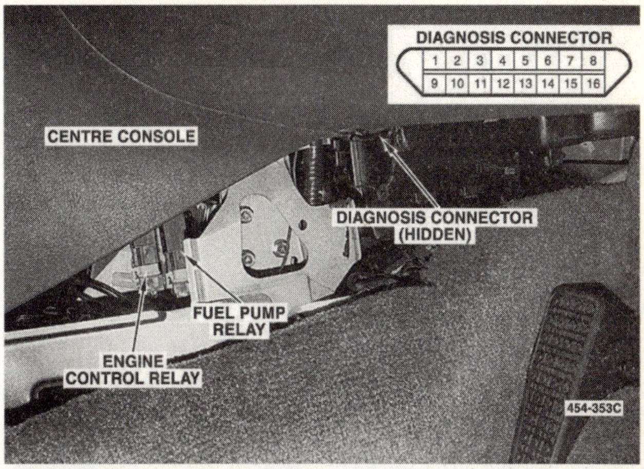

Location of the diagnosis connector. Inset shows terminal identification.

EXHAUST SYSTEM

Operations relating to the removal and installation of the exhaust system are covered under the Exhaust System heading at the end of this chapter.

To Inspect

1 Raise the front of the vehicle with a jack and support the vehicle with chassis stands under the recommended points. Refer to General Information (chapter 1) for the correct trolley jack and chassis stand positions to prevent damage or injury.

2 Check the alignment of the exhaust system, ensuring it has sufficient operating clearance.

3 Inspect the condition of the mounting rubbers. Any rubber which is broken or oil soaked should be replaced and the source of the oil leak identified and repaired.

4 Have an assistant hold a thick wadded cloth at the opening of the tail pipe for a short period (approx. 5 seconds), while the engine is running at idle speed. Check the system for excessive leaks, which will be heard as a hissing noise.

5 Check the condition of the gaskets. Any gasket excessively protruding from the joints, should be renewed.

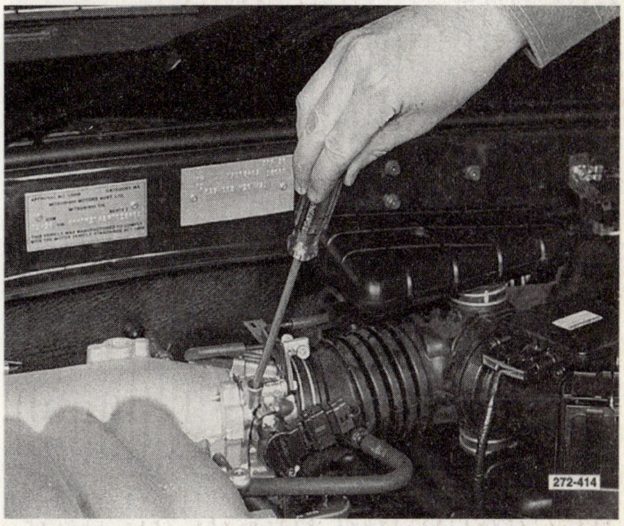

Adjusting the engine idle speed.

Check the exhaust system gaskets for leakage.

6 Inspect the catalytic converter, the resonator and rear muffler for corrosion, dents or damage.

7 Lightly bump the resonator and rear muffler with a rubber mallet. They should not have any loose parts inside.

> *Do not strike the catalytic converter, as the core may shatter, causing irrepairable damage to the converter.*

8 Inspect the exhaust system heat shields for damage and correct installation. Renew the heat shields as necessary.

9 Check the tightness of the nuts and bolts retaining the joints.

VALVE CLEARANCES

The valve clearance is not adjustable as a service operation. Valve clearance is normally adjusted automatically by the hydraulic tappets.

Trouble Shooting

ENGINE MISFIRES AT IDLING SPEED

> *Engine misfire can be caused by ignition, fuel or mechanical faults. If the misfire cannot be rectified by performing the procedures described at the start of this chapter, refer to the Trouble Shooting heading in Fuel and Engine Management (chapter 7) prior to checking the following possible causes.*

1 Air leak at the inlet manifold or plenum chamber gasket(s): Tighten the manifold bolts or renew the gasket(s) as necessary.

2 Blown head gasket: Check the cylinder compressions and renew the head gasket as necessary.

3 Burnt valves or seats in the cylinder head: Check the cylinder compressions and overhaul the cylinder head(s) as necessary.

4 Weak or broken valve springs: Remove the rocker gear and check the condition of the valve springs.

5 Broken or worn piston rings: Check the cylinder compressions and renew the piston rings as necessary.

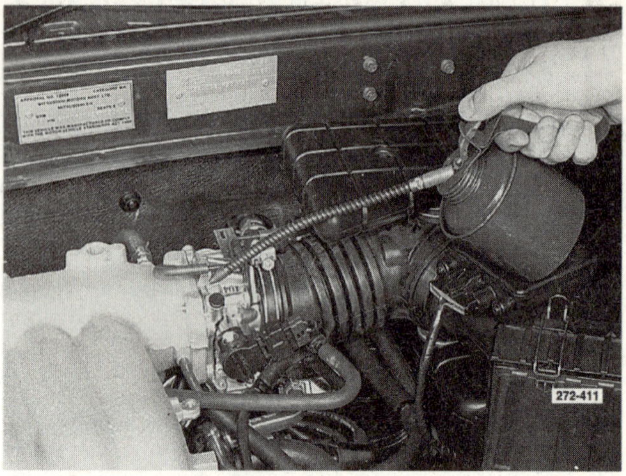

Check for vacuum leaks using engine oil.

> TIP
>
> *Check for air leaks at the inlet manifold and plenum chamber by applying engine oil or dewatering fluid around the manifold joints while the engine is idling. If the engine speed alters or smoke begins to emit from the exhaust system, there is an air leak at the inlet manifold.*
>
> *Cylinder compressions should be checked using a compression gauge as described later in this chapter.*
>
> *Broken valve springs will usually be obvious by excessive valve (tappet) noise.*

NOISY VALVE OPERATION

1 Faulty or blocked hydraulic tappets: Remove the rocker gear and clean or renew the tappets.

2 Excessive valve clearances: Check for worn tappets and/or camshaft lobes, and valve tips. Renew faulty components as necessary.

3 Worn rocker arms or shafts: Remove the rocker gear and check the components for wear.

4 Low oil pressure: Refer to the Low Oil Pressure heading.

5 Weak or broken valve springs: Remove the rocker cover and check the condition of the valve springs.

6 Worn valve guides: Overhaul the cylinder head.

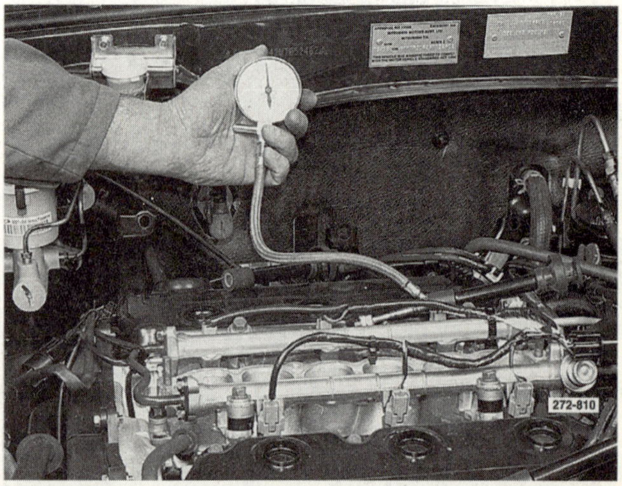

Using a compression gauge to measure the compression pressure.

CONNECTING ROD BEARING NOISE

1. Low oil pressure: Refer to the Low Oil Pressure heading.
2. Excessive bearing clearance: Renew the bearing shells. Renew the crankshaft if the crankpin journals are oval, tapered or scored.
3. Misaligned connecting rod bearings: Align the connecting rods and renew the connecting rod bearing shells.

TIP *Connecting rod bearing noise is indicated by a metallic knock which is usually loudest at approximately 60 km/h with the throttle closed. Before dismantling the engine to inspect the connecting rods, check the engine oil for correct level and dilution on the dipstick. Also remove the oil pressure sender unit and connect an oil pressure gauge into the oil gallery to check the oil pressure readings.*

MAIN BEARING NOISE

1. Low oil pressure: Refer to the Low Oil Pressure heading.
2. Loose flywheel/driveplate: Install new flywheel/driveplate bolts and tighten to the specified torque.
3. Excessive crankshaft end float: Renew the thrust washers and inspect the main bearings.
4. Excessive bearing to journal clearance: Renew the bearing shells. Renew the crankshaft if the journals are oval, tapered or scored.

TIP *Main bearing noise is indicated by a heavy but dull knock when the engine is under load. A loose flywheel/driveplate is indicated by a thud or dull click when the engine stops and is usually accompanied by vibration. Crankshaft end float noise is indicated by a sharp rap at idle speed. The crankshaft can be checked for excessive end float by levering the crankshaft backwards and forwards. If the oil pressure is satisfactory, consult an engine reconditioning specialist or remove the main bearing cap assembly and check the bearing clearance using Plastigage. Ovality and wear on the main bearing journals can only be checked with a micrometer after the crankshaft has been removed.*

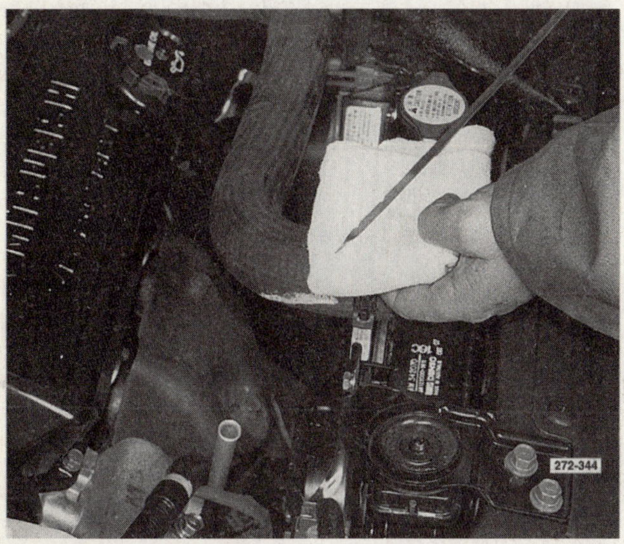

Checking the engine oil level.

EXCESSIVE OIL CONSUMPTION

1. Incorrect engine oil: Drain the engine sump and fill with the correct grade of oil.
2. Oil leaks: Check that the relevant retaining bolts or nuts are tight or renew the gasket or seal as necessary.
3. Faulty positive crankcase ventilation (PCV) system operation: Check the PCV system as described in Emission Control (chapter 8).
4. Damaged or worn valve stem oil seals: Renew the valve stem oil seals.
5. Excessive valve stem to valve guide clearance: Have a cylinder head specialist renew the valves and/or guides as necessary.
6. Worn or broken piston rings: Renew the piston rings and/or pistons as necessary.
7. Rings too tight or stuck in grooves: Renew the rings and clean out the ring grooves.
8. Excessive wear in the cylinders, pistons and rings: Recondition the cylinders and install oversize pistons and rings.
9. Compression rings incorrectly installed or the oil control rings blocked or broken: Renew the piston rings.

TIP *Damaged or worn valve stem oil seals which allow oil to be drawn down past the valve stems into the combustion chambers can be diagnosed by allowing the engine to idle for a few minutes and then opening the throttle. If oil is being drawn down past the valve stems, a heavy discharge of blue smoke will be seen at the tailpipe.*

Piston, ring and cylinder bore faults are normally accompanied by a loss of compression. Cylinder compressions can only be accurately checked using a compression gauge as described later in this chapter.

LOW OIL PRESSURE

1. Low engine oil level: Check and replenish the oil to the full mark on the dipstick.
2. Incorrect or diluted oil: Drain the engine sump and fill with the correct grade of oil. If the oil is diluted, check for the source of the water or fuel. Ensure that the engine is not overheating.
3. Blocked oil filter: Renew the oil filter.
4. Oil pump pressure relief valve stuck open or the spring broken: Free up the valve or renew the spring.
5. Split or obstructed oil pump pickup pipe: Remove the pickup pipe and clear the obstruction or renew the pickup pipe.
6. Excessive bearing clearance: Renew the bearing shells and check the condition of the crankshaft journals.
7. Excessive wear of the oil pump components: Dismantle the oil pump and renew the components as necessary.

TIP *Remove the oil pressure switch and connect a pressure gauge into the oil gallery. Check the oil pressure with the engine cold and hot. If the oil pump or pressure relief valve are faulty, low pressure will be indicated with the engine both hot and cold. However, if the bearings are at fault, a fairly high oil pressure will be indicated when the engine is cold but a marked fall in pressure will be noted when the engine is hot.*

Checking the condition of the timing belt.

ENGINE ROTATES BUT WILL NOT START

1 Ignition system fault: Refer to Roadside Trouble Shooting (chapter 3).

2 Fuel system fault: Refer to Fuel and Engine Management (chapter 7).

ENGINE WILL NOT ROTATE

1 Broken timing belt: Remove the front timing belt cover and examine the timing belt. If broken, also inspect the valves and pistons. Dismantling of the engine may be necessary.

2 Starter motor drive jammed: Remove the starter motor, check and renew the damaged drive and/or the flywheel ring gear.

3 Engine overheated and seized: Remove and dismantle the engine, check and renew the damaged components.

4 Coolant in the cylinders due to a blown head gasket or cracked cylinder block or head: Remove the cylinder head and if the gasket is blown, check the cylinder block and head for distortion. Renew the cylinder head and/or the cylinder block if cracked or excessively distorted.

5 Broken crankshaft, connecting rod or piston due to overheating, excessive engine speed, fatigue, etc: Remove and dismantle the engine, examine and renew the components as necessary.

6 Valve head broken off due to overheating, excessive engine speed, fatigue, etc: Remove the cylinder heads, check the heads, pistons and cylinders for damage and repair or renew as necessary.

Invariably when an engine seizes because of overheating due to lack of oil and/or water, damage is done to the bearings, pistons etc. Although there may be instances where an engine will restart and run after it has cooled and the oil and coolant have been replenished, it will generally be found that oil consumption increases, oil pressure drops and the engine will be noisier, depending on the extent of damage.

Sufficient coolant can enter a cylinder due to a blown gasket, cracked cylinder block or head to prevent the engine rotating. Normally this is preceded by difficult starting, misfiring, excessive steam from the exhaust and loss of coolant from the radiator cap or tank. With the engine fast idling, excessive discharge may be noticed from the reserve tank overflow pipe. When a cylinder head gasket blows allowing coolant into the cylinders, or compression loss between cylinders, it is essential to check the gasket faces on the cylinder block and head for distortion.

Cylinder Compression Test

To Test Compression — Compression gauge

A compression test is one of the best methods of assessing the mechanical condition of the upper part of the engine (pistons, rings, valves and head gasket). It is also the best place to start looking for poor performance, a misfire or bad fuel economy. Unless cylinder compressions are even, renewing other components to try to get an engine running better will be a waste of time.

Compression testing is also performed if an engine is overheating, or loosing coolant and no leaks externally have been found. If compression pressure is getting into the coolant, unsatisfactory cooling system performance will result.

The accuracy of compression gauges vary, so the important thing to look for in a test is that all cylinders are even, rather than the actual result.

1 Ensure that the battery is fully charged.

2 Run the engine until it reaches normal operating temperature. On 3.5 litre models, remove the engine cover.

3 To gain access to the rear cylinders, remove the plenum chamber as described later in this chapter.

4 Ensure that the area around the top of the rocker cover is clean to prevent foreign matter entering the spark plug tubes and/or cylinders. Remove the spark plug leads.

5 Remove fuse No 15 (engine control relay) from the engine compartment fuse and relay box.

Failure to remove the fuse can result in damage to the ignition system or an explosion.

6 Install a compression gauge to No 1 cylinder. Refer to the instrument manufacturers instructions if necessary.

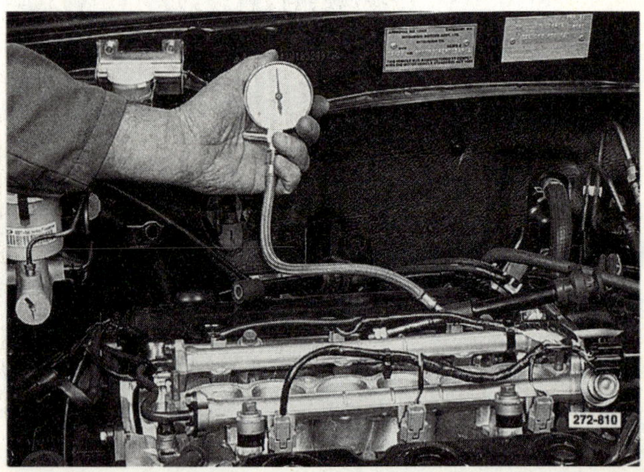
Using a compression gauge to measure the compression pressure.

7 Have an assistant operate the starter motor. Observe the compression gauge and stop the engine when the gauge has reached the highest reading, usually after three or four revolutions.

8 Record the reading and check the compression of the remaining cylinders in the same manner.

9 Compare all the readings taken. Refer to Specifications for the minimum acceptable pressure. The difference between the highest and lowest readings should not exceed 98 kPa.

10 If a low reading is recorded on one or more cylinders, the trouble source may be isolated as follows:

a Inject a small amount of engine oil into the spark plug hole of the cylinder and ensure that the oil is evenly distributed by rotating the engine.

b Repeat the compression test on the cylinder concerned.

A substantial increase of compression pressure indicates faulty or worn piston rings, piston or cylinder bore.

Little or no increase of compression pressure indicates burnt, obstructed or sticking valves, or a leaking or blown cylinder head gasket.

Engine Assembly

To Remove and Install Engine—Suitable lifting tackle, trolley jack

TO REMOVE AND INSTALL

The engine and transaxle can be removed from the vehicle as an assembly, however the vehicle must be raised quite high to allow the engine and transaxle to pass underneath. In most situations, it may not be possible to raise the vehicle to the required height unless a vehicle hoist or suitable long chassis stands are available. The minimum clearance required to remove the engine and the transaxle from below the vehicle is 50 cm, measured from the engine drain plug.

As most people do not have this equipment, the following procedure describes the engine only being removed from a vehicle in the conventional manner, using overhead lifting tackle.

Front three quarter view of the engine. Models with air conditioning.

1 Place suitable protection around the engine compartment to prevent damage to the vehicle body. On 3.5 litre models, remove the engine cover.

2 Reduce the fuel system pressure and disconnect the fuel supply and return hoses, referring to Fuel and Engine Management (chapter 7). Discard the O rings. Seal the hoses to prevent the entry of dirt.

TIP *Loosen the fuel filler cap, to prevent pressure build up in the tank and subsequent fuel leakage while the engine is removed.*

3 Disconnect the negative battery terminal.

4 Disconnect the windscreen washer hose on the drivers side of the bonnet. While an assistant supports the bonnet, remove the retaining bolts and carefully remove the bonnet.

5 Remove the transaxle, referring to the relevant transaxle chapter.

On manual transaxle models a pull type clutch is fitted, making it necessary to disconnect the release bearing clip during transaxle removal. Always refer to the procedure in Manual Transaxle (chapter 10) to avoid damaging components.

6 Drain the cooling system and remove the top and bottom radiator hoses. Disconnect the heater hoses, located below the throttle body. Refer to Cooling System (chapter 5), if necessary.

7 Remove the power steering pump, referring to Steering (chapter 12), if necessary.

8 Remove the drive belt referring to Routine Maintenance at the beginning of this chapter, if necessary.

9 Remove the bolts retaining the air conditioning compressor to the bracket on the cylinder block. Suspend the compressor with the hoses still attached using wire or rope, clear of the work area.

The top compressor retaining bolts are close to and virtually identical in appearance to the pressure hose flange retaining bolts. Do not remove the pressure hose flange bolts as the escaping refrigerant could cause serious injury.

10 Suitably mark and disconnect the wiring from the TPS, crank angle sensor, alternator, temperature gauge sender, coolant temperature sensor, ISC motor, distributor, radio suppressor, oil pressure switch and fuel injector wiring connector.

11 Remove the bolts retaining the wiring harness to the alternator and to the thermostat housing and place the wiring clear of the work area. Remove the bolt retaining the oil pressure switch wiring bracket to the cylinder block.

12 Mark the installed position of the throttle cable retaining bracket, remove the bolts, disconnect the earth cable and disconnect the inner throttle cable from the throttle lever. Place the earth cable clear of the work area.

Rear three quarter view of the engine.

13 Suitably mark and disconnect the vacuum hoses from the charcoal canister, the brake booster and the pressure regulator. If applicable, disconnect the vacuum supply hose from the cruise control diaphragm.

14 Remove the bolts retaining the windscreen washer bottle and partially withdraw the bottle. Disconnect the wiring connector(s) and the hose(s) and remove the washer bottle.

15 Attach a suitable length chain or rope to the engine lifting hooks and allow the lifting equipment to take the weight of the engine.

Remove the oil filler cap if the chain/rope interferes with the cap. Place a clean cloth into the opening to prevent dirt entering the engine. It may be necessary to place a piece of timber between the engine lifting hook and the front rocker cover to prevent damage to the cover.

16 Remove the through bolt from the right hand side engine mounting. Place the power steering supply hose over the engine mounting.

17 Ensuring that all electrical wiring and hoses are clear and disconnected, carefully raise the engine and remove it from the vehicle.

Installation is a reversal of the removal procedure with attention to the following points:

1 Tighten all bolts securely.

2 Ensure that the wiring and vacuum hoses are correctly routed and securely connected.

3 Connect the fuel supply and return hoses, using new O rings.

4 Renew the engine exhaust pipe gaskets.

5 Securely tighten the fuel filler cap.

6 Fill the engine with the correct grade and quantity of oil. Fill the cooling system as described in Cooling System (chapter 5).

If the cooling system is not filled as described in the Cooling System chapter, engine overheating may result.

7 Check for external fuel leakage before starting the engine. This procedure is described in Fuel and Engine Management (chapter 7).

Removing the engine assembly from the vehicle.

8 Start and run the engine until it reaches normal operating temperature and check for fuel, oil and coolant leaks. Rectify as necessary.

9 Allow the engine to cool and check that the nuts retaining the exhaust engine pipe to the manifold flanges are tight and that all hoses and connections are secure.

10 If new components have been installed to the engine, for example pistons, rings, bearings etc, the engine must be run-in at reduced speeds and loads for the first 1 000 km or so. Do not operate the engine at full throttle or allow it to labour in any gear during this period. Follow the component manufacturers advice and it is recommended that the engine oil and filter are changed at the end of this period.

Plenum Chamber

TO REMOVE AND INSTALL

1 Disconnect the negative battery terminal.

2 On 3.5 litre models, remove the bolts retaining the engine cover and withdraw the cover from the engine.

3 Remove the air cleaner inlet hose, the airflow sensor and the air cleaner element, referring to Fuel and Engine Management (chapter 7), if necessary.

4 Drain the cooling system, referring to Cooling (chapter 5), if necessary. Disconnect the coolant bypass hoses from the throttle body.

5 Mark the installed position of the throttle cable retaining bracket, disconnect the earth cable, remove the bolts and disconnect the throttle cable from the throttle lever. Place the throttle and earth cables clear of the work area.

6 Suitably mark and disconnect the vacuum hoses from the charcoal canister, the brake booster and the pressure regulator. If applicable, disconnect the vacuum supply hose from the cruise control diaphragm.

7 Suitably mark and disconnect the wiring connectors from the TPS and the ISC motor. Remove the fuel injector and the crank angle sensor wiring connectors from the retaining brackets. Release the injector wiring connector from the front retaining bracket.

8 Remove the bolts retaining the plenum chamber to the rear support brackets.

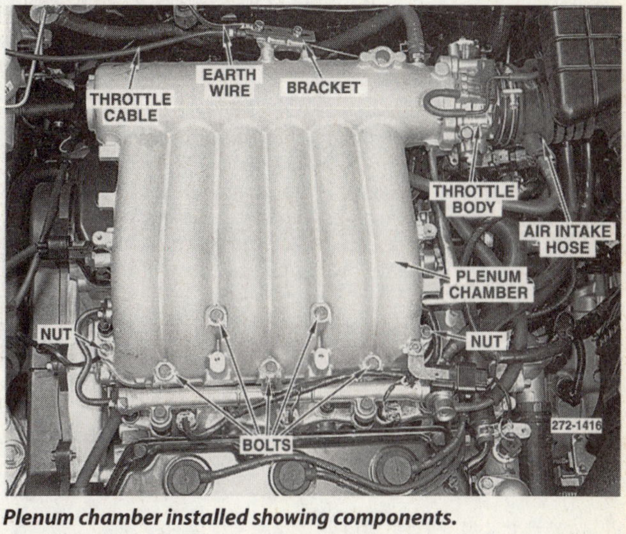

Plenum chamber installed showing components.

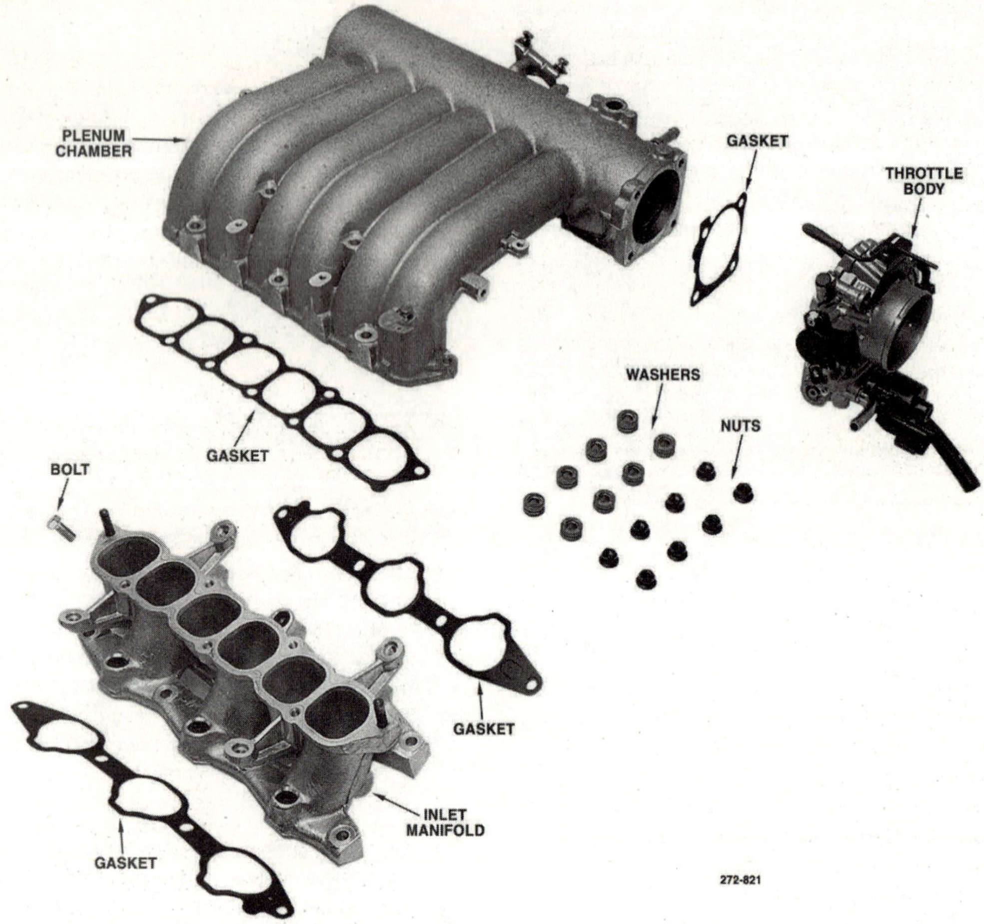

Plenum chamber and inlet manifold components.

9 Remove the bolts and nuts retaining the plenum chamber to the inlet manifold and manoeuvre the plenum chamber from the engine compartment. Discard the gasket.

10 If necessary, remove the bolts and separate the throttle body from the plenum chamber. Discard the gasket.

Installation is a reversal of the removal procedure with attention to the following points:

1 Ensure that all traces of old gasket material have been removed from the gasket surfaces.

2 If removed, install the throttle body using a new gasket.

TIP *Whenever the plenum chamber is removed, check that the throttle body is free from carbon deposits. Clean if necessary, using a clean rag soaked with carburettor cleaner. If excessively dirty, remove the fast idle air valve and the ISC motor as described in Fuel and Engine Management (chapter 7). Clean the ISC motor ensuring that carburettor cleaner is not used on the fast idle air valve*

3 Install a new gasket, ensuring that the holes on the gaskets are aligned with the holes on the plenum chamber. Place the plenum chamber onto the inlet manifold.

4 Tighten the retaining nuts and bolts to the specified torque in a spiral pattern, starting with the centre bolts, .

5 Securely connect the wiring connectors and the vacuum hoses as marked during removal.

6 Connect the throttle cable, install the earth cable and tighten the bolts retaining the bracket.

7 Install the air inlet hose and the airflow sensor. Connect the wiring connector. Install the air cleaner element.

Inlet Manifold

TO REMOVE AND INSTALL

1 Remove the plenum chamber as previously described.

2 Remove the fuel rail, referring to Fuel and Engine Management (chapter 7), if necessary.

3 Remove the bolts retaining the front timing belt cover and remove the manifold retaining bolt, located above the water pump.

TIP *Place a piece of clean cloth over the water pump pulley to prevent the bolt falling inside the timing belt covers.*

4 Remove the inlet manifold retaining nuts and washers and withdraw the manifold from the engine. Discard the gaskets and the washers.

5 Remove the fuel injector O rings from the manifold.

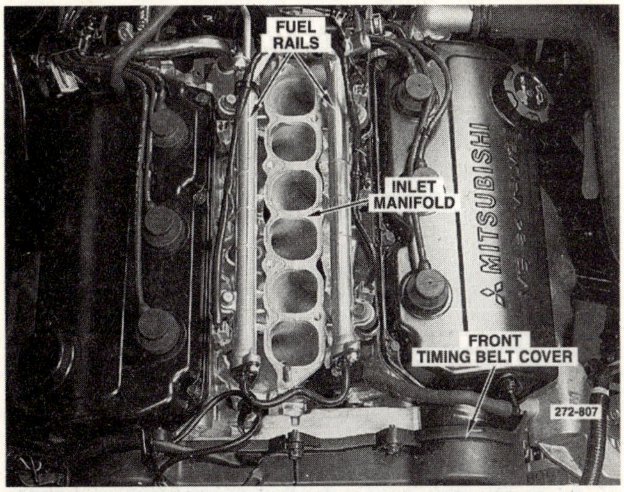

Plenum chamber removed showing the fuel rail installed.

Installation is a reversal of the removal procedure with attention to the following points:

1. Using new gaskets, install the inlet manifold to the cylinder heads, ensuring that the holes on the gaskets are aligned with the holes on the inlet manifold.
2. Place the new spring washers and nuts onto the studs. Tighten the front row of retaining nuts to 7 Nm.
3. Tighten the rear row of retaining nuts to 23 Nm.
4. Tighten the front and then the rear row of retaining nuts to 23 Nm.
5. Install the retaining bolt above the water pump and tighten securely. Remove cloth and install the timing belt cover.
6. Install the fuel rail, using new O rings.
7. Install the plenum chamber, using a new gasket as previously described.

Exhaust Manifold

TO REMOVE AND INSTALL

1. Raise the front of the vehicle and support on chassis stands under the recommended points. Refer to General

Inlet manifold retaining nuts and bolt.

Information (chapter 1) for the correct trolley jack and chassis stand positions to prevent damage or injury.

2. Remove the retaining bolts and remove the heat shield from the relevant exhaust manifold.
3. If removing the front exhaust manifold, disconnect the wiring connectors and remove the alternator. Refer to Engine Electrical (chapter 6), if necessary.
4. Remove the nuts retaining the engine pipes to the exhaust manifolds, lower the engine pipe assembly and support using a piece of wire tied to the crossmember of the vehicle.

> *Do not allow the pipe to hang unsupported, as damage to the flexible joint may occur.*

5. Remove the nuts and washers retaining the relevant exhaust manifold to the cylinder head and withdraw the manifold from the vehicle. Discard the gasket.

Installation is a reversal of the removal procedure with attention to the following points:

1. Check the condition of the cylinder head and exhaust manifold mating surfaces. Ensure that all traces of old gasket material have been removed from the gasket surfaces.
2. Check the condition of the studs in the cylinder head. If they are corroded or the threads are damaged, renew the studs.

> *To remove the studs from the cylinder head, refer to General Information (chapter 1).*

3. Apply a small amount of Permatex Anti Seize paste or equivalent to the threads of the exhaust manifold and engine pipe retaining nuts, before installing the nuts. Place a new gasket onto the cylinder head, ensuring that the gasket does not block the openings of the cylinder head or the exhaust manifold.
4. Tighten the retaining nuts progressively to the specified torque.
5. Run the engine and check for exhaust leaks.

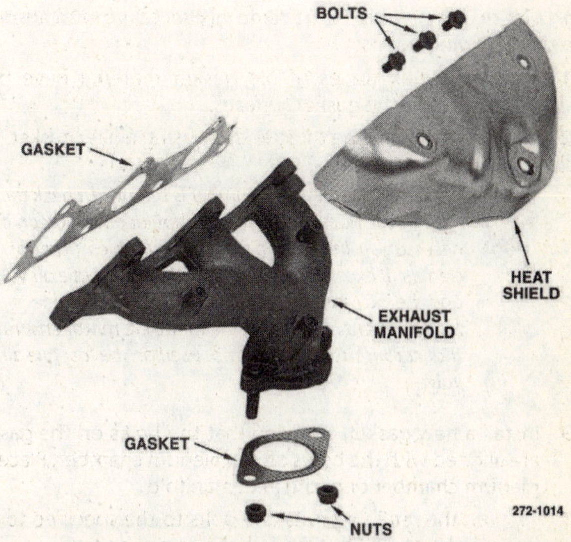

Exhaust manifold components.

Rocker Covers

To Remove and Install

1. Disconnect the negative battery terminal. On 3.5 litre models, remove the bolts retaining the engine cover and withdraw the cover from the engine.
2. Clean around the top of the rocker cover to prevent dirt, etc. entering the spark plug tubes or the engine. Disconnect the high tension leads from the spark plugs. Pull only on the rubber boot at the end of each lead to prevent damage.
3. If removing the rear rocker cover, remove the plenum chamber, as previously described.
4. Remove the bolts retaining the front and rear timing belt covers and withdraw the covers from the engine compartment.

TIP *If removing the front rocker cover only, do not remove the rear timing belt cover.*

5. Disconnect the PCV vacuum hoses from the rocker covers.
6. Remove the bolts retaining the rocker cover and withdraw the cover from the engine. Discard the gasket.

TIP *The rocker cover may be tightly seated, due to the spark plug tube seals in the cover. Use a flat bladed screwdriver wrapped in cloth to evenly separate the cover from the cylinder head.*

Installation is a reversal of the removal procedure with attention to the following points:

1. Clean the rocker cover in solvent. Check that the gasket face is not distorted or cracked. Repair or renew the rocker cover if necessary.
2. Check the condition of the spark plug tube seals. Renew the spark plug tube seals if excessively hard, soft or damaged.
3. Install a new gasket and progressively tighten the rocker cover retaining bolts to the specified torque.
4. Connect the spark plug leads securely.
5. If removed, install the plenum chamber, referring to the appropriate heading in this chapter.
6. Install the timing belt cover(s). Tighten the retaining bolts securely.

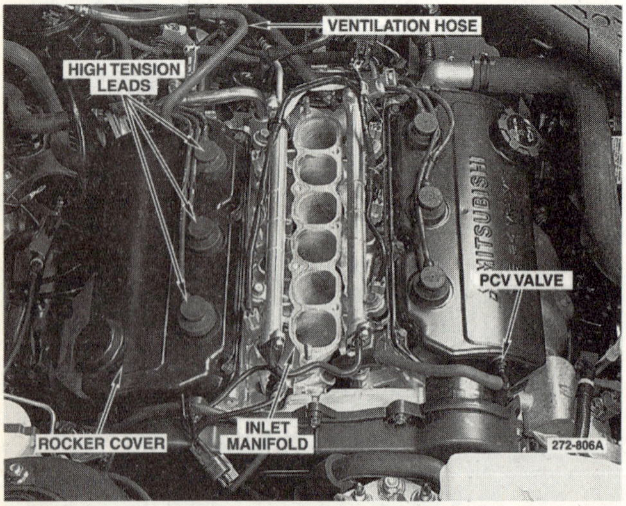

Rocker covers installed, plenum chamber removed.

Timing Belt

To Remove Crankshaft Pulley — Harmonic balancer puller
To Set Timing Belt Initial Tension — Small torque wrench, fabricated setting tool

TO REMOVE

1. Raise the front of the vehicle and support it on chassis stands. Refer to General Information (chapter 1) for the correct jacking and support points. Remove the bolts and scrivets retaining the right hand side splashguard and withdraw it from the vehicle.
2. Disconnect the negative battery terminal.
3. Support the engine under the sump, using a floor jack with a piece of timber placed between the head of the jack and the sump.
4. Remove the bolts retaining the windscreen washer bottle and partially withdraw the bottle. Disconnect the wiring connector(s) and the hose(s) and remove the washer bottle.
5. Remove the through bolt from the right hand side engine mounting.
6. Remove the nuts and bolt retaining the right hand side engine mounting to the support bracket and withdraw the mounting from the engine compartment.

Note the direction of the arrows on the rubber pads, prior to removing the pads.

7. Remove the power steering and the alternator drive belts, referring to Routine Maintenance heading in this chapter, if necessary.
8. Remove the power steering drive belt tensioner pivot bolt and the locknut. Withdraw the assembly from the engine compartment.
9. Remove the alternator, referring to Engine Electrical (chapter 6).

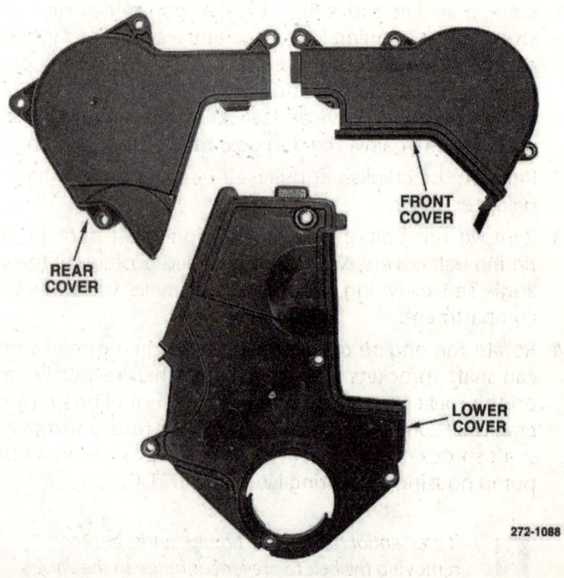

Timing belt covers.

Timing belt covers and other components installed prior to removing the timing belt.

10 Noting their installed position, loosen the bolts retaining the right hand side engine mounting support bracket and withdraw the bracket from the engine compartment.

> **TIP** *The bolts cannot be withdrawn separately from the bracket as they interfere with the vehicle body.*

11 Using a socket and a long breaker bar, remove the crankshaft pulley retaining bolt. On manual transaxle models, the engine can be locked by selecting fourth gear and applying the brakes. On automatic transaxle models, remove the starter motor and have an assistant hold a large screwdriver or tyre lever against the ring gear teeth to lock the engine.

12 Remove the crankshaft pulley. If necessary, use a harmonic balancer puller.

13 Remove the bolts retaining the front, rear and the lower timing belt covers. Noting the installed position of the crank angle sensor wiring, withdraw the covers from the engine compartment.

14 Rotate the engine clockwise until the timing marks on the camshaft sprockets are aligned with their respective marks on the rocker covers. If the marks are not aligned, turn the crankshaft pulley clockwise an other full turn, until the crankshaft sprocket mark is aligned with the protrusion on the oil pump housing, indicating No 1 piston TDC.

> **i** *It is essential to align the timing marks prior to removing the belt to prevent damage to the valves or pistons.*

15 Remove the bolts retaining the crank angle sensor to the oil pump housing. Carefully lift the sensor away from the plate and place it clear of the work area.

16 If the timing belt has done little work and is to be reused, mark the timing belt with an arrow to show the direction of rotation.

> **TIP** *It is important to return a used timing belt to the original direction of rotation or adverse wear could result.*

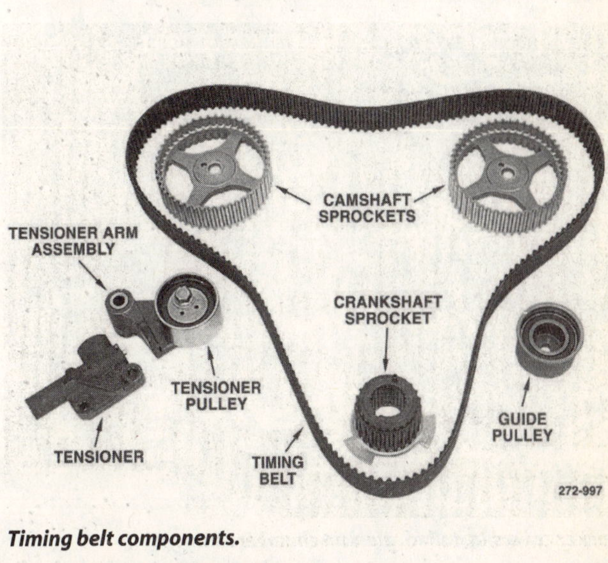

Timing belt components.

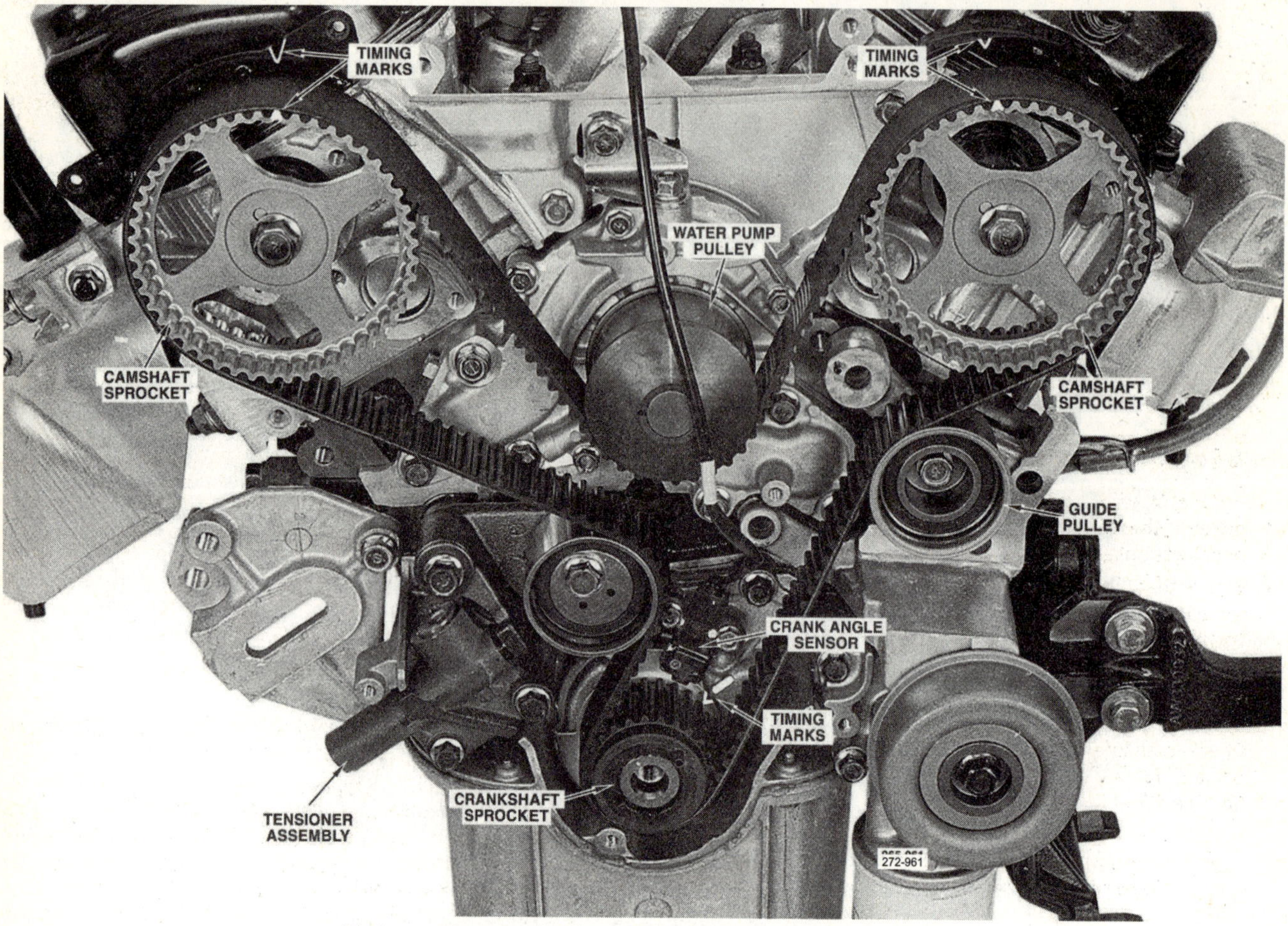

Timing marks correctly aligned prior to removing the timing belt.

17 Loosen the tensioner pulley lock bolt.

18 Remove the retaining bolts and withdraw the timing belt tensioner assembly.

19 Carefully remove the timing belt from the sprockets and pulleys. Withdraw the belt from the vehicle.

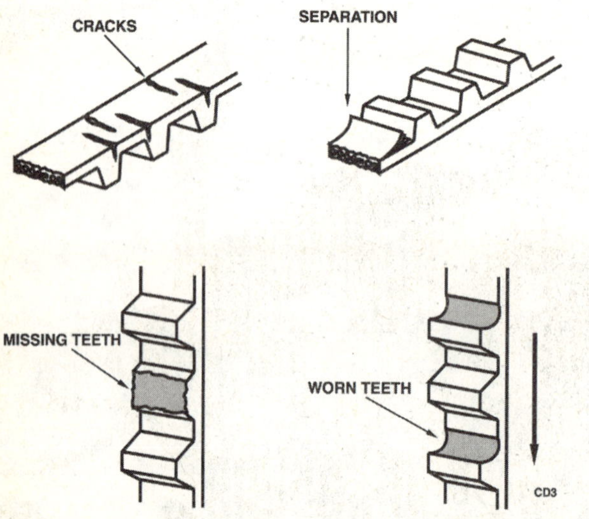

Check the timing belt for the illustrated faults.

The rear camshaft sprocket may rotate due to valve spring tension.

TO CLEAN AND INSPECT

Do not wash the timing belt, sprockets or tensioner pulley. Keep these components free of oil and grease. Wipe the components with a clean cloth if required.

1 Inspect the crankshaft and camshaft sprockets for foreign material on the teeth or wear and damage. If necessary, ease the crankshaft sprocket from the end of the crankshaft using two large screwdrivers. Ensure that the key in crankshaft is not misplaced.

2 Check the condition of the crank angle sensor blades. If necessary, separate the blade assembly from the sprocket and renew.

3 If the camshaft sprockets are to be removed, retain the relevant camshaft sprocket with a screwdriver or similar, and remove the camshaft sprocket retaining bolt.

Ensure that the cylinder head or rocker cover is not damaged while removing the sprocket.

4 Turn the tensioner and guide pulleys by hand and check for wear and roughness in the bearing. Check the contact

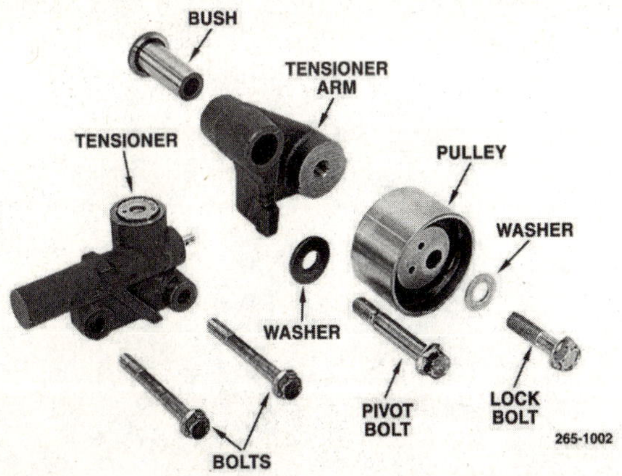

Tensioner components dismantled.

surface of each pulley for wear and damage. Renew the pulley if the contact surface is worn or damaged or if the bearing is unserviceable.

5 Ensure that the tensioner arm is pivoting freely. If necessary, remove the pivot bolt and withdraw the arm. Check the condition of the bush and sparingly lubricate or renew as necessary.

6 Inspect the timing belt for damage or wear. Check the belt thoroughly for peeling, cracking or hardening of the rubber. Check to ensure that the belt is not contaminated with oil or coolant. Renew the belt and rectify the leaks if contamination is evident.

TIP *When inspecting the timing belt do not bend, twist or turn the belt inside out as damage could occur.*

7 If the timing belt is worn or damaged on one side only, check the alignment of the sprockets, the guide and the tensioner pulley.

TO INSTALL

Installation is a reversal of the removal procedure with attention to the following points:

1 If removed, install the camshaft sprockets and tighten the retaining bolts to the specified torque. Hold the sprockets using the same method as removal.

2 Place the tensioner assembly in a vice fitted with soft jaws and slowly compress the piston rod until the holes in the tensioner body and the piston rod are aligned.

3 Insert a 1.5 mm thick wire into the hole in the tensioner body, ensuring that the wire is pushed through the piston rod and is engaging the other side of the tensioner body.

4 Ensure that the timing marks are correctly aligned.

5 Place the timing belt over the crankshaft sprocket. If installing the original belt, ensure that the arrow marked on the belt is facing the direction of rotation. While applying pressure to the belt, place it over the guide pulley.

6 With an assistant retaining the front camshaft sprocket, place the belt over the front camshaft sprocket, ensuring that the belt correctly engages the teeth of the sprocket and that the timing marks are aligned.

7 Guide the belt under the water pump pulley and over the rear camshaft sprocket. Have the assistant retain the rear sprocket to prevent the sprocket moving.

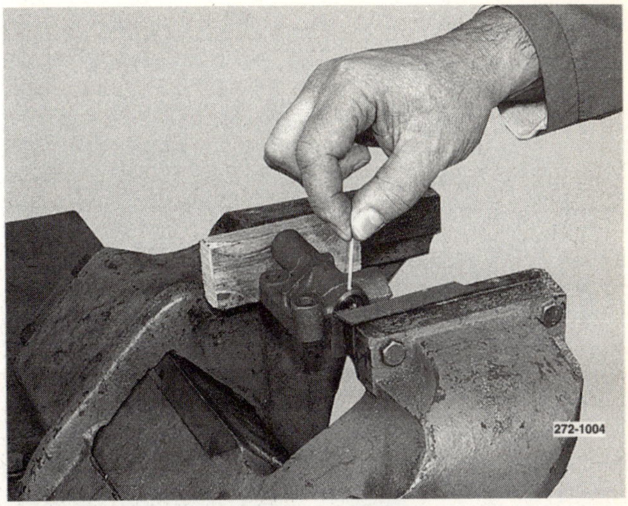

Method of resetting the timing belt tensioner.

8 While keeping the belt tight between the rear camshaft sprocket and the crankshaft sprocket, install the belt over the tensioner pulley.

9 Press the tensioner pulley against the timing belt and temporarily tighten the lock bolt. Ensure that all timing marks are correctly aligned.

10 Install the tensioner assembly and tighten the retaining bolts to the specified torque. Do not remove the wire from the assembly at this stage.

11 Install the crankshaft pulley retaining bolt with a thick washer (10 mm) on the shank to the crankshaft. Turn the engine a quarter of a turn anticlockwise and then a quarter of a turn clockwise, aligning the timing marks at the end of the turn. Loosen the tensioner pulley lock bolt. If the timing marks do not align, the belt is not installed correctly. Remove the belt and start again.

Installing the front camshaft sprocket.

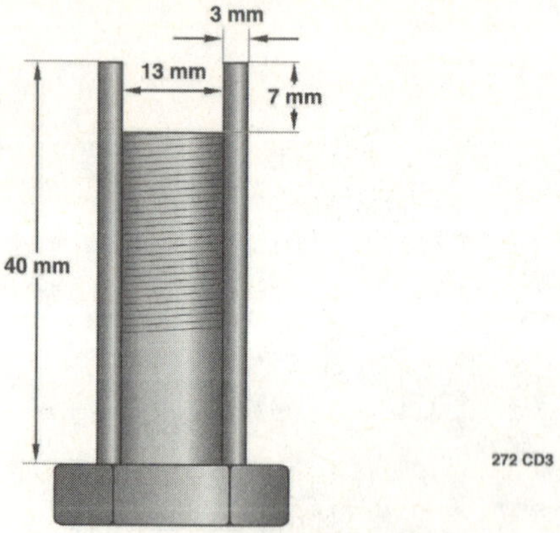

Diagram of the setting tool necessary to adjust the initial timing belt tension. The 3 mm steel rod should be brazed or silver soldered to the bolt shank.

> *The thick washer must be installed on the crankshaft pulley retaining bolt prior to installation, to prevent damage to the crankshaft, through the bolt bottoming out.*

12 To correctly set the timing belt initial tension, fabricate a tool to the dimensions shown on the illustration. Alternatively obtain the genuine tool part No EMB 998767. Install the tool to the holes in the tensioner. Ensure that the lock bolt is in the uppermost position.

13 Apply 5 Nm tension to the tool and while holding the tool under tension, tighten the lock bolt to the specified torque.

> *The tensioner pulley must be turned anticlockwise, towards the water pump pulley, with the lock bolt in the uppermost position and the tool inserted at the bottom, as shown on the illustration.*

14 Turn the engine two turns in a clockwise direction, ensuring that the timing marks are still correctly aligned. Let the engine stand for approximately 5 minutes, then withdraw the wire from the tensioner body.

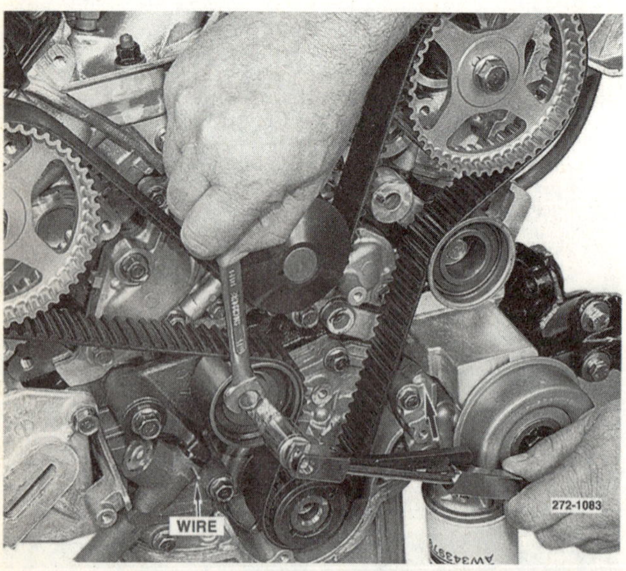

Using the setting tool to initially adjust the timing belt tension. Note that the wire is still installed to the tensioner.

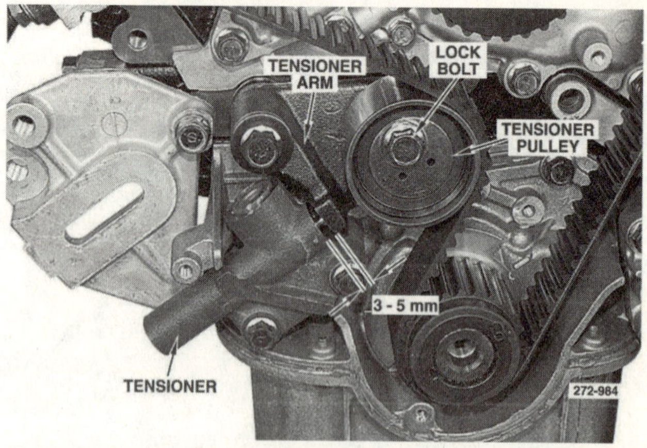

Check the distance between the tensioner and the pulley arm.

15 If the wire can not be withdrawn from the adjuster using the fingers, repeat steps 13 and 14, until the wire can be easily removed. Do not use pliers to remove the wire.

16 Check the distance between the tensioner and the tensioner arm. The distance should be 3-5 mm.

17 Install the crank angle sensor and tighten the retaining bolts securely.

18 Install the timing belt covers and tighten the bolts securely. Ensure that the crank angle sensor wiring is correctly routed as noted on removal.

19 Install the crankshaft pulley and tighten the retaining bolt to the specified torque. Use the same method used on removal to prevent the crankshaft turning.

20 Install the right hand side engine mounting support bracket and tighten the retaining bolts to the specified torque.

21 Install the power steering drive belt tensioner assembly and tighten securely.

22 Install the alternator, referring to Engine Electrical (chapter 6), if necessary.

23 Install the drive belts, referring to Routine Maintenance heading in this chapter, if necessary.

24 Install the engine mounting and tighten the retaining bolt and nuts to the specified torque, ensuring that the arrows on the rubber pads are pointing towards the right hand side of the vehicle.

25 Road test the vehicle and check for correct engine performance.

Rocker Arms, Shafts and Hydraulic Tappets

TO REMOVE

1 Raise the front of the vehicle and support on chassis stands under the recommended points. Refer to General Information (chapter 1) for the correct trolley jack and chassis stand positions to prevent damage or injury.

2 Remove the timing belt as previously described in this chapter.

3 Rotate the front and rear camshaft sprockets anticlockwise until the dowel pins aligned as shown on the illustration. In this position the camshaft lobes will exert equal force on the rocker assembly.

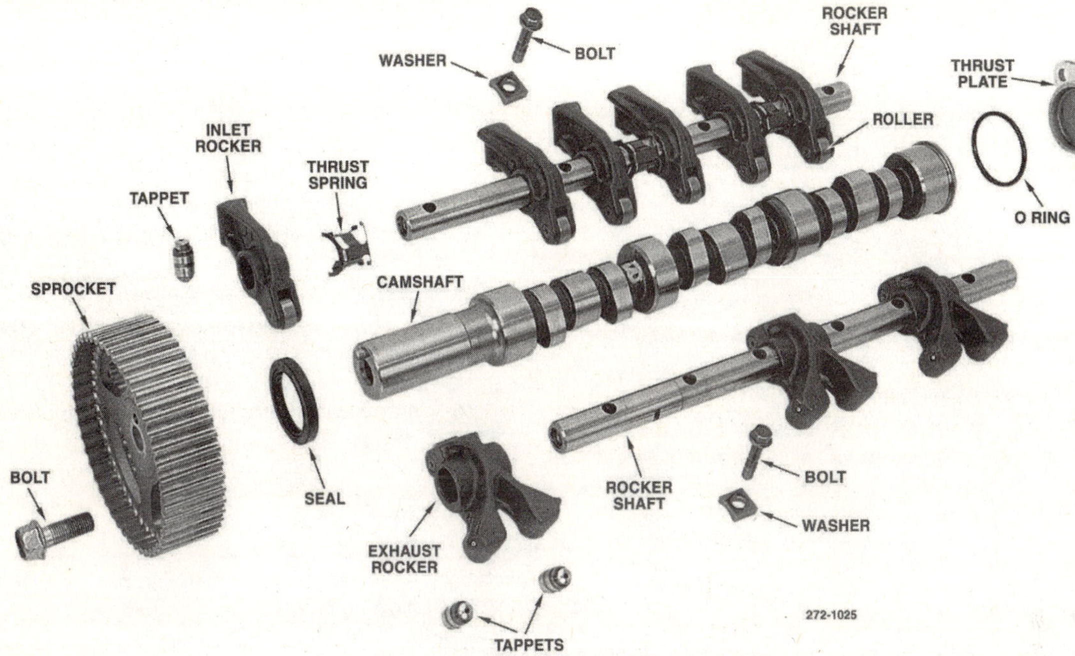

Camshaft and rocker shafts partially dismantled.

Do not turn the crank and/or camshafts clockwise, or damage to the engine may occur.

4 Remove the plenum chamber and the rocker covers, referring to the relevant sections in this chapter.

If only working on the front cylinder head, it is not necessary to remove the plenum chamber.

5 Block the cylinder head oil drain holes with clean lint free cloth to prevent the hydraulic tappets being dropped into the engine while removing the rocker assembly.

6 Gradually and progressively loosen the rocker shaft retaining bolts, one turn at a time, working from each end towards the centre, until all pressure is released from the assembly. Noting their installed direction remove the washers.

7 Wrap electrical tape or similar around the hydraulic tappets, to prevent the tappets falling from the rocker arms. Carefully lift the rocker arms and shafts from the cylinder head and place them on a clean flat surface.

Position of the camshaft dowel pins prior to removing the rocker assembly.

TIP *Retaining clips are available from the manufacturer to prevent the hydraulic tappets falling from the rocker arms.*

TO DISMANTLE AND INSPECT

TIP *Prior to dismantling, mark the rocker arms to ensure they are returned to their original position on assembly.*

Suitably mark the installed position of each rocker shaft.

1 Slide the rocker arms and thrust washers from the rocker shafts keeping all components in order as an aid to assembly. Withdraw the tappets from the rocker arms, noting their installed position.

Keep each rocker arm and tappet assembly together if these components are to be reused.

Rocker arms and shafts installed showing major components.

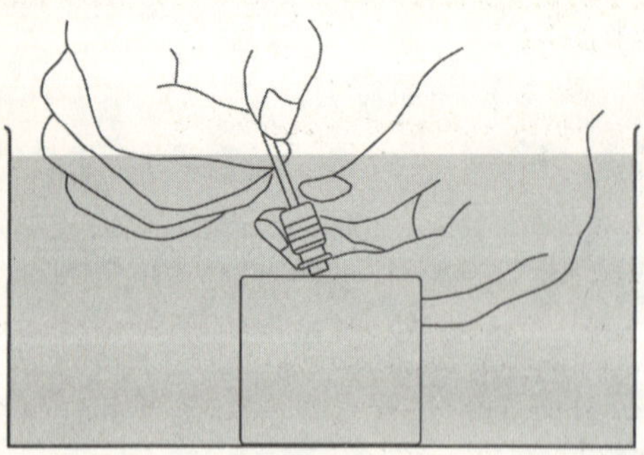

Method of bleeding the hydraulic tappets in clean diesel fuel.

2 Clean the components in cleaning solvent.

3 Inspect the rocker shafts for excessive wear and damage, evidenced by scoring or excessive rocker arm movement when the rocker arm is placed on the shaft and twisted diagonally to the shaft. If excessive wear is suspected, or the engine has been dismantled for a noisy top end, measure the outside diameter of the shaft and the inside diameter of the arm and compare to Specifications. Renew the shafts and/or rocker arms if necessary.

4 Ensure that the roller on the camshaft end of each rocker arm rotates freely and has an even wear pattern. Renew any rocker arms with worn or damaged rollers.

If a roller is damaged, carefully check its corresponding camshaft lobe for wear and damage. Renew components as necessary.

5 Inspect each hydraulic tappet for wear, especially on the valve stem contact face. If the valve stem face of the hydraulic tappet is lightly worn it can be carefully dressed using an oil stone. Renew the hydraulic tappet if the wear is excessive.

Dismantling of the tappet assembly is not recommended by the manufacturer.

6 Remove each tappet in turn, keeping it with its rocker arm. Clean the hydraulic tappets thoroughly in diesel fuel and bleed, using the following procedure:

a Fill a clean container with clean diesel fuel and place a solid block into the container.

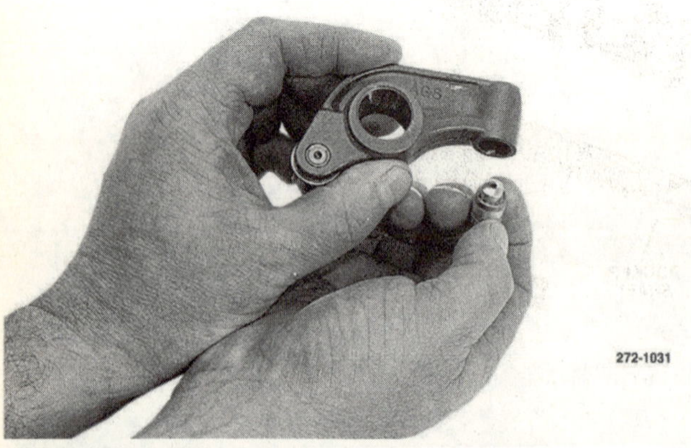

Ensure that the tappet oil hole is facing the roller arm when installing.

b Place the tappet into the container and support the base of the tappet on the block. Ensure that the tappet is completely submerged in the diesel fuel.

c Insert a 1.0 - 1.5 mm diameter wire or sewing needle through the oil bleed hole in the top of the tappet and push down lightly on the spring loaded ball inside the hydraulic tappet.

d While depressing the spring loaded ball, pump the tappet up and down five to ten times.

e Remove the wire and attempt to pump the tappet again. The tappet should not move. If it moves, repeat steps c and d several times.

f If the tappet still moves when pumped, renew the tappet.

g Install the tappets to the rocker arms, ensuring that the oil hole on the tappet assembly is facing the rocker arm.

TIP *Apply a dab of petroleum jelly to the body of each tappet to retain it in the rocker arm.*

TO ASSEMBLE AND INSTALL

Assembly is a reversal of the dismantling procedure with attention to the following points:

1 Ensure that all passages and holes are clear, especially the fine holes in the rocker arms.

2 Lubricate all components with clean engine oil before assembly. Do not allow the oil to enter the rocker shaft assembly bolt holes in the cylinder head.

3 One end of each rocker shaft has a notch as an identification mark. Ensure that the notched end of the shaft on the front cylinder head is facing the left hand side, while on the rear cylinder head it is towards the right hand side. The oil feed holes on all rocker shafts must face down, towards the cylinder head.

4 Install the rocker arms and thrust springs to the shafts in the position from which they were removed. If marks were not made or the components have been mixed up since disassembly, assemble all the components as shown in the illustration.

The dished side of the washer must face towards the rocker shaft.

Thrust springs are only installed to the inlet rocker shafts.

5 Install all components to the cylinder head ensuring that the hydraulic tappets remain in each rocker arm.
6 Install the washers, ensuring that the dished side of the washer engages the rocker shaft. Install the retaining bolts finger tight.
7 Ensure that the camshaft dowel pins are aligned, as on removal to avoid damaging the pistons or the valves.
8 Progressively tighten the rocker shaft bolts, working in a spiral sequence from the centre outwards, to the specified torque.

It is important that the rocker shafts be kept level as they are pulled down into position by the retaining bolts. Damage to or breakage of the rocker arm and/or cylinder head could occur if the rocker shafts are tilted during installation.

9 Thoroughly clean the mating surfaces of the rocker cover and the cylinder head and remove the cloth plugs from the cylinder head drain holes.
10 Install the rocker cover retaining bolts and progressively tighten them to the specified torque. Install the spark plug leads.
11 Install the timing belt as previously described.
12 Install the plenum chamber, referring to the Plenum Chamber heading if necessary.
13 Run the engine and check for oil leaks. Rectify as necessary.
14 If the tappets are noisy, slowly increase the engine speed from idle to 3 000 rpm and back to idle again once over a one minute period. Repeat this operation several times until the air is bled from the hydraulic tappets.

Camshafts

To Check Camshaft — Micrometer

TO REMOVE

1 Remove the cylinder heads, closely following the procedure, as described later in this chapter.
2 Remove the bolt retaining the sprocket and withdraw the sprocket from the camshaft.
3 When working on the front cylinder head, remove the bolts retaining the thrust plate, carefully prise the plate from the head and discard the O ring.
4 Carefully guide the camshaft from the cylinder head, ensuring that the cam lobes do not damage the camshaft bearing running surface in the head.
5 Noting the installed depth of the oil seal, carefully prise the seal from the front of the head.

TO CHECK AND INSPECT

1 Thoroughly wash and dry each camshaft.
2 Check the camshaft lobes, journals and thrust faces for pitting and wear. If wear is evident, measure the lobes with a micrometer and compare with Specifications. Renew the camshaft if pitting or wear is excessive.

If the camshaft lobes are worn, check the corresponding rocker arm roller for wear and damage and renew as necessary.

3 Check the bearing running surfaces inside the cylinder head for signs of scoring or overheating. If a bearing surface is worn or damaged, renew the cylinder head or consult an engine reconditioning specialist.

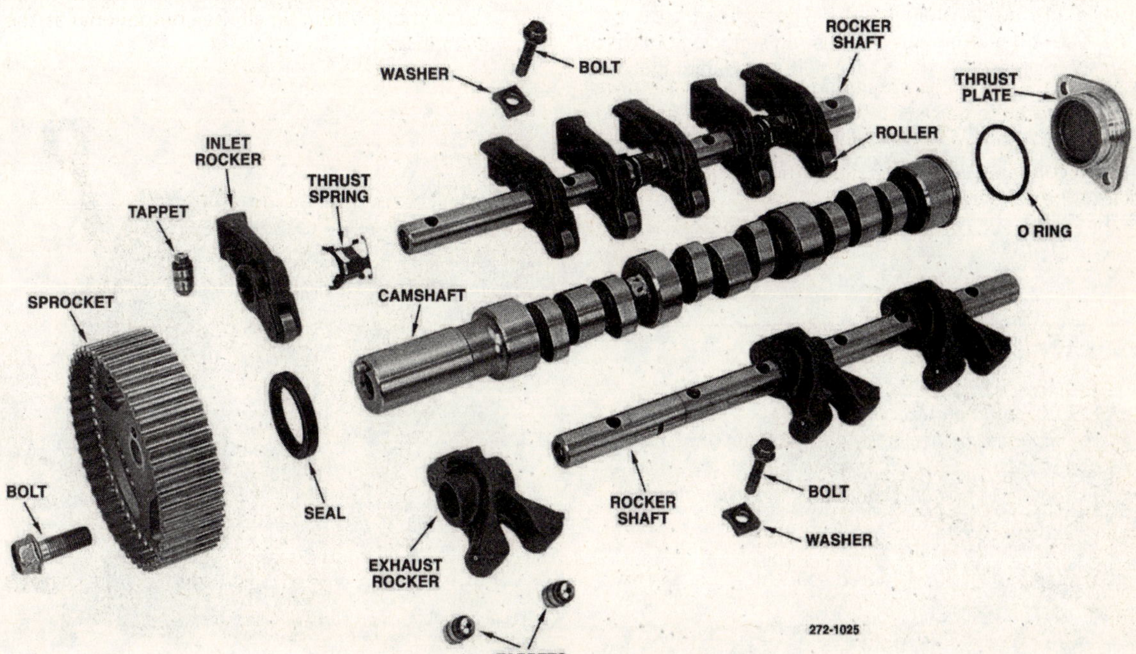

Camshaft and rocker shafts partially dismantled.

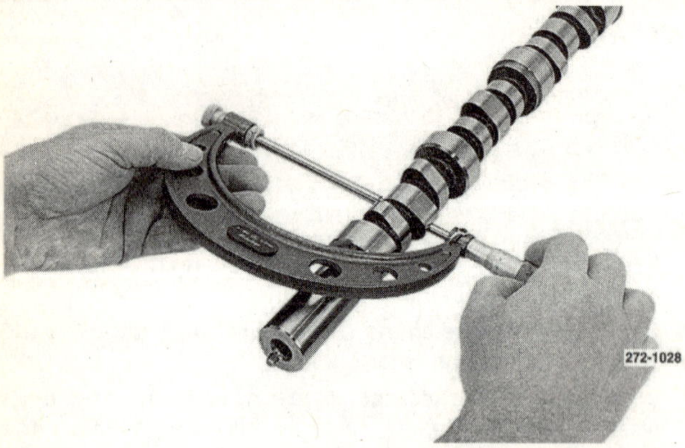

Measuring the cam lobe height using a micrometer.

4 If it is suspected that the camshaft bearings are worn, or the engine has been dismantled for a noisy top end, measure the camshaft bearing journals and the camshaft bearing bores, using inside and outside micrometers. Subtract the diameter of the bearing journal from the diameter of the bearing bore to calculate the oil clearance. Compare to Specifications and if worn consult an engine reconditioning specialist or renew the camshaft and/or the cylinder head as necessary.

TO INSTALL

Installation is a reversal of the removal procedure with attention to the following points:

1 Lubricate the journals and cam lobes with clean engine oil prior to installation.
2 Install a new camshaft oil seal to the cylinder head using a suitable tube or socket that pushes on the outer circumference of the seal, to the depth noted on removal. When installed the seal lips must face inwards and the seal must be square. Lubricate the lips of the seal with multipurpose grease.
3 Carefully guide the camshaft into the cylinder head, ensuring that the lobes do not damage the bearing surfaces, and that the camshaft slides into the oil seal without damaging the seal lips.
4 Install the thrust plate using a new O ring.
5 Install the rocker assembly, using the procedure previously described.
6 Install the cylinder head, referring to the procedure described later in this chapter.

Cylinder Heads

To Remove – 10mm Allen key socket
To Dismantle and Assemble — Valve spring compressor and adaptor
To Inspect — Straight edge, inside and outside micrometers

TO REMOVE AND INSTALL

1 Remove the rocker covers, the inlet manifold, the exhaust manifolds, and the timing belt as previously described in this chapter.

Do not rotate the camshafts after the timing belt is removed, as damage to the valves and/or pistons may occur.

2 Remove the valley cover plate retaining bolts and withdraw the plate from the right hand side of the engine.
3 Release clips retaining the spark plug leads and remove the screws retaining the distributor cap. Withdraw the distributor cap with the leads attached from the engine.
4 Disconnect the distributor wiring connector, remove the retaining nuts and withdraw the distributor from the cylinder head. Refer to Fuel and Engine Management (chapter 7), if necessary.

There is no need to mark the installed position of the distributor, as the shaft is offset and can only be installed one way.

5 Drain the cooling system, referring to Cooling (chapter 5), if necessary.
6 Remove the bolts retaining the steel heater pipes and withdraw the pipes from the thermostat housing. Discard the O rings from the end of the pipes.
7 Release the clamp retaining the top radiator hose to the thermostat cover and remove the hose.
8 Suitably mark and disconnect the wiring from the thermostat housing at the left hand end of the cylinder head. Place the wiring clear of the work area.
9 Remove the bolts retaining the thermostat housing to the cylinder heads and withdraw the housing from engine. Remove and discard the O ring from the transfer pipe. Discard the gaskets from between the thermostat housing and the cylinder heads.

It is good mechanical practice to check the thermostat whenever the cylinder head gaskets are renewed. Refer to Cooling System (chapter 5) for the correct procedure.

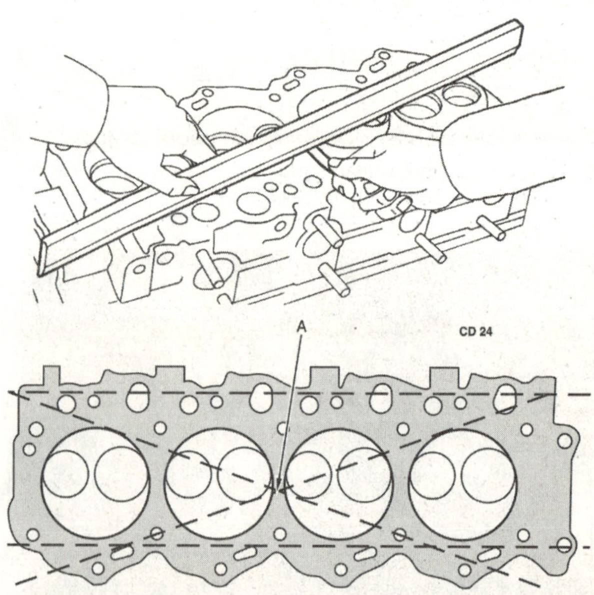

Check the cylinder head for distortion using a feeler gauge and straight edge. It is common to find a depression in area A.

10 Remove the bolt retaining the transfer pipe to the front cylinder head.

11 Remove the power steering pump retaining bolts and place the pump clear of the work area. If necessary, refer to Steering (chapter 12). Remove the power steering pump retaining bracket from the cylinder head.

12 Remove the alternator, referring to Engine Electrical (chapter 6), if necessary.

13 Using a 10 mm Allen key socket, progressively loosen the cylinder head retaining bolts in a spiral sequence, starting at each end and working towards the centre.

Do not loosen the cylinder head bolts while the engine is hot. Allow the engine to cool before cylinder head removal.

14 Lift each cylinder head away from the cylinder block and remove the cylinder heads from the vehicle. If it is difficult to remove a cylinder head from the block, insert a lever between the side of the block and the cylinder head at the timing belt end of the engine to assist in breaking the gasket adhesion. Do not lever on the gasket surface.

Do not rotate the camshafts after the cylinder heads are removed, as damage to the valves and/or pistons may occur when installing the head. If removing the rocker assembly, closely follow the procedure described in the Rocker Arms, Shafts and Hydraulic Tappets heading.

15 Using a sharp scraper, clean the cylinder block and the cylinder head, ensuring that both surfaces are free from gasket material, using care to prevent foreign material entering the cylinders.

TIP *Permatex Gasket Remover or equivalent will assists in removing any traces of old gasket material.*

16 Check the cylinder block face for distortion diagonally and lengthwise using a straight edge and feeler gauges. If the cylinder block is distorted more than specified, it will be necessary to remove and dismantle the engine and consult an engine reconditioning specialist.

17 Inspect the cylinder block and pistons for damage.

18 Check the cylinder head surfaces for distortion as described under the heading To Clean and Inspect.

Cylinder head gasket correctly positioned on the cylinder block.

Installation is a reversal of the removal procedure with attention to the following points:

1 Ensure that the cylinders and the cylinder head bolt threads are free of foreign materials. Remove the oil residual from the cylinder head bolt holes. If necessary, clean the thread holes using a suitable tap.

2 Use new gaskets and seals during assembly.

3 Ensure that No 1 piston is on TDC.

4 Install new cylinder head gaskets to the cylinder block. The gasket holes, especially the oil passage, should be aligned with the corresponding holes in the cylinder block. Refer to the illustration for the correct positioning of the gasket. Carefully lower the cylinder head onto the cylinder block and engage the dowel pins.

Do not slide the cylinder head across the dowels as this may damage the cylinder head face.

5 Lightly oil the threads of the retaining bolts and the washers. Place the washers onto the bolts, ensuring that the chamfered edge of the washer is facing the head of the bolt.

Never put oil directly in the bolt holes in the cylinder head as damage or incorrect torque settings could result.

6 Progressively tighten the cylinder head bolts in three stages, to the specified torque, following the illustrated sequence.

7 Install the rocker covers, using new gaskets.

8 Place a new O ring on the end of the transfer pipe and install the thermostat housing using new gaskets. Place new O rings on the heater pipes and install the pipes. Tighten the retaining bolt securely.

9 Install the distributor and the spark plug leads. Note that the distributor drive tang is offset, and the distributor will only go in one way.

10 Install the inlet manifold and the plenum chamber, as previously described.

11 Refill the cooling system, following the procedure in Cooling System (chapter 5).

If the cooling system is not filled as described in the Cooling System chapter, engine overheating may result.

Cylinder head bolt tightening sequence. Loosen in the reverse order. The sequence is the same for the other cylinder head.

12 Securely tighten the fuel filler cap.

13 Start and run the engine and check for oil, water and fuel leaks. Rectify as necessary.

TO DISMANTLE

Overhauling the cylinder heads requires the use of specialist equipment. Buying or hiring this equipment is usually too expensive to be considered when overhauling a pair of cylinder heads. Overhauling the cylinder heads is best left to a cylinder head specialist with the necessary equipment. Some specialists offer a cylinder head exchange service which greatly reduces the amount of time the vehicle is off the road.

1 Place the cylinder head down on a clean cloth to prevent scratching the cylinder head face.

2 Suitably mark the valves to ensure that all the components of the cylinder head are returned to their original locations.

3 Remove the rocker arms and shafts as previously described and guide the camshaft from the cylinder head.

4 Drill 12 holes in two pieces of board to receive the valves removed from the cylinder heads. Number the holes to ensure that the valves can be replaced in their original positions in the head.

5 Position the cylinder head on its side and using a valve spring compressor, compress a valve spring. With some types of valve spring compressor an adaptor maybe necessary to allow clearance from the rocker cover sealing surface. An adaptor can be fabricated by cutting an opening in a short length of metal tubing. Remove the collets from the valve spring retainer.

TIP *If the collets are stuck to the valve spring retainer, apply light pressure to the retainer and tap the edge of the retainer using a brass punch and hammer.*

6 Release the spring compressor and remove the retainer, spring and seat from the cylinder head.

7 Remove all burrs from the end of the valve stem using a small file to ensure that the valve guide is not damaged as the valve is withdrawn.

Removing the valve collets using a valve spring compressor and adapter.

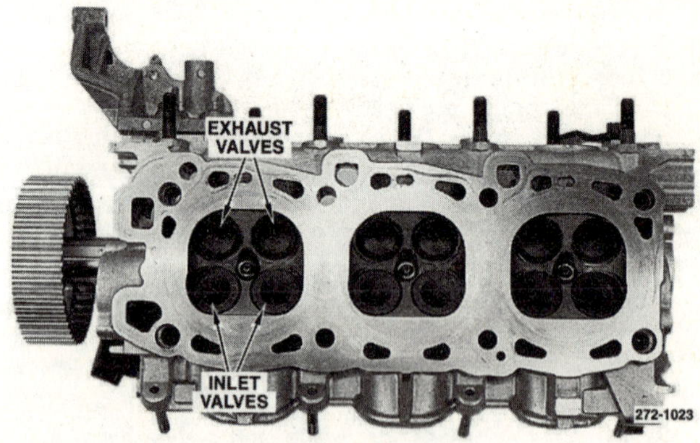

Cylinder head showing the valve arrangement.

8 As each valve is withdrawn, insert it into its appropriate hole in the piece of board to ensure correct installation to the original position during assembly.

9 Repeat the removal procedure with the remaining valves.

10 Remove the oil seal from the valve guide using long nosed pliers.

TO CLEAN AND INSPECT

1 Thoroughly clean the cylinder head in solvent and dry it with compressed air, if available.

2 Inspect the cylinder head for cracks or damage. Check the camshaft bearing surfaces for scoring or pitting.

3 Thoroughly clean the valves of carbon deposits, using a wire brush and electric drill. Do not mark the valve seats. Clean the valve guides using a piece of wadded cloth and a suitable solvent.

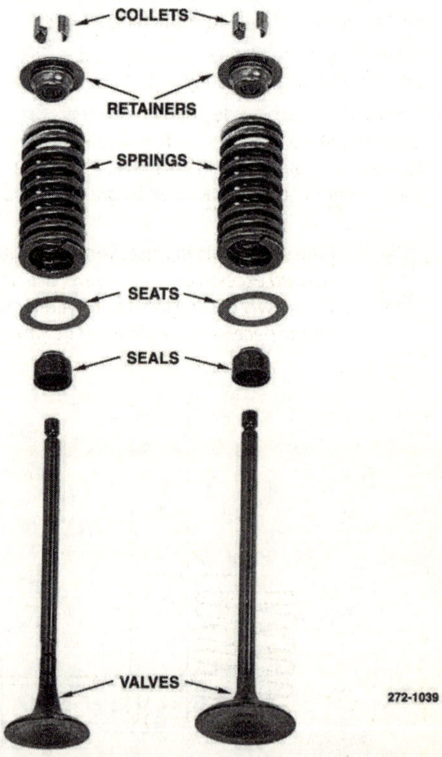

Inlet and exhaust valve components.

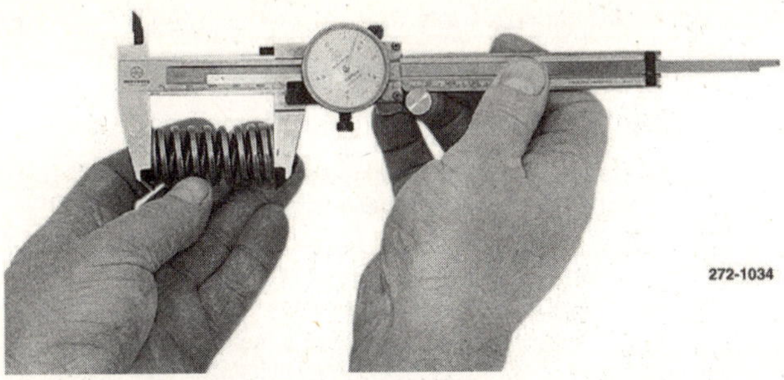

Measuring the valve spring free length.

4 Inspect each valve for burning, distortion or cracking. A valve with any one or more of these characteristics should be discarded and a new valve fitted.

5 Using inside and outside micrometers, measure the valve stem and the valve guide bores. Subtract the outside diameter of the valve stem from the inside diameter of the valve guide to calculate the oil clearance. Compare to Specifications. Renew the valve and/or the valve guide, as necessary. This work is best referred to a cylinder head specialist.

6 Check each valve spring free length as shown on the illustration and compare to Specifications.

7 Stand the valve springs upright and check their squareness using a set square. Rotate each spring slowly and observe the distance between the top coil of the spring and the square. Renew the spring if it is more than 4° out of square.

8 To check the tension of a valve spring if a spring tester is not available, check the springs by comparison with a new spring as follows:

a Install one of the springs and the new spring in a vice, end to end, with a steel plate in between.

b Tighten the vice until the new spring is compressed to a length of 44.2 mm and measure the length of the original spring.

c If the length of the original spring is within five percent of the new spring, it indicates that the tension is acceptable.

9 Remove all carbon deposits from the combustion chambers, cylinder head face and valve throats. Check the condition of the valve seats and if necessary, recut the valve seats with a valve seat cutter to the specified angle and width.

If the valve seats in the cylinder head are worn, cracked or recessed excessively, new inserts will have to be installed. This work is best referred to a cylinder head specialist who has the necessary specialised equipment.

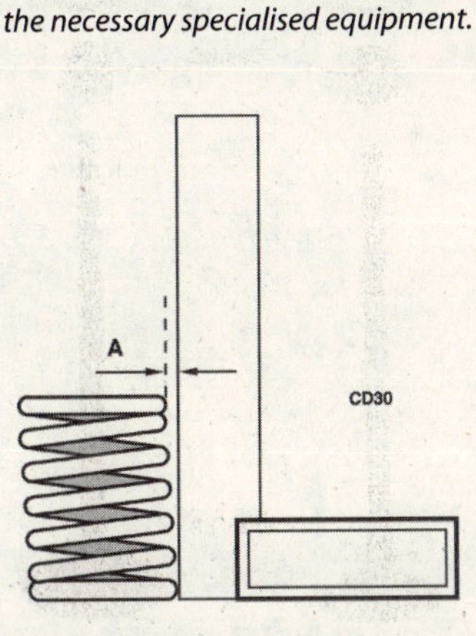

Renew the valve spring if the distortion marked A exceeds 4°.

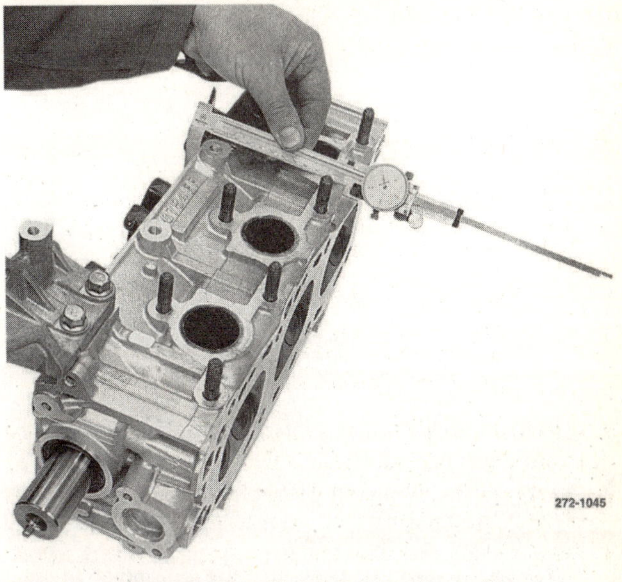

Measuring the cylinder head height using vainer calipers. Measure at several points and average the results.

10 Using a steel straight edge and feeler gauges, check along and across the cylinder head face for distortion. If the cylinder head is distorted, it must be machined. If it is necessary to machine more than 0.20 mm from the cylinder head, the cylinder head should be renewed.

To determine if the cylinder head gasket face has sufficient thickness to allow machining, measure the cylinder head height between the rocker cover sealing surface and the head gasket face. From this measurement subtract the distortion measurement. If the figure is less than 119.7 mm, the cylinder head should be renewed.

Keep in mind also, that if the block is distorted and requires machining, that the total of cylinder head and cylinder block machining is not to exceed 0.20 mm.

Machining the cylinder head or cylinder block faces beyond the specified limit will increase the compression ratio. This may cause engine detonation or pinging, result in loss of performance and could increase exhaust emission levels.

11 Check the cylinder head face and combustion chambers for cracks and burn marks. Inspect the coolant transfer holes in the cylinder head for corrosion. Check the spark plug tubes for looseness and leaks. If necessary, renew the head or consult an engine reconditioning specialist.

TO REFACE VALVES AND VALVE SEATS

1 Reface the valves to the specified angle on a suitable valve refacing machine. If a valve head margin thickness is less than specified limit after the grinding operation, the valve should be renewed.

2 Inspect the valve stem tips and if pitted, they should be reground taking care to remove only enough material to restore tip flatness.

3 Using a valve seat cutter, cut the valve seats to the specified angle. When cutting the valve seats it is important that the correct size valve guide pilot is used for the cutting tool. A true and complete surface must be obtained.

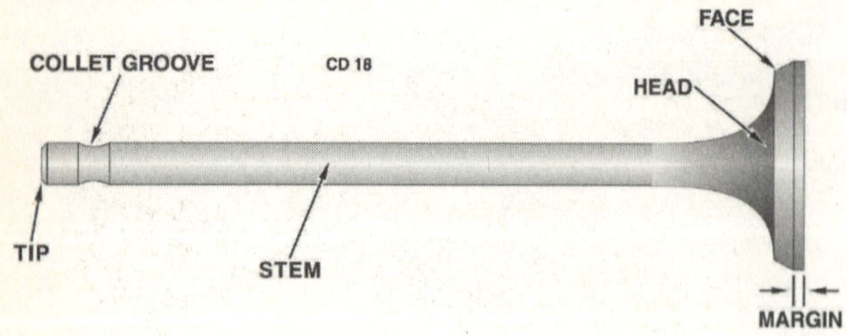

Various parts of the valve.

4 Lap the valves to the valve seats with a suitable oil or water mixed lapping compound until a true and concentric seating has been obtained.

5 Check the valve to seat contact using the following the procedure:

a Apply a smear of Prussian Blue to the valve face and rotate the valve approximately 1/6 of a turn while lightly holding the valve head against the seat.

b Withdraw the valve and check that the marking on the seat is complete and even around the seat, indicating full seat contact. Remove all Prussian Blue from the valve and seat.

c If the valve is not seating fully, install the valve and rotate it one complete revolution.

d Withdraw the valve and check the marking on the seat. If full seat contact is indicated, the valve requires refacing. If partial seat contact is indicated, recut the seat.

6 Install the valves and measure the installed height of each valve, measuring from the spring seat on the cylinder head to the valve stem tip and compare with Specifications. If the installed height on any valve exceeds Specifications, the cylinder head will require new valve seat inserts.

TO ASSEMBLE

Assembly is a reversal of the dismantling procedure with attention to the following points:

1 Clean the cylinder head thoroughly removing all traces of abrasive material.

2 Place the cylinder head face down on a thick, clean cloth on the workbench.

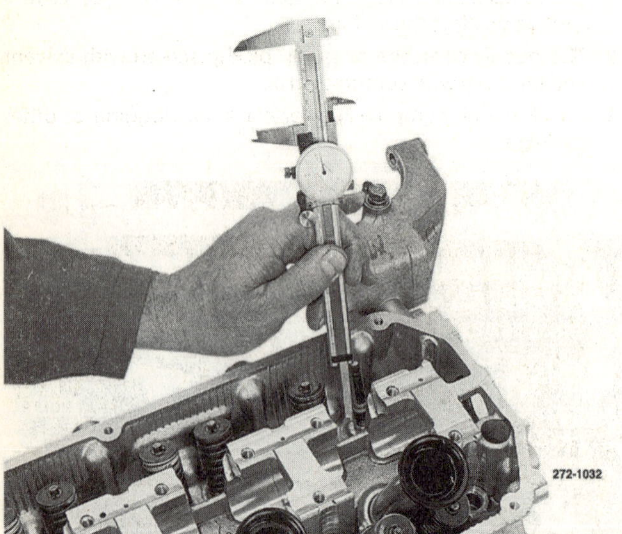

Measuring the installed height of the valve.

3 Using a suitable size tube, that pushes evenly on the outer edge of the seal, install new seals on each valve stem until they are fully seated.

> *At the time of publication the exhaust valve seals had a light green coloured band around the top of the seal, while the inlet valve seals had no colour markings. Consult your parts supplier prior to installing the seals.*

4 With the cylinder head supported on its side, lubricate the valve stems and guides sparingly with clean engine oil and insert each valve into its respective guide, ensuring that the valves are installed in their original position.

5 Install the valve spring seats and the valve springs. Note that the valve springs are installed with the close coils at the bottom.

6 Place the retainer over the valve, compress the assembly using the valve spring compressor and install the collets. Slowly release the valve spring compressor and check that the collets are correctly engaged with the valve stem grooves.

> TIP *A dab of multipurpose grease applied to the inside surface of the collets will hold the collets in the correct position on the valve stem during installation.*

7 Lightly tap the end of the valve stem using a soft faced hammer to seat the collets in the valve stem groove.

8 Repeat these operations with the remainder of the valves.

9 Install the camshaft and the rocker assembly to the cylinder head as previously described, ensuring that the camshaft dowel is positioned as on removal. Progressively tighten the retaining bolts to the specified torque in a spiral sequence working from the centre.

10 Install the cylinder head as previously described.

Sump and Pickup Pipe

The sump is made of two parts. The upper sump is aluminium alloy and provides additional stiffness to the lower part of the cylinder block. The lower sump is conventional pressed metal. Separating the two sumps is not normally necessary.

Sump installed. Models with air conditioning. The large arrows indicate the sump forcing holes.

TO REMOVE AND INSTALL

1 Raise the front of the vehicle and support on chassis stands under the recommended points. Refer to General Information (chapter 1) for the correct trolley jack and chassis stand position to prevent damage or injury.

2 Remove the bolts and scrivets retaining the front and right hand side splashguards and withdraw the guards from the vehicle.

3 Drain the engine oil into a suitable container, referring to the Routine Maintenance heading in this chapter, if necessary. Install a new O ring to the drain plug and tighten the plug to the specified torque.

4 Remove the nuts retaining the engine pipes to the exhaust manifolds, lower the pipe assembly and support away from the work area, using a piece of wire tied to the body of the vehicle.

Do not allow the pipe to hang unsupported, as damage to the flexible joint may occur.

5 On models with air conditioning, proceed as follows:

a Remove the bolts retaining the air conditioning compressor to the bracket on the cylinder block. Suspend the compressor with the hoses still attached using wire or rope, clear of the work area.

The top compressor retaining bolts are close to and virtually identical in appearance to the pressure hose flange retaining bolts. Do not remove the pressure hose flange bolts as the escaping refrigerant could cause serious injury.

b Remove the bolts retaining the air conditioning compressor support bracket to the cylinder block, and withdraw the bracket from the engine compartment.

6 Remove the bolt retaining the dipstick tube to the cylinder block and withdraw the tube from the cylinder block. Discard the O ring from the tube.

7 Remove the starter motor referring to Engine Electrical (chapter 6), if necessary.

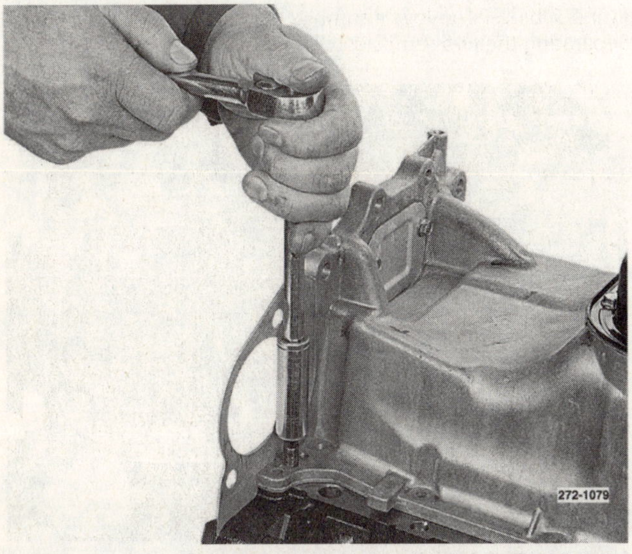

Separating the sump from the cylinder block using a suitable bolt in the forcing holes provided.

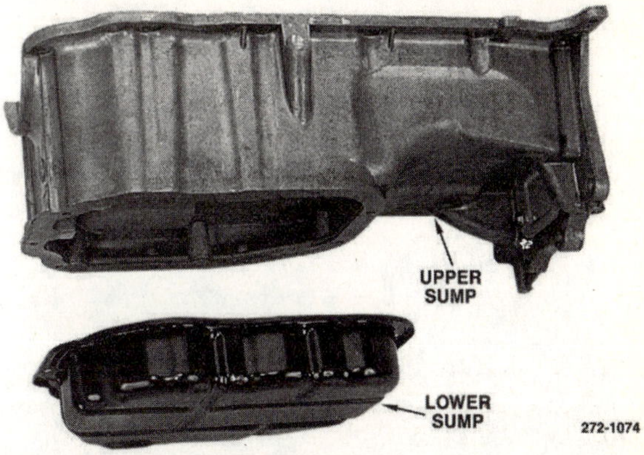

Sump components dismantled.

8 Remove the bolts retaining the sump to the cylinder block.

9 Insert two suitable length Metric 10x1.25 mm bolts into the forcing holes at the transaxle end of the sump and evenly tighten both bolts until the sump is separated from the cylinder block. Remove the sump from the engine compartment.

Do not use a screwdriver or similar to separate the sump from the cylinder block as damage to the sump and/or cylinder block could occur.

10 If necessary, remove the retaining bolts and withdraw the baffle plate from the sump.

11 If necessary, remove the bolts retaining the steel lower sump to the alloy upper sump and, using a sharp knife or cutter, separate the steel sump.

12 If required, remove the bolts retaining the oil pump pickup pipe and withdraw the pipe from the engine compartment. Discard the gasket.

Installation is a reversal of the removal procedure with attention to the following points:

1 Thoroughly clean the mating surfaces of the cylinder block and the sump(s). Use a scraper or razor blade to remove the old sealant. When all the sealant has been removed, clean the sump mating face with solvent which will not leave a residue or affect the surface.

2 Thoroughly clean the oil pump pickup screen with solvent and blow dry with compressed air.

3 Check the oil pump pickup for cracking, clogging or other damage.

Sump removed showing the oil pump pickup pipe.

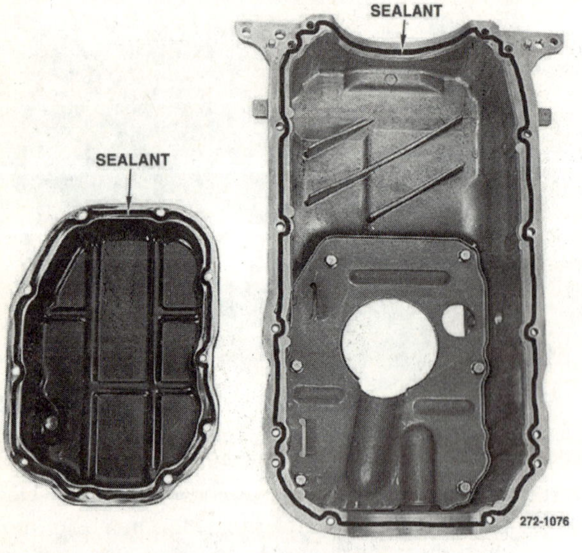

Apply a 4mm wide bead of sealant as shown prior to installing the sump.

4 Check the sump for cracks and distortion or damage to the sealing surface. Renew components as necessary.

5 Apply a 4 mm wide bead of Permatex Ultra Blue or equivalent sealant to the sump mating surface.

6 Place the sump onto the cylinder block, install the retaining bolts and tighten the bolts to the specified torque in the order shown on the illustration.

> *Remove all traces of the sealant squeezed out on the flywheel side of the sump.*

7 If removed, apply sealant to the lower sump mating surface, install the lower sump and progressively tighten the retaining bolts to the specified torque, in a spiral sequence, starting with the centre bolts.

8 Install the sump and tighten the retaining bolts progressively to the specified torque as shown on the illustration.

9 Install a new O ring to the dipstick tube and install the tube to the sump. Tighten the retaining bolt securely.

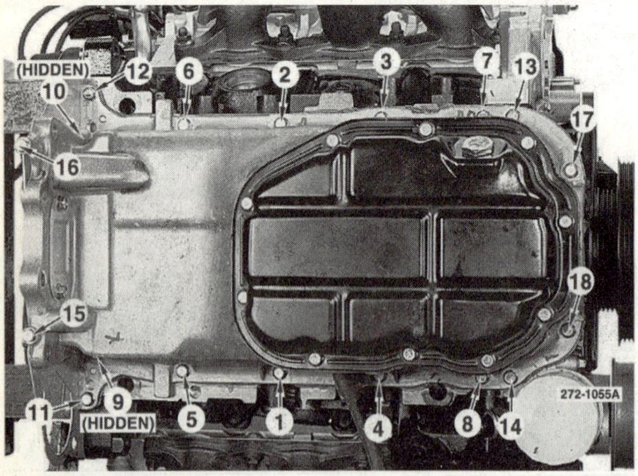

Sump bolt tightening sequence. The two long bolts, numbers 15 and 16 have a lower toque specification than the other bolts.

10 Install the air conditioning compressor bracket and the compressor.

11 Refill the engine, using the specified quality and quantity of engine oil.

> *Wait at least 30 minutes, to allow the sealant to cure before filling the sump with engine oil.*

12 Start the engine and ensure that the oil pressure warning lamp extinguishes after a short period. Check for oil leaks and rectify as necessary.

Oil Pump

TO REMOVE AND INSTALL

1 Remove the sump and pickup pipe as previously described.

2 Remove the drive belts referring to the Routine Maintenance heading in this chapter.

3 Remove the timing belt and withdraw the crankshaft sprocket, as described in this chapter under the Timing Belt heading.

4 Raise the timing belt tensioner assembly and secure it using a wire or similar, away from the oil pump. Alternatively, remove the pivot bolt and withdraw the tensioner assembly from the engine.

5 Remove the bolt retaining the alternator/air conditioning drive belt tensioner and withdraw the assembly.

6 Disconnect the oil pressure switch wiring connector.

7 Remove the bolts retaining the oil pump housing to the cylinder block. Using a rubber mallet or similar, carefully tap the oil pump housing to separate it from the cylinder block.

8 Using a soft faced hammer, gently tap the oil pump to separate it from the cylinder block. Manoeuvre the oil pump from the end of the crankshaft. Discard the O ring.

9 Without turning the pump inner rotor, mark the position of the rotor in relation to the pump cover, to assist installation of the pump. Not rotating the crankshaft while the oil pump is removed, will make installation easier.

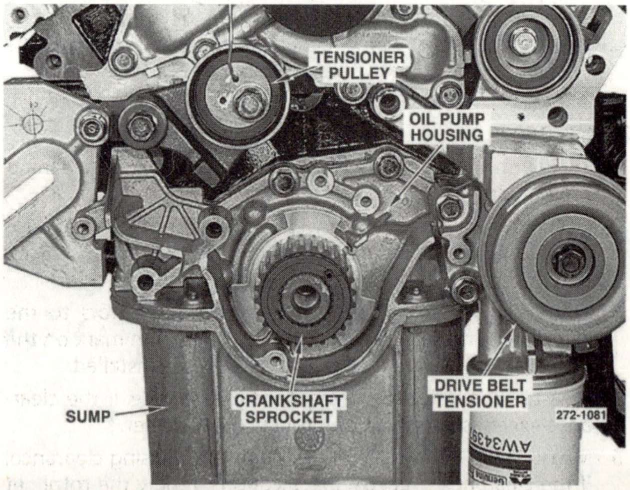

Oil pump and associated components installed.

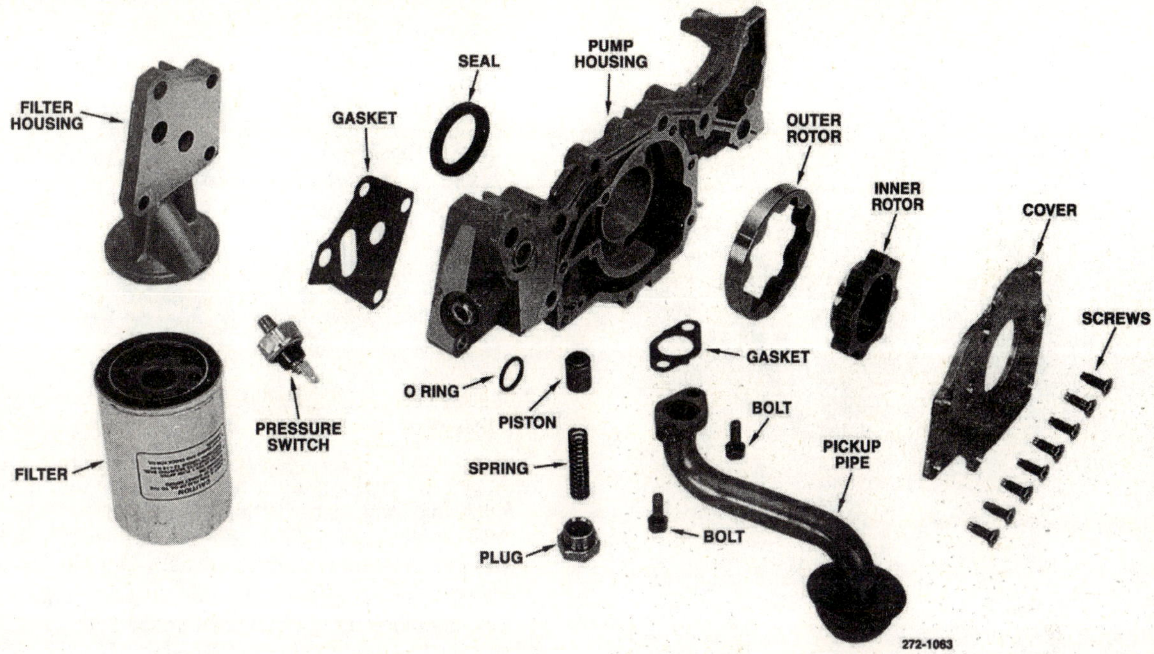

Oil pump components dismantled.

TO DISMANTLE

1. Remove the bolts retaining the pickup pipe and discard the gasket. Remove the filter from the housing and discard.
2. Using an impact driver, remove the screws retaining the oil pump cover to the housing. Ease the oil pump cover from the housing.
3. Withdraw the inner rotor and, noting the mark on the surface, remove the outer rotor from the oil pump housing.
4. Remove the plug retaining the pressure relief valve in the oil pump housing and, noting their installed position, withdraw the spring and piston.
5. Noting the installed depth, prise the crankshaft oil seal from the oil pump housing using an oil seal removal tool or a screwdriver. Use care not to scratch the seal bore. Discard the oil seal.
6. If necessary, remove the bolts retaining the filter housing and separate the filter housing from the pump housing. Discard the gasket. Remove all traces of old gasket from the oil filter housing and the oil pump housing.

TO INSPECT

1. Thoroughly clean the oil pump components in cleaning solvent, dry with compressed air if available.
2. Check the pump cover for wear and scoring, the rotors for pitting and galling, and the housing for scoring and cracks. Renew components as necessary.
3. Temporarily assemble the inner and outer rotors to the pump housing, ensuring that the identification mark on the outer rotor is towards the pump cover when installed.
4. Measure the inner to outer rotor tip clearance. If the clearance exceeds Specifications, renew the rotor set.
5. Measure the outer rotor to the oil pump housing clearance. If the clearance exceeds Specifications, renew the rotor set and/or the oil pump housing as necessary.
6. Hold a straight edge across the end face of the pump and measure the end clearance between the rotors and the straight edge with a feeler gauge. The clearance should not exceed Specifications. Renew the rotor set and/or the oil pump housing as necessary.
7. Check the pressure relief valve spring for breakage or loss of tension. Renew the spring if its condition is suspect.
8. Check the relief valve piston for wear and scoring. Apply a light film of oil to the valve and check that it slides freely in the oil pump housing by its own weight. Check that the oil passages are clear. Check the condition of the sealing surface of plug. Renew if necessary.

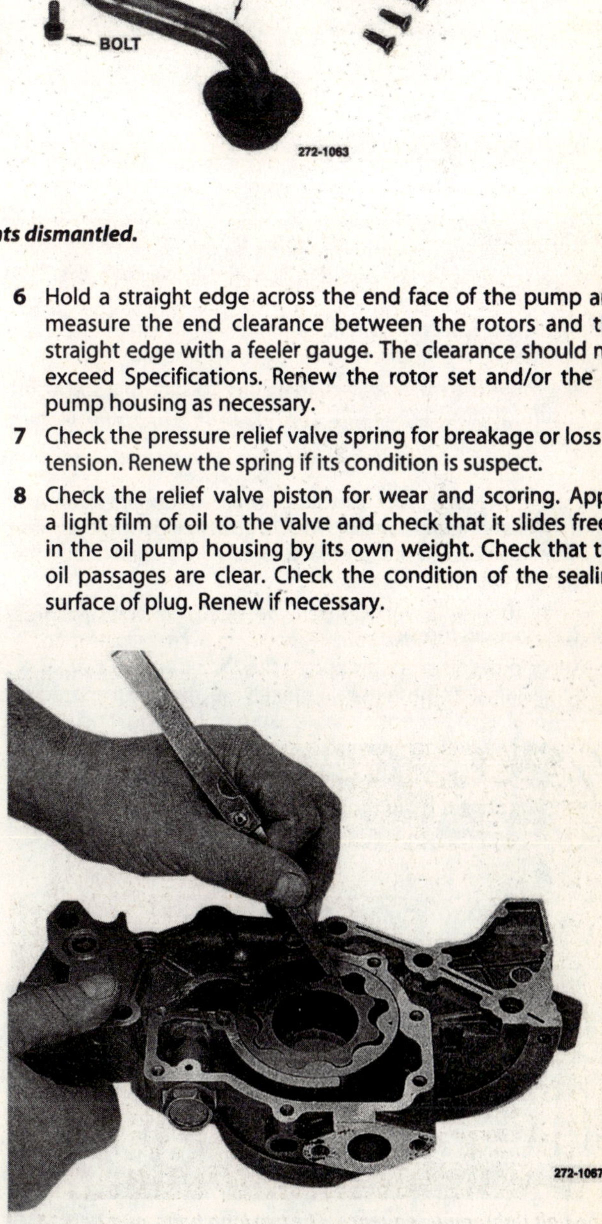

Measuring the inner to outer rotor tip clearance.

Measuring the outer rotor to oil pump housing clearance.

9 Thoroughly clean the oil pump pickup and screen with a wire or bottle brush.

10 Blow the oil pump pickup and screen clean with compressed air.

TO ASSEMBLE

Assembly is reversal of the dismantling procedure with attention to the following points:

1 Ensure that all components are thoroughly clean and liberally coat the components with clean engine oil.

2 Assemble the inner and outer rotors to the pump housing ensuring that the identification mark on the outer rotor is towards the pump cover.

> TIP *It is good practice to pack the space between the rotors with petroleum jelly. This will achieve rapid increase of oil pressure when starting the engine.*

3 Install the pump cover and tighten the retaining screws to the specified torque.

4 Apply multipurpose grease to the lip of a new crankshaft oil seal. Install the oil seal squarely, to the depth noted on removal, into the front of the oil pump housing. Ensure that the seal lip is facing inward.

5 Install the relief valve piston and spring to the oil pump housing. Tighten plug to the specified torque.

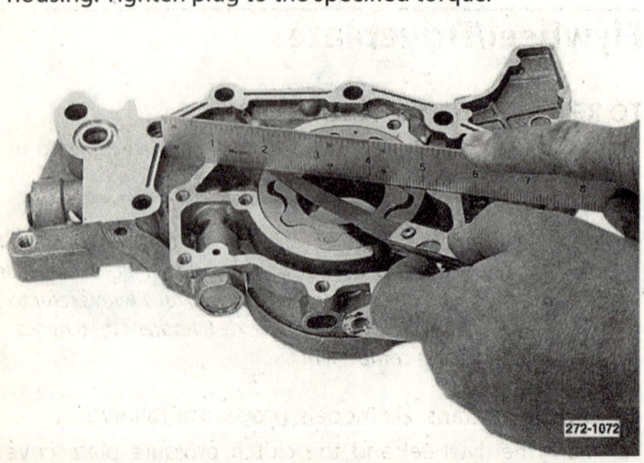

Measuring the end clearance of the rotors.

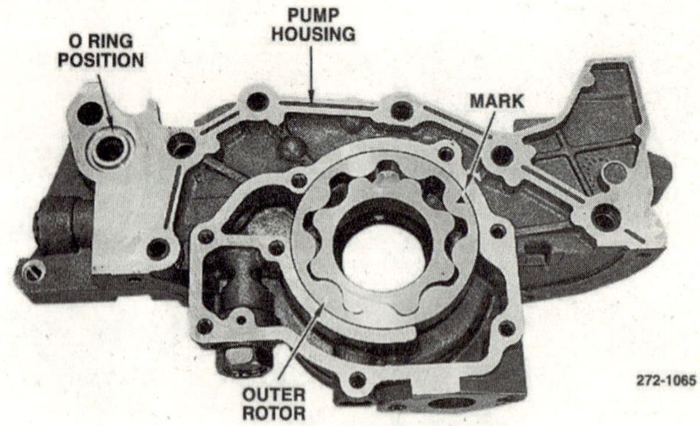

The mark on the outer rotor must be towards the end cover.

6 If petroleum jelly was not used to fill the space between the rotors, the oil pump must be primed with the correct grade of clean engine oil before it is installed to the engine. Fill the inlet port with oil and rotate the inner rotor to distribute the oil within the pump body.

7 Using a new O ring, install the oil pump housing to the cylinder block, using the procedure that follows.

TO INSTALL

Installation is a reversal of the removal procedure with attention to the following points:

1 Remove all traces of gasket material from the cylinder block.

> *It is recommended, that the front crankshaft oil seal is removed whenever the oil pump is removed.*

2 Align the marks made during removal on the inner rotor and the pump cover.

3 Apply a 3 mm wide bead of Permatex Ultra Blue or equivalent sealant to the oil pump mating surface, around the inside of the bolt holes, and place a new O ring in the groove. Install the oil pump.

> TIP *If difficulty experienced sliding the oil pump onto the crankshaft, rock the pump slightly to the left and to the right, until the inner rotor engages the flat portion of the crankshaft.*

4 Install the oil pump retaining bolts and tighten to the specified torque.

5 Install the sump, the timing belt and the drive belts, referring to the appropriate headings in this chapter.

6 Fill the engine with the recommended quantity and grade of engine oil. Run the engine and check that the oil lamp goes out within a few seconds. Also check for oil leaks and rectify as necessary.

Crankshaft Oil Seals

FRONT CRANKSHAFT OIL SEAL

To Remove and Install

1 Remove the timing belt as previously described.

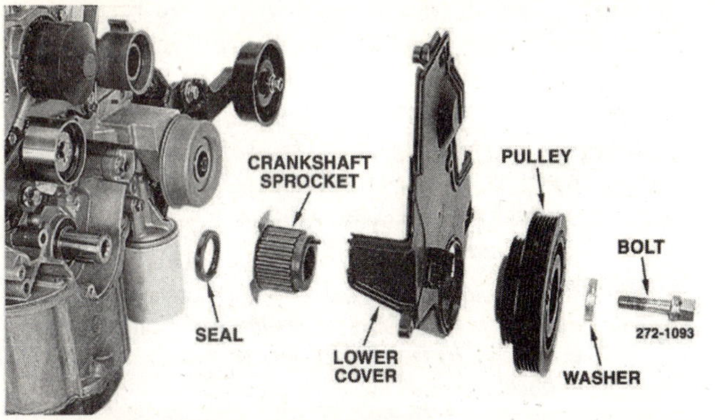

Crankshaft pulley and seal components.

2 Remove the crankshaft sprocket from the crankshaft. Do not remove the key.
3 Using a centre punch, tap a small hole into the front face of the oil seal. Use care not to make the hole too deep and damage the oil pump housing.
4 Screw a self tapping screw into the oil seal. Grip the head of the screw with a pair of pliers and withdraw the seal from the oil pump housing.
5 Check the seal contact surface on the crankshaft for scoring or damage. If necessary, refer the engine to an engine reconditioning specialist.
6 Apply multipurpose grease to the oil seal lip and install the seal squarely into the oil pump housing using a mandrel or tube that pushes on the outer edge of the seal only. The oil seal lip must face towards the engine and the outer face of the oil seal must be recessed approximately 1mm from the face of the oil pump housing.
7 Install the crankshaft sprocket and the timing belt as previously described.

REAR CRANKSHAFT OIL SEAL

To Remove and Install

1 Remove the transaxle from the vehicle as described in the relevant transaxle chapter.

On manual transaxle models a pull type clutch is fitted. Because of this, it is necessary to disconnect the release bearing clip during transaxle removal. Always refer to the procedure in Manual Transaxle (chapter 10) to avoid damaging components.

Removing the front crankshaft oil seal, using a self tapping screw.

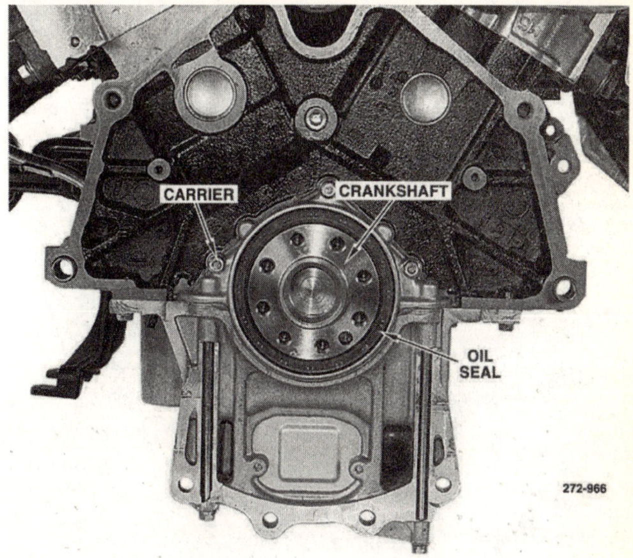

Rear crankshaft oil seal.

2 Remove the flywheel/driveplate as described later in this chapter.
3 Using a centre punch, tap a small hole into the front face of the oil seal. Use care not to make the hole too deep and damage the carrier.
4 Screw a self tapping screw into the oil seal. Grip the head of the screw with a pair of pliers and, noting the installed depth of the seal, withdraw the seal from the carrier.
5 Check the seal contact surface on the crankshaft for scoring or damage. If scored or damaged, refer the engine to an engine reconditioning specialist.

Installation is a reversal of the removal procedure with attention to the following points:

1 Apply multipurpose grease to the oil seal lip and install the oil seal to the carrier using a mandrel or tube that pushes on the outer edge of the seal only. The oil seal lip must face towards the engine. Ensure that the outer face of the seal is installed to the depth noted during removal.
2 Install the flywheel/driveplate as described later in this chapter.
3 Install the transaxle to the engine as described in the relevant transaxle chapter.

Flywheel/Driveplate

TO REMOVE AND INSTALL

1 Remove the transaxle from the vehicle as described in the relevant transaxle chapter.

On manual transaxle models a pull type clutch is fitted. Because of this, it is necessary to disconnect the release bearing clip during transaxle removal. Always refer to the procedure in Manual Transaxle (chapter 10) to avoid damaging components.

2 On manual transaxle models, proceed as follows:
a Mark the flywheel and the clutch pressure plate cover to ensure correct assembly.

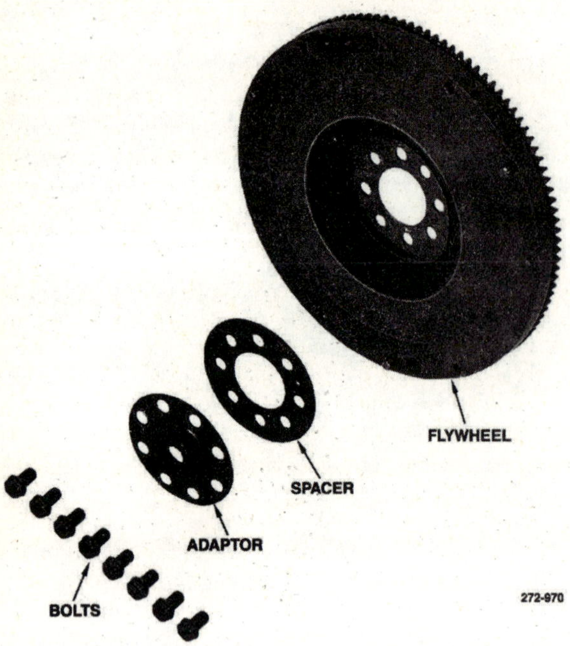

Flywheel dismantled.

b Progressively loosen and remove the pressure plate retaining bolts.

c Lift the pressure plate from the flywheel and remove the clutch driven plate.

d Using a dial gauge, check the runout of the flywheel clutch face contacting surface. If the runout exceeds 0.1 mm, renew the flywheel.

> *On automatic transaxle models the manufacturer does not specify the driveplate runout.*

3 Suitably mark the flywheel or driveplate so that the unit can be installed in the original position on the crankshaft.

4 Remove the retaining bolts. On automatic transaxle models, note the installed position and remove the spacer. On manual transaxle models, mark the installed position and installed direction of the adaptor and spacer plates and withdraw them from the flywheel. Lift the flywheel or driveplate from the crankshaft.

Driveplate installed.

TO INSPECT

1 Where applicable, inspect the flywheel clutch face contacting surface for cracks or excessive scoring.

> *The retaining bolts on the reverse side of the flywheel must not be removed, as the balance of the flywheel may be affected. If these bolts have been disturbed, renew the flywheel.*

2 Inspect the ring gear teeth for damage, cracks or wear. The ring gear can not be separated from the torque converter or the flywheel, and the whole assembly must be renewed if damaged.

3 Inspect the driveplate bolt holes for elongation, caused by loose retaining bolts.

4 Inspect the driveplate for cracks.

5 Renew components as necessary.

TO INSTALL

Installation is a reversal of the removal procedure with attention to the following points:

1 Align the marks made when removing the flywheel/driveplate. On automatic transaxle models, install the spacer with the raised lip facing away from the driveplate. On manual transaxle models, install the spacer and the adaptor plates, in the position and direction noted on removal. Refer to the illustration if necessary.

2 Install and tighten the retaining bolts evenly, and progressively to the specified torque.

3 On manual transaxle models, align the clutch driven plate with the adaptor centre hole using an aligning tool. Tighten the pressure plate bolts progressively in a diagonal sequence. Refer to Clutch (chapter 9) for more information on the aligning tool and the pressure plate bolt torque.

4 Install the transaxle to the vehicle as described in the relevant transaxle chapter.

Engine Overhaul

Engine overhaul refers to major work on the pistons, connecting rods, crankshaft bearings and cylinder block as well as checking the items already covered in this chapter.

OPTIONS

If the engine is worn and the rest of the vehicle is in good enough condition to warrant overhauling the engine, there are four common steps that can be taken:

1 Install a reconditioned long engine. A long engine usually includes the cylinder block and heads assembled with camshafts rocker covers and sump etc. ready to bolt on your manifolds and other components such as starter motor, alternator, power steering pump etc.

Installing a long engine from an engine reconditioning specialist is usually the best policy if intending to keep a vehicle that is otherwise in good condition for a long period.

Usually it is possible to install the engine yourself or the specialist will do it for you for an agreed fee.

2 Install a reconditioned short engine. A short engine is only the cylinder block, pistons and crankshaft. All other components come from your old engine.

A short engine is really only a proposition if the cylinder heads have been recently been reconditioned or the engine has done only a relatively low mileage and something has gone wrong with the lower part of the engine. If the other components are well worn the restored compression of the short engine will soon show up weaknesses in other areas.

3 Install a good secondhand engine from a wrecker or specialist. This is usually the most economical option and may be the best option to keep and old vehicle on the road for a while longer. These engines often come with some accessories installed but great care must be taken to ensure that the engine chosen, is actually suitable for the vehicle.

Imported engines can vary a lot from the local product originally installed in this vehicle so the best policy is to try and get an engine from the same type of vehicle as yours.

Ensure that some sort of warranty comes with the engine and if possible, test run it at the wreckers etc. before purchase.

4 Overhaul the engine yourself. This is really only economically and practically viable for those with previous experience and quality measuring equipment. It must be remembered that the modern engine runs tighter clearances and is less tolerant of assembly glitches than older style engines. If taking this path the following information may help you.

Systematically mark components or note the manufactures marks as the engine is progressively dismantled. It will still be necessary to take certain components to a specialist for operations like reboring, crankshaft checking and piston to connecting rod installation.

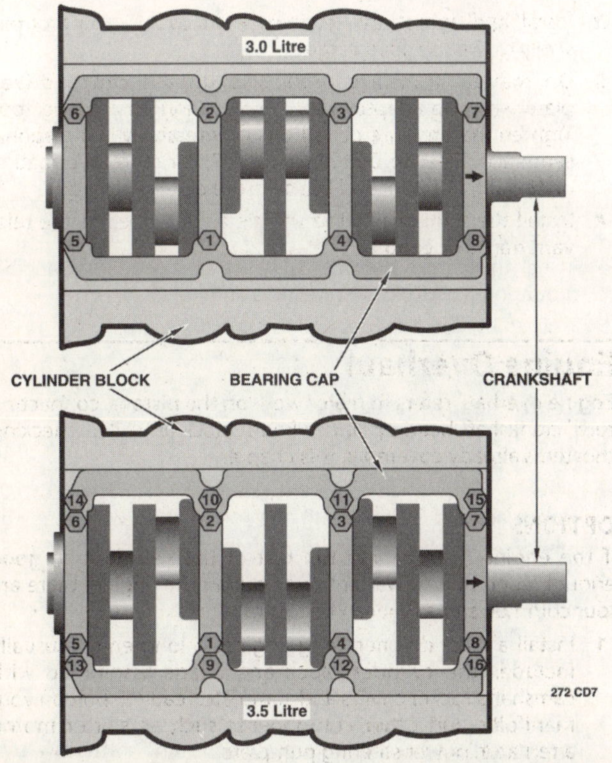

Main bearing cap tightening sequence for the 3.0 and 3.5 litre engines.

The specialist should check clearances for you however, take nothing for granted and check everything with the comprehensive specifications listed at the start of this chapter. Cleanliness is one of the most important aspects of engine assembly along with checking that all components are installed correctly. If there is a fault with the engine when it goes back together it can cost you a lot of money and time.

Overhauling this engine, even though of advanced design follows standard workshop practice, with attention to the following points:

a The main bearing shells have a plain half and a grooved half. The grooved half with the oil hole must be installed in the cylinder block.

b Crankshaft end float is adjustable at the No 3 main bearing using thrust washers. The thrust washers must be installed with the grooves facing outwards.

c The crankshaft and main bearings are graded. Consult your parts supplier about the correct size replacement crankshaft bearings.

d Follow the illustration when tightening the main bearing cap bolts. 3.0 litre engines are equipped with two bolt main bearings and 3.5 litre engines have four bolt main bearings. Both engines have the caps united by a cradle arrangement.

GENERAL PROCEDURES

When installing an overhauled or secondhand engine certain items should be inspected, renewed or cleaned to prevent problems occurring.

1 The radiator should be flushed out and if suspect should be taken to a radiator specialist for through cleaning and checking. It is advisable to renew the thermostat and water pump as a matter of course. Thoroughly inspect the radiator cap and renew if, it is suspect. Always use the correct coolant in the cooling system.

The temperature gauge, radiator fan and coolant temperature sensor must all be functioning correctly.

2 The timing belt should be renewed.

3 Install new spark plugs of the correct type. Thoroughly check the distributor, rotor, cap and high tension leads and renew if necessary.

4 Install a new air cleaner element and fuel filter. Remove the injectors and have them cleaned by a specialist.

5 All hoses in the engine compartment must be thoroughly checked for splits, perishing and hardening. New hoses should be fitted unless the old hoses are perfect. Fuel hoses are a vital safety item and only high quality hoses suitable for fuel injection systems are to be used.

6 Inspect the engine mountings and the exhaust system for wear and deterioration and renew components as necessary.

7 Use engine oil of the correct grade and a new oil filter to prevent excessive wear on start up. Ensure that the oil pressure warning lamp system is functioning correctly and repair if necessary.

8 Refer to the engine installation procedure for more information on how to treat an overhauled engine.

Engine Mountings

TO REMOVE AND INSTALL

Engine mountings that are oil soaked should be renewed and the cause of the oil leak rectified to ensure a normal working life for the new mounting.

Right Hand Engine Mounting

1 Raise the front of the vehicle and support on chassis stands under the recommended points. Refer to General Information (chapter 1) for the correct trolley jack and chassis stand positions to prevent damage or injury.

2 Support the engine under the sump, using a floor jack with a piece of timber placed between the head of the jack and the sump.

3 Remove the bolts retaining the windscreen washer bottle and partially withdraw the bottle. Disconnect the wiring connector(s) and the hose(s) and remove the washer bottle.

4 Remove the through bolt from the right hand side engine mounting.

5 Remove the nuts and bolt retaining the right hand side engine mounting to the support bracket and manoeuvre the mounting from the engine compartment.

Note the direction of the arrows on the rubber pads, prior to removing the pads.

6 If necessary, remove the support bracket as follows:

a Remove the alternator, referring to Engine Electrical (chapter 6).

b Noting their installed position, remove the bolts retaining the support bracket and manoeuvre the bracket from the engine compartment.

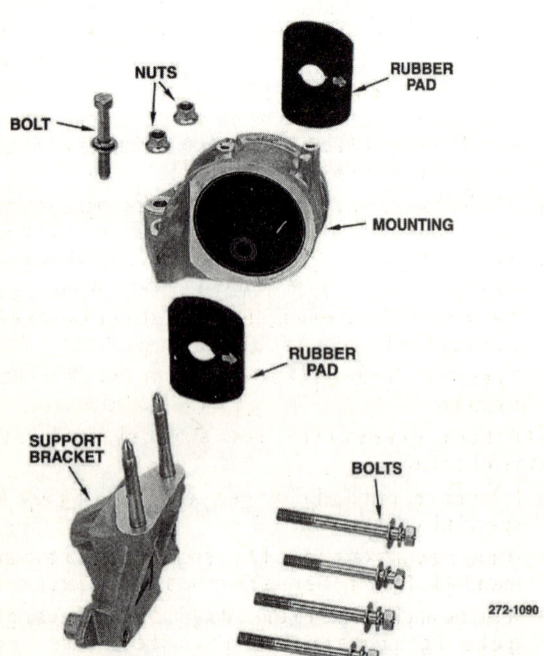

Right hand side engine mounting dismantled.

TIP: The bolts cannot be withdrawn separately from the bracket as they interfere with the vehicle body.

Installation is a reversal of the removal procedure, ensuring that the arrows on the rubber pads are pointing towards the right hand side of the vehicle.

Left Hand Engine Mounting

1 Raise the front of the vehicle and support on chassis stands under the recommended points. Refer to General Information (chapter 1) for the correct trolley jack and chassis stand positions to prevent damage or injury.

2 Using a floor jack, support the transaxle ensuring that a timber block is placed between the head of the jack and the transaxle.

3 Remove the air cleaner inlet hose, the airflow sensor and the air cleaner assembly, referring to Fuel and Engine Management (chapter 7), if necessary.

4 Remove the nuts and washers retaining the front and rear engine mountings. Noting their installed position, withdraw the through bolts from the mountings.

5 Release the clip retaining the A/C pipe to the centre member. Remove the retaining bolts and withdraw the centre member from the vehicle.

6 Loosen the transaxle mounting retaining nuts and remove the nut from the through bolt. Noting the installed direction of the through bolt, withdraw the bolt and, on models with automatic transaxle, the damper, Carefully lower the floor jack, until sufficient clearance is obtained to remove the mounting and withdraw the mounting from the engine compartment.

Installation is a reversal of the removal procedure with attention to the following points:

1 Ensure, that the arrows on the rubber pads are pointing downwards prior to installation. On models with automatic transaxle, place the damper on the through bolt.

2 Place the mounting onto the studs and loosely install the retaining nuts.

3 Raise the transaxle and install the through bolt and nut.

4 Install the centre member and the front and rear engine mounting through bolts, washers and nuts.

5 Tighten the retaining nuts to the specified torque.

Front and Rear Mountings

1 Raise the front of the vehicle with a jack and support the vehicle with chassis stands under the recommended points. Refer to General Information (chapter 1) for the correct trolley jack and chassis stand positions to prevent damage or injury.

2 Using a floor jack, support the transaxle ensuring that a timber block is placed between the head of the jack and the transaxle.

3 Remove the nuts and washers and, noting their installed position, withdraw the through bolts from the mountings.

4 If fitted, release the clip retaining the A/C pipe to the centre member. Remove the retaining bolts and withdraw the centre member from the vehicle.

5 Remove the retaining bolts and manoeuvre the mounting from the engine compartment.

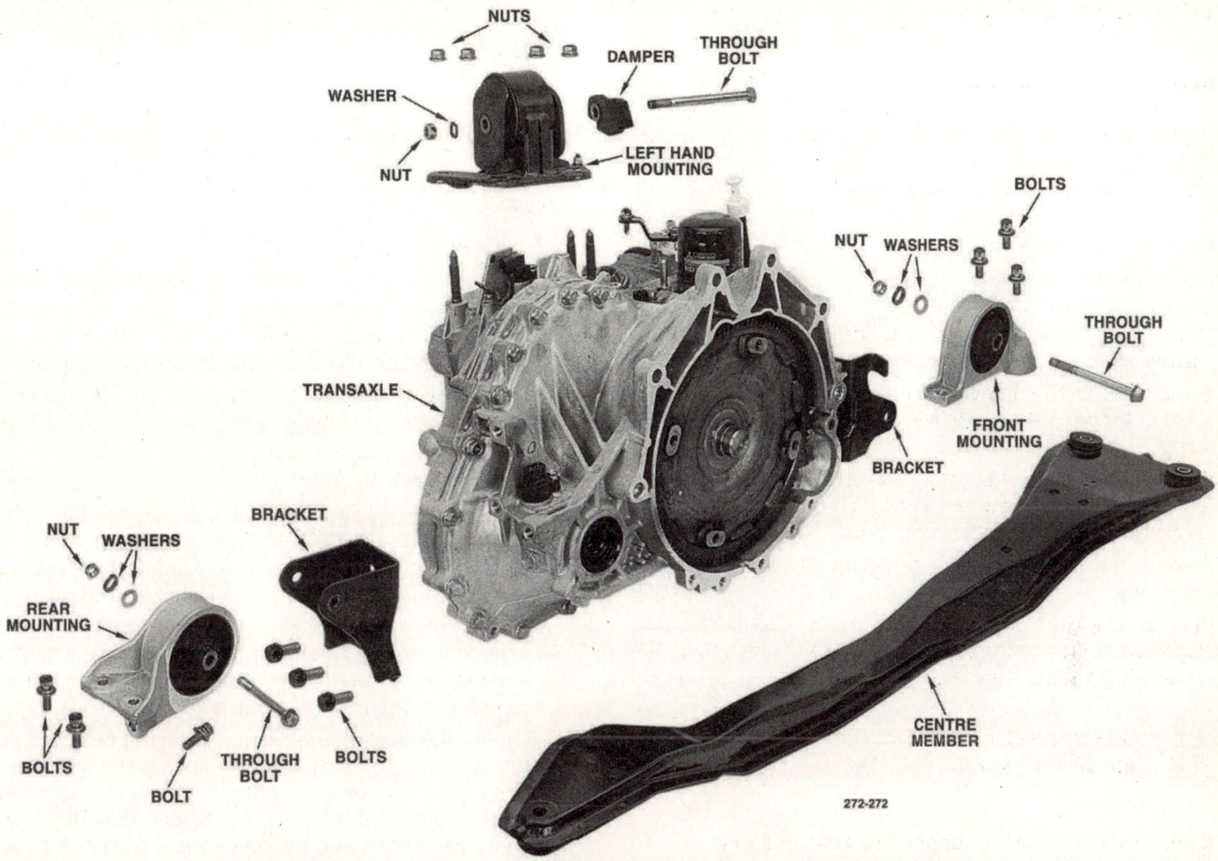

Front, rear and left hand side engine mountings. Automatic transaxle model shown.

6 If necessary, remove the bolts retaining the bracket and withdraw the bracket from the vehicle.

Installation is a reversal of the removal procedure, ensuring that the retaining nuts and bolts are tightened to the specified torque.

Exhaust System

TO REMOVE AND INSTALL

TIP *It is advisable to soak all nuts and bolts necessary to remove the exhaust system with a suitable penetrating oil to prevent the bolts and studs breaking. It is good mechanical practice when working on the exhaust system, to wrap the end of the tail pipe in a piece of soft cloth to prevent damage to the rear bumper.*

Engine Pipe

1 Disconnect the oxygen sensor wiring connector as follows:

a Working inside the vehicle, on the passenger side, remove the clip retaining the carpet to the centre tunnel and fold the carpet down.

b Remove the screws retaining the metal bracket and withdraw the bracket.

c Disconnect the oxygen sensor wiring connector.

2 Raise the front of the vehicle and support on chassis stands under the recommended points. Refer to General Information (chapter 1) for the correct trolley jack and chassis stand positions to prevent damage or injury.

3 Remove the bolts retaining the engine pipe to the catalytic converter. Discard the gasket.

4 Carefully prise the rubber grommet surrounding the oxygen sensor wiring from the floor and withdraw the wiring from the interior of the vehicle.

5 While supporting the rear of the engine pipe, remove the bolts retaining the engine pipe to the rubber mountings.

6 Remove the nuts retaining the engine pipe to the exhaust manifolds and carefully manoeuvre the engine pipe from the vehicle. The oxygen sensor must not be knocked or damaged during removal. Discard the gaskets.

7 If necessary, remove the oxygen sensor, using a suitable size spanner.

Installation is a reversal of the removal procedure, with attention to the following:

1 If removed, install the oxygen sensor and tighten to the specified torque.

2 Using new gaskets, install the engine pipe to the exhaust manifolds. Tighten the retaining nuts to the specified torque.

3 Join the engine pipe to the catalytic converter, using a new gasket. Tighten the retaining nuts to the specified torque.

4 Feed the oxygen sensor wiring through the opening in the floor and install the rubber grommet securely.

Engine pipe installed.

5 Install the bolts retaining the engine pipe to the rubber mountings and tighten the retaining bolts securely.

6 Connect the oxygen sensor wiring connector securely and replace the carpet inside the vehicle.

Catalytic Converter

1 Raise the front of the vehicle and support on chassis stands under the recommended points. Refer to General Information (chapter 1) for the correct trolley jack and chassis stand positions to prevent damage or injury.

2 Remove the bolts retaining the catalytic converter flange to the engine pipe and discard the gasket.

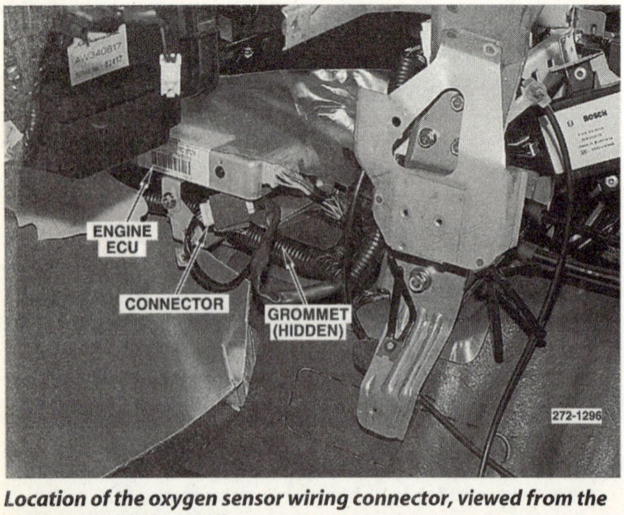

Location of the oxygen sensor wiring connector, viewed from the passenger side footwell.

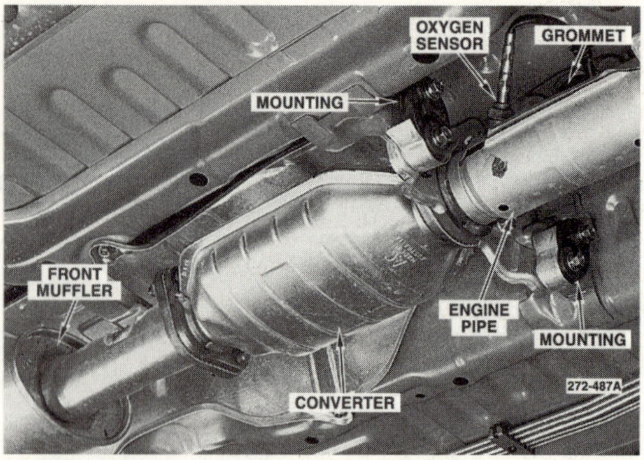

Catalytic converter and oxygen sensor installed.

3 While supporting the catalytic converter, remove the bolts retaining the catalytic converter flange to the front muffler assembly. Noting the directional arrow at the rear of the catalytic converter, withdraw the converter and discard the gasket.

Installation is a reversal of the removal procedure, with attention to the following points:

1 Install the catalytic converter, using new gaskets, ensuring that the arrow is pointing in the direction of flow (rearwards), as noted on removal.

2 Tighten the retaining bolts to the specified torque.

Front Muffler and Intermediate Pipe

1 Raise the rear of the vehicle with a jack and support the vehicle with chassis stands under the recommended points. Refer to General Information (chapter 1) for the correct trolley jack and chassis stand positions to prevent damage or injury.

2 Remove the bolt retaining the earth cable to the front muffler assembly.

3 Noting the installed position of the heat shield adjacent to the fuel tank, remove the bolts retaining the intermediate pipe flange to the rear muffler flange. Discard the gasket.

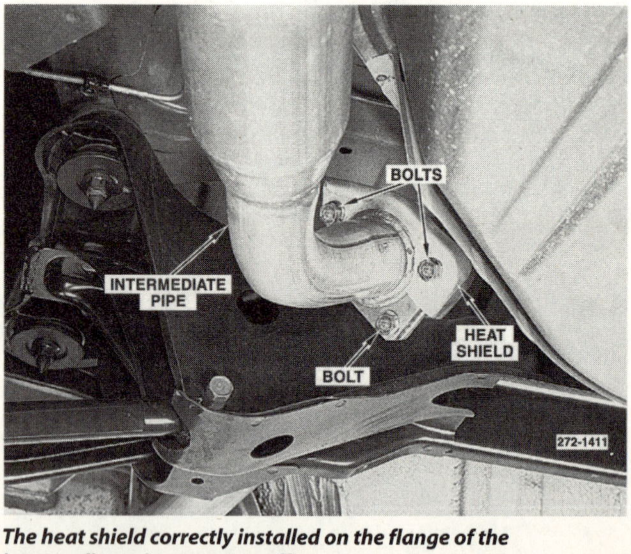

The heat shield correctly installed on the flange of the intermediate pipe to rear muffler.

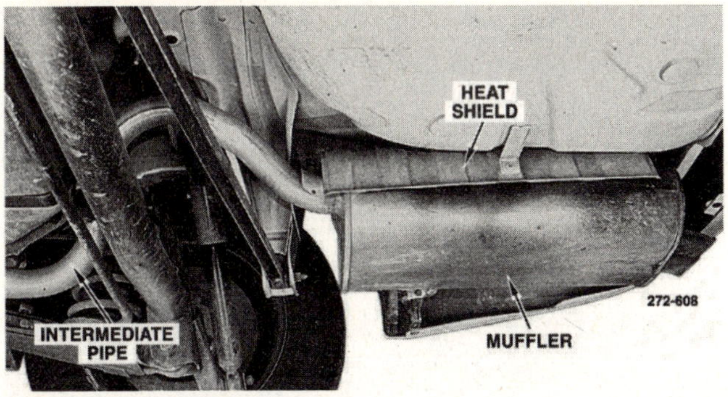

Rear muffler installed.

4 Remove the bolts retaining the front muffler flange to the catalytic converter and discard the gasket.

5 Spray soapy water onto the rubber mountings and while supporting the assembly, prise the mountings from the studs. Withdraw the front muffler and intermediate pipe assembly from the vehicle.

Installation is a reversal of the removal procedure, with attention to the following points:

1 Spray soapy water onto the studs and the rubber mountings and, while supporting the front muffler in position, press the rubber mountings onto the studs.

2 Place a new gasket between the front muffler flange and the catalytic converter and install the retaining bolts.

3 Install the heat shield to the rear flange of the intermediate pipe as noted on removal, and using a new gasket, join the flanges. Tighten the retaining bolts securely.

4 Install the bolt retaining the earth cable to the front muffler assembly and tighten securely.

Rear Muffler

1 Raise the rear of the vehicle and support on chassis stands under the recommended points. Refer to General Information (chapter 1) for the correct trolley jack and chassis stand positions to prevent damage or injury.

2 Noting the installed position of the heat shield adjacent to the fuel tank, remove the bolts retaining the rear muffler flange to the intermediate pipe flange. Discard the gasket.

3 Spray soapy water onto the rubber mountings and while supporting the assembly, prise the mountings from the studs. Withdraw the rear muffler assembly from the vehicle.

Installation is a reversal of the removal procedure, with attention to the following points:

1 Spray soapy water onto the studs and the rubber mountings and, while supporting the rear muffler in position, press the rubber mountings onto the studs.

2 Using a new gasket, install the flange retaining bolts.

3 Install the heat shield to the flange of the rear muffler as noted on removal, and using a new gasket, join the flanges. Tighten the retaining bolts securely.

Chapter 5
Cooling System

Contents

Specifications

Coolant:
- Capacity ... Approximately 9.5 litres
- Inhibitor concentration ... 30 – 50 % ethylene glycol and clean, demineralised water

Water pump type ... Centrifugal impeller

Thermostat:
- Opening temperature ... 88 °C
- Valve lift at 95° C ... More than 8 mm

Radiator fan:
- Type ... Electric, four blade
- Motor current draw ... 12.2 ± 1.5 amps

Radiator type ... Vertical flow corrugated fin

Radiator cap opening pressure ... 74 – 103 kPa

TORQUE WRENCH SETTINGS

Cylinder block drain plug ... 39 Nm

Thermostat cover bolts ... 18 Nm

Water pump retaining bolts:
- 8 mm bolts ... 24 Nm
- 10 mm bolt ... 41 Nm

Description

The cooling system is of the thermo siphon type with a water pump driven by the timing belt and an electric cooling fan providing assistance. The system is pressurised in order to raise the boiling point of the coolant and so increase the efficiency of the engine.

The radiator tanks are manufactured in a resin material to reduce weight and resist corrosion. To facilitate coolant drainage, a drain tap is located in the lower radiator tank and drain plugs are located in each bank of the cylinder block.

On automatic transaxle models, the lower radiator tank also houses the transaxle oil cooler.

The radiator overflow hose is connected to a reserve tank which is mounted adjacent to the battery. As the coolant volume expands due to an increase in temperature, the pressure valve in

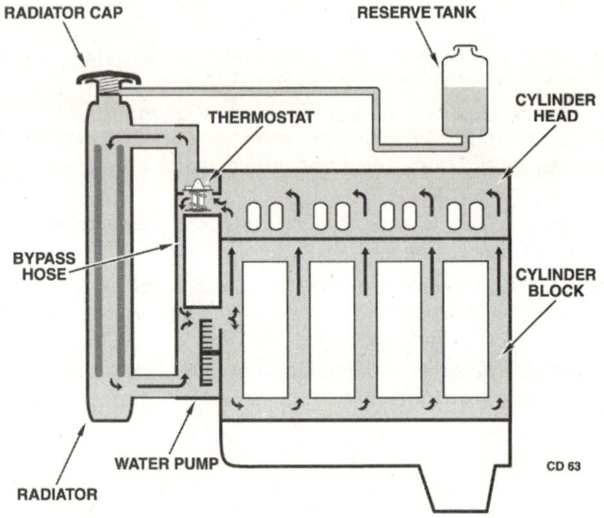

Schematic showing coolant flow through a typical engine.

the radiator cap opens allowing the excess coolant to flow into the reserve tank. When the engine is stopped and the coolant temperature falls, the vacuum valve in the radiator cap opens, allowing the excess coolant in the reserve tank to siphon back into the radiator. This makes frequent topping up of the coolant unnecessary.

Temperature within the cooling system is controlled by a thermostat, which has an additional valve to block the bypass circuit when the coolant temperature rises and the thermostat opens. The function of the thermostat is to prevent the circulation of the coolant through the radiator until the engine has reached operating temperature. This restricted circulation allows the engine to warm up quickly, which improves driveability and fuel economy.

The water pump, driven by the timing belt from the crankshaft pulley, is of the centrifugal impeller type. The body of the water pump is made from an aluminium alloy to assist in engine weight reduction and contains a prelubricated bearing and a spring loaded seal assembly. A drain hole, located between the seal and the bearing, prevents bearing lubricant contamination if coolant leaks past the seal. The water pump is a one piece unit and if defective, must be renewed as a unit.

A variable speed electric cooling fan is mounted on the rear of the radiator and is controlled by the engine ECU using information from the coolant temperature sensor. Power is supplied by relays and fuses which are located in the engine compartment relay box.

On models with air conditioning, an additional variable speed electric fan is installed beside the radiator fan to assist in cooling the condenser. This fan is controlled by relays and fuses which are located in the engine compartment relay box.

Different brands and types of coolants may not be compatible. Mixing some coolants may lead to overheating or corrosion of engine components. It is important that when topping up, the same type of coolant is used, and at the same concentration. If changing the coolant, the cooling system must first be thoroughly flushed with clean water.

To avoid accidental scalding, when working on the cooling system, exercise caution when releasing the radiator cap of an engine that is at normal operating temperature. Allow the engine to cool prior to releasing the radiator cap.

Routine Maintenance

COOLANT

To Check Level

1 With the engine cold, check the coolant level in the reserve tank using the dipstick provided in the cap. The level should be between the Full and Low marks.

2 If the level is on or below the Low mark, remove the radiator cap and check that the coolant is level with the overflow hose orifice. If necessary, add coolant to the radiator until the level is at the overflow hose orifice.

> i *When adding coolant, always add the same type of coolant that is in the system. Not all coolants are compatible. If in doubt, drain and flush the entire cooling system and refill using the manufacturers or a good quality, brand name coolant concentrate.*

3 Install the radiator cap and add coolant to the reserve tank until the level is at the Full mark.

4 Run the engine and check for coolant leaks. If necessary, pressure test the radiator cap to check for serviceability, if coolant loss occurs and leaks are not apparent.

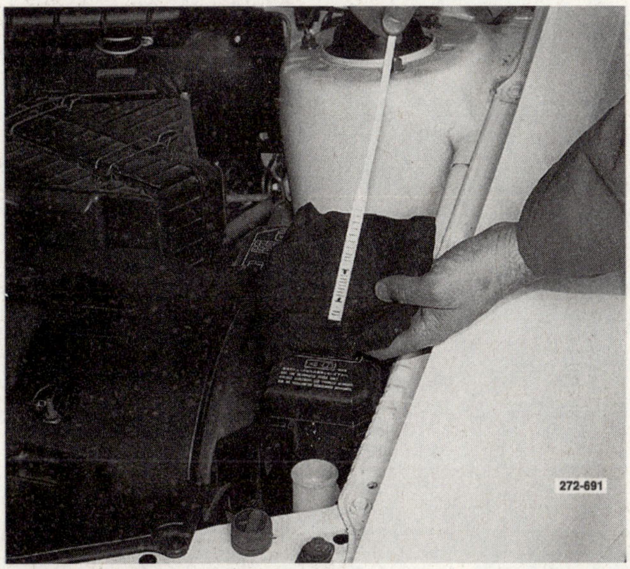

Checking the coolant level in the reserve tank using the dipstick in the cap.

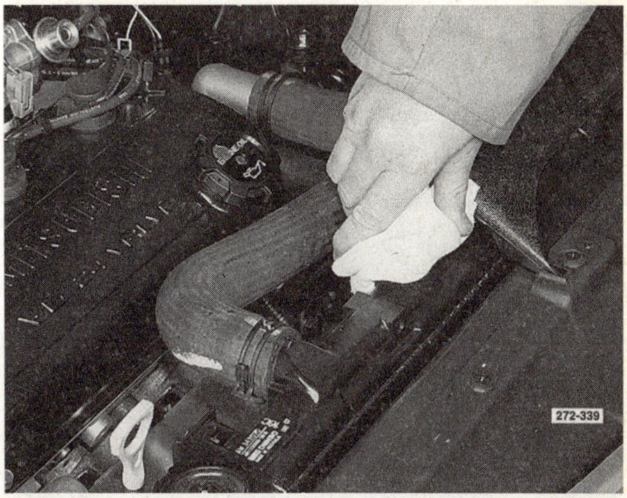

If removing the radiator cap on a warm engine, allow the engine to cool for at least 30 minutes and cover the cap with a thick cloth before removing. If the upper radiator hose can be squeezed together, the system is not pressurised and it is safe to remove the cap.

> i *If there is a problem with the secondary seal on the radiator cap, or a blocked or split overflow hose, it is possible for the reserve tank to indicate satisfactory coolant level, but the radiator coolant level to be low.*

To Drain, Flush and Refill

Follow the procedure below carefully and use a brand name coolant inhibitor concentrate, which you mix yourself rather than the premixed coolant. This allows the mixture to be adjusted to compensate for the remaining water in the cylinder block/heater core.

The procedure below applies to cooling systems, which are regularly serviced and where the recommended, high quality coolant is used. On vehicles, with a blocked radiator core or a neglected cooling system, the radiator may have to be removed and the radiator and/or cooling system reverse flushed as described in the Radiator heading.

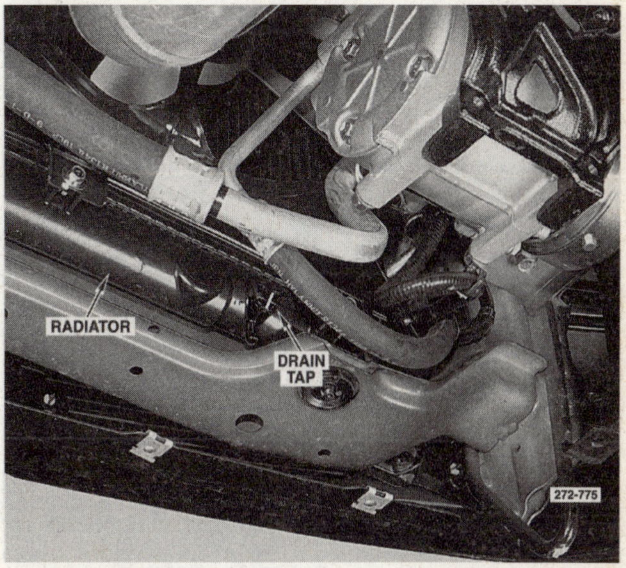

Location of the radiator drain tap.

Cylinder block drain plug located at the front of the engine.

Rear view of the throttle body showing the coolant hose locations.

1 Raise the front of the vehicle and support on chassis stands. Refer to General Information (chapter 1), if necessary for the position of the jacking and support points. A hole is provided in the plastic splashguard for draining the coolant, if more clearance is required, remove the fasteners and withdraw the splashguard from below the radiator.

2 Position a container under the radiator drain tap and the cylinder block drain plugs. One drain plug is located behind the lower part of the dipstick and the other one is behind the rear engine pipe.

3 With the engine cool, remove the radiator cap.

Do not remove the cap from a hot engine, as boiling coolant could cause serious injury. Allow the engine to cool for at least half an hour.

4 Open the radiator drain tap, remove the cylinder block drain plugs and drain the coolant into a container.

5 When the coolant has drained, insert a garden hose into the radiator hose and, using low water pressure, flush the system until the water flowing out of the drain holes and radiator filler neck is clean and contains no trace of old coolant. Disconnect one of the coolant hoses from the throttle body. Insert the garden hose into the throttle body hose and repeat the flushing procedure.

Do not use high water pressure to flush the cooling system, as damage to various cooling system components may occur.

6 Clean the reserve tank using water pressure and a bottle brush, if necessary. Note that the hose can be disconnected from the bottom of the reserve tank to make cleaning out easier. Ensure that the hose is reconnected once the tank is clean.

7 Close the radiator drain tap and tighten securely.

8 Clean the openings for the cylinder block drain plugs and apply a small amount of Permatex No 3 or equivalent sealant to the threads of the plugs. Install and tighten the cylinder block plugs to the specified torque.

9 Following the instructions on the container, add the required quantity of high quality coolant concentrate to the radiator.

10 Add clean, clear water (demineralised in areas with hard water) to the radiator until coolant flows out of the open end of the throttle body hose. Reconnect the hose and secure with the hose clamp.

11 Fill the radiator to the overflow hose orifice and fill the reserve tank halfway between the Low and Full marks.

Cylinder block drain plug located at the rear of the engine.

Splashguard partially removed from the left hand front wheel arch showing the coolant reserve tank.

TIP *It is recommended to add approximately a cup full of coolant concentrate to the reserve tank.*

12 Run the engine for approximately 6 minutes at 1 500 rpm. While the engine is running, keep the radiator full by adding clean water.

13 Stop the engine and install the radiator cap. Top up the reserve tank to the Full mark.

14 Run the engine and check for correct operating temperature. Check the system for leaks and check the coolant level as previously described.

HOSES

To Inspect

1 Inspect all hoses for signs of cracking, perishing, excessive softness and bulging. Usually, the first signs of a deteriorating hose is the ends of the hose turning up.

2 Squeeze each hose in turn. No hose should feel hard and brittle.

3 Bend the heater and throttle body hoses and check for surface cracking.

4 Check the engine and transaxle for oil leaks. Oil leaking onto heater and radiator hoses could severely weaken the hose.

Renew any suspect hoses detected in the above checks.

5 Check that the hose clamps are secure, and renew any that have lost their tension.

Inspecting the radiator hoses for deterioration.

To Renew

1 Drain the cooling system as previously described.

2 Remove the necessary components to gain access to the radiator, throttle body or heater hoses.

3 Noting the installed position of the hose, release the hose clamps and using a twisting motion, remove the hose.

TIP *If the hose is difficult to remove, it is advisable to carefully cut the hose lengthwise to ease removal. Avoid cutting a groove in the pipe or tank fittings.*

4 Check the condition of the hose fitting surface and, if necessary, clean the metal fitting using a wire brush or emery paper.

TIP *The radiator tank and heater fittings are made of plastic. Use a piece of cloth to clean the fittings.*

5 Check the condition of the hose clamps. Renew if necessary.

6 Install the hose clamps to the hose, apply a small amount of water to the hose fitting and install the hose in the position noted on removal.

7 Fill the cooling system, as previously described under this heading.

8 Run the engine and check for leaks.

RADIATOR

To Inspect

1 Check the condition of the tanks, hose fittings and the radiator core.

To check the condition of the front of the radiator core on models with air conditioning, the radiator must be removed, as described later in this chapter.

2 Check the radiator retaining bolts for tightness. Ensure, that the mountings are in serviceable condition.

3 Check the condition of the coolant in the reserve tank. Flush the cooling system as previously described if the condition of the coolant is suspect.

4 Allow the engine to cool and remove the cap from the radiator.

Do not remove the cap from a hot engine, unless the top radiator hose can be easily squeezed together. Boiling coolant could cause serious injury.

5 Check the condition of the radiator filler neck. The radiator cap will not hold the recommended pressure, if the filler neck is faulty. Renew the radiator or consult a radiator specialist if necessary.

6 Using running water, clean the rubber seals and the vacuum valve on the radiator cap.

Check the condition of the seals, spring and the valve. If necessary, renew the radiator cap with one of the specified opening pressure.

Check the condition of the radiator cap seals.

7 If available, a pressure tester can quickly determine the condition of the radiator cap. Moisten the main rubber seal with water and apply the pressure indicated on the cap. Renew the cap if the pressure cannot be maintained using the tester.

Trouble Shooting

ENGINE OVERHEATING

1 Insufficient coolant: Replenish the coolant, check the cooling system for leaks and repair as necessary.
2 Obstructed air flow through the radiator: Remove the obstruction from the radiator core with compressed air or water pressure.
3 Faulty thermostat: Test and renew the thermostat if necessary.
4 Faulty cooling fan: Check the operation of the cooling fan as described later in this chapter.
5 Poor coolant circulation: Check and rectify as described under Coolant Circulation Faulty.
6 Restricted exhaust system: Remove the restriction or renew the restricted components.
7 Low engine oil level: Stop the engine immediately and replenish the oil to the correct level.
8 Dragging brakes: Check the operation of the brake system and handbrake adjustment. Refer to Brakes (chapter 15).
9 Slipping clutch: Repair the clutch as necessary.
10 Leaking cylinder head gasket: Check and rectify as described under Coolant Leakage — Internal.
11 Engine tight after overhaul: Stop the engine and allow it to cool. Check that the engine is not unduly tight.

In addition to indication on the temperature gauge, engine overheating is usually accompanied by steam emitting from the coolant overflow pipe and loss of engine power.

In some cases, engine overheating can result in cylinder head gasket failure due to distortion of the cylinder head or total engine seizure.

If there are no symptoms of engine overheating, check for a faulty temperature gauge sender or a faulty temperature gauge.

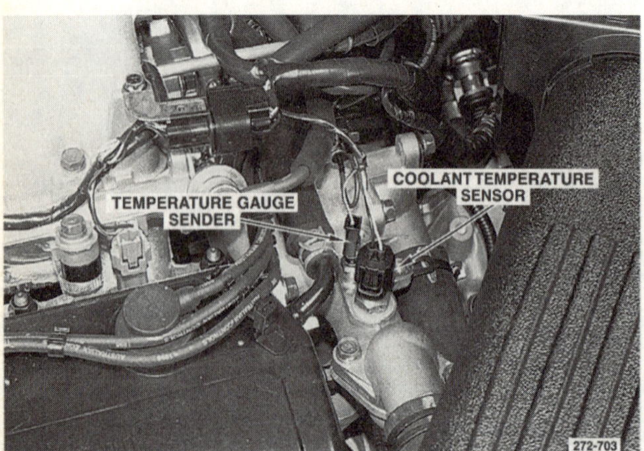

Location of the temperature gauge sender.

COOLANT LEAKAGE — EXTERNAL

1 Loose hose clamp or faulty hose: Renew the hose clamp or the faulty hose.
2 Leaking radiator core or tanks: Repair or renew the radiator.
3 Worn water pump causing leakage at the water pump seal: Check the water pump bearings for excessive play and renew the pump if necessary.
4 Loose or damaged coolant drain plugs: Tighten or renew the faulty component.
5 External crack in the cylinder block or cylinder head: Repair or renew the faulty component.
6 Leaks at the thermostat cover, water pump or coolant pipe fitting: Tighten or renew components as necessary.
7 Leaking heater core or heater valve: Repair or renew the faulty component.
8 Corroded or damaged welch plugs: Renew the faulty welch plugs.

Check the cooling system for external leakage by running the engine to operating temperature over a dry floor and checking for the leak source. Also check the vehicle interior below the heater unit for moisture.

On air conditioned models, it is normal for water to be seen draining under the vehicle from the evaporator after the vehicle has been operated with the air conditioning on.

COOLANT LEAKAGE — INTERNAL

1 Cylinder head gasket leak due to distorted cylinder head or cylinder block gasket faces: Check the heads and block for distortion and if necessary renew the cylinder heads or cylinder block. Renew the cylinder head gaskets.
2 Crack in the cylinder head or cylinder block: Repair or renew the faulty components.
3 Corroded or cracked throttle body: Renew the throttle body.

Check the engine for internal leakage by withdrawing the dipstick and checking for emulsified oil. Run the engine and check for excessive steam at the exhaust pipe which would indicate water leakage into the combustion chamber. Check the compression as described in Engine (chapter 4) under the Routine Maintenance heading, to locate a leaking cylinder head gasket.

A leaking head gasket is sometimes indicated by bubbles in the radiator when the engine is running.

COOLANT CIRCULATION FAULTY

1 Insufficient coolant in the system: Replenish the coolant and check for leaks.
2 Faulty thermostat: Check and renew the thermostat.
3 Partial blockage of the radiator core tubes: Reverse flush or renew the radiator.
4 Sludge deposits in the engine coolant passages: Clean and reverse flush the cylinder block.
5 Faulty water pump: Renew the water pump.

TIP *This condition is easily checked by removing the radiator cap and running the engine until it reaches the normal operating temperature. Check for coolant turbulence in the radiator. A normal system should show turbulence at part throttle. As most adverse conditions can be caused by more than one fault, diagnosis can only be made after checking the items by process of elimination in the order shown. If blocked radiator core tubes are detected which cannot be cleared by reverse flushing, it is recommended that the radiator be taken to a reliable specialist for repair.*

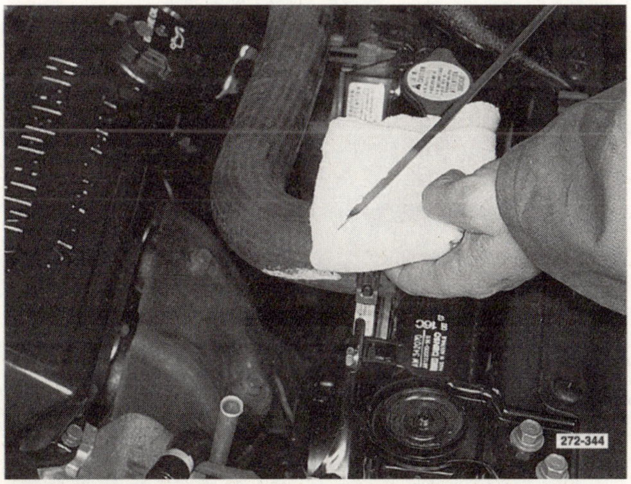

Check the engine dipstick for emulsified oil indicating internal coolant leakage.

Radiator

TO REMOVE

1 Drain the cooling system as previously described.
2 Remove the fasteners and withdraw the air intake from above the radiator. Separate the wiring connectors for the electric cooling fan and if fitted, the air conditioning condenser fan.
3 On automatic transaxle models, disconnect the oil cooler hoses from the lower radiator tank. Plug the hoses and the fittings to prevent the loss of fluid and the entry of dirt.
4 Disconnect the upper and lower hoses from the radiator.
5 Disconnect the overflow hose from the radiator filler neck.
6 Remove the bolts retaining the radiator support brackets to the top of the radiator support panel. Withdraw the brackets from the vehicle.
7 Manoeuvre the radiator and electric cooling fan assembly from the engine compartment.

Reverse flush the radiator using water pressure.

8 If necessary, remove the mounting bushes from the pegs on the lower radiator tank.
9 If necessary, remove the bolts retaining the fan assembly and separate the assembly from the radiator.

TIP *To prevent internal deposits drying and possibly blocking the radiator tubes and causing overheating, seal the hose fittings and fill the radiator with coolant if it is to be removed from the vehicle for any length of time.*

TO FLUSH AND CLEAN

1 With the radiator removed as previously described, install the radiator cap securely and carefully invert the radiator.
2 Insert a hose in the radiator outlet fitting, seal the gap around the hose with suitable waste cloth and reverse flush the radiator until the water flowing from the radiator inlet fitting is clean.

i *If excessive back pressure or diminished flow indicates severely blocked radiator core tubes, which cannot be cleaned by reverse flushing, it is recommended that the radiator be taken to a reliable radiator specialist for repair.*

3 Apply a stream of water or compressed air to the radiator core from the rear to the front.

Cleaning the radiator core.

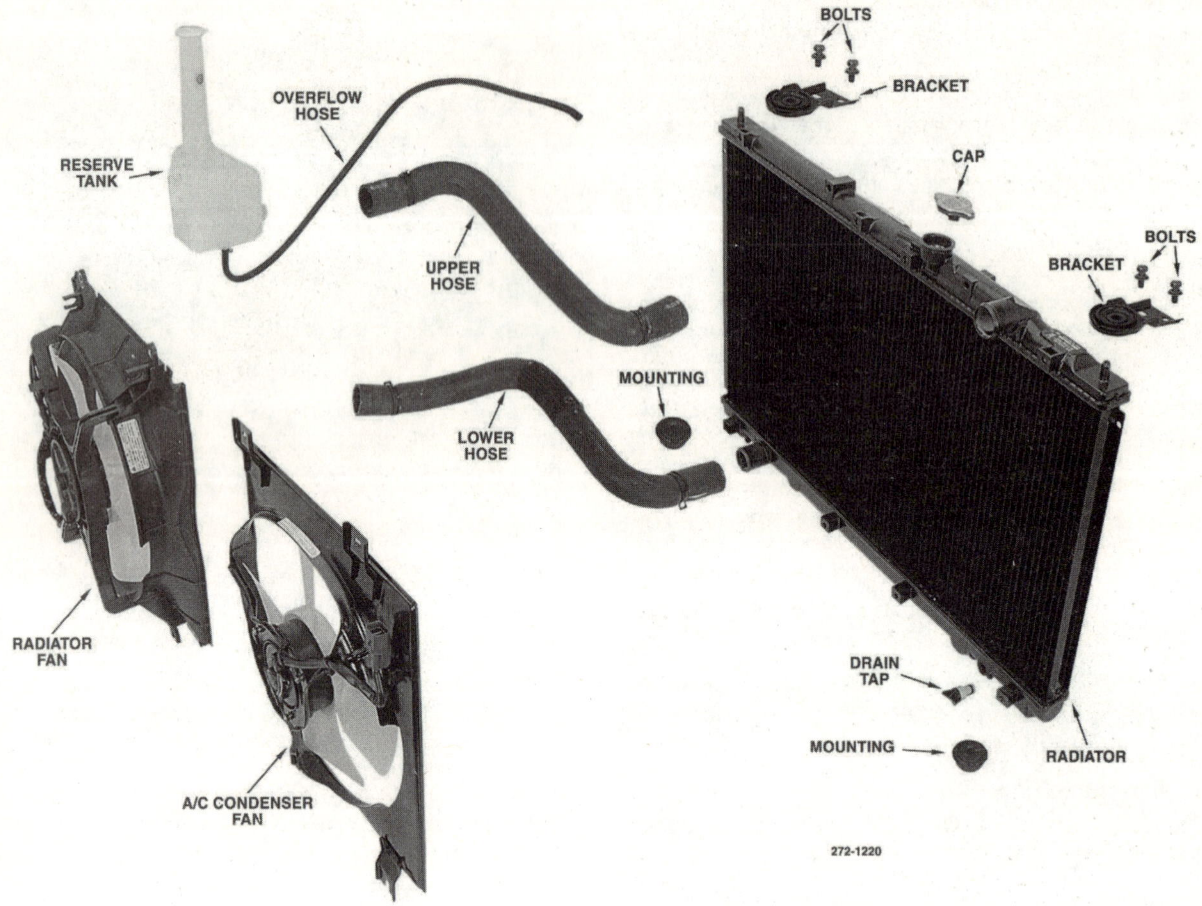

Radiator and associated components.

If using high pressure cleaning equipment, maintain a distance of 40 - 50 cm between the cleaning nozzle and the radiator core to avoid damage to the radiator core.

Continue the cleaning procedure until all dirt and foreign matter is removed from the radiator core.

TO INSTALL

Installation is a reversal of the removal procedure with attention to the following points:

1. If removed, install the fan assembly to the radiator.
2. If removed, install the mounting bushes to the pegs on the lower radiator tank.
3. Carefully install the radiator, ensuring that the mounting bushes on the bottom of the radiator are fully inserted in the lower mounting brackets.
4. On automatic transaxle models, connect the oil cooler hoses to the lower radiator tank.
5. Refill the cooling system as previously described in Routine Maintenance.
6. On automatic transaxle models check and if necessary top up the transaxle fluid to the correct level. Refer to Automatic Transaxle (chapter 11) for the correct procedure.
7. Check the cooling system for leaks and rectify as necessary.
8. Road test the vehicle and check for correct operating temperature.

Cooling Fan and Relay

TO TEST COOLING FAN SYSTEM ON VEHICLE

1. With the engine cold and the air conditioning switched Off, start the engine and disconnect the coolant temperature sensor wiring connector.

   If the cooling fan system is functioning correctly, the radiator fan and air conditioning condenser fan should rotate at high speed.
2. If the cooling fans rotate with the coolant temperature sensor connected, check the coolant temperature sensor as described in Fuel and Engine Management (chapter 7).

Avoid leaving the coolant temperature sensor disconnected for any longer than necessary to prevent the engine ECU detecting a fault. If the Check Engine Lamp remains illuminated when the engine is running after the above tests, clear the fault codes as described in Fuel and Engine Management (chapter 7).

3. If the cooling fan does not operate when the coolant temperature sensor wiring connector is disconnected, use the wiring diagrams as a guide to performing the following procedures:

a. Disconnect the cooling fan motor wiring connector and using a voltmeter, ensure that power is reaching the two

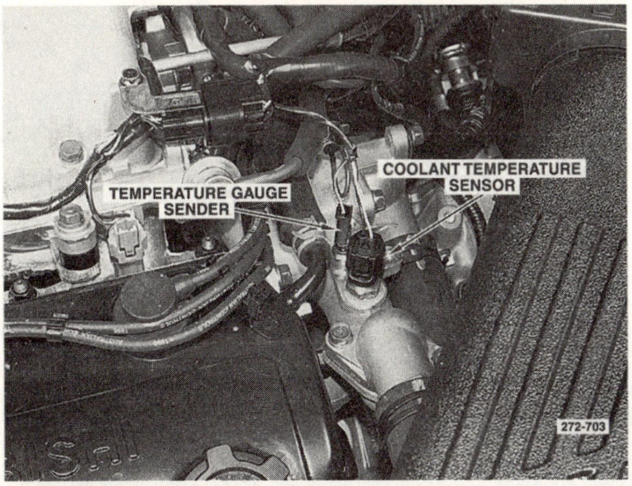

Location of the coolant temperature sensor.

power supply wires of the motor connector. If not, check the appropriate fusible link and fuse for serviceability. Also check the relays by substituting with a known good relay of the same type from one in the relay box.

b If the fuses and relays are serviceable but no power is reaching the motor, check the wiring for breaks or damage and repair as necessary. If the wiring is serviceable but still no power is reaching the fan connector a problem may exist with the engine ECU refer to Fuel and Engine Management (chapter 7) for more information.

c If power is reaching the cooling fan, switch the engine Off and check for a good connection to earth at the black wires. Note that both black wires are not always connected to earth at all times, on Verada models.

d Using jumper leads connected to battery power, connect the positive jumper lead to the blue wire of the fan motor and the negative jumper lead to the black wire and then the yellow wire. The fan should operate at low speed when the negative lead contacts each earth wire.

Connect the positive jumper lead to the green wire and repeat the procedure.

e As a final check, connect the positive jumper lead to both the green and blue wires and the negative jumper lead to both the yellow and black earth wires. The fan should now operate on high speed.

Renew the fan motor if it does not operate as described.

TO REMOVE AND INSTALL COOLING FAN MOTOR

1. Disconnect the negative battery terminal.
2. Drain the cooling system sufficiently to allow removal of the upper radiator hose. If necessary, refer to the procedure described previously.
3. Remove the fasteners and withdraw the air intake from above the radiator. Separate the wiring connectors for the electric cooling fan and if fitted, the air conditioning condenser fan.
4. Release the hose clamp and disconnect the upper radiator hose from the radiator. Disconnect the overflow hose from the radiator shroud.
5. On models with air conditioning, remove the upper condenser fan to radiator bolts and loosen the lower ones. If necessary the condenser fan can be manoeuvred from the radiator.

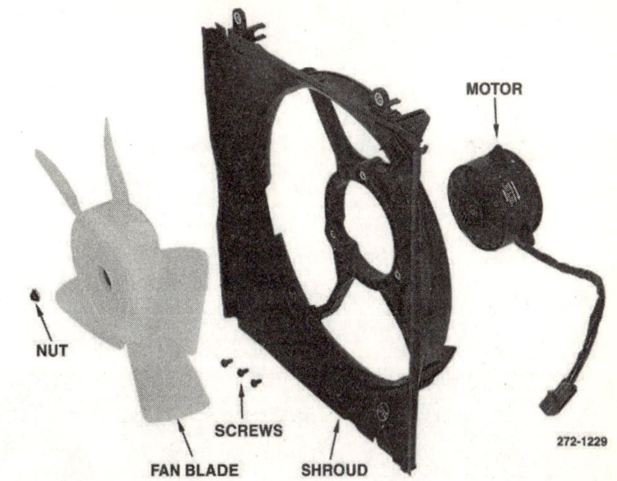

Cooling fan, shroud and motor.

6. Remove the upper bolts retaining the fan assembly to the radiator. Loosen the lower retaining bolts.
7. Manoeuvre the fan motor and shroud as an assembly from the vehicle being careful not to damage the radiator.
8. Remove the nut retaining the fan blade to the motor shaft.
9. Remove the fan blade from the motor shaft.
10. Disengage the clips retaining the fan wiring and the connector to the fan shroud. Remove the retaining screws and remove the fan motor from the shroud.

Installation is a reversal of the removal procedure with attention to the following points:

1. Install the fan motor to the fan shroud, ensuring that the drain pipe on the fan motor points downwards.
2. Ensure that the fan blade does not contact the shroud when rotating.
3. Tighten all screws and bolts securely.
4. Securely connect the fan motor wiring connector and install the overflow hose to the retainers on the fan shroud.
5. Test the cooling fan operation as previously described, prior to road testing the vehicle.

Thermostat

To Test Thermostat — Thermometer

TO REMOVE AND INSTALL

1. With the engine cool, remove the radiator cap.
2. Loosen the radiator drain tap and drain the cooling system into a container to a level just below the thermostat cover.
3. Remove the fasteners and withdraw the air intake from above the radiator.
4. Release the hose clamp and using a twisting motion disconnect the upper radiator hose from the thermostat cover.
5. Remove the bolts retaining the thermostat cover to the thermostat housing and lift the thermostat cover from the housing.
6. Withdraw the thermostat from the thermostat housing.

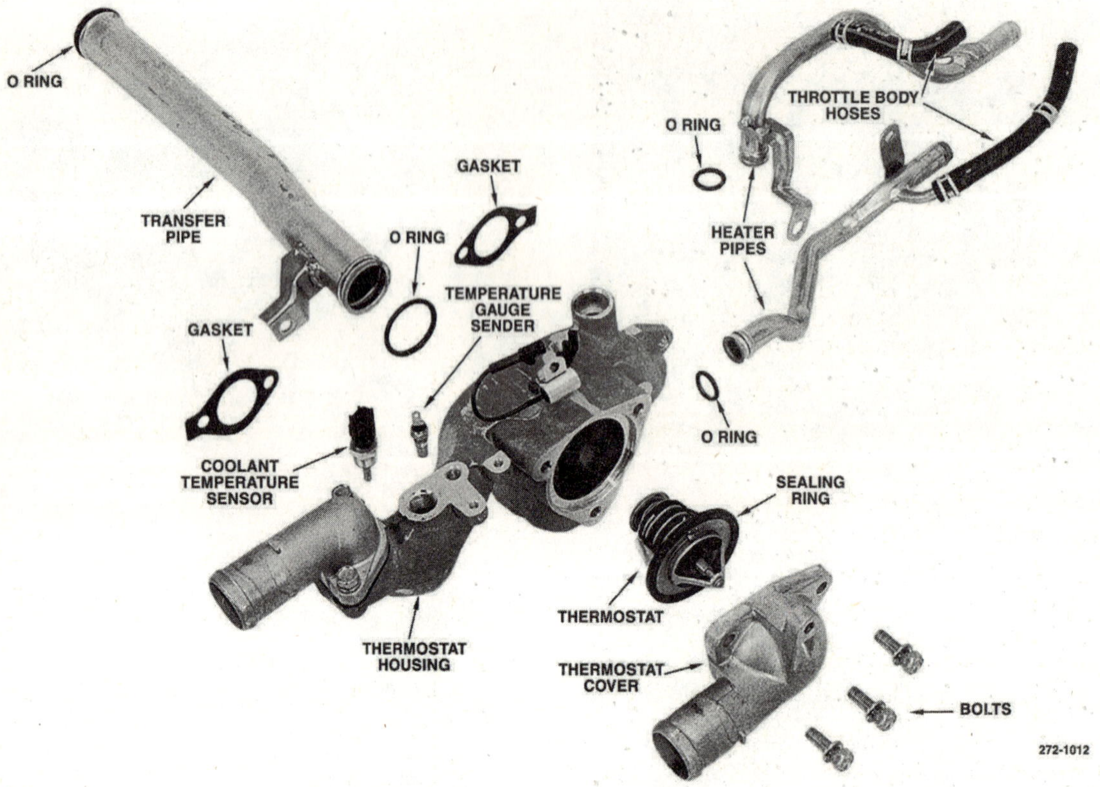

Thermostat and related components.

A visual examination of the thermostat will often determine its serviceability and make further testing unnecessary. For example, a thermostat with its valve fully open when removed from a cold engine is obviously faulty and should be renewed.

Installation is a reversal of the removal procedure with attention to the following points:

1 Ensure that the mating surfaces of the thermostat housing and cover are clean and flat.

TIP *If necessary, a distorted thermostat housing gasket face can be corrected by rubbing the face of the cover in a figure eight motion on fine emery paper placed on a flat surface.*

2 Install the thermostat using a new sealing ring, ensuring that the jiggle valve is uppermost, as shown on the illustration.

3 Install the thermostat cover and tighten the bolts to the specified torque.

4 Fill the cooling system as previously described.

5 Road test the vehicle and check for correct operating temperature and for coolant leaks. Rectify as necessary.

TO TEST

1 Remove the thermostat as previously described.

2 Check that the thermostat is closed when cold.

3 Suspend and immerse the thermostat, together with a reliable thermometer in a vessel of cold water ensuring that neither the thermostat nor the thermometer are touching the sides or bottom of the vessel.

4 Progressively heat the water noting the temperature reading on the thermometer as the thermostat valve commences to open. Refer to Specifications for the correct opening temperature. Heat the water to 95°C and check that the thermostat valve opens more than 8 mm.

A thermostat which does not commence to open at the specified temperature or does not open more than 8 mm at 95 °C, should be renewed.

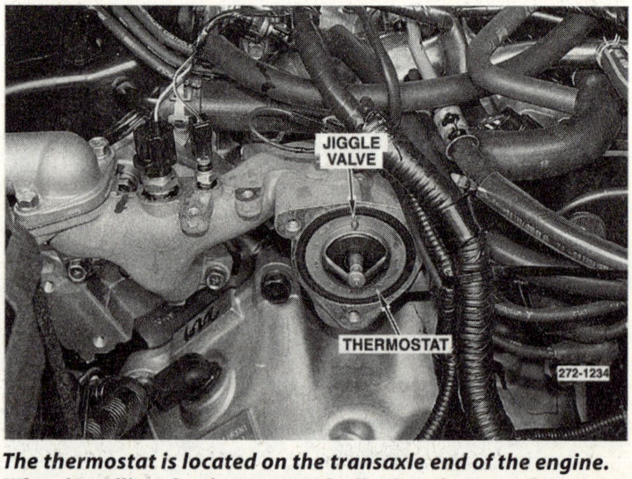

The thermostat is located on the transaxle end of the engine. When installing the thermostat the jiggle valve must be uppermost.

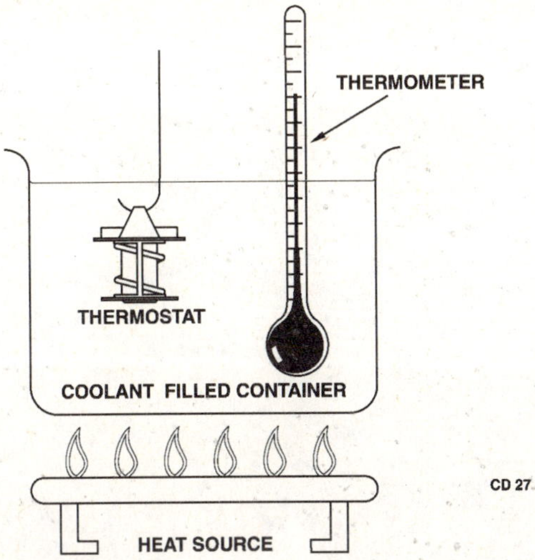

Testing the thermostat.

Water Pump

TO REMOVE AND INSTALL

1. Disconnect the negative battery terminal.
2. Drain the cooling system as previously described.
3. Remove the engine timing belt as described in Engine (chapter 4).
4. Remove the bracket retaining bolts and bracket.
5. Remove the water pump retaining bolts noting the length of each bolt as it is removed. There is one larger diameter bolt fitted but it can only go in one place.
6. Remove the water pump using a twisting motion to free the pump from the transfer pipe. Remove and discard the O ring.
7. Using a sharp scraper remove all traces of old gasket material from the cylinder block.
8. Check the water pump drain hole for signs of coolant leakage which indicates a leaking seal and/or worn bearing.

Timing belt removed showing water pump and retaining bolts (arrowed) and the position of the short bolt.

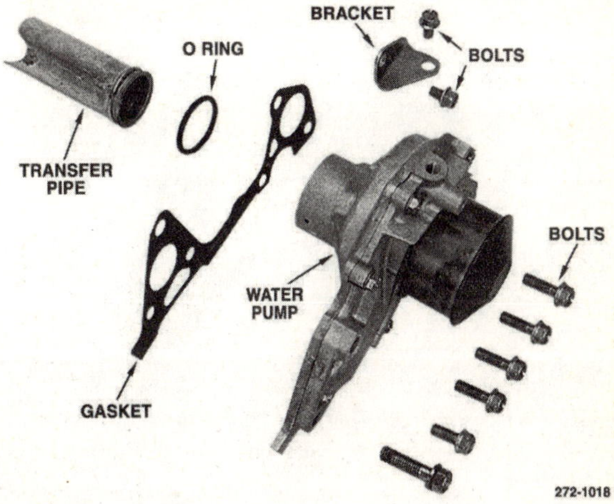

Water pump components.

TIP *If the water pump has been leaking always renew the timing belt.*

9. Rotate the water pump pulley by hand and check the bearing for rough rotation, wear or looseness in the water pump body. Check the impeller and water pump body for wear and damage.

Installation is a reversal of the removal procedure with attention to the following points:

1. Install a new transfer pipe O ring and lubricate the O ring with water just prior to installation.
2. Using a new gasket, install the water pump to the cylinder block and water pipe. Ensure that the short bolt goes into the right position. Refer to the illustration if necessary.
3. Tighten the water pump retaining bolts to the specified torque.
4. Ensure that the timing belt is correctly installed and adjusted. Refer to Engine (chapter 4) for the correct procedure.
5. Install and adjust the drive belts as described under the Routine Maintenance heading in Engine (chapter 4).
6. Fill the cooling system as previously described.
7. Road test the vehicle and check for correct operating temperature and for coolant leaks. Rectify as necessary.

Welch Plugs

To facilitate the casting of the cylinder block and cylinder head passages and water jackets, a number of openings were made during manufacture. These openings are sealed with welch plugs.

The welch plugs will have to be renewed when cleaning the water jackets or when they become defective due to corrosion.

i *If a welch plug is found to be corroded, it is advisable to renew all the welch plugs.*

Location of the welch plugs at the rear of the cylinder block.

TO RENEW

1 Drain the cooling system as previously described.

2 Remove the necessary engine components or accessories to gain ample working space around the damaged plug.

> *It will be necessary to remove the transaxle to access the welch plug at the rear of the engine. Leaks from this welch plug will drip from the bottom of the clutch housing or the automatic transaxle torque converter housing.*

3 Using a punch and hammer, tap the welch plug on its outer circumference into its opening.

4 Grasp the edge of the welch plug with a pair of multigrip pliers and using the shoulder of the pliers as a fulcrum, lever the plug out of its opening.

5 Thoroughly clean and dry the welch plug opening.

6 Lightly smear the edge of the new welch plug and the opening with a suitable jointing compound, such as Permatex No 2.

7 Place the new welch plug onto the opening and using a large socket or piece of tubing fitting neatly inside the rim of the welch plug, tap the plug squarely into the opening.

Method of removing a welch plug using multigrip pliers.

> *The welch plug must be entered squarely into its opening or leakage may occur.*

8 Install the components which were removed to gain access to the welch plug.

9 Refill the cooling system as previously described. With the radiator cap installed, run the engine until it reaches normal operating temperature and check for coolant leaks. Rectify as necessary.

Drive Belts

The renewal of the drive belts is fully covered under the Routine Maintenance heading in Engine (chapter 4).

Chapter 6
Engine Electrical

Contents

Specifications

BATTERY

Type ..... 12 volt lead acid
Polarity ..... Negative (-) earth
Specific gravity:
Fully charged ..... 1.240 – 1.260
Fully discharged ..... Below 1.130

ALTERNATOR

Make ..... Bosch KC-A
Maximum output ..... 110 amps
Voltage ..... 14.25 – 14.55 volts
Brush length:
Standard ..... 12 mm
Minimum ..... 3.8 mm
Slip ring:
Minimum diameter ..... 26.7 mm
Maximum runout ..... 0.06 mm
Rotor resistance ..... 3 – 5 ohms
Stator winding resistance ..... 0.015 ohms ± 10%

STARTER MOTOR

Make ..... Bosch DV
Type ..... Four pole, four brush, with planetary reduction gears
Output ..... 1.2 kW
Armature:
Maximum runout ..... 0.5 mm
End float ..... 0.05 – 0.30 mm
Commutator:
Maximum runout ..... 0.1 mm
Minimum diameter ..... 31.2 mm
Depth of undercut —
Standard ..... 0.8 mm
Minimum ..... 0.5 mm
Brush length, minimum ..... 7 mm
Pinion end gap ..... 0.5 – 2.0 mm

TORQUE WRENCH SETTINGS

Alternator pulley retaining nut ..... 54 – 68 Nm
Starter motor retaining bolts ..... 27 – 34 Nm

Routine Maintenance

BATTERY

Keep naked flame and spark away from the battery as the battery gases are highly explosive.

1 Keep the battery and the surrounding area clean and dry. In particular, ensure that the top of the battery is free from dirt and moisture to prevent voltage flow between the terminals.

If necessary, clean the battery case as follows:

a Remove the battery from the vehicle as described later in this chapter.

b Pour hot water over the battery being careful not to allow any water to enter the battery through the filler cap vent holes.

c Wipe the battery case clean.

2 Clean the battery terminals and posts using a battery post cleaner or a wire brush. Remove all dirt and corrosion.

3 Remove the filler caps and check that the vent holes are clear.

4 Check the electrolyte level, as described later, and replenish with distilled water if necessary. The correct level is marked on the case on some batteries. Alternatively, bring the level to the bottom of the well in the filler hole ensuring that the plates are submerged. Do not overfill the battery.

5 If the battery electrolyte requires frequent replenishing, check for overcharging by measuring the alternator output as described later in this chapter. Check the battery case for cracks. If the case is damaged, renew the battery.

Ensure that the battery terminals are clean and secure.

6 If the electrolyte has overflowed and contaminated the battery carrier and surrounding body panels, proceed as follows:

a With the battery removed from the vehicle, sprinkle bicarbonate of soda over the corroded area.

b Pour hot water over the bicarbonate of soda. The bicarbonate of soda will react with the water and begin to foam.

c When the bicarbonate of soda has stopped foaming, rinse the area with clean water and allow to dry.

d Remove any blistered paint from the contaminated area.

e Paint the contaminated area with a corrosion inhibiting paint.

7 Install the battery as described later in this chapter.

Terminals and Leads

1 Clean the battery terminals and posts using a battery post cleaner or a wire brush. Remove all dirt and corrosion.

2 Apply petroleum jelly to the battery terminals to prevent corrosion.

3 Check the condition of the battery leads and terminals.

4 Ensure, that the positive lead is securely connected to the positive battery terminal and to the starter motor and that the negative lead is securely connected to the negative battery terminal and to the transaxle.

To Test and Charge

1 Using a hydrometer, test the specific gravity of the battery electrolyte.

Refer to the following chart which shows charge conditions for various specific gravity readings:

SPECIFIC GRAVITY	CONDITION
1.110 – 1.130	Fully discharged
1.140 – 1.160	Nearly discharged
1.170 – 1.190	Approximately one quarter charged
1.200 – 1.220	Half charged
1.230 – 1.240	Approximately three quarters charged
1.240 – 1.260	Fully charged

Using a hydrometer to check the battery specific gravity.

These readings are for electrolyte at a temperature of 20°C. For every 10° above 20°C, add 0.007. For every 10 below 20°C, subtract 0.007.

An accurate specific gravity reading cannot be obtained if distilled water has recently been added to the electrolyte.

If the electrolyte level is below the battery plates, replenish with distilled water and charge the battery before testing the specific gravity. Do not transfer electrolyte from one cell to another.

2 If a low state of charge is indicated, the battery should be charged using a commercially available battery charger with attention to the following points:

Specific gravity variations of more than 0.025 between cells indicate defective cells in the battery.

a Disconnect the negative and positive battery terminals.

b Remove the filler caps from the battery.

c Check the battery electrolyte level and replenish as necessary.

d Follow the battery charger manufacturers instructions.

TIP *Where practical, use a slow charge rate in preference to a fast charge rate.*

3 Should the battery fail to operate satisfactorily after charging, it is recommended that it be taken to an authorised dealer or auto electrician with the necessary equipment to perform a load test. Load testing, when performed in conjunction with the specific gravity tests described above, will give an accurate indication of the serviceability of the battery.

ALTERNATOR

The alternator itself does not require maintenance under normal conditions, apart from checking that the drive belt is serviceable and correctly tensioned as described in Engine (chapter 4). Also check that oil or coolant is not allowed to enter the alternator cooling slots, as both these could cause the alternator to malfunction.

Occasionally check the alternator mounting bolts and wiring to ensure that they are secure.

Battery and Charging System Trouble Shooting

BATTERY UNDERCHARGED

1 Loose or broken alternator drive belt: Adjust or renew the alternator drive belt as described in Engine (chapter 4).

2 Dirty or loose battery terminals: Clean and check the battery terminals.

3 Faulty charge warning lamp bulb or circuit: Renew the bulb or repair the circuit as necessary.

4 Faulty battery: Renew the battery.

5 Fault in the charging circuit wiring or connections: Check and renew or repair as necessary.

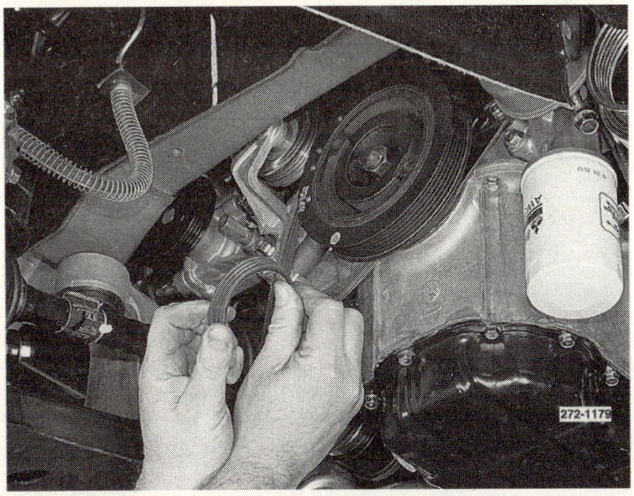

Checking the drive belt for cracks, glazing or wear.

6 Fuse or fusible link blown: Rectify the fault and renew as necessary.
7 Faulty alternator or voltage regulator: Refer the alternator to an authorised repairer.

Check the state of charge (specific gravity) of the battery as described under the Battery heading in this chapter. If the battery is undercharged, check the possible causes in the order given.

BATTERY OVERCHARGED

1 Faulty charging circuit wiring or connections: Check and renew any faulty components.
2 Faulty battery: Install a new battery of the correct type and capacity.
3 Faulty alternator or voltage regulator: Refer the alternator to an authorised repairer

TIP *An overcharged battery is indicated by continual loss of water through boiling. This is usually accompanied by discolouration of the electrolyte. It is important that the shroud surrounding the battery is not discarded as this keeps the temperature of the battery at acceptable levels.*

CHARGE INDICATOR LAMP REMAINS ON

1 Broken drive belt: Renew the drive belt. Refer to the Engine chapter.
2 B+ lead from alternator broken: Repair or renew the B+ lead.
3 S lead from alternator broken: Repair or renew the S lead.
4 Short to earth in the warning lamp circuit: Check and repair the circuit.
5 Battery overcharged: Refer to the previous heading.
6 Faulty alternator or voltage regulator: Refer the alternator to an authorised repairer.

CHARGE INDICATOR LAMP DOES NOT OPERATE

1 Faulty voltage regulator or worn alternator brushes: Overhaul or renew the alternator.
2 Charge indicator lamp bulb blown: Check and renew the faulty bulb.

Spin the alternator pulley by hand to check the bearings for noise.

3 Charge indicator lamp bulb holder faulty: Check the bulb holder and renew if necessary.
4 Open circuit in the wiring: Check and rectify the open circuit.
5 Open circuit in the rotor windings: Overhaul or renew the alternator.

NOISE IN DRIVE BELT OR ALTERNATOR

1 Loose or frayed drive belt: Renew the drive belt, as described in Engine (chapter 4).
2 Loose alternator pulley: Tighten the pulley retaining nut to the specified torque.
3 Faulty alternator bearings: Overhaul or renew the alternator.

TIP *To check if the noise is in the alternator or drive belt, remove the drive belt and run the engine for a short period. If the noise ceases, check the serviceability of the drive belt and the alternator components. Also, check the bearings of other components driven by the engine drive belts.*

Battery and Starter Trouble Shooting

STARTER LACKS POWER TO CRANK ENGINE

1 Battery terminals loose or corroded: Clean and tighten the terminals.
2 Battery undercharged: Check the charging system and rectify as necessary.
3 Battery fully charged but will not crank engine: Check the starter circuit, also check for a locked drive and ring gear, an internal starter fault or a seized engine.
4 Battery will not hold charge: Check and renew the battery.
5 Faulty or dirty solenoid switch or contacts: Check the solenoid as described later in this chapter.
6 Faulty starter motor: Check and overhaul the starter motor.

Check the state of charge of the battery and check all terminals for cleanliness and security. A battery terminal that is not making good contact will usually become warm when attempting to crank the engine. If necessary, test the starter motor as described later in this chapter.

STARTER WILL NOT ATTEMPT TO CRANK ENGINE

1 Immobiliser system activated: Refer to the Engine Immobiliser System heading in part 1 of Fuel and Engine Management (chapter 7).

2 Battery discharged: Check for a fault or short circuit in the electrical system.

3 Faulty automatic transaxle inhibitor switch: Check and adjust or renew as necessary. Refer to Automatic Transaxle (chapter 11).

4 Fusible link blown: Rectify the cause and renew the fusible link.

5 Faulty starter motor: Check and overhaul the starter motor.

6 Faulty ignition switch: Renew the ignition switch.

7 Open circuit in the starting system: Check for dirty or loose terminals and rectify as necessary.

8 Battery fully charged and all components serviceable but engine will not crank: Check for a locked drive pinion and ring gear and/or a seized engine.

Rotate the engine manually to ensure that the starter motor drive pinion is not locked with the ring gear and that the engine is not seized. Ensure that the ignition is switched Off before rotating the engine.

Battery

TO REMOVE AND INSTALL

Original equipment sound systems are fitted with a security system. Disconnecting the battery terminal(s) will require the operator to enter the security code.

If the number has been lost, it may be obtained by providing proof of ownership to an authorised dealer.

Battery packs which plug into the cigarette lighter and maintain security code and radio tuning memory while the battery is disconnected, are commercially available.

1 Open the battery cover and prise out the clips retaining the upper part of the shroud to the radiator support panel and to the battery clamp. Manoeuvre the upper part of the shroud from the vehicle.

2 Disconnect the negative and positive battery terminals in that order.

3 Remove the battery clamp retaining nuts, and withdraw the clamp from the vehicle.

4 While holding the battery securely, lift it out of the vehicle.

5 If necessary, remove the bolts retaining the various components and the lower shroud and manoeuvre the shroud from the vehicle.

Installation is a reversal of the removal procedure with attention to the following points:

1 If removed, install the lower battery shroud to the vehicle and install all bolts securely.

2 Install the battery, ensuring that the terminals and posts are clean.

3 When connecting the leads to the battery, ensure that the polarity is correct and connect the positive lead first.

Under no circumstances should the battery terminals be crossed over as severe damage will occur to the electronic circuitry.

4 Do not overtighten the terminal clamp nuts.

5 Apply petroleum jelly to the terminals to prevent corrosion.

Alternator

To Test — Voltmeter 0 – 20 volts DC, 120 – 0 – 120 ammeter

DESCRIPTION

The internal fan Bosch alternator has the voltage regulator and brush holder unit mounted directly to the slip ring end bracket of the alternator. The battery voltage sensitive regulator ensures that the alternator charging rate is directly determined by the condition of the battery.

The alternator consists of a fixed coil winding or stator, a shaft wound coil or rotor, a rectifier or diode pack and the voltage regulator/brush holder unit.

A magnetic field is produced in the rotor by electrical current fed into the rotor windings by brushes bearing on slip rings mounted on the rotor shaft. Three phase alternating current (AC) is produced in the star wound stator by the rotation of the magnetic field.

As it is not possible to charge a storage battery with alternating current (AC), it is necessary to rectify the output of the stator windings to direct current (DC). This is done using a bank of diodes or diode pack, mounted within the alternator slip ring end bracket.

The current output of the alternator is governed by the inbuilt characteristics of the alternator. The voltage output of the alternator is governed by the integrated circuit regulator. The voltage regulator is not adjustable.

An electrical cut out unit is not necessary as the diodes prevent a reverse current flow through the alternator.

SERVICE PRECAUTIONS

1 Ensure that the battery is connected with the negative terminal to earth.

2 Do not short circuit or ground any terminals common to the charging circuit.

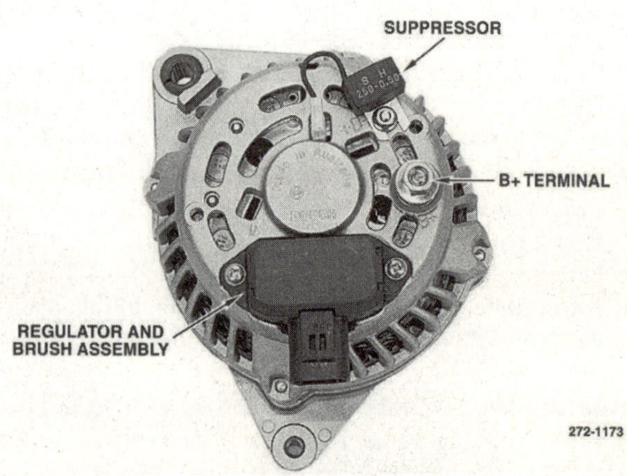

End view of alternator.

3 Always disconnect the battery before connecting a battery charger.
4 If a booster battery is used, always connect it in a parallel circuit, that is positive to positive (+ to +) and negative to negative (- to -) to maintain a 12 volt supply.
5 Never disconnect the battery or terminals in the charging circuit while the engine is running.
6 Regularly check the drive belt.
7 Keep the battery terminals clean and all electrical connections tight.
8 Disconnect the battery, alternator and any electronic items when electric welding on the vehicle. Refer to Electronic Components under the Vehicle Safety heading in General Information (chapter 1).

TO TEST ON VEHICLE

If the warning lamp stays on after the engine has been started and run at approximately 1200 rpm or the charge rate is suspected of being outside the specified range, indicated by a discharged or overcharged battery, perform the following tests.

1 Stop the engine. Do not open circuit any parts of the charging circuit while the engine is running.
2 Check the condition and tension of the alternator drive belt.
3 Clean the battery terminals and tighten all the electrical connections. Check the wiring for short circuits to earth and/or bridged circuits. Ensure that the battery is fully charged. If the battery is in a low state of charge, either recharge it or replace it with a fully charged battery.
4 Start the engine, allow it to run until normal operating temperature is reached and then stop the engine.

While performing the following test on the alternator, avoid attaching the test leads with alligator clips.

All connections should be made with clamp or bolted type connectors. This will prevent leads becoming detached during the test and open circuiting the alternator which would result in failure of the diodes in the alternator.

5 Disconnect the negative battery terminal.
6 Remove the nut from the alternator output terminal (B+) and disconnect the alternator output lead.

Alternator installed with washer bottle, and upper timing belt cover removed.

7 Connect one lead of a 120 - 0 - 120 ammeter to the output terminal (B+) and the other lead of the ammeter to the lead removed from the output terminal.
8 Connect the negative battery terminal and the voltmeter to the battery.
9 Switch the headlamps onto high beam. Leave the headlamps on for approximately five minutes to reduce the charge of the battery and to apply a load to the circuit.
10 Start the engine and gradually increase the speed to approximately 2 500 rpm. The reading on the ammeter should be approximately 75% of the rated output of the alternator listed in the Specifications.
11 Switch off all the lamps and run the engine at 2 500 rpm until the indicated charge is below 10 amps. The voltage reading should be the same as listed in Specifications.

Should the alternator fail to reach the specified output, the unit will have to be removed and overhauled, or a replacement unit installed. Should the voltage reading be above or below Specifications, the regulator will have to be renewed.

TO REMOVE AND INSTALL

1 Disconnect the negative battery terminal.
2 Remove the bolts, disconnect the hose(s) and wiring connectors and withdraw the washer bottle from the vehicle.
3 Remove the output terminal nut (B+) and wiring from the rear of the alternator.
4 Disconnect the wiring connector from regulator assembly.
5 Loosen and remove the alternator drive belt. If necessary refer to Engine (chapter 4) for the correct procedure.
6 Remove the upper alternator mounting bolt from the bracket.
7 Remove the bolts retaining the front upper timing belt cover and withdraw the cover.

TIP *Use care not to allow anything to fall into the timing cover or to damage/contaminate the belt.*

8 Remove the wiring bracket retaining bolt and place the wiring clear of the work area.
9 Remove the bolts retaining the upper alternator bracket to the cylinder head and manoeuvre the bracket from the vehicle.
10 Remove the lower alternator mounting bolt nut and washers and slide the through bolt towards the passenger side towards the passenger side of the vehicle.
11 On models with air conditioning, place a small jack with a suitable block of wood under the air conditioning compressor bracket and just take the weight of the engine. Remove the front engine mounting through bolt and raise the jack slightly until the alternator can be manoeuvred from the engine compartment.

Avoid damaging the timing belt and air conditioning pipes with the alternator on removal.

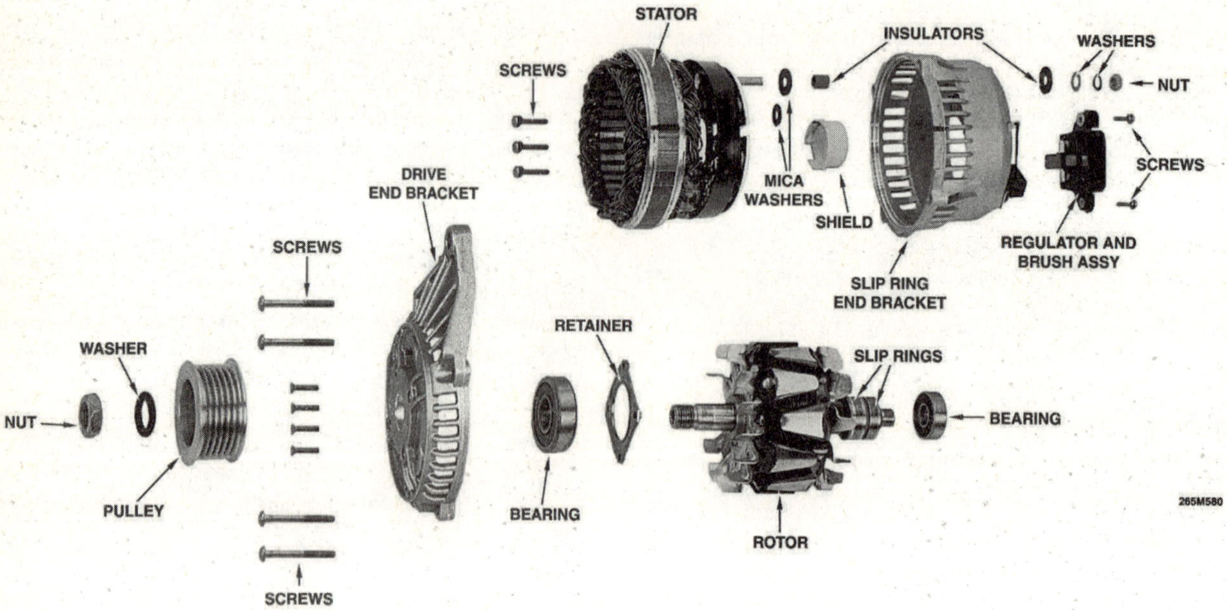

Alternator components.

Installation is a reversal of the removal procedure with attention to the following points:

1 Do not overtighten the alternator drive belt. Refer to the Routine Maintenance heading in Engine (chapter 4), for the correct procedure.

2 Ensure that the washer bottle hose(s) are securely connected on installation.

3 Ensure that all alternator connections are providing good electrical contact.

TO OVERHAUL

Extensive knowledge and specialised equipment is required to overhaul the alternator, with the cost of the equipment normally exceeding the repair charges of an auto electrician or the cost of an exchange unit. A dismantled view is shown to better understand the internal working of the alternator and help those who have the equipment and the knowledge to carry out the overhaul procedure.

Starter Motor

DESCRIPTION

The Bosch starter motor is of the induced pole type with four brushes and four pole shoes with field coils, using a planetary gear arrangement to increase torque delivered from the starter motor.

A solenoid switch is attached to the starter drive end bracket. The solenoid plunger is connected to the overrunning clutch and drive pinion assembly through a link and lever assembly.

When the solenoid windings are energised, the plunger acting on the lever and pivot causes the drive pinion to engage with the flywheel ring gear. At the same time, the solenoid switch closes, supplying power from the positive battery terminal to the starter motor field coils and to the armature, causing the armature to rotate.

As the engine fires, the overrunning clutch of the drive pinion assembly prevents high speed rotation of, and possible damage to, the starter armature if the solenoid windings are not immediately de-energised by releasing the ignition key.

TO TEST ON VEHICLE

Should the starter fail to operate when the ignition switch is moved to the Start position, refer to Roadside Trouble Shooting (chapter 3) to determine if the fault is in the starter motor, the security system or the battery and cables.

TO TEST OFF VEHICLE

When testing the starter motor off the vehicle with jumper leads exercise caution not to allow the positive leads to short on the vice or starter motor body.

1 Remove the starter motor, as described later in this chapter and secure it in a vice fitted with soft jaws.

2 Connect a positive battery jumper lead to the starter motor solenoid B+ terminal.

3 Connect the negative jumper lead to the drive end bracket of the starter motor, ensuring it is securely connected.

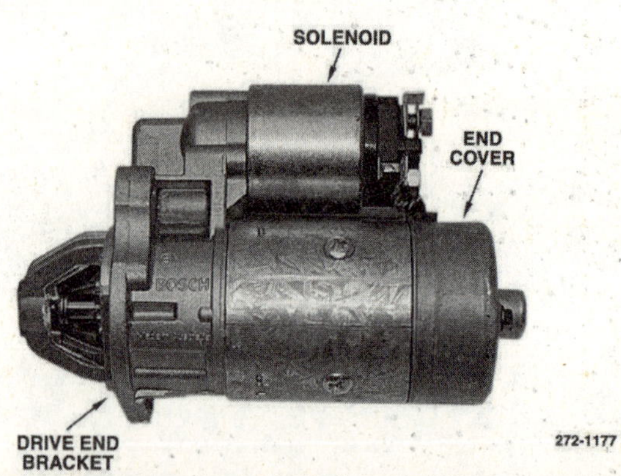

Starter motor removed from the vehicle.

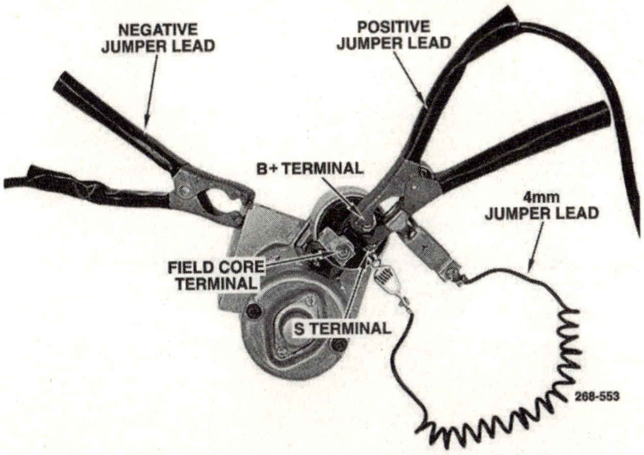

Method of testing the starter removed from the vehicle.

4 Connect the other end of the jumper leads to a fully charged 12 volt battery, ensuring that the positive lead is connected to the positive battery terminal and the negative lead is connected to the negative battery terminal.

5 Connect a 4 mm jumper lead from the B+ terminal to the S terminal of the solenoid.

If the starter fails to operate, proceed as follows:

a Remove the positive jumper leads from the solenoid B+ and S terminals.

b Remove the 4 mm jumper lead from the positive jumper lead.

c Touch the positive jumper lead to braided field core wire terminal on the starter motor solenoid.

If the starter motor operates, the fault is in the starter motor solenoid. Renew the solenoid, using the illustration as a guide.

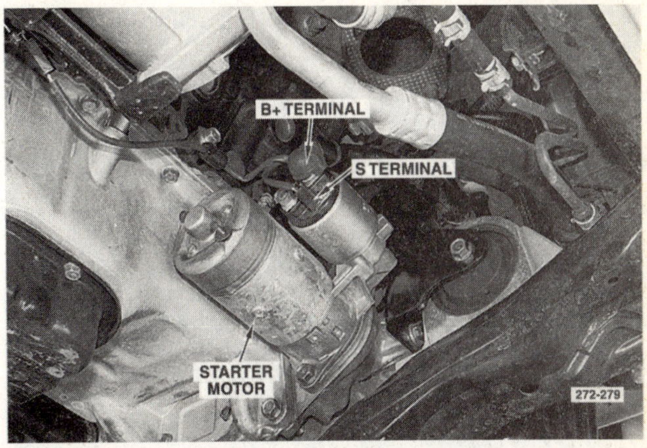

Starter motor installed with heat shield removed.

If the starter motor does not operate, dismantle and overhaul, or install an exchange starter motor.

TO REMOVE AND INSTALL

1 Disconnect the negative battery terminal.

2 Raise the front of the vehicle and support it on chassis stands. Refer to General Information (chapter 1) for the position of the and chassis stands.

3 Remove the splashguard bolts and scrivets and remove the splashguard from the vehicle.

4 Where fitted, remove the retaining nut and bolt and remove the heat shield from the starter motor.

5 Disconnect the wiring from the starter motor solenoid and remove the starter motor retaining bolts.

6 Manoeuvre the starter motor from below the engine compartment.

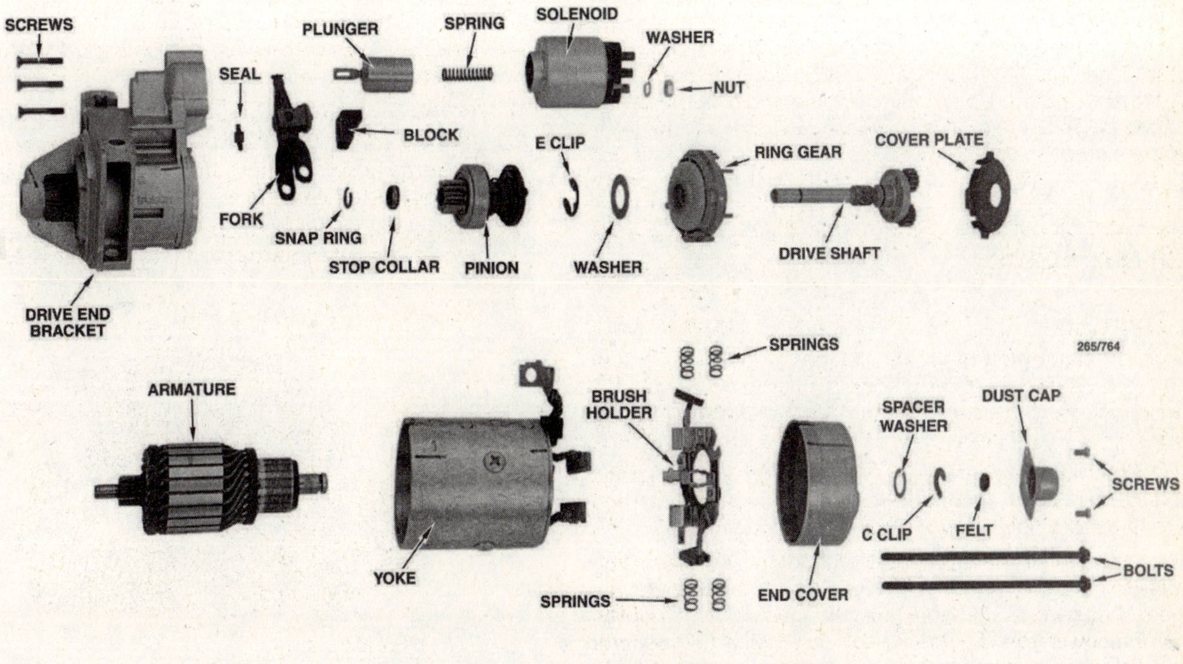

Starter motor components.

Installation is a reversal of the removal procedure with attention to the following points:

1 Tighten the starter motor retaining bolts to the specified torque.
2 Ensure that the wiring is securely connected to the starter motor solenoid.

TO OVERHAUL

Extensive knowledge and specialised equipment is required to overhaul the starter motor, with the cost of the equipment normally exceeding the repair charges of an auto electrician or the cost of an exchange unit. A dismantled view is shown to better understand the internal working of the starter motor and help those who have the equipment and the knowledge to carry out the overhaul procedure.

Ignition System

The ignition system is part of the engine management system.

The specifications, testing and the overhaul procedures are covered in Fuel and Engine Management (chapter 7).

Chapter 7
Fuel and Engine Management

Part 1. General Servicing and Repair

Contents

Specifications

Type ... Electronic multipoint fuel injection with ignition control

Idle speed:
- Kerb ... 700 ± 100 rpm
- Base ... 700 ± 50 rpm

[1] Ignition timing:
- 3.0 litre ... 15° ± 3° BTDC
- 3.5 litre ... 10° ± 3° BTDC

CO concentration at idle ... 0 – 0.5%

Spark plug, 3.0 litre:
- Front bank —
  - Champion ... RC8YCC4
  - NGK ... BKR6E–11
- Rear bank (platinum) —
  - Champion ... RC8PYP4
  - NGK ... PFR6J–11

Spark plug, 3.5 litre:
- Front bank —
  - Champion ... RC10YCC4
  - NGK ... BKR5E–11
- Rear bank (platinum) —
  - Champion ... RC10PYP4
  - NGK ... PFR5J–11

[2] Spark plug gap ... 1.0 – 1.1 mm

High tension lead resistance ... Less than 22 k/ohms

Ignition coil resistance:
- Primary ... 0.5 – 0.7 ohms
- Secondary ... 9 – 13 k/ohms

Fuel pressure:
- At idle ... Approx 265 kPa
- With vacuum hose disconnected from regulator ... 324 – 343 kPa

Injector resistance at 20°C ... 13 – 16 ohms

Idle speed control (ISC) motor resistance at 20°C ... 28 – 33 ohms

[1] Non adjustable

[2] Platinum plugs are pre gapped and should not have the gaps adjusted

TORQUE WRENCH SETTINGS

Coolant temperature sensor	29 Nm
Fuel filter banjo bolt	36 Nm
Fuel pressure regulator bolts	9 Nm
Fuel rail retaining bolts	12 Nm
Fuel rail supply pipe bolts	9 Nm
Fuel rail pressure pipe bolts	9 Nm
Fuel tank retaining bolts (sedan)	25 Nm
Plenum chamber to inlet manifold nuts and bolts	17 Nm
Throttle body bolts	11 Nm
Throttle pedal bracket nuts	13 Nm

Description

To prevent severe electrical shock, extreme care must be taken when working on or near the electronic ignition systems as dangerous voltages are produced.

The fuel and engine management system employs sensitive electronic components. Refer to the Service Precautions and Procedures heading in this chapter before attempting any service procedures.

The engine management system on the range of vehicles covered by this manual, control the operation of the fuel injection and ignition systems.

The central component of the engine management system is the electronic control unit (ECU), which is a microcomputer used to control the ignition system and the amount of fuel injected according to signals received from various sensors. As changes are detected in engine load and speed, coolant temperature, air temperature, throttle position and exhaust emissions, the ECU alters the ignition timing and the fuel injection to achieve optimum engine efficiency and fuel economy under those conditions.

When the engine is running, the ECU monitors the signals it is receiving from the various sensors. If the ECU detects that an abnormal signal has continued for an extended period, a fault code will be recorded and the ECU will substitute an alternate signal for the abnormal signal, enabling near normal driveability except in the most severe failure situations. When the self diagnosis mode is activated, the ECU will generate coded signals to indicate the circuit that the fault has been detected.

When the ECU is set in self diagnosis mode, the coded signals stored in memory can be read to identify the area in which the fault has occurred. For people with the confidence and knowledge to identify and undertake fault code retrieval and repair procedures, refer to Part 2 of this chapter.

In the multipoint fuel injection system, a metered amount of fuel is injected into the air stream at the entrance to each inlet port of the cylinder head. The air/fuel mixture then enters the combustion chamber via the inlet valve. The fuel is supplied by a high pressure electric fuel pump, which is mounted inside the fuel tank. The fuel is maintained at a constant pressure, relative to variances in manifold vacuum, by the fuel pressure regulator. Excess fuel is returned to the fuel tank via the fuel return line.

The fuel pressure and the duration that the injectors are open regulate the injection quantity. The basic time that the injectors remain open is controlled by the ECU which uses information from the distributor, air flow sensor and intake air temperature sensor. The basic injection time is then modified by the ECU which uses information from the coolant temperature sensor, throttle position sensor, vehicle speed sensor and oxygen sensor to give optimum engine efficiency and exhaust emissions for the relevant conditions.

The electronic ignition system consists of a distributor with an integral ignition coil and an igniter. The distributor has two functions, to produce and distribute secondary voltage to the spark plugs and to provide the ECU with information on engine speed and piston position. The ECU in turn commands the igniter to interrupt the ignition coil primary earth circuit, causing the secondary voltage to be produced.

The ECU constantly adjusts the ignition timing to achieve optimum engine efficiency and the lowest exhaust emissions for the relevant conditions.

The idle speed control (ISC) motor is mounted to the throttle body and maintains the engine idle speed at a set value when the engine is cold or various loads are placed upon the engine. The throttle body is located on the inlet manifold plenum chamber and controls the amount of air entering the engine by action of the throttle valve.

Efficient operation of the engine management system depends on controlled intake air flow, correct fuel supply and accurate signals being received by the ECU. Therefore, it is essential that all air, fuel and electrical connections are clean and tight.

Part 1 of this chapter covers common servicing and repair procedures, which are considered within the scope of the average home mechanic.

Part 2 provides more detailed specifications and test procedures. These procedures should only be attempted by mechanics or those with a good understanding of electrical components and circuits.

Reference should be made to the Service Precautions and Procedures heading before performing any repair procedures in this chapter.

Routine Maintenance

The following routine maintenance items are covered under the Routine Maintenance heading in Engine (chapter 4).

Air cleaner element.

Fuel filter.

Spark plugs.

Distributor cap and leads.

The ignition timing is non adjustable and can only accurately be checked using the manufacturers dedicated test equipment.

IDLE SPEED

1 Run the engine until it reaches normal operating temperature.

2 Ensure that all accessories are turned off and, on automatic transaxle models, ensure that the transaxle is in Neutral or Park.

3 Stop the engine and connect the tachometer. Refer to Part 2 of this chapter if necessary.

Engine compartment showing various engine management system components.

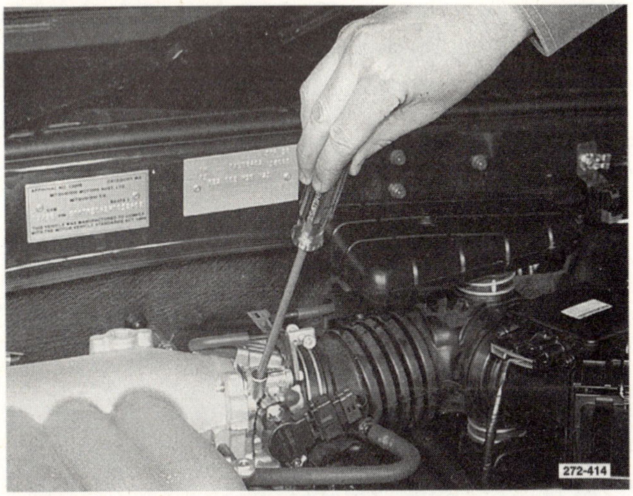

Adjusting the engine idle speed.

4 Start the engine and check the idle speed. If the kerb idle speed is not to Specifications, proceed as follows:

a Stop the engine and bridge the diagnosis connector terminals 1 and 4 using a paper clip or split pin.

b Start the engine and check the base idle speed. Adjust the idle speed screw if the base idle speed is not within Specifications. If the idle speed is high and cannot be adjusted using the idle adjusting screw, check for vacuum leaks, an incorrectly adjusted throttle stop screw or a faulty fast idle air valve. The throttle stop screw adjustment is covered under the Throttle Body heading.

c Remove the jumper lead from the diagnosis connector. After the engine has been idling for 10 minutes, check that the kerb idle speed is to Specifications.

HOSES, PIPES AND CONNECTIONS

1 Raise the vehicle and check the fuel tank and connections for leaks. Check the fuel tank for damage.

2 Check that the fuel supply and return lines are not leaking between the fuel tank and the engine compartment. Check that the fuel supply and return pipes are not damaged or unsecured. Check that the rubber hoses are not cracked or chafing.

3 Lower the vehicle and check that the fuel supply and return lines in the engine compartment are not leaking. Check that there are no leaks at the fuel rail connections to the fuel hoses, the pressure regulator and the injectors.

4 Check that all vacuum hoses in the engine compartment are secure at each connection. Check that the hoses are not split, cracked or perished.

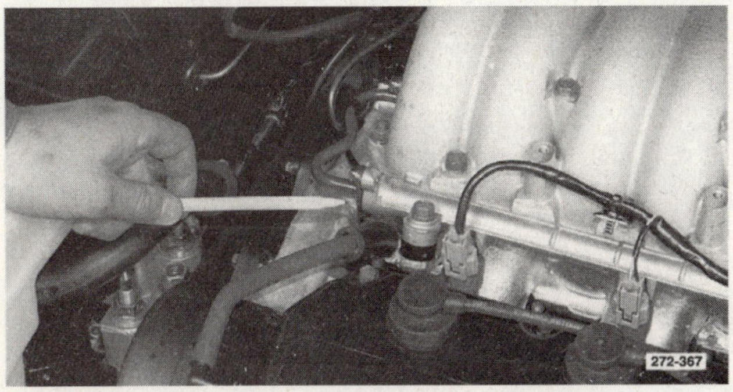

Check all fuel system connections for signs of leakage.

AIR INLET HOSE

1 Check that the inlet hose connections between the air cleaner, air flow sensor and the throttle body are secure. Check that the hose is not split or cracked.

2 Check that the air cleaner inlet hose is not obstructed.

THROTTLE LEVER

1 Operate the throttle body linkage by hand and check that the movement is smooth and free. If necessary, lubricate the throttle lever and shaft. When the throttle is slowly released, check that it closes fully.

If the throttle does not close fully, check for an incorrectly adjusted throttle cable as described later in this chapter.

2 Check that the inner throttle cable is not frayed at the ends of the cable.

Trouble Shooting

The following trouble shooting procedures are basic checks only and assume that the checks in Roadside Trouble Shooting (chapter 3) have been performed to confirm that the problem is in the fuel and engine management system. Prior to performing any of the following checks, refer to the Service Precautions and Procedures heading.

If the following checks fail to locate the fault, refer to Part 2 for more thorough testing, or refer the vehicle to an authorised dealer or fuel injection specialist.

ENGINE HARD TO START OR WILL NOT START

1 Faulty No 23 and 24 fusible links: Check the fusible links and renew as necessary.

2 Faulty fuel pump relay or engine control relay: Test the relays and renew if necessary.

3 Fault in the fuel supply: Test the fuel pump and the fuel pressure.

4 Faulty engine management component wiring connections: Check that all wiring connections are clean and secure, especially the distributor connections.

5 Faulty ignition coil: Check the ignition coil resistances. Check the ignition system.

6 Faulty injectors: Test the injector circuit.

7 Faulty engine immobiliser system: Check the immobiliser system.

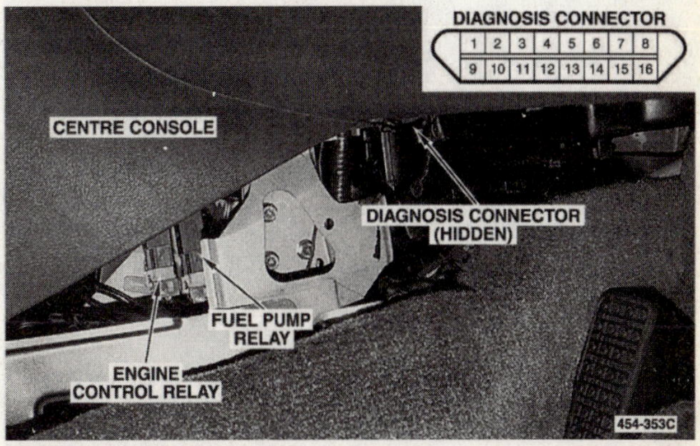

Location of the engine control relay and the fuel pump relay. Inset shows the diagnosis connector terminals.

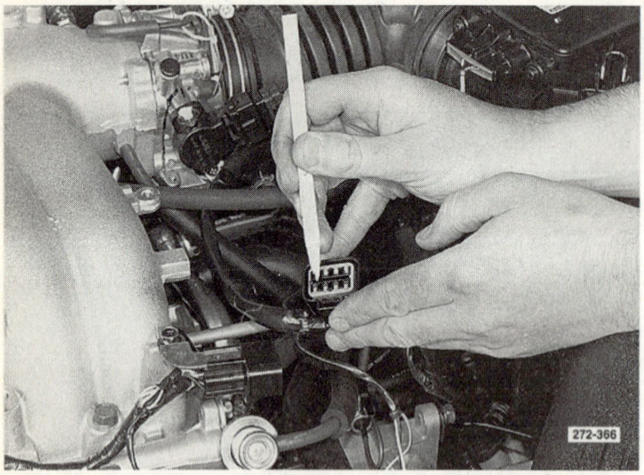

Ensure that all wiring connections in the engine management system are clean and secure.

8 Engine flooded: Depress the throttle pedal fully while cranking the engine until it starts.

ENGINE STARTS THEN STALLS

1 Moisture under the distributor cap: Remove the cap and check for moisture, cracks and tracking.

2 Incorrect idle speed: Check for air leaks around the intake manifold and all vacuum hoses.

3 Faulty engine management component wiring connections: Check that all wiring connections are clean and secure.

4 Faulty idle speed control (ISC) motor: Check the ISC motor operation.

5 Fault in the fuel supply: Check the fuel pressure.

ENGINE MISFIRES

1 Faulty, dirty or incorrectly adjusted spark plugs: Renew or clean and adjust the spark plugs.

2 Faulty high tension leads: Check the resistance of the leads and renew as necessary.

3 Dirt or moisture under distributor cap: Remove the distributor cap and clean the cap and the rotor. Renew if excessively dirty or cracked.

TIP *It is recommended that the rotor and distributor cap be renewed together.*

Check the distributor cap for cracks, deterioration or tracking.

Service the air cleaner element regularly.

4 Water in the fuel: Drain the fuel from the system and renew the fuel filter.

5 Faulty ignition coil: Check the coil resistances and renew if necessary.

6 Faulty fuel injectors: Check the electrical connections and test the injectors.

ENGINE LACKS POWER

1 Blocked air cleaner element: Check the element and renew as necessary.

2 Incorrectly adjusted throttle cable: Adjust the throttle cable.

3 Low fuel pressure: Check the fuel pressure. If low, renew the fuel filter and check the lines for blockage. Check the fuel pressure regulator.

4 Water in the fuel: Drain the fuel from the system and renew the fuel filter.

5 Exhaust system blockage: Check the exhaust system, including the catalytic converter for damage and remove any restrictions.

6 Timing belt has jumped a tooth: Check that the valve timing is correctly set as described under the Timing Belt heading in Engine (chapter 4).

EXCESSIVE FUEL CONSUMPTION

1 External leakage in the fuel supply system: Check the connections and components for leakage.

2 Blocked air cleaner element: Check the element and renew as necessary.

3 Engine running below correct operating temperature due to a faulty thermostat: Check and renew the thermostat as described in Cooling System (chapter 5).

4 High fuel pressure: Test the fuel pressure. If high, check for a faulty fuel pressure regulator or a blocked fuel return line.

5 Exhaust system blockage: Check the exhaust system, including the catalytic converter for damage and remove any restrictions.

i *Excessive fuel consumption can also be caused by driver compensation for loss of engine power. Check the cylinder compressions as described under the Routine Maintenance heading in Engine (chapter 4).*

Service Precautions and Procedures

To avoid injury or damage to system components, note the following precautions and procedures before commencing any work on the engine management system.

Electronic ignition systems can produce dangerously high voltages. For this reason, extreme care must be taken when working on or near the ignition system.

Do not start the engine if the battery terminals are not clean and tight.

Do not disconnect the battery terminals while the engine is running.

Do not use the fast charge or boost settings on a battery charger to start the engine.

Disconnect the battery terminals before charging the battery.

Do not subject the ECU to temperatures above 80° C.

Ensure that all wiring connectors are clean and tight.

Ensure that the ignition is Off or the battery terminals are disconnected before disconnecting any engine management component.

Reduce the fuel system pressure before disconnecting any fuel supply components. The fuel system maintains a high residual pressure at all times.

Check the fuel system for external leakage whenever a connection in the fuel supply system has been disturbed.

Disconnect the battery terminals and the ECU wiring connector and remove the ECU from the vehicle before attempting any welding operations.

Do not direct a steam cleaning or pressure washing nozzle at engine management components when cleaning the engine.

Refer to the illustration in Part 2 of this chapter for information on how to read insets.

Handle all engine management components with care, particularly the oxygen sensor.

Do not attempt to dismantle any of the electrical components If faulty, these components must be renewed as complete assemblies.

Do not directly apply battery voltage to the injectors.

USING ELECTRICAL TEST EQUIPMENT

Multimeter

It is essential that in all tests where voltage or resistance is to be measured, a digital or analogue display multimeter with a minimum 10 megohm impedance be used.

All meter readings are to be performed with the component wiring connectors installed unless otherwise instructed. This can be done by inserting the meter test probes from the wiring harness side of the connector or backprobing. If the test probes are not thin, they should be modified by attaching a thin piece of wire to the end of the probe, giving a fine point that will not damage the connector. Ensure that the wire is making good contact with the probe.

If sealing plugs or boots are removed from the wiring connectors to facilitate backprobing, ensure that they are correctly installed at the conclusion of the test procedure.

Test Lamp

When a test lamp is specified, it is essential that the test lamp does not draw a current of more than 0.3 amps to avoid damage to the electronic components.

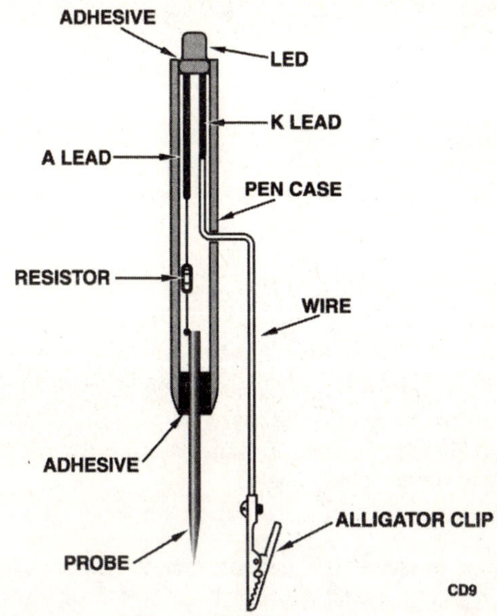

Schematic of an LED test lamp. Ensure that the connections to the K (cathode) lead and A (anodde) lead of the LED are as illustrated.

To check the current draw of the test lamp, connect an accurate ammeter, such as the multimeter described previously, in series with the test lamp. Test the assembly across the vehicle battery.

If the ammeter reads less that 0.3 A or 300 mA, the test lamp is suitable.

If the test lamp is not suitable, fabricate an LED (light emitting diode) test lamp as described in General Information (chapter 1).

An LED test lamp will only allow current flow in one direction. Therefore, the lamp will only operate with the lead connected to negative (earth) and the probe to positive.

PRELIMINARY CHECKS

Prior to investigating a fault or referring the vehicle to an authorised dealer or fuel injection specialist, the following checks should be made.

1. Check for adequate fuel in the fuel tank.
2. Check the wiring connectors and earth points of all engine management components for clean and secure connections.
3. Check the condition of the battery. Refer to the Routine Maintenance heading in Engine Electrical (chapter 6) for checking procedures. Rectify any faults as necessary.
4. Check the air cleaner element for restrictions.
5. Check the fuel pressure as described later in this chapter.
6. Ensure that the engine is in satisfactory mechanical condition and is in tune. Refer to the Routine Maintenance heading in Engine (chapter 4).
7. Check that all engine vacuum and emission control hoses are serviceable and securely connected.
8. Check that there are no air leaks at the inlet manifold, throttle body and plenum chamber.
9. Check the operation of the positive crankcase ventilation (PCV) valve as described in Emission Control (chapter 8).
10. Check the coolant level and ensure that the engine is reaching normal operating temperature.

Fuel System

To Check Fuel Pressure — 600 kPa pressure gauge

REDUCING FUEL SYSTEM PRESSURE

The pressure in the fuel system should be reduced before disconnecting any fuel lines or removing any components of the fuel supply system.

1 Gain access to the fuel pump relay. Refer to the Fuel Pump and Engine Control Relay heading later in this chapter.

2 With the engine idling, disconnect the fuel pump relay wiring connector.

3 When the engine stalls, operate the starter 2 or 3 times to relieve any remaining pressure.

4 Turn the ignition Off and install the fuel pump relay.

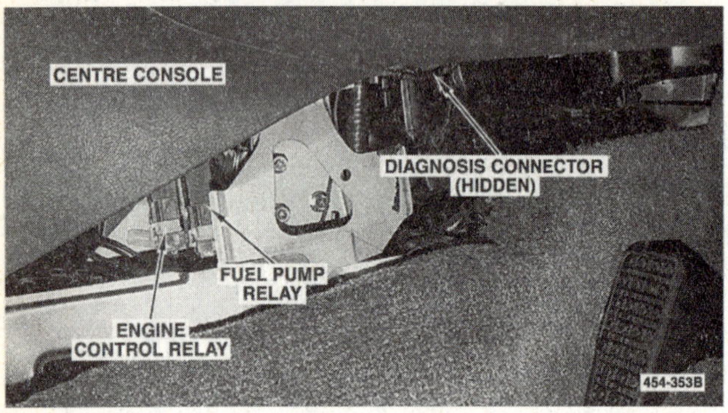

Location of the engine control relay and the fuel pump relay.

TO CHECK FOR EXTERNAL FUEL LEAKAGE

1 Using a jumper lead, apply battery voltage to terminal 2 (black/blue wire) of the engine check connector. Hold the fuel return hose at the fuel pressure regulator and check that fuel can be felt flowing through the hose. In quite conditions it should also be possible to hear the fuel pump operating inside the fuel tank.

2 Check all fuel line connections between the fuel tank and injectors for leakage.

Location of the engine check connector. Inset shows the terminal identification. Looking into the connector, terminal 2 is the fuel pump test and terminal 3 is the tachometer terminal.

3 If repair of any leak requires loosening a fuel line or removal of a component, reduce the fuel system pressure as previously described.

TO TEST FUEL PRESSURE

1 Reduce the fuel pressure and install a pressure gauge to the fuel line forward of the fuel filter.

If suitable fittings are not available, the pressure gauge can be installed by cutting the fuel supply hose and installing a T piece using clamps. On completion of the test procedure, remove the T piece and join the fuel hose using a double ended barbed fitting and clamps. Ensure that all connections are capable of withstanding pressures up to 600 kPa.

2 Using a jumper lead, apply battery voltage to terminal 2 (black/blue wire) of the engine check connector and check for fuel leaks. Disconnect the jumper lead from the check connector.

3 Start the engine and check that the fuel pressure at idle is to Specifications.

If the pressure is high, check that the pressure regulator vacuum hose is connected and that the fuel return line is not restricted. If no fault can be found, renew the fuel pressure regulator.

If the fuel pressure is low, check the fuel filter and supply lines for restriction or blockage. Also check for a blocked fuel pump pickup strainer.

4 Disconnect the vacuum hose from the pressure regulator and check that the pressure increases instantly to Specifications.

If the pressure is low or slow to increase when the vacuum hose is disconnected, check for a faulty pressure regulator or fuel pump.

5 Connect the vacuum hose to the pressure regulator and stop the engine.

Using a bridging wire to power up the fuel pump and check the fuel pressure.

If the fuel pressure decreases within two minutes, pinch the return line closed. If the pressure now remains steady, renew the pressure regulator. If the pressure continues to drop, check the fuel system for internal and external leaks. If no leaks are present, renew the fuel pump.

FUEL PUMP

To Test

1 Switch the ignition On but do not start the engine.

2 Using a jumper lead, apply battery voltage to terminal 2 (black/blue wire) of the engine check connector and check that the fuel pump can be heard operating. It should also be possible to feel a slight vibration at the fuel return line, indicating that the fuel pump is operating.

3 If the pump is not operating, check the power supply and earth circuits at the pump wiring connector. Access the fuel pump wiring connector as follows:

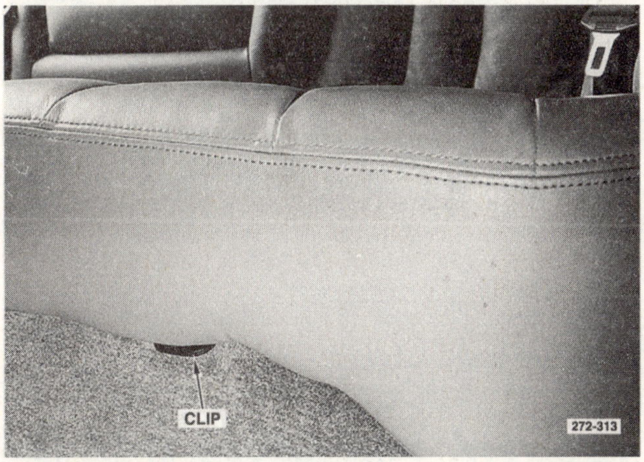

Location of the rear seat retaining clip. A second clip is located on the other side of the seat.

a On sedan models, remove the rear seat cushion by locating the two retaining clips at the front of the seat cushion. Pull the clip toward the front of the vehicle and lift the seat cushion upward. Repeat this on the second retaining clip and remove the seat cushion from the vehicle.

Remove the screws retaining the fuel pump cover and lift it from the floor of the vehicle.

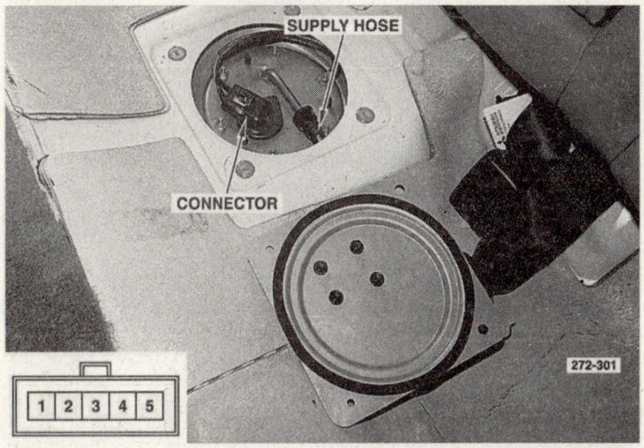

Rear seat cushion and cover removed showing the fuel pump wiring connector. Sedan models. Inset shows the terminal identification.

b On station wagon models, raise the rear of the vehicle and support on chassis stands.

4 Disconnect the fuel pump wiring connector and measure the resistance between the fuel pump terminals 4 and 5 (black/blue wire and black wire). Renew the pump if an open circuit is indicated.

5 Connect the fuel pump wiring connector, start the engine and disconnect the bridge wire from the check connector. If the engine stalls shortly after disconnecting the bridge wire, check the fuel pump relay and the wiring between the fuel pump relay and the ECU. If no fault can be found, renew the ECU.

To Remove and Install

1 Reduce the fuel system pressure as previously described.

2 On sedan models, proceed as follows:

a Remove the rear seat cushion by locating the two retaining clips at the front of the seat cushion. Pull the clip toward the front of the vehicle and lift the seat cushion upward. Repeat this on the second retaining clip and remove the seat cushion from the vehicle.

b Remove the screws retaining the fuel pump cover and lift it from the floor of the vehicle.

3 On station wagon models, raise the rear of the vehicle and support on chassis stands. Refer to General Information (chapter 1) for the positions of the jacking points.

4 Disconnect the wiring connector from the fuel pump/gauge sender unit.

5 Squeeze the tangs together on each side of the fuel supply hose fitting and pull the hose from the fuel pump/gauge sender unit.

6 Mark the relationship between the sender unit and fuel tank to ensure that it is returned to the same position.

7 Remove the retaining nuts, on sedans or screws, on station wagons and then remove the fuel pump/gauge sender unit from the fuel tank.

8 If necessary, separate the fuel pump from the sender unit assembly as follows:

a On station wagon models, remove the screw retaining the fuel strainer to the sender unit bracket.

b Slide the bottom of the fuel pump and the strainer assembly from the sender unit bracket and remove the rubber support from the base of the pump.

c Slide the supply hose clamp away from the fuel pump and separate the hose from the pump.

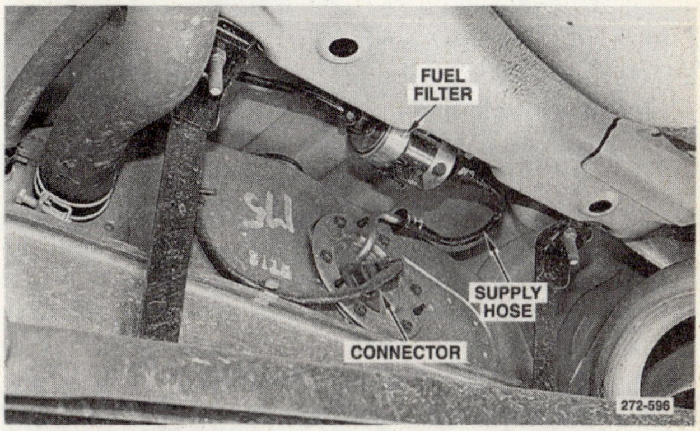

Station wagon fuel pump connector location. The terminal identification is the same for sedan models

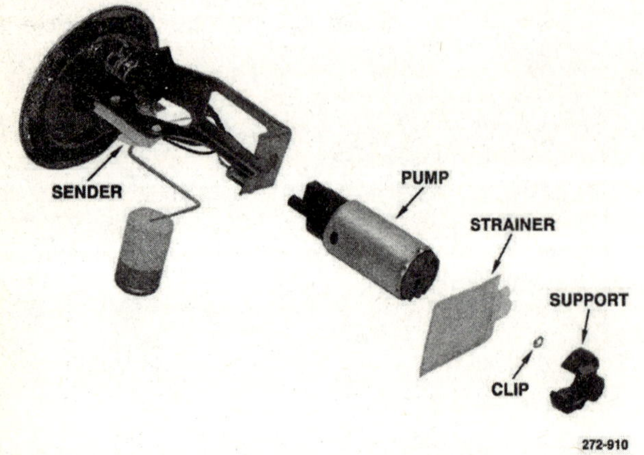

Fuel pump components. Sedan shown, wagon similar.

d Disconnect the wiring connector from the top of the fuel pump.

e Remove the clip retaining the strainer to the bottom of the pump and remove the strainer from the pump.

9 Check all components for deterioration and renew as necessary. It is advisable to renew the strainer as the material ages over time.

Installation is a reversal of the removal procedure with attention to the following points:

1 If dismantled, assemble the pump as follows:

a Position the strainer onto the pump and install the retaining clip securely.

b Connect the pump to the supply hose and install the clamp and the wiring connector.

c Install the rubber support to the base of the pump and slide the base of the pump into the sender unit bracket.

d On station wagon models, install the strainer bracket securing screw.

2 Install the sender unit to the position noted during removal using a new gasket. On sedan models, ensure that the wiring clip is installed to the sender unit stud closest to the rear of the vehicle. Tighten the retaining nuts, on sedans, or screws on station wagons evenly and securely.

3 Connect the wiring and the fuel supply hose to the fuel pump/gauge sender unit.

4 On sedan models, install the cover and tighten the retaining screws securely and install the rear seat cushion, ensuring that the retaining clips lock into position.

FUEL PRESSURE REGULATOR

To Remove and Install

1 Reduce the fuel system pressure as previously described.

2 On Verada models, remove the cover from the top of the engine.

3 Disconnect the vacuum hose and the fuel return hose from the fuel pressure regulator.

4 Remove the pressure regulator retaining bolts and withdraw the regulator from the fuel rail. Remove and discard the O ring.

Installation is a reversal of the removal procedure with attention to the following points:

1 Install a new O ring to the pressure regulator and lubricate the O ring with clean fuel.

2 Ensure that all connections are tightened securely.

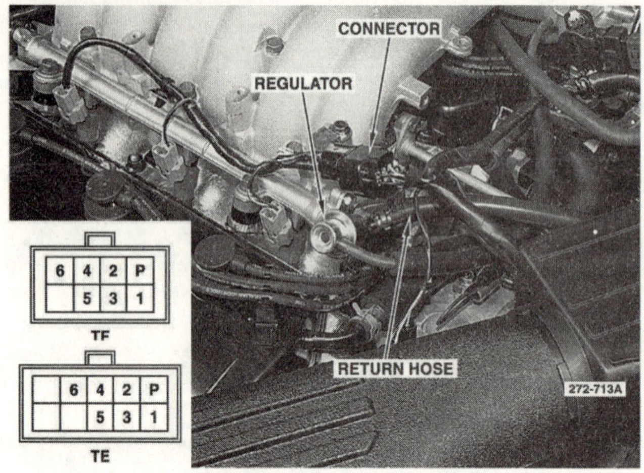

Fuel pressure regulator and injector block wiring connector. Inset shows the terminal identification for the connector.

3 Ensure that the vacuum hose is serviceable and is connected securely.

4 Test the fuel system for external leakage as previously described.

FUEL RAILS AND INJECTORS

To Test Injector Circuit

1 On Verada models, remove the engine cover from the top of the engine.

2 Disconnect the block wiring connector for the injectors from the passenger side of the engine.

3 Using the inset, measure the resistance between the power supply terminal (P) and each injector terminal at the block connector.

If any of the injector resistances are not to Specification, it will be necessary to disconnect the wiring from that injector and measure the resistance between the injector terminals.

Renew any injectors that are not to Specifications.

> i *It will be necessary to remove the plenum chamber from the inlet manifold as described in Engine (chapter 4) to access the rear bank of injectors.*

4 Connect a test lamp between the power supply terminal (P) of the block wiring connector and earth. Check that the test lamp illuminates with the ignition On.

If the test lamp does not illuminate, check the engine control relay and the 20 amp No 15 fuse located in the engine compartment fuse box.

5 With the test lamp connected between the power supply terminal (P) and one of the injector trigger wire terminals, check that the test lamp flashes when the starter motor is operated. Repeat this on each injector trigger wire terminal.

The power supply terminal (P) has battery power applied to it when the ignition is switched On. Each numbered terminal goes to an injector, (No1 terminal goes to No 1 injector and so on). These are called trigger wires and each trigger wire is pulsed to earth through the ECU. When these wires are pulsed to earth, a circuit through the injector is completed, and the injector opens, allowing fuel to pulse into the engine.

If the test lamp does not flash at one or more of the terminals, check the wiring between the connector and the ECU. Also check that the ECU terminals are clean and securely making contact. If not fault can be found, a fault is indicated in the ECU. Using the wiring diagram as a guide, check the ECU power supply and earth circuits. Also refer to Part 2 of this chapter and check for any stored fault codes or refer the vehicle to an authorised dealer or fuel injection specialist.

TIP *If no cranking signal is being received by the ECU, the test lamp will not flash. Refer to part 2 of this chapter or refer the vehicle to an authorised dealer or fuel injection specialist.*

If the test lamp flashes when connected to each injector trigger wire terminal, the circuit can be considered serviceable.

6 To check that each injector is functioning, proceed as follows:

a Connect the injector block wiring connector and start the engine.

b While the engine is idling, disconnect the wiring from each injector in turn for no more than five seconds. The engine should misfire each time the injector is disconnected.

This can only be done on No 1, 2, 4 and 6 cylinders, as access to No 3 and 5 cylinders is restricted by the plenum chamber. Do not attempt to start the engine with the plenum chamber removed.

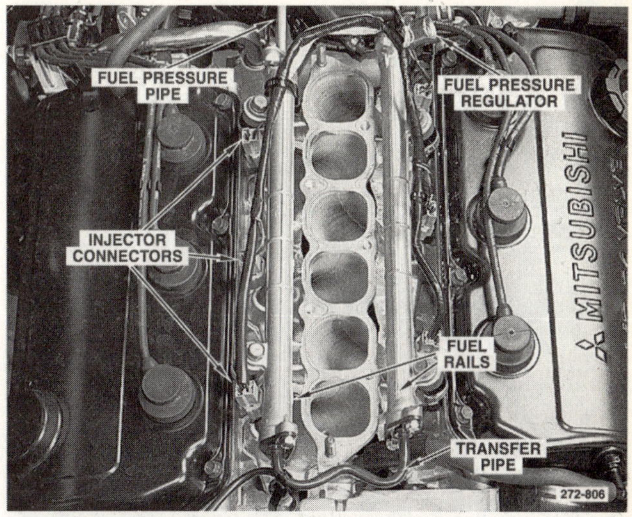

Plenum chamber removed showing the fuel rails.

c If the engine does not misfire, the injector may be blocked or sticking. However, before removing the injectors for testing and cleaning, check the ignition high tension circuit as previously described in Roadside Trouble Shooting (chapter 3).

Special equipment is required to clean and test the injectors. Refer to an authorised dealer or a fuel injection specialist.

To Remove and Install

1 Reduce the fuel system pressure as previously described.

2 Remove the plenum chamber as described in Engine (chapter 4).

3 Remove the bolts retaining the fuel pressure pipe to the fuel rail. Remove the pipe and discard the O ring. Remove the bolts retaining the fuel pressure regulator to the fuel rail and

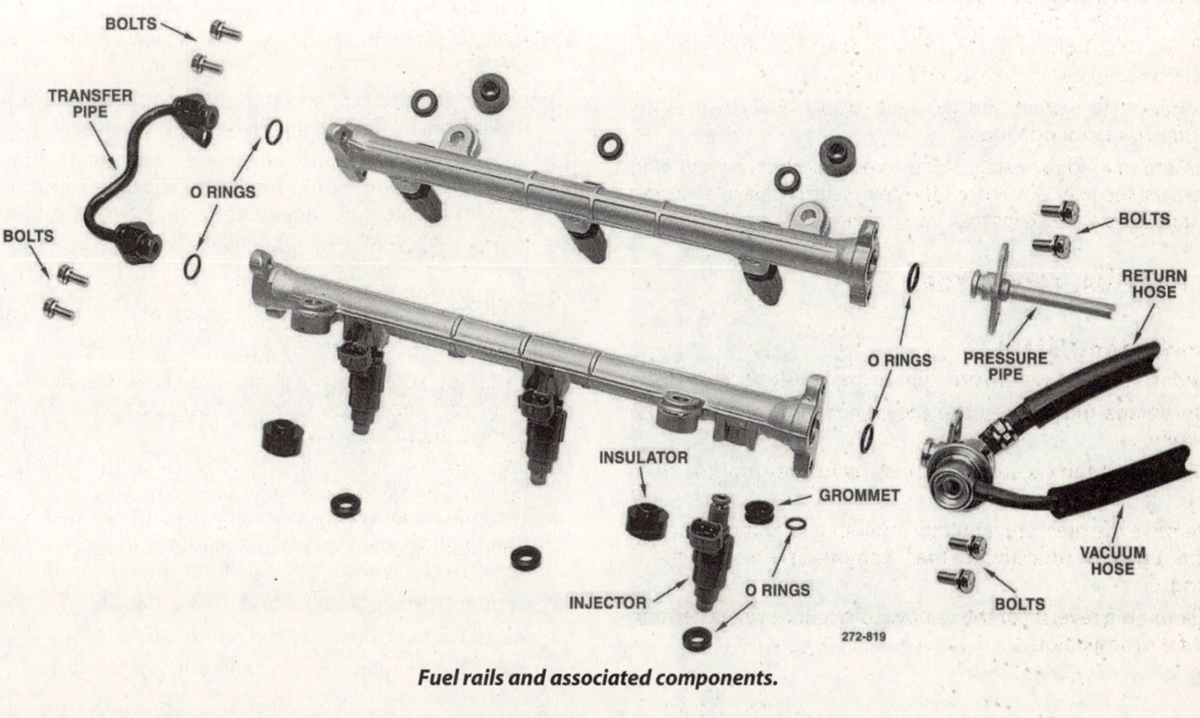

Fuel rails and associated components.

remove the regulator. Discard the O ring from the fuel pressure regulator.

4 Remove the bolts retaining the transfer pipe to the fuel rails. Manoeuvre the transfer pipe from the fuel rails and discard the O rings.

TIP *Loosen the fuel filler cap several turns to prevent pressure build up in the fuel tank from causing a fuel leak.*

5 Suitably mark and disconnect the wiring connectors from the fuel injectors.

6 Remove the retaining bolts and remove the fuel rail and injectors from the engine ensuring that the injectors do not fall from the fuel rail. Remove the spacers.

7 Pull the injectors out of the fuel rail and remove the O rings and grommets from the injectors. Discard the O rings and the grommets.

8 Remove the insulators from the cylinder head.

Installation is a reversal of the removal procedure with attention to the following points:

1 Renew the grommets and the O rings sealing the injectors in the fuel rail and the cylinder head.

2 Lubricate the O rings sparingly with fuel and install the injectors squarely into the fuel rail. Ensure that the injectors can be rotated smoothly in the fuel rail.

3 Ensure that the spacers are correctly installed to the fuel rail retaining bolts and that the injectors are correctly positioned.

4 Tighten the fuel rail retaining bolts to the specified torque and verify that the injectors can be rotated smoothly.

5 Ensure that the wiring is securely connected to the injectors.

6 Install new O rings to the fuel rail transfer pipe, fuel pressure regulator and fuel pressure pipe prior to installation. Tighten the retaining bolts to the specified torque.

7 Renew the plenum chamber gasket and tighten the plenum chamber bolts to the specified torque.

8 Test the fuel system for external leakage as previously described.

Rear seat cushion and cover removed showing the fuel pump wiring connector. Sedan models.

FUEL GAUGE SENDER UNIT

The fuel gauge sender unit is combined with the fuel pump. Refer to the Fuel Pump heading for service operations.

FUEL TANK

To Remove and Install — Sedan

1 Reduce the fuel system pressure as previously described.

2 Raise the rear of the vehicle and support on chassis stands. Refer to General Information (chapter 1) for the positions of the jacking points.

3 Disconnect the fuel pump/sender unit wiring connector located at the rear of the fuel tank.

4 Squeeze the tangs on each side of the fuel supply hose fitting and remove it from the fuel filter inlet side.

5 Working beneath the vehicle, disconnect the fuel return hose from the pipe on the front left corner of the fuel tank and the fuel vapour hose at the centre rear of the tank.

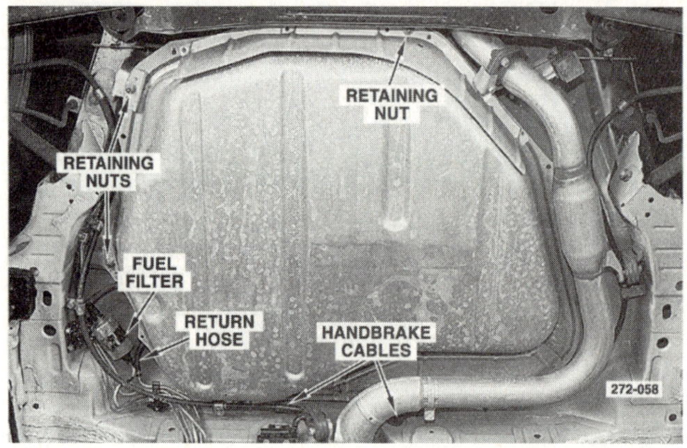

View of the fuel tank showing some of the retaining nuts. Sedan models.

6 Drain the fuel from the fuel tank using a hand pump or a suitable siphon. Do not siphon fuel using the mouth.

7 Remove the bolts retaining the handbrake cables to the fuel tank.

8 Loosen the hose clamps and remove the filler and breather hoses from the rear of the fuel tank above the crossmember.

9 Remove the nuts and bolts and separate the intermediate exhaust pipe from between the tailpipe and catalytic converter. Refer to Engine (chapter 4).

10 Remove the bolts retaining the exhaust bracket to the left of the tunnel in the floor and just forward of the fuel tank.

11 With the aid of an assistant, support the fuel tank and remove the nuts and bolts retaining the fuel tank to the floor.

12 Lower the fuel tank and manoeuvre it toward the rear of the vehicle to clear the handbrake cables and from the vehicle.

Installation is a reversal of the removal procedure with attention to the following points:

1 With the aid of an assistant, partially raise the tank and position the handbrake cables beneath the corners of the tank. Install the tank onto the retaining studs and install the nuts.

2 Connect the vapour hose and the fuel supply hose.

3 Install the filler, breather and return hoses.

4 Install the handbrake cable bolts and tighten securely.

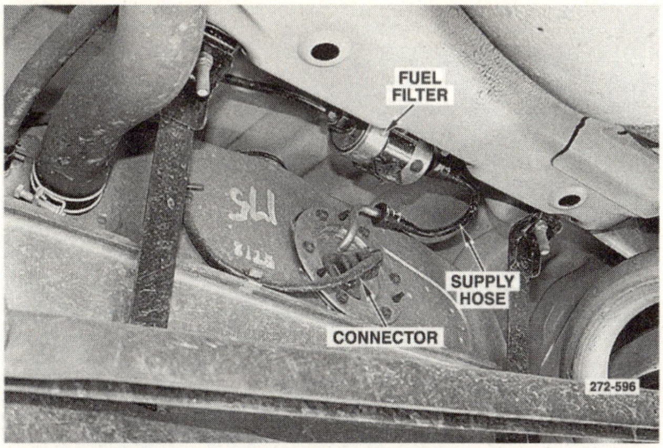

Station wagon fuel pump connector location.

5 Install the exhaust intermediate pipe, ensuring that the heat shield is correctly positioned.

To Remove and Install — Station Wagon

1 Reduce the fuel system pressure as previously described.
2 Raise the rear of the vehicle and support on chassis stands. Refer to General Information (chapter 1) for the positions of the jacking points.
3 Disconnect the fuel pump/sender unit wiring connector.
4 Squeeze the tangs on each side of the fuel supply hose fitting and remove it from the fuel pump/sender unit.
5 Disconnect the fuel return hose from the pipe on the front left corner of the fuel tank.
6 Drain the fuel from the fuel tank using a hand pump or a suitable siphon. Do not siphon fuel using the mouth.
7 Remove the bolts retaining the handbrake cables to the fuel tank.
8 Loosen the hose clamps and remove the filler and breather hoses from the rear of the fuel tank.
9 Support the fuel tank and remove the fuel tank strap rear retaining nuts.
10 Lower the fuel tank, unclip the vapour hoses, manoeuvre the tank toward the rear of the vehicle to clear the handbrake cables and straps and remove it from the vehicle.

Installation is a reversal of the removal procedure with attention to the following points:

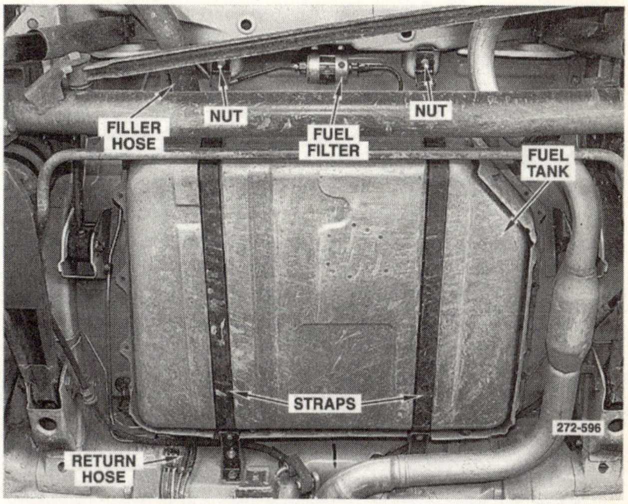

Fuel tank installed showing the retaining straps. Station wagon models.

1 Partially raise the tank and position the handbrake cables beneath the corners of the tank. Install the vapour hose to the tank.
2 Raise the tank into position and install the straps on the studs and install the retaining nuts. Ensure that the wiring and supply hose can be accessed and that the tank is sitting centrally on the straps and tighten the nuts securely.
3 Install the filler, breather and return hoses.
4 Install the handbrake cable bolts and tighten securely.
5 Place some fuel into the fuel tank and check for external fuel leaks as previously described, prior to installing the cover over the fuel pump/sender unit and the rear seat cushion.

Air Intake System

AIR CLEANER

To Remove and Install

1 Release the retaining clips, separate the air cleaner front cover and rear housing and slide the air cleaner element from between the cover and housing.

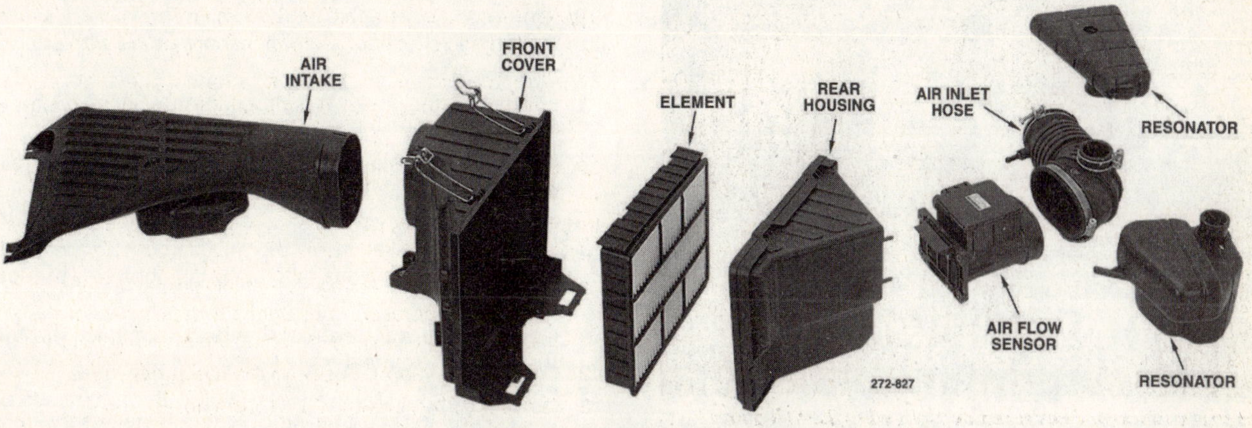

Air cleaner components.

2 Loosen the air inlet hose clamps, disconnect the PCV valve hose and remove the air inlet hose from the vehicle.
3 Disconnect the wiring from the airflow sensor, pivot the rear of the airflow sensor down, disengaging the rear air cleaner housing from the front cover. Remove the rear cover and air flow sensor as an assembly from the vehicle.
4 Remove the plastic screws from the clips retaining the cover over the radiator support panel. When the screws are removed, raise the cover up, gathering the clips from the support panel. Remove the panel from the vehicle.
5 Pull the air intake toward the front of the vehicle, disconnecting it from the front cover and remove the intake from the vehicle.
6 Remove the nut and bolts retaining the front cover to the vehicle and remove the front cover from the vehicle.

One of the bolts is located on the inside the front cover.

7 Clean all the components except the air cleaner element with a damp cloth.
8 Check the air cleaner front cover, rear housing and the air inlet hose for splits and damage. Ensure that the housing and cover is not distorted and that it is sealing completely around the air cleaner element.
9 Ensure that the rear housing and air flow sensor are correctly sealed. Renew the gasket if necessary.

Installation is a reversal of the removal procedure with attention to the following points:

1 Tighten the air cleaner mounting bolts securely.
2 Ensure that the element is fully seated between the housing and cover.
3 Connect the wiring connector to the air flow sensor.
4 Tighten the inlet hose clamps securely.

THROTTLE BODY

To Remove and Install

1 Disconnect the PCV valve hose from the air inlet hose.
2 Loosen the hose clamps and remove the air intake hose and resonator as an assembly from between the throttle body and air flow sensor.

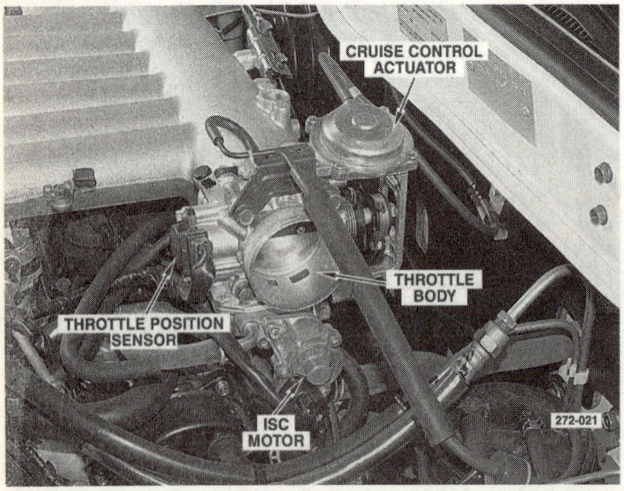

Throttle body and associated components. Cruise control model illustrated.

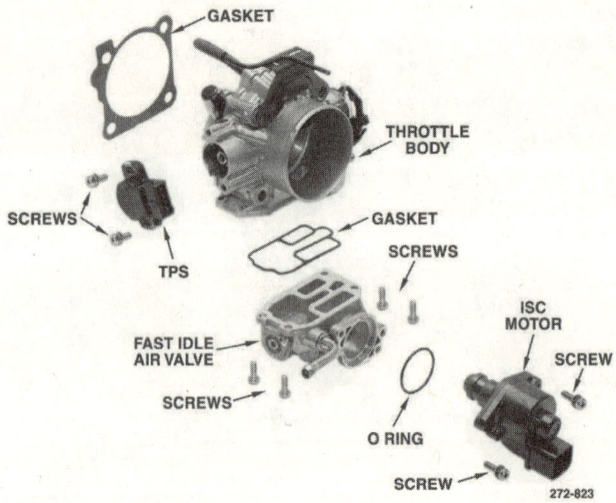

Throttle body components.

3 Disconnect the throttle position sensor and idle speed control (ISC) motor wiring connectors.
4 On models with cruise control, proceed as follows:
a Disconnect the vacuum hose from the vacuum actuator.
b Remove the screws retaining the actuator bracket to the throttle body.
c Hold the throttle valve open and twist the actuator linkage until it can be removed from the throttle valve.
5 Disconnect the vacuum and coolant hoses from the throttle body. Seal the coolant hoses to prevent loss of coolant.
6 Open the throttle lever by hand and disconnect the inner throttle cable.
7 Remove the retaining bolts and remove the throttle body from the plenum chamber. Discard the gasket.
8 Remove the bolts retaining the fast idle air valve to the bottom of the throttle body. Separate the two components and discard the gasket.
9 Thoroughly clean the throttle body using suitable solvent (carburetor cleaner or similar) with particular attention to the throttle valve and idle passages. Do not use this type of cleaner to clean the fast idle air valve as the chemicals within the cleaners can destroy the wax pellet inside the fast idle valve.
10 Operate the throttle lever and check for smooth operation. Ensure that the throttle closes fully when slowly released. Lubricate the throttle shaft with a small amount of machine oil.

Installation is a reversal of the removal procedure with attention to the following points:

1 Ensure that the plenum chamber, throttle body and fast idle air valve mating faces are free of old gasket material.
2 Install a new gasket and then install the fast idle air valve onto the throttle body and tighten the retaining bolts securely.
3 Using a new gasket, install the throttle body to the plenum chamber. Install the retaining bolts and tighten to the specified torque.
4 Open the throttle lever by hand and connect the inner throttle cable to the throttle lever.
5 Check that the coolant and vacuum hoses are not soft, cracked or deteriorated and renew if necessary before connecting them to the throttle body.

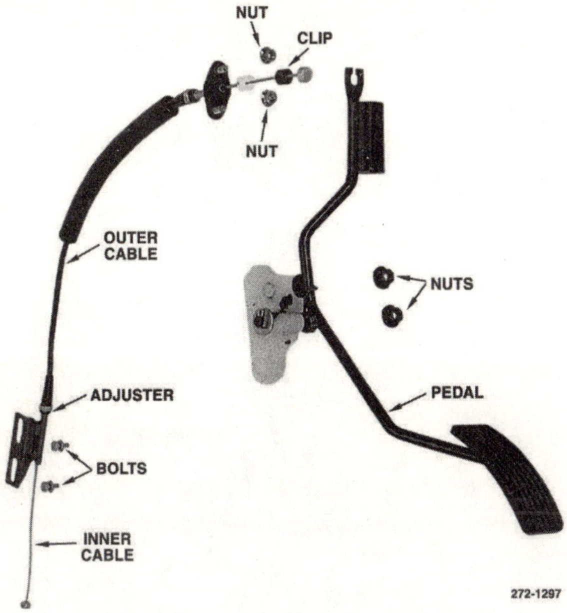

Throttle cable and pedal removed from the vehicle.

6 Connect the throttle position sensor and idle speed control valve wiring connectors.

7 Check the throttle cable adjustment as described later in this chapter.

8 Check and if necessary top up the coolant level.

THROTTLE CABLE

To Remove and Install

1 On models with cruise control, proceed as follows:

a Disconnect the vacuum hose from the vacuum actuator.

b Remove the screws retaining the actuator bracket to the throttle body.

c Hold the throttle valve open and twist the actuator linkage until it can be removed from the throttle valve.

2 Open the throttle lever by hand and disconnect the inner throttle cable.

3 Remove the bolts retaining the outer cable to the plenum chamber. Remove the earth lead and position the remove the outer throttle cable from the engine.

4 Working inside the vehicle, partially depress the throttle pedal and disconnect the inner throttle cable from the pedal.

5 Remove the bolts retaining the outer throttle cable to the firewall and then pull the throttle cable through the firewall and into the engine compartment.

Installation is a reversal of the removal procedure with attention to the following points:

1 Pass the throttle cable from the engine compartment side through the firewall.

2 Connect the inner cable to the throttle pedal, ensuring that the retaining clip is securely attached to the pedal.

3 Install the bolts retaining the outer cable to the firewall and tighten securely.

4 Place the outer cable onto the plenum chamber. Install the retaining bolts ensuring that the earth wire is placed onto the bolt closest to the driver side of the vehicle.

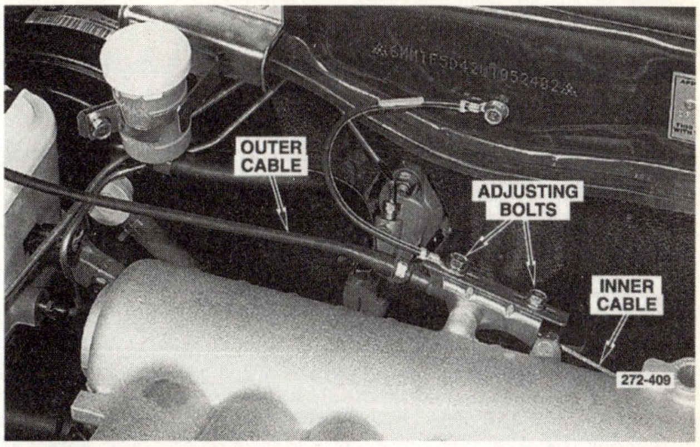

Throttle cable adjusting point.

5 Open the throttle lever by hand and connect the inner throttle cable to the throttle lever.

6 On models with cruise control, install the vacuum actuator linkage to the throttle valve and then position the actuator body into position onto the throttle body. Tighten the retaining screws securely.

7 Check the throttle cable adjustment, as described later in this chapter.

To Adjust

1 Have an assistant slowly depress the throttle pedal while observing the throttle lever at the throttle body.

2 When the throttle pedal is fully depressed, the throttle lever should be fully open. If necessary, loosen the outer cable adjusting bolts on the plenum chamber and slide the outer cable along the bracket until the correct adjustment is achieved.

If the throttle cable does not open the throttle fully, the engine will have a noticeable decrease in engine power. If the throttle pedal opens the throttle lever before the pedal is against the floor, wear and possible breakage of the throttle cable is possible. Ensure the throttle cable is adjusted so as to open the throttle fully only when the pedal touches the floor of the vehicle.

3 Tighten the adjusting bolts against the bracket and check the throttle cable adjustment. If the adjustment has altered, repeat the procedure.

Location of the distributor. Inset shows the terminal identification.

Electrical system

IGNITION SYSTEM

To Test

1 Check for spark at the high tension leads as described in Roadside Trouble Shooting (chapter 3) under the heading Ignition and Electrical System Checks.

If problems are indicated, refer to Engine (chapter 4) and thoroughly check the spark plugs, high tension leads and under the distributor cap for dirt, cracks etc.

2 Check that the wiring connections at the distributor are clean, dry and secure.

3 Disconnect the wiring connector from the distributor and check that battery voltage is available at terminal 1 (black/white wire) of the connector with the ignition On.

If battery voltage is not present, check the wiring between the harness connector and the ignition switch for an open circuit or a short to earth. Also check the wiring between the ignition switch, the capacitor and the fuel pump relay for a short to earth.

4 Measure the coil primary resistance between the distributor terminals 1 and 2 and the secondary resistance between the coil high tension terminal and the distributor connector terminal 1. Renew the distributor if the coil resistances are not to Specifications.

5 If no faults can be found and the ignition system is still faulty, consult an authorised dealer or a fuel injection specialist. For those with a good understanding of electrical components and circuits, refer to Part 2 of this chapter under the headings TDC Sensor (Code 23) and Ignition Circuit (Code 44, 52, 53).

DISTRIBUTOR AND IGNITION COIL

At the time of writing, the only serviceable components of the distributor are the cap, cap gasket, rotor and housing O ring. The O ring can be renewed after removing the distributor, as described below.

As the vehicle ages, specialist automotive computer companies may provide an exchange distributor on a change over basis and this avenue should be checked before renewal of the distributor.

The ignition timing is non adjustable, as there is no elongation in the distributor housing on the lower retaining stud hole.

The rotor position with the engine at TDC (top dead centre) No 1 cylinder compression.

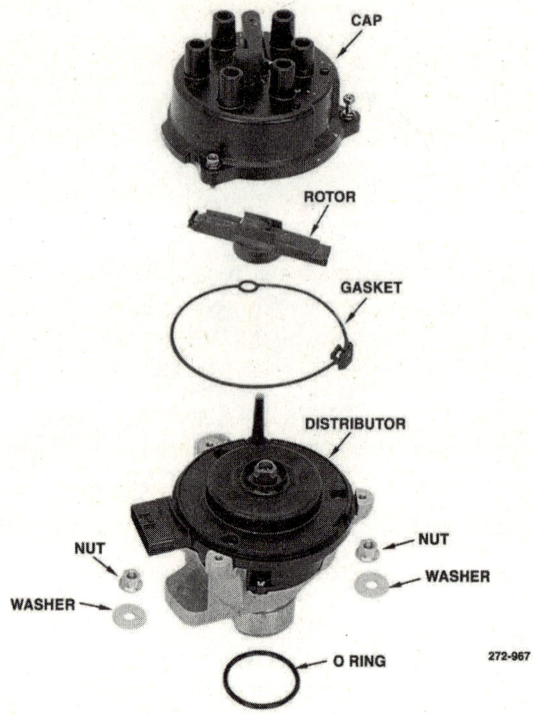

Distributor components.

To Remove and Install

1 Disconnect the negative battery terminal.

2 Remove the air inlet hose from between the throttle body and the air flow sensor. Refer to the Air Cleaner heading in this chapter if necessary.

3 Remove the distributor cap retaining screws and remove the cap and gasket from the distributor. To ease installation, mark the position of the rotor arm and avoid rotating the crankshaft while the distributor is removed.

TIP *It is not necessary to set the engine on No 1 cylinder TDC as the distributor is driven by an offset tang and will only fit in one position.*

4 Disconnect the wiring connector from the distributor.

5 Remove the retaining nuts and washers and slide the distributor from the cylinder head.

Installation is a reversal of the removal procedure with attention to the following points:

1 Install a new O ring to the distributor and apply a smear of engine oil to it.

2 Rotate the distributor until the mark made on removal aligns and slide the distributor into the cylinder head. If no marks were made, slowly rotate the distributor shaft until it engages with the camshaft.

3 Connect the wiring connector to distributor and install the washers and nuts retaining the distributor and tighten the nuts securely.

4 Install the distributor cap and secure with the retaining screws.

5 Start and run the engine to check it is running correctly. If timing problems are suspected, have an authorised dealer check the ignition timing and make repairs as necessary.

Location of the coolant temperature sensor.

COOLANT TEMPERATURE SENSOR

To Test

Testing the coolant temperature sensor is a straightforward operation and is covered in Part 2 of this chapter.

To Remove and Install

1 Loosen the radiator drain plug and partially drain the cooling system to a level below the coolant temperature sensor.

2 Disconnect the wiring from the coolant temperature sensor.

3 Remove the sensor from the thermostat housing.

Installation is a reversal of the removal procedure with attention to the following points:

1 Apply a smear of Permatex Ultra Grey or equivalent to the threads of the sensor.

2 Install the sensor and tighten to the specified torque.

3 Ensure that the wiring is securely connected and making good electrical contact.

4 Fill the cooling system as described in Cooling System (chapter 5).

IDLE SPEED CONTROL (ISC) MOTOR

To Remove and Install

1 Remove the throttle body as previously described under the Throttle Body heading.

Location of the ISC motor and throtttle stop screw. Inset shows the terminal identification.

2 Remove the screws retaining the ISC motor to the fast idle air valve and remove the valve and gasket. Discard the gasket.

Installation is a reversal of the removal procedure with attention to the following points:

1 Ensure that the mating faces between the throttle body and the ISC motor are clean.

2 Using a new gasket install the ISC motor and tighten the retaining screws securely.

3 Install the throttle body as previously described.

To Test

1 Switch the ignition On and confirm that the ISC motor can be heard operating, indicating normal operation.

2 If the ISC motor does not operate, disconnect the wiring from the ISC motor and check for battery voltage at the connector terminals 2 and 5 (red wires) with the ignition On.

If battery voltage is not available, check the wiring between the engine control relay and the ISC motor for continuity and shorts.

3 Measure the resistance between the ISC motor terminal 5 and terminals 4 and 6, and between terminal 2 and terminals 1 and 3. Renew the ISC motor if any resistance is not to Specifications.

4 Check the wiring between the ISC motor and the ECU for continuity and shorts to earth.

ENGINE IMMOBILISER SYSTEM

The engine immobiliser consists of a key reader surrounding the ignition key lock assembly, the ignition key incorporating a transmitter within the key and a module located behind the centre console. The system interacts with the engine ECU and will not allow the engine to start until verification that the correct ignition key was used to start the vehicle.

When renewing or adding an extra key or renewing the immobiliser module, a key training procedure will need to be carried out by an authorised dealer. In this case, all keys being used will have to be taken with the vehicle to an authorised dealer.

Due to the need of specialised equipment to diagnose faults within the immobiliser system, the following limited procedures will assist the operator in a few basic checks prior to referring the vehicle to an authorised dealer.

To Check

1 If the vehicle will not start, attempt to start the engine using another ignition key.

> *If more than one ignition key is on the key ring, the key reader can pick up signals from both keys causing a "no start" situation.*

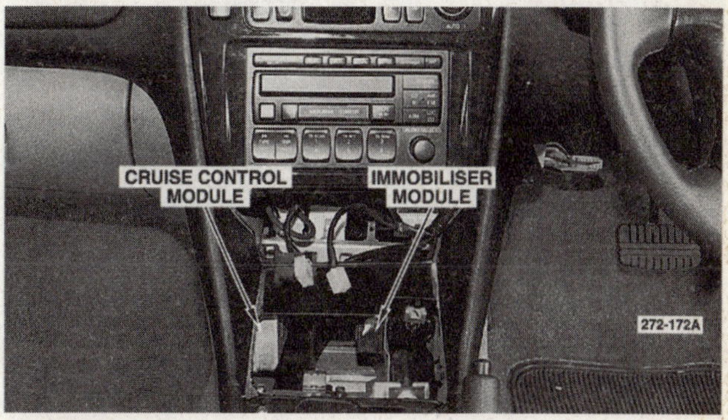

Immobiliser and cruise control module location.

2 If the vehicle now starts, the immobiliser module doesn't recognise the ignition key. Refer the vehicle to an authorised dealer to have the key coded to the immobiliser module.

If the vehicle still will not start, check the key reader as follows:

a Remove the screws retaining the bonnet lever to the lower dash panel and position the lever clear of the work area.

b Remove the screw cover from the top of the driver side lower dash panel. Remove the remaining screws from the lower dash panel, disconnect the panel from the top left side of the dashboard and partially remove the panel. Where fitted, rotate the footwell lamp bulb holder anti clockwise and remove the panel from the vehicle.

c Remove the screws from the bottom of the ignition lock surround panel and manoeuvre the panel from the dashboard. Disconnect the key reader wiring connector.

d Using an ohmmeter, check the resistance between the key reader terminals. If the resistance is not 8 ± 3 ohms, renew the key reader assembly.

3 Remove the console fascia panel as follows:

a On manual transaxle models, remove the gear lever knob.

b On Verada models, open the coin tray lid and firmly pull the coin tray from the console. Disconnect the wiring connectors from the base of the accessory socket and then pull the cup holder from the front of the console.

c Grasp the rear edges of the console fascia panel and pull upward. Working toward the front of the console, raise the fascia panel, releasing the retaining clips. When all of the clips are released, manoeuvre the fascia panel from the console.

d On Magna models, remove the ashtray from the vehicle. Pull the ashtray surround panel partially from the dashboard, disengaging the clips retaining the top of the surround to the dashboard. Disconnect the cigarette lighter wiring from the base of the socket and remove the surround.

e On Verada models, pull the ashtray assembly, disengaging the clips retaining the top of the ashtray assembly to the dashboard. Disconnect the cigarette lighter wiring from the base of the socket and remove the ashtray.

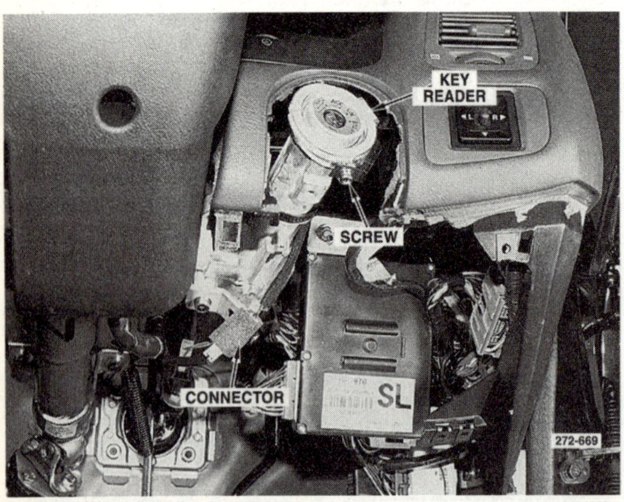

Ignition lock surround panel removed showing the key reader screw.

4 Remove the bolt retaining the immobiliser module. Disengage the module from the front of the console and disconnect the wiring connector.

5 Check for continuity to earth at the two black wires. If continuity is not available, repair the earth circuit.

6 Check for battery power with the ignition On at the black/white wire, and the two red wires. Use the wiring diagrams to trace the fault if necessary.

7 Check that the brown wire and the grey/red wire running between the immobiliser module and the engine ECU is not damaged or broken. Ensure that all connections are clean and secure. Use the wiring diagrams for a guide.

8 With the key reader wiring reconnected, check that the key reader resistance obtained previously is the same when checked between the black/red wire and the white/red wire. If the resistance has altered, check the wiring and connections between the key reader and immobiliser module.

If the above checks fail to isolate the fault, refer the vehicle to an authorised dealer.

FUEL PUMP AND ENGINE CONTROL RELAYS

To Remove and Install

1 Pull the carpet down from beneath the driver side of the console.

2 Remove the ashtray surround panel. Refer to the Sound System heading in Body Electrical (chapter 16) if necessary.

3 Remove the relay bracket retaining bolts and pass the relays and wiring through the side of the console.

4 Disconnect the wiring connectors and remove the relays from the vehicle.

Installation is a reversal of the removal procedure.

To Test

1 Check that continuity exists between relay terminals 2 and 4.

2 Check that continuity does not exist between terminals 1 and 3.

3 Apply 12 volts between terminals 2 and 4 and check that continuity exists between terminals 3 and 1.

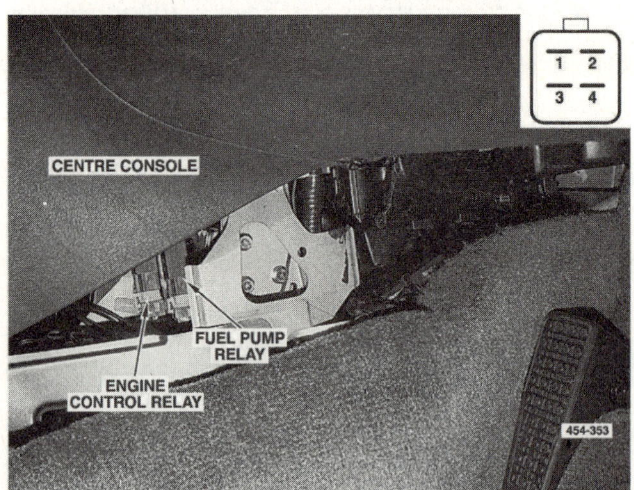

Location of the engine control relay and the fuel pump relay. Inset shows the terminal identification.

Part 2. System Testing

Contents

Important Note

This chapter consists of additional checks and specifications for diagnosing faults within the fuel injection and ignition systems. The following procedures should only be attempted by mechanics or persons with a good understanding of electrical components and circuits.

To diagnose fuel and engine management problems in the minimum amount of time it is important not to overlook the obvious and complete tasks in a logical order. If unsure of the vehicle, the best place to start is with a compression test which is then followed with a complete tune-up as described in Engine (chapter 4).

Next it is most important to check fuel pressure and the injector circuit as described in Part 1 of this chapter.

TIP *Trying to diagnose an engine which has poor compression or incorrect fuel pressure can cause misleading results.*

Before starting work, read the Service Precautions and Procedures heading in Part 1 of this chapter prior to carrying out repair procedures in this chapter.

Reduce the fuel system pressure as described in Part 1 of this chapter before disconnecting any fuel lines.

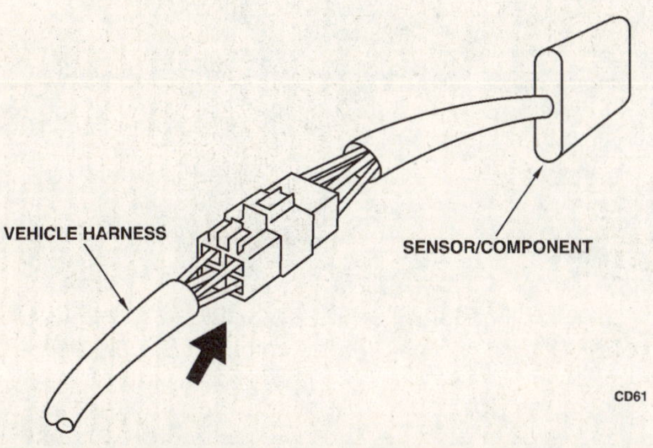

Large arrow indicates how insets used in this manual are viewed unless otherwise stated.

Specifications

Intake air temperature sensor resistance between terminals 5 and 6:
- At 20°C....... 2.3 – 3.0 k/ohms
- At 80°C....... 0.30 – 0.42 k/ohms

Coolant temperature sensor resistance:
- At 20°C....... 2.1 – 2.7 k/ohms
- At 80°C....... 0.26 – 0.36 k/ohms

Throttle position sensor resistance between terminals 1 and 4....... 3.5 – 6.5 k/ohms

Oxygen sensor heater element resistance at 20°C....... 7 – 40 ohms

ECU VOLTAGES

Manual Transaxle Models

Terminal 1, No 1 injector:
- Ignition On....... Battery V

Terminal 2, No 3 injector:
- Ignition On....... Battery V

Terminal 3, No 5 injector:
- Ignition On....... Battery V

Terminal 4, ISC motor:
- Just after starting engine at operating temperature....... Fluctuates 0.3 – 14 V

Terminal 5, ISC motor:
- Just after starting engine at operating temperature....... Fluctuates 0.3 – 14 V

Terminal 8, A/C relay control:
- Engine idling as A/C is switched On....... Above 6 volts decreasing to 0 – 3 V

Terminal 10, power transistor control:
- Engine running at 3 000 rpm....... 0.3 – 3.0 V

Terminal 12, ECU power supply:
- Ignition On....... Battery V

Terminal 14, No 2 injector:
- Ignition On....... Battery V

Terminal 15, No 4 injector:
- Ignition On....... Battery V

Terminal 16, No 6 injector:
- Ignition On....... Battery V

Terminal 17, ISC motor:
- Just after starting engine at operating temperature....... Fluctuates 0.3 – 14 V

Terminal 18, ISC motor:
- Just after starting engine at operating temperature....... Fluctuates 0.3 – 14 V

Terminal 19, air flow sensor reset signal:
- Engine at idle speed....... 0 – 1 V
- Engine at 3 000 rpm....... 6 – 9 V

Terminal 20, high speed radiator fan relay:
- Radiator fan not operating....... Battery V
- Radiator fan operating....... 0 – 3 V

Terminal 21, low speed radiator fan relay:
Radiator fan not operating .... Battery V
Radiator fan operating .... 0 – 3 V
Terminal 22, fuel pump relay control:
Ignition On .... Battery V
Engine idling .... 0 – 3 V
Terminal 25, ECU power supply:
Ignition On .... Battery V
Terminal 36, Check Engine lamp:
Ignition On, lamp On .... 0 – 3 V
Ignition On, lamp Off .... 9 – 13 V
Terminal 37, power steering pressure switch:
Engine idling, steering wheel stationary .... Battery V
Engine idling, steering wheel turning .... 0 – 3 V
Terminal 38, engine control relay:
Ignition Off .... Battery V
Ignition On .... 0 – 3 V
Terminal 45, A/C On signal:
Engine idling, A/C Off .... 0 – 3 V
Engine idling, A/C On .... Battery V
Terminal 71, Start signal:
Engine cranking .... Above 8 V
Terminal 72, intake air temperature sensor:
At 0°C .... 3.2 – 3.8 V
At 20°C .... 2.3 – 2.9 V
At 40°C .... 1.5 – 2.1 V
At 80°C .... 0.4 – 1.0 V
Terminal 76, oxygen sensor:
Engine at 2 500 rpm and at
operating temperature .... Fluctuates 0 – 0.8 V
Terminal 80, ECU constant power supply:
Ignition Off .... Battery V
Terminal 81, sensor power supply .... 4.5 – 5.5 V
Terminal 82, ignition switch:
Ignition On .... Battery V
Terminal 83, coolant temperature sensor:
At 0°C .... 3.2 – 3.8 V
At 20°C .... 2.3 – 2.9 V
At 40°C .... 1.3 – 1.9 V
At 80°C .... 0.3 – 0.9 V
Terminal 84, throttle position sensor:
Ignition On, throttle closed .... 0.3 – 1.0 V
Ignition On, throttle fully opened .... 4.5 – 5.5 V
Terminal 86, vehicle speed sensor:
Ignition On, vehicle moving
slowly .... Fluctuates 0 – 5 V
Terminal 87, TPS idle switch:
Ignition On, throttle closed .... 0 – 1 V
Ignition On, throttle slightly opened .... Above 4 V
Terminal 88, TDC sensor:
Engine cranking .... 0.4 – 3.0 V
Engine idling .... 0.5 – 2.0 V
Terminal 89, crank angle sensor:
Engine cranking .... 0.4 – 4.0 V
Engine idling .... 1.5 – 2.5 V
Terminal 90, air flow sensor:
Engine idling and at 2 500 rpm .... 2.2 – 3.2 V

Automatic Transaxle Models

Terminal 1, No 1 injector
Ignition On .... Battery V
Terminal 2, No 4 injector
Ignition On .... Battery V
Terminal 9, No 2 injector
Ignition On .... Battery V
Terminal 10, No 5 injector:
Ignition On .... Battery V
Terminal 11, power transistor control:
Engine running at 3 000 rpm .... 0.3 – 3.0 V
Terminal 14, ISC motor:
Just after starting engine at
operating temperature .... Fluctuates 0.3 – 14 V
Terminal 15, ISC motor:
Just after starting engine at
operating temperature .... Fluctuates 0.3 – 14 V
Terminal 17, high speed radiator fan relay:
Radiator fan not operating .... Battery V
Radiator fan operating .... 0 – 3 V
Terminal 18, low speed radiator fan relay:
Radiator fan not operating .... Battery V
Radiator fan operating .... 0 – 3 V
Terminal 19, air flow sensor reset signal:
Engine at idle speed .... 0 – 1 V
Engine at 3 000 rpm .... 6 – 9 V
Terminal 20, A/C relay control:
Engine idling as A/C is
switched On .... Above 6 volts decreasing to 0 – 3 V
Terminal 21, fuel pump relay control:
Ignition On .... Battery V
Engine idling .... 0 – 3 V
Terminal 22, Check Engine lamp:
Ignition On, lamp On .... 0 – 3 V
Ignition On, lamp Off .... 9 – 13 V
Terminal 24, No 3 injector
Ignition On .... Battery V
Terminal 25, No 6 injector
Ignition On .... Battery V
Terminal 28, ISC motor:
Just after starting engine at
operating temperature .... Fluctuates 0.3 – 14 V
Terminal 29, ISC motor:
Just after starting engine at
operating temperature .... Fluctuates 0.3 – 14 V
Terminal 41, ECU power supply:
Ignition On .... Battery V
Terminal 44, coolant temperature sensor:
At 0°C .... 3.2 – 3.8 V
At 20°C .... 2.3 – 2.9 V
At 40°C .... 1.3 – 1.9 V
At 80°C .... 0.3 – 0.9 V
Terminal 45, crank angle sensor:
Engine cranking .... 0.4 – 4.0 V
Engine idling .... 1.5 – 2.5 V
Terminal 46, sensor power supply .... 4.5 – 5.5 V
Terminal 47, ECU power supply:
Ignition On .... Battery.V
Terminal 49, engine control relay:
Ignition Off .... Battery V
Ignition On .... 0 – 3 V
Terminal 52, power steering pressure switch:
Engine idling, steering wheel stationary .... Battery V
Engine idling, steering wheel turning .... 0 – 3 V
Terminal 56, TDC sensor:
Engine cranking .... 0.4 – 3.0 V
Engine idling .... 0.5 – 2.0 V
Terminal 58, A/T inhibitor switch:
Ignition On, Selector in P or N .... 0 – 3 V

Ignition On, Selector not in P or N	8 – 14 V
Terminal 59, Start signal:	
Engine cranking	Above 8 V
Terminal 64, intake air temperature sensor:	
At 0°C	3.2 – 3.8 V
At 20°C	2.3 – 2.9 V
At 40°C	1.5 – 2.1 V
At 80°C	0.4 – 1.0 V
Terminal 65, air flow sensor:	
Engine idling and at 2 500 rpm	2.2 – 3.2 V
Terminal 66, ECU constant power supply:	
Ignition Off	Battery V
Terminal 71, oxygen sensor:	
Engine at 2 500 rpm and at operating temperature	Fluctuates 0 – 0.8 V
Terminal 78, throttle position sensor:	
Ignition On, throttle closed	0.3 – 1.0 V
Ignition On, throttle fully opened	4.5 – 5.5 V
Terminal 79, TPS idle switch:	
Ignition On, throttle closed	0 – 1 V
Ignition On, throttle slightly opened	Above 4 V
Terminal 80, vehicle speed sensor:	
Ignition On, vehicle moving slowly	Fluctuates 0 – 5 V
Terminal 83, A/C On signal:	
Engine idling, A/C Off	0 – 3 V
Engine idling, A/C On	Battery V
Terminal 98, ignition switch:	
Ignition On	Battery V

Self Diagnosis

The self diagnosis codes (fault codes) are recorded in the ECU memory as they occur and the Check Engine lamp on the instrument panel will illuminate if any faults are detected in primary circuits. If the fault is intermittent, the lamp will illuminate each time the fault reoccurs. However, the fault code will be stored in memory until the codes are cleared.

It is normal for the Check Engine lamp to illuminate for 5 seconds after the ignition is switched On.

FAULT CODES

11	Oxygen sensor
12	Air flow sensor
13	Intake air temperature sensor
14	Throttle position sensor
21	Coolant temperature sensor
22	Crank angle sensor
23	TDC (top dead centre) sensor
24	Vehicle speed sensor
41	Injectors
44	Ignition circuit
52	Ignition circuit
53	Ignition circuit
54	Immobiliser system
61	A/T communication failure (renew the ECU)

The codes are displayed by the Check Engine lamp after switching the ignition On and earthing terminal 1 of the diagnosis connector. Read the codes as follows:

1 The codes will be set as a series of pulses and will be displayed as a series of flashes of the Check Engine lamp.

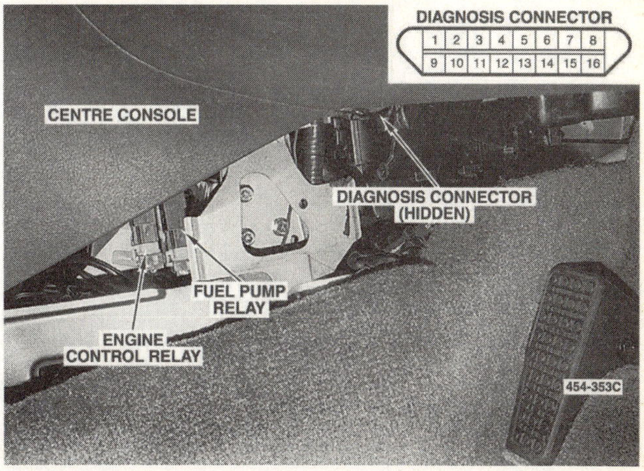

Location of the diagnosis connector. Inset shows terminal identification.

2 Each code is represented by two series of flashes separated by a pause.

3 The first series are long flashes and represent tens. The second series are short flashes and represent units. For example, four long flashes followed by a pause, followed by one short flash would indicate code 41.

4 If more than one code is in memory, each code will be displayed in sequence.

5 Record each code for confirmation.

6 If no fault codes are in memory, the Check Engine lamp will flash continually at 0.5 second intervals

CONFIRMING CODES

The code/s should be confirmed prior to further diagnosis.

1 Clear the codes.

2 Road test the vehicle and recheck the codes.

3 If the code is not set, an intermittent fault is indicated.

CLEARING CODES

1 Switch the ignition Off.

2 Disconnect the negative battery terminal for 20 seconds.

3 Confirm that no fault codes are present.

4 Bring the engine to normal operating temperature and run it at idle speed for 10 minutes.

DIAGNOSTIC CIRCUIT CHECK

1 Switch the ignition On and confirm that the Check Engine warning lamp illuminates for approximately 5 seconds.

2 If the lamp does not illuminate, check for a blown globe or fuse or an open circuit between the ECU, the Check Engine lamp and the fuse panel.

3 If the lamp remains illuminated, check the circuit between the ECU and the warning lamp for a short to earth.

Components and System Tests

TACHOMETER CONNECTION

1 Connect the tachometer positive lead to terminal 3 (white wire) of the engine check connector.

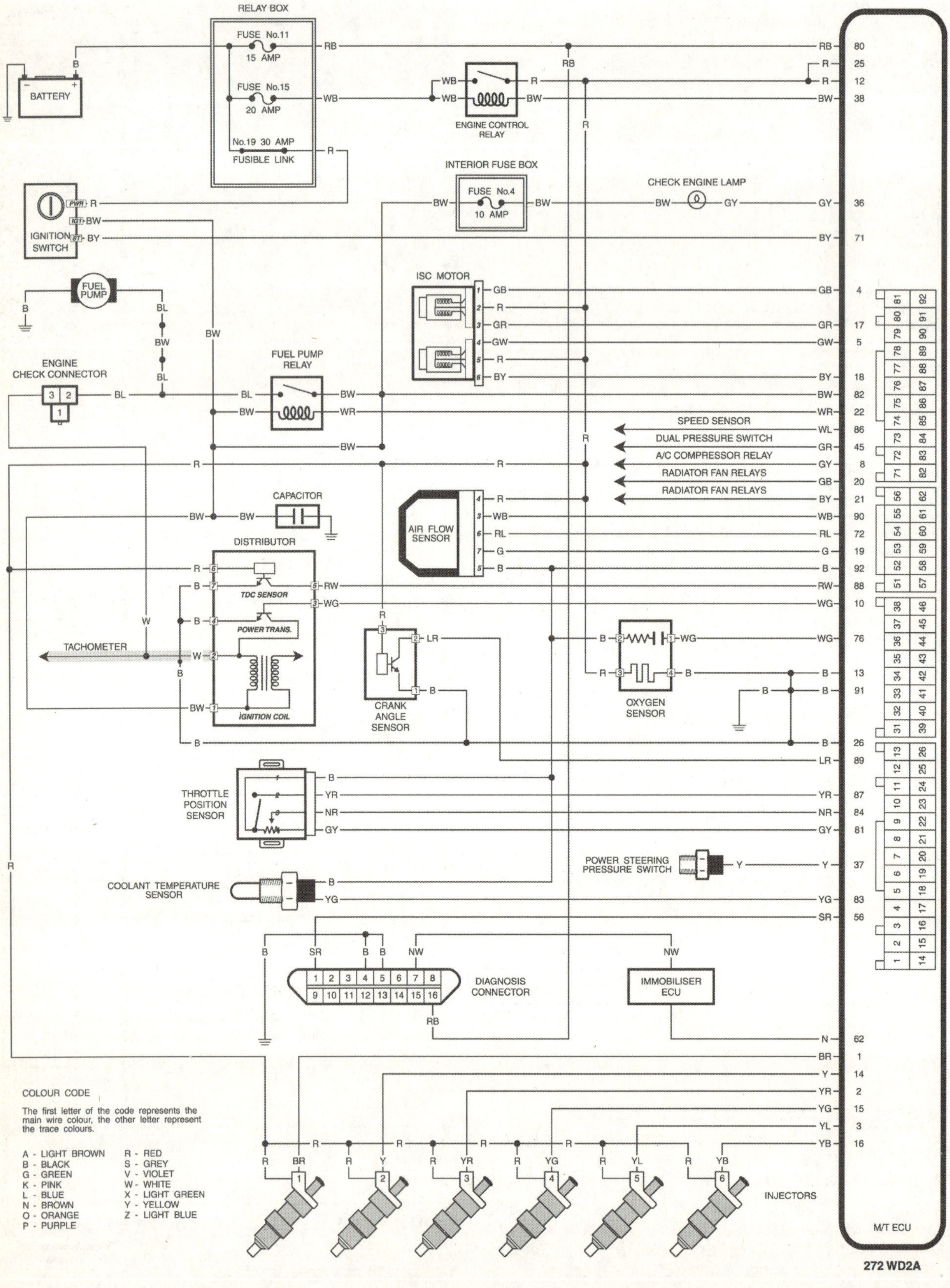

Fuel and engine management system circuit diagram – manual transaxle. TE/TF/TH & KE/KF/KH models. Refer to Body Electrical chapter for wiring diagrams for other models.

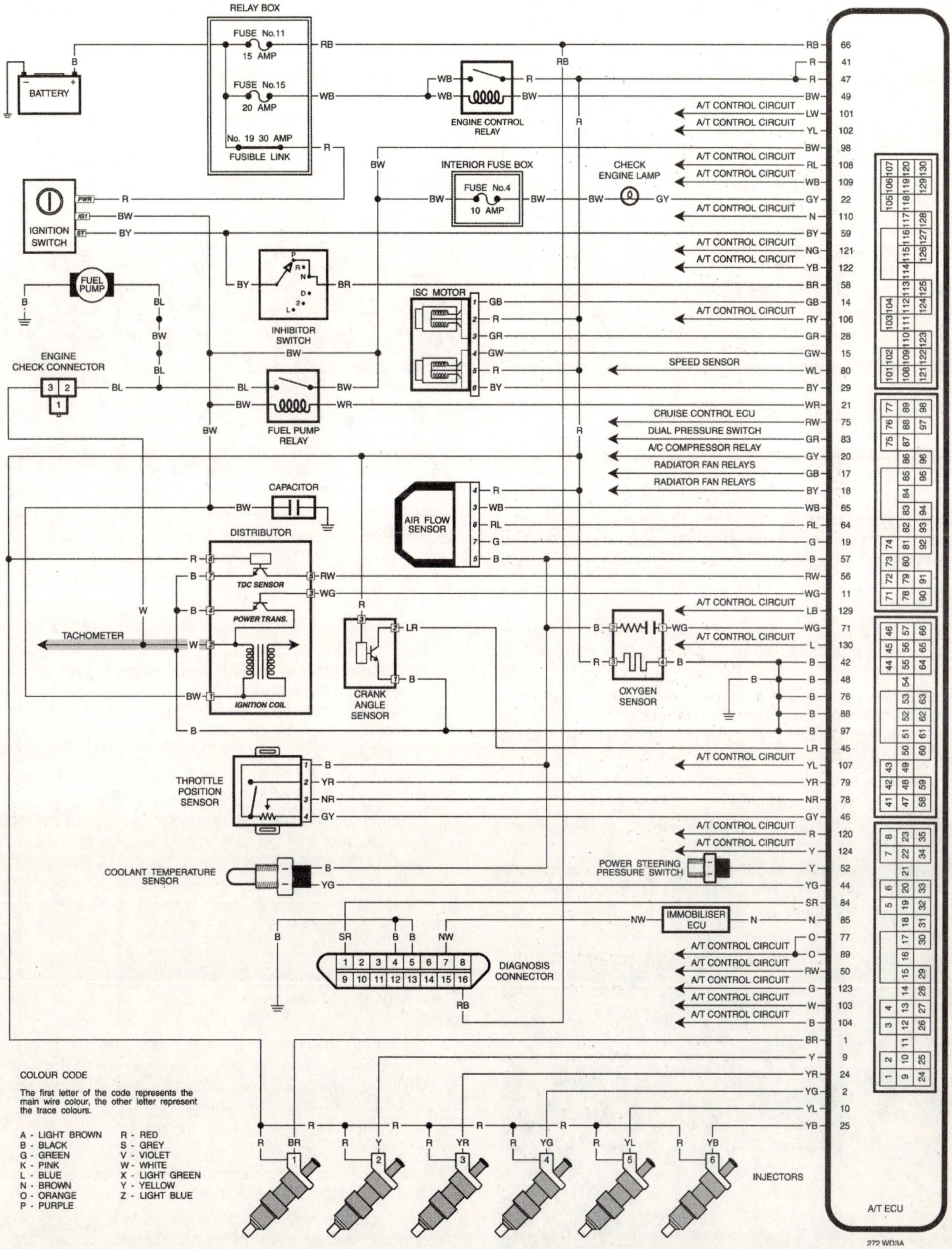

Fuel and engine management system circuit diagram – automatic transaxle. TE/TF/TH & KE/KF/KH models. Refer to Body Electrical chapter for wiring diagrams for other models.

Location of the engine check connector. Inset shows the terminal identification looking into the connector. Terminal 2 is the fuel pump test and terminal 3 is the tachometer terminal.

2 Connect the tachometer negative lead to a good earth point.

OXYGEN SENSOR (CODE 11)

Because the oxygen sensor detects the oxygen content in the exhaust gas, any conditions that prevent complete combustion of the air/fuel mixture will affect the operation of the oxygen sensor. A faulty oxygen sensor may be incorrectly diagnosed due to incorrect fuel pressure, a misfiring cylinder, leaking injector or a faulty coolant temperature sensor.

1. Disconnect the oxygen sensor connector and check the resistance between the heater element terminals 3 and 4. Renew the sensor if the resistance is not to Specifications.
2. Using jumper leads, connect terminals 3 and 4 to the battery terminals, positive to terminal 3. Connect a voltmeter to terminals 1 and 2
3. Start the engine and bring to operating temperature. Repeatedly race the engine. The voltmeter should indicate 0.6 – 1.0 volts. Renew the oxygen sensor if the voltage is not correct.
4. Check that battery voltage is available at the oxygen sensor harness connector terminal 3 with the ignition On.
5. Check for continuity to earth at the oxygen sensor harness connector terminals 2 and 4.
6. Check the wiring between the oxygen sensor connector terminal 1 and the ECU for continuity and shorts to earth.
7. If no faults can be found and the fault code continues, renew the ECU.

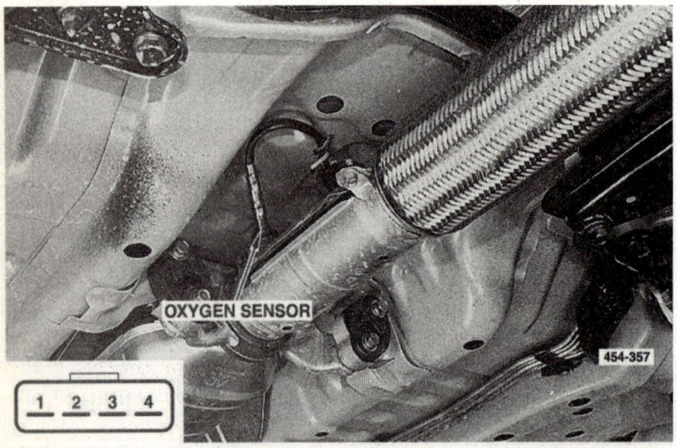

Location of the oxygen sensor. The wiring connector is located below the centre console.

AIR FLOW SENSOR (CODE 12)

1. Measure the voltage at the air flow sensor connector terminal 3 with the engine idling and at 3 000 rpm. The voltage should be 2.2 – 3.2 volts.
2. Measure the voltage at the air flow sensor connector terminal 7 with the engine idling and at 3 000 rpm. The voltage should be 0 – 1 volt and 6 – 9 volts respectively

Location of the air flow sensor. Inset shows the terminal identification.

3. If the voltages in steps 1 and 2 are correct and the fault code continues, renew the ECU.
4. Switch the ignition Off and disconnect the wiring from the air flow sensor.
5. Switch the ignition On and measure the voltage at the air flow sensor connector terminals 4 and 3. Battery voltage should be available at terminal 4, and 4.8 – 5.2 volts at terminal 3.

   If the voltage at terminal 4 is not correct, check the wiring between the air flow sensor connector and the engine control relay for continuity and shorts.

   If the voltage at terminal 3 is not correct, check the wiring between the air flow sensor connector and the ECU for continuity and shorts. If no fault can be found, renew the ECU.
6. Check for continuity to earth at the at the air flow sensor connector terminal 5.

INTAKE AIR TEMPERATURE SENSOR (CODE 13)

The intake air temperature sensor is integral with the air flow sensor.

1. Disconnect the wiring from the air flow sensor and check for 4.5 – 5.5 volts at the connector terminal 6 with the ignition On. If the voltage is not correct, check the ECU power supply and the wiring between terminal 6 and the ECU.

   If no fault can be found, renew the ECU.
2. Check for continuity to earth at the sensor connector terminal 5. If there is no continuity, check the wiring and the ECU earth circuits. If no fault can be found, renew the ECU.
3. Connect an ohmmeter between the air flow sensor terminals 5 and 6 and measure the resistance while heating the air inside the air cleaner using a hair dryer or heat gun. Renew the air flow sensor assembly if the resistances are not to Specifications.

THROTTLE POSITION SENSOR (CODE 14)

1 Confirm that the throttle lever is contacting the throttle stop screw when the throttle is closed. Check the throttle cable or throttle stop screw adjustment. Adjustment of the screw is covered under the Throttle Body heading.

Location of the throttle position sensor. Inset shows terminal identification.

2 Check that continuity is displayed between the TPS terminals 1 and 2 when the throttle is closed and an open circuit indicated when the throttle is in any position other than closed.

If a fault is indicated, adjust the TPS as follows:

a On models without cruise control, check that the throttle lever is contacting the throttle stop screw and install a 0.45 mm feeler gauge between the lever and the screw.

b On models with cruise control, insert a 1.4 mm feeler gauge between the throttle lever and the cruise control lever to a depth of no more than 3 mm.

c Loosen the TPS retaining bolts and rotate the sensor in an anticlockwise direction until there is continuity between terminals 1 and 2.

d Slowly rotate the sensor in a clockwise direction until an open circuit is indicated and tighten the retaining bolts securely.

e Connect the wiring to the TPS and check the voltage at the TPS connector terminal 3 with the ignition On. The voltage should be 0.3 – 1.0 volt with the throttle closed, and progress smoothly to 4.5 – 5.5 volts as the throttle is opened fully.

3 Measure the resistance between the TPS terminals 1 and 4. Renew the TPS if the resistance is not to Specifications.

4 Check that the resistance between the TPS terminals 3 and 4 increases steadily as the throttle is slowly opened.

5 With the ignition On, check for approximately 5 volts at terminals 2 and 4 at the TPS wiring connector.

If the voltage is not correct, check the wiring between the TPS and the ECU for continuity and shorts. Test the ECU power supply and earth circuits. If no fault can be found, renew the ECU.

COOLANT TEMPERATURE SENSOR (CODE 21)

1 Remove the coolant temperature sensor from the inlet manifold and check that the resistances are to Specifications.

2 Check that 4.5 - 5 volts is available between the sensor connector terminals with the ignition On.

If the voltage is not as specified, check the ECU power supply and earth circuits and the wiring between the sensor connector and the ECU for continuity and shorts. If the circuits are satisfactory, renew the ECU.

Location of the coolant temperature sensor.

CRANK ANGLE SENSOR (CODE 22)

1 Disconnect the wiring from the crank angle sensor and switch the ignition On.

2 Check that battery voltage is available at the crank angle sensor harness connector terminal 3. If battery voltage is not present, check the engine control relay.

3 Check that 4.8 – 5.2 volts is available at the crank angle sensor harness connector terminal 2.

Location of the crank angle sensor wiring connector. Inset shows the terminal identification.

If the voltage is not as specified, check the ECU power supply circuits and the wiring between the sensor connector terminal 2 and the ECU for continuity and shorts. If the circuits are satisfactory, renew the ECU.

4 Check for continuity to earth at the crank angle sensor harness connector terminal 1.

5 If the previous checks do not indicate a fault, renew the crank angle sensor.

TDC SENSOR (CODE 23)

1 Disconnect the wiring connector from the distributor and switch the ignition On.

2 Check that battery voltage is available at the distributor harness connector terminal 6. If battery voltage is not present, check the engine control relay.

3 Check that 4.8 – 5.2 volts is available at the crank angle sensor harness connector terminal 5.

Location of the distributor. Inset shows the terminal identification.

If the voltage is not as specified, check the ECU power supply circuits and the wiring between the sensor connector terminal 5 and the ECU for continuity and shorts. If the circuits are satisfactory, renew the ECU.

4 Check for continuity to earth at the crank angle sensor harness connector terminal 1.

5 If the previous checks do not indicate a fault, renew the distributor.

VEHICLE SPEED SENSOR (CODE 24)

1 If the speedometer operates, check the voltage at the ECU connector terminal 80 (A/T), or terminal 86 (M/T). If the voltage is to Specifications and the fault code continues, renew the ECU.

If the voltage is not to Specifications, check for a loose or dirty connection at the ECU or a open circuit in the wiring between the speed sensor and the ECU.

2 If the speedometer does not operate, disconnect the wiring from the speed sensor and check for battery voltage at the speed sensor connector terminal 1 with the ignition On. Check the 10 amp No 13 fuse in the interior fuse box.

3 Check for continuity to earth at the speed sensor harness connector terminal 2.

4 Check that approximately 5 volts is available at the speed sensor harness connector terminal 3.

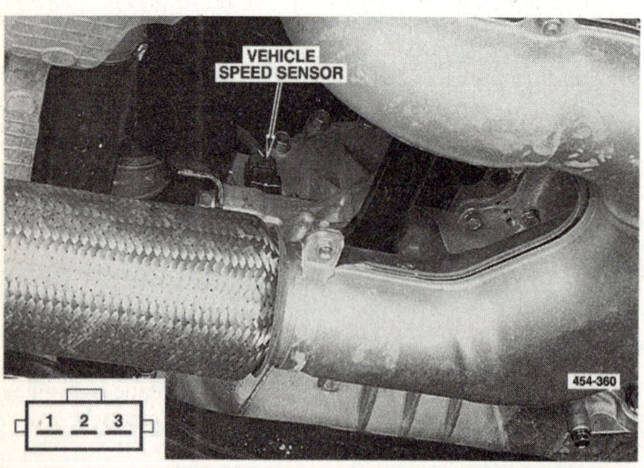

Location of the vehicle speed sensor. Inset shows the terminal identification.

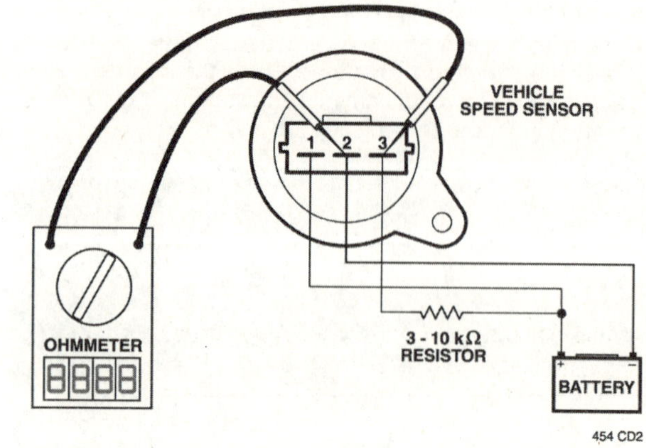

Method of testing the speed sensor.

If the voltage is not as specified, check the ECU power supply circuits and the wiring between the sensor connector terminal 3 and the ECU for continuity and shorts. If the circuits are satisfactory, renew the ECU.

5 If the previous checks do not indicate a fault, test the speed sensor as follows:

a Connect the speed sensor to a 12 volt battery using a 3 – 10 k/ohm resistor as shown in the illustration.

b Connect an voltmeter between the speed sensor terminals 2 and 3 and rotate the speed sensor shaft. The voltage should fluctuate four times per revolution

FUEL INJECTORS (CODE 41)

1 Disconnect the block wiring connector from the injectors, measure the resistance of each injector and compare to Specifications. Refer to Part 1 of this chapter.

2 Connect a test lamp between the terminals on an injector wiring connector and crank the engine. The test lamp should flash. Repeat the procedure on the other connectors. If the test lamp remains constantly illuminated, check the wiring for shorts to earth. If the wiring is serviceable, renew the ECU.

3 If the test lamp does not illuminate, check for battery voltage at the injector connector red wire (P terminal) with the ignition On. If battery voltage is not displayed, check the engine control relay.

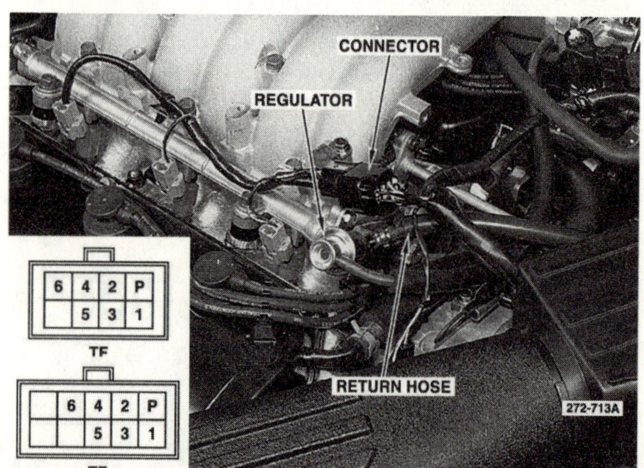

Fuel pressure regulator and injector block wiring connector. Inset shows the terminal identification for the connector.

4 Check the wiring harness for continuity between the injector connectors and the ECU.

5 Check the ECU power supply and earth circuits.

6 Check the crank angle and TDC sensors.

IGNITION CIRCUIT (CODES 44, 52, 53)

1 Reduce the fuel system pressure as previously described but do not install the fuel pump relay on completion. If the engine will not start, disconnect the wiring from the injectors.

2 While cranking the engine, check that strong, constant spark is capable of jumping a 4 mm gap. To prevent possible electronic component failure, the gap should not exceed 4 mm.

3 Disconnect the wiring connector from the distributor.

4 Check for battery voltage at the distributor harness connector terminal 1 with the ignition On.

Location of the distributor. Inset shows the terminal identification.

5 Check the ignition coil primary resistance between the distributor terminals 1 and 2, and the secondary resistance between terminal 1 or 2 and the high tension terminal. Renew the ignition coil if a fault is indicated.

6 Connect the positive lead of an analogue ohmmeter to terminal 4 and the negative lead to terminal 2.

7 Connect a 1.5 volt battery to terminals 3 and 4 with positive connected to terminal 3. The ohmmeter should display continuity only when the battery is connected. Renew the power transistor if a fault is indicated.

8 Check the wiring for continuity to earth at the distributor harness connector terminal 4.

9 Check the wiring between the distributor harness connector terminal 3 and the ECU for continuity and shorts to earth.

10 Connect the wiring to the distributor and while cranking the engine, check the voltage at the distributor terminal 3. If there is no voltage, check the fault codes. If code 22 or 23 is not displayed, renew the ECU.

IMMOBILISER SYSTEM (CODE 54)

The immobiliser ECU is located below the centre console next to the fuel pump and engine control relays. Part 1 of this chapter contains a description of how the immobiliser system operates, as well as some preliminary checks.

1 Ensure that any other ignition keys are not in close proximity to the ignition switch when starting the vehicle.

2 Check the engine ECU and the immobiliser ECU wiring connectors for clean and secure terminals.

3 Check the wiring for continuity and shorts between the immobiliser and the engine ECU terminal 85 (A/T) or terminal 62 (M/T).

4 Check that battery voltage is available at the red wires of the immobiliser ECU wiring connector.

5 Check for continuity to earth at the black wires of the immobiliser ECU wiring connector.

6 If the fault code continues, refer the vehicle to an authorised dealer to have the immobiliser system diagnosed.

A/T COMMUNICATION (CODE 61)

During transaxle up shifts, the operation of the fuel injectors is momentarily interrupted to reduce transmission torque shock. This code is set when a fault is caused by an internal communication error in the ECU between the engine and the transaxle control processing units. Renew the ECU.

THROTTLE BODY

1 Check the throttle body for damage and wear. Ensure that the throttle valve opens and closes smoothly without binding.

2 Ensure that the throttle valve bore and ports are clean.

3 Check that the throttle lever contacts the throttle stop screw when the throttle is closed slowly.

4 Adjust the throttle stop screw as follows:

a Loosen the locknut and unscrew the throttle stop screw until it is not contacting the throttle lever.

b Adjust the throttle stop screw until it is just contacting the throttle lever and then turn the screw in an additional 1 1/4 turns.

c Hold the throttle stop screw in this position and tighten the locknut securely.

Location of the ISC motor and the throttle stop screw.

ECU POWER SUPPLY AND EARTH CIRCUITS

1 Check that battery voltage is available with the ignition Off at the ECU terminal 66 (A/T), or terminal 80 (M/T). If the voltage is not as specified, check fuse 13 in the engine compartment fuse and relay box.

2 With the ignition On, check that battery voltage is available at the ECU terminals 41 and 47 (A/T) or terminals 12 and 25 (M/T).

3 Check for continuity to earth at the ECU terminals 42, 48, 76, 88 and 97 (A/T), or 13, 26 and 91 (M/T).

START SIGNAL CIRCUIT

1 Disconnect the wiring from the starter motor solenoid.
2 On models with an automatic transmission, ensure that the selector lever is in the Park or Neutral position.
3 With the ignition switch in the Start position, measure the voltage at the ECU connector terminal 59 (A/T) or terminal 71 (M/T).
4 If battery voltage is not present, check the wiring between the starter circuit and the ECU for continuity and shorts. Rectify as necessary.

POWER STEERING PRESSURE SWITCH

1 Disconnect the wiring connector from the pressure switch and check that battery voltage is available at the harness connector terminal with the ignition On.
2 If battery voltage is not displayed, check the wiring between the pressure switch and the ECU for continuity and shorts to earth. Also check the ECU power supply and earth circuits.
   If no fault can be found, renew the ECU.
3 Connect an ohmmeter between the pressure switch terminal and earth.
4 With the engine idling, turn the steering wheel and check that continuity exists. If continuity does not exist, renew the pressure switch.

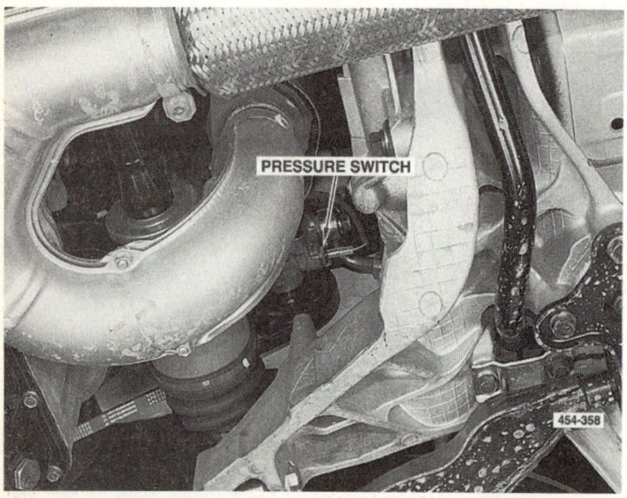

Location of the power steering pressure switch.

THROTTLE PEDAL POSITION SENSOR

The throttle pedal position sensor (APS) is a variable resistor that works much like the throttle position sensor and is mounted on the end of the throttle shaft on the throttle body. The traction control system uses this signal to determine the position of the throttle.

To Check

1 Disconnect the throttle pedal position sensor electrical connector.

Location of the accelerator pedal position sensor.

2 Measure the resistance between terminals 1 and 4. The resistance should be 3.5–6.5 k/ohms.
3 Reconnect the sensor electrical connector and backprobe terminal 2 of the connector.
4 Have an assistant depress the throttle pedal slowly and check as follows:
a Ensure that the voltage varies smoothly in relation to throttle movement.
b Ensure that the voltage is between 0.3-1.0 volts when the throttle is fully closed.
c Ensure that the voltage is approximately 4.5 volts when the throttle is fully opened.
5 If the sensor does not perform as indicated, renew it with a new one.

To Renew

1 Disconnect the cable from the negative battery terminal.
2 Disconnect the electrical connector from the throttle pedal position sensor.
3 Detach the wiring harness clamp, sensor mounting screws and remove the sensor from the throttle body.
4 Refitting is the reverse of removal, ensuring that the sensor is correctly adjusted as follows.

To Adjust

1 Connect an ohmmeter between terminals 3 and 4 of the sensor electrical connector.
2 Insert a 0.85 mm feeler gauge between the throttle lever and the cruise control lever, but no more than 3 mm.
3 Loosen the accelerator pedal position sensor mounting screws, and rotate the sensor body until the ohmmeter reading switches between indicating continuity and indicating open circuit.
4 Tighten the throttle pedal position sensor mounting screws at the point where the ohmmeter changes to open circuit, indicating the idle contacts are open.

Chapter 8
Emission Control

Contents

Description

To reduce the output levels of the three primary automotive emissions causing air pollution carbon monoxide (CO), hydrocarbons (HC) and oxides of nitrogen (NOx) several emission control systems are used in the range of vehicles covered by this manual. The systems will be discussed under the headings Positive Crankcase Ventilation (PCV) System, Evaporative Control System and Exhaust Control System.

The engine electronic control unit (ECU) regulates the amount of fuel injected to suit the engine operating conditions to maintain optimum performance and minimal emissions.

Refer to Fuel and Engine Management (chapter 7) for further information on the engine management system.

Positive Crankcase Ventilation (PCV) System

The crankcase ventilation system is of the closed type and is designed to reduce the emission of hydrocarbons and carbon monoxide. Clean air is drawn from within the induction system by a hose connecting a port on the air intake duct to the rocker cover. The air passes through the engine and combines with crankcase vapours which contain blow-by gas that has escaped from the combustion chambers past the piston rings.

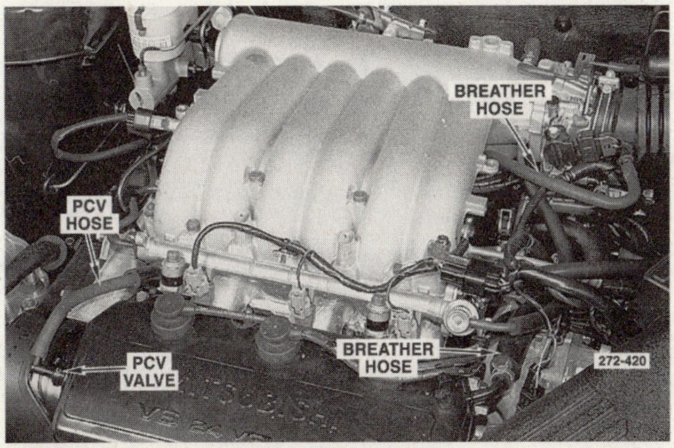

Positive crankcase ventilation components.

The air is then drawn into the inlet manifold through the PCV valve and the gases are consumed by the combustion process.

The engine oil dipstick and the oil filler cap are sealed to prevent crankcase vapours escaping into the atmosphere.

This system is most effective at part throttle operation where a particularly high vacuum is present in the plenum chamber allowing the PCV valve to open and for all the crankcase vapours to be scavenged from the crankcase. At full throttle operation the manifold vacuum is not sufficient to draw all the crankcase vapours through the PCV valve and into the plenum chamber. Under these conditions the crankcase ventilation flow may reverse, with the vapours being drawn into the throttle body instead of the plenum chamber.

If the engine is excessively worn and blow-by is at a high level, a certain amount of crankcase vapour will recycle through the rocker covers and be drawn into the air intake duct, regardless of the throttle position.

ROUTINE MAINTENANCE

1. Regularly check the system hoses for deterioration, clogging and poor connections. Also check the port in the throttle body and clean out if necessary. Renew the hoses as necessary.
2. Check the seals on the oil filler cap and dipstick. Renew components as necessary.
3. With the engine idling, remove the PCV valve from the front rocker cover.
4. Block the end of the valve using a finger.

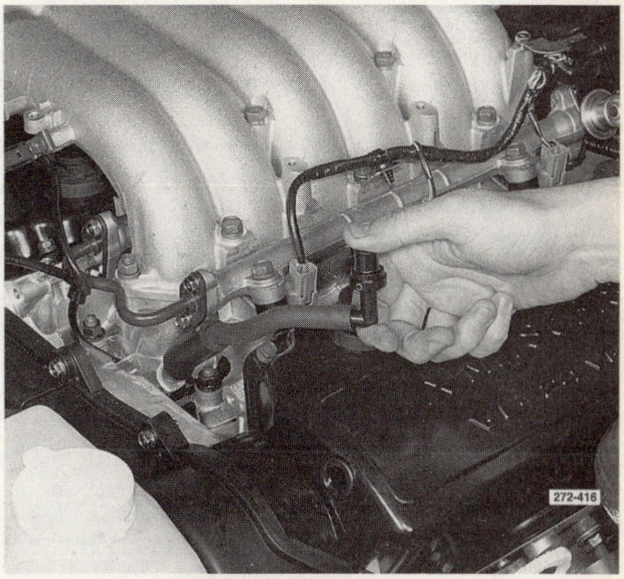

Checking the operation of the PCV valve.

5 Suction should be felt and the valve operation should be heard as the end of the valve is blocked and opened. If necessary clean or renew the valve and hose.

6 Switch the engine Off, disconnect the hose from the PCV valve and remove the valve from the front rocker cover.

7 Attach a clean piece of hose to the rocker cover end of the PCV valve.

8 Blow through the PCV valve from the rocker cover end and ensure that air passes easily through the valve.

Do not suck air through the valve as the inhalation of substances within the valve could be harmful.

9 Disconnect the hose and attach it to the plenum chamber end of the PCV valve.

10 Blow through the PCV valve from the inlet manifold chamber end and ensure that the valve restricts the passage of air. Should the PCV valve fail the above tests and cleaning the valve does not cure the problem, it must be renewed.

Evaporative Control System

The evaporative control system is used to reduce the level of hydrocarbons released to the atmosphere from the fuel tank.

The main components of the system are the charcoal canister incorporating a breather valve, a sealed fuel tank, the purge port on the throttle body and vaccuum hoses and pipes. A check valve is fitted to the purge hose going to the throttle body.

When the engine is stationary, fuel vapour caused by evaporation gradually fills the air space in the fuel tank. As the fuel tank is equipped with a sealed filler cap, vapour pressure builds up within the system. Excess vapour from the fuel tank is forced into and absorbed by the charcoal in the canister.

When the engine is running and the throttle is opened sufficiently, the stored vapour in the charcoal canister is drawn into the throttle body and burnt with the air/fuel mixture.

ROUTINE MAINTENANCE

1 Regularly check the system hoses for deterioration and poor connections. Renew the hoses as necessary.

2 Check the fuel tank filler cap seal for deterioration.

3 If it is felt that the system is not functioning correctly, check that the purge port is not blocked and clean it out using carburetor cleaner or similar.

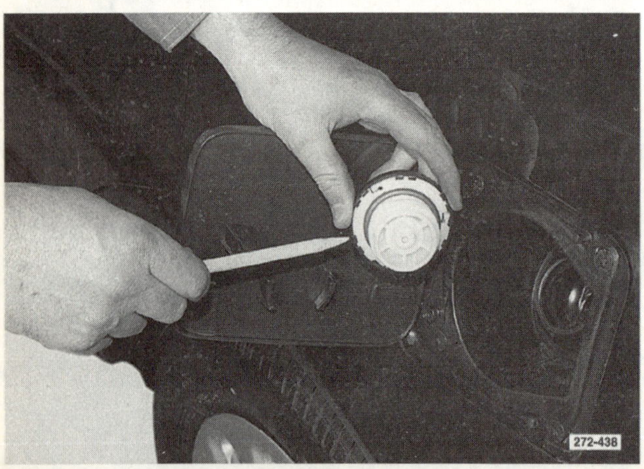

Check the fuel filler cap seal for deterioration.

Location of the charcoal canister.

TO CHECK CHARCOAL CANISTER

1 Remove the air inlet duct from the engine compartment. Refer to Fuel and Engine Management (chapter 7) if necessary.

2 Disconnect the hoses from the charcoal canister. Using a pair of pliers, carefully bend the retaining tang clear of the top of the canister and slide the canister from the bracket on the firewall.

3 Check the charcoal canister for cracks, dents or other damage, do not attempt to wash the canister. Renew if necessary.

4 Reverse the removal procedure and install the charcoal canister to the vehicle.

Exhaust Control System

The exhaust system is the greatest source of pollutant emission from the internal combustion engine. The level of emissions are controlled by the engine management system and a catalytic converter.

The engine ECU constantly monitors the exhaust gases via an oxygen sensor located in the engine pipe. The ECU uses the signals produced by the oxygen sensor to adjust the fuel injection and the ignition timing to ensure that the correct air/fuel ratio and complete combustion is maintained, reducing exhaust emissions.

For more information, refer to Fuel and Engine Management (chapter 7).

CATALYTIC CONVERTER

The three element catalytic converter is located at the front of the exhaust system. The converter contains platinum, palladium and rhodium which convert hydrocarbons and carbon monoxide, present in the exhaust gases to carbon dioxide, water and heat and the oxides of nitrogen present in the exhaust gases to oxygen and nitrogen.

The catalytic converter does not require periodic maintenance or renewal under normal circumstances. However, it can be damaged or poisoned by excessively high exhaust gas temperatures, incorrect air/fuel ratio or the use of leaded fuel. To prevent the use of leaded fuels, a small diameter fuel filler neck with a spring loaded trap door is used.

The converter can also be damaged by hard impacts, which may shatter the core. If the converter overheats or is contaminated with leaded fuel the core will melt and become blocked or restricted, resulting in a severe loss of performance.

To prevent damage to the catalytic converter, the following precautions should be observed.

Do not operate the vehicle on leaded fuel.

Do not push or tow start the vehicle.

Do not drop the catalytic converter as internal damage could result.

Do not switch the ignition Off while the vehicle is in motion and the transaxle is in gear.

Do not allow oil, coolant or unburnt fuel to enter the converter as this will cause contamination of the catalyst.

Do not operate the vehicle for long periods if the engine is misfiring.

Avoid running the vehicle out of fuel.

ROUTINE MAINTENANCE

1 Inspect the catalytic converter for dents and damage. If the converter is excessively damaged, it should be renewed.

2 Inspect the converter heat shield for damage and ensure that it is positioned correctly. Renew the heat shield if necessary.

TO REMOVE, INSPECT AND INSTALL

The removal and installation procedure for the catalytic converter is covered under the Exhaust System heading in Engine (chapter 4).

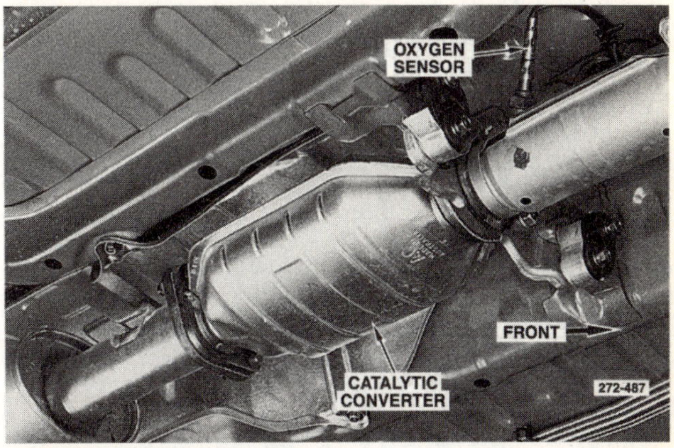

Location of the catalytic converter.

The only way to check the servicability of the catalytic at home, is to remove it from the exhaust system, shine a torch through one end and ensure that the core is not damaged.

Chapter 9
Clutch

Contents

Specifications

Type	Single dry plate
Operation	Hydraulic, self adjusting
Pressure plate spring	Diaphragm
Release bearing	Ball thrust type prelubricated and sealed
Fluid type	Dot 4 brake fluid
Driven plate:	
Outside diameter	225 mm
Wear limit	0.3 mm above rivet heads
Run out limit	0.8 mm
Pressure plate:	
Finger height	31.9 ± 0.7 mm
Flywheel runout limit	0.13 mm
Master cylinder bore diameter	15.87 mm
Slave cylinder bore diameter	20.5 mm
Clutch pedal:	
Height	168 – 173 mm
Pushrod freeplay	1 – 3 mm
Height from floor, clutch disengaged	35 mm or more

TORQUE WRENCH SETTINGS

Bleeder valve	11 Nm
Clutch damper bolt	18 Nm
Flywheel bolts	98 Nm
Hydraulic pipe unions	15 Nm
Master cylinder mounting nuts	13 Nm
Pressure plate cover to flywheel bolts	25 Nm
Release fork shaft lock bolt	11 Nm
Slave cylinder mounting bolts	18 Nm
Slave cylinder banjo bolt	22 Nm
Transaxle to engine bolts	50 Nm

Description

The clutch fitted to this range of vehicles is a pull type clutch rather than the more conventional push type clutch. For this reason the clutch release bearing is clipped to the pressure plate diaphragm spring and a special procedure must be used when removing the manual transaxle. This procedure is fully described in Manual Transaxle (chapter 10).

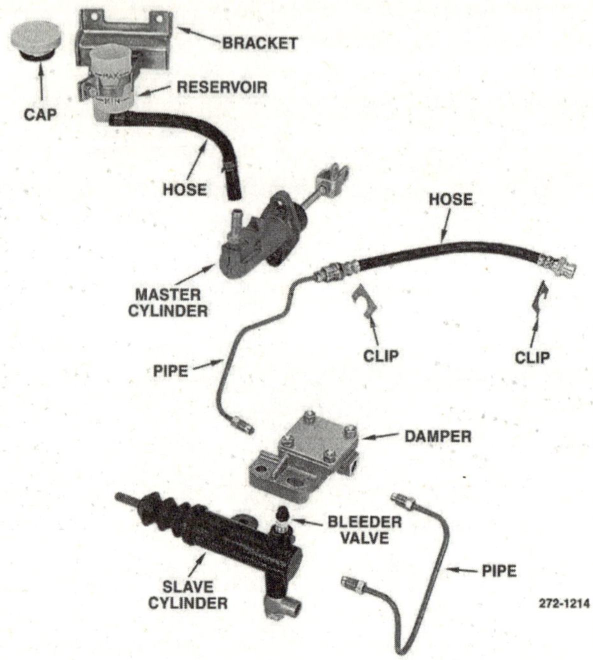

Clutch hydraulic components.

Most of the clutch components are similar to a conventional clutch with the main differences being the release lever pivot, release bearing and pressure plate. The clutch consists of a single dry driven plate assembly, splined to slide on the transmission input shaft, and a diaphragm type pressure plate assembly which is bolted to the engine flywheel.

The driven plate is clamped between the pressure plate and the flywheel and transmits the drive from the engine to the transaxle.

Cushion springs and rubbers are positioned between the hub of the driven plate assembly and the plate facings to cushion the drive.

The release mechanism is activated hydraulically by the clutch pedal via a clutch master cylinder and slave cylinder. The release mechanism consists of the release bearing, release lever and the diaphragm spring in the pressure plate assembly.

Operation of the clutch pedal moves the release lever, which pulls the release bearing and the centre of the diaphragm towards the transaxle. When the diaphragm deflects, the pressure is relieved and the clutch disengages.

The slave cylinder piston is spring loaded towards the release lever, which keeps the pushrod in constant contact with the release lever, allowing the slave cylinder a self adjusting capability. As the driven plate wears, the slave cylinder piston repositions itself, thus keeping the clutch pedal engagement height consistent.

A damper is installed in the hydraulic line between the master and slave cylinders to provide smooth clutch operation.

Pedal freeplay and height is adjusted at the pedal stop bolt and the master cylinder pushrod.

Routine Maintenance

TO CHECK CLUTCH FLUID

The clutch fluid level should be checked regularly ensuring that the level is between the Minimum and Maximum marks on the side of the reservoir. If the level is low, clean any dirt around the cap. Remove the cap and top up the reservoir with the specified type of fluid. If the clutch fluid is very low, check for any signs of fluid leakage from the clutch master cylinder, slave cylinder, pipes and fittings.

To ensure a long service life for the master and slave cylinders, the clutch fluid should be renewed at the same time as the brake fluid.

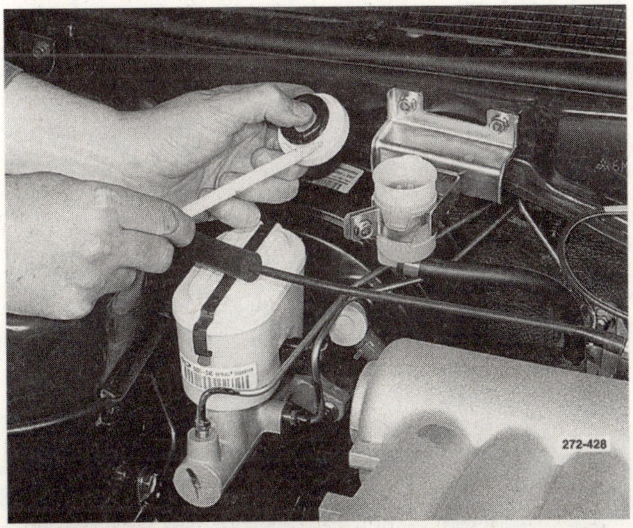

Check the reservoir cap breather hole for restrictions when topping up the master cylinder.

LUBRICATION

There are no items which require regular lubrication. Lubrication of the release lever mechanism and input shaft splines is only necessary during clutch overhaul.

PEDAL PAD

The clutch pedal pad should be regularly inspected for splitting and other signs of wear. The pedal pad should be renewed if split or the tread is worn.

TO RENEW CLUTCH FLUID

The clutch fluid should be renewed at regular intervals and it is suggested that it is done when renewing the brake fluid. Renewing the clutch fluid follows the same basic procedure as bleeding the clutch fluid so refer to the relevant heading later in this chapter with attention to the following points:

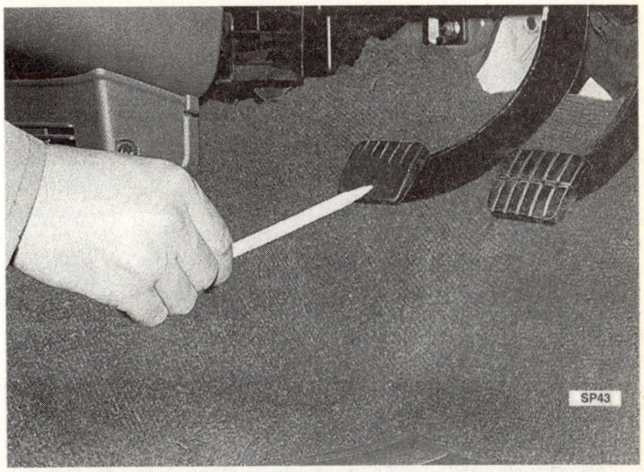

The clutch pedal pad should be regularly inspected for splitting and other signs of wear.

1 Before commencing to change the fluid, withdraw as much old fluid as possible from the clutch master cylinder using a syringe.
2 Continue bleeding until a constant stream of new fluid, without any air bubbles is being discharged from the slave cylinder.

Trouble Shooting

CLUTCH SLIPPING

1 Insufficient clutch pedal free travel: Check and adjust the free travel to Specifications.
2 Worn driven plate facings: Check and renew the clutch driven plate.
3 Weak or broken pressure plate diaphragm spring: Check and renew the pressure plate assembly.
4 Worn or scored pressure plate face: Check and renew the pressure plate assembly.
5 Worn or scored flywheel face: Check and renew or machine the flywheel as necessary.
6 Oil on the driven plate facings: Check and renew the clutch driven plate. Rectify the oil leak.

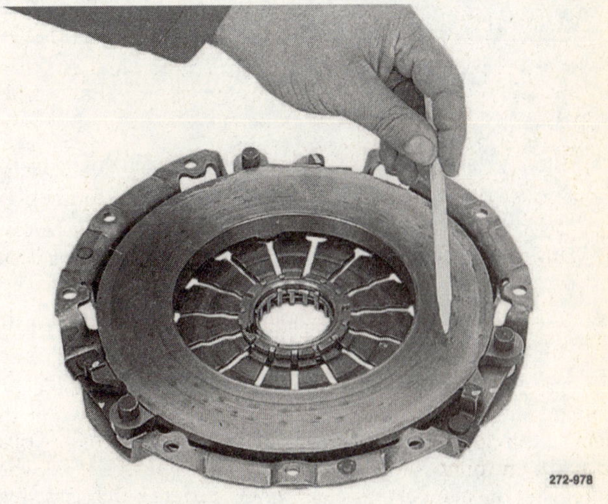

Check the pressure plate face for scoring, cracks and wear.

TIP *Usually clutch slippage is first indicated by a marked increase in engine speed, for no apparent reason, when driving up a steep hill. The clutch condition can be positively identified as follows: With the handbrake firmly applied and the wheels chocked, place the transmission in top gear. With the engine speed at approximately 1 500 rpm, slowly release the clutch pedal. Clutch slippage is evident if the engine does not stall. Perform the test as quickly as possible to prevent any further clutch damage. Ensure that the clutch pedal free travel is correct prior to performing this test.*

CLUTCH SHUDDER

1. Oil on the driven plate facings: Check and renew the clutch driven plate. Rectify the oil leak.
2. Scored pressure plate or flywheel face: Renew the pressure plate assembly and machine or renew the flywheel.
3. Loose or damaged engine mountings: Check and renew the engine mountings as necessary.
4. Loose or damaged driven plate hub: Check and renew the clutch driven plate.
5. Loose driven plate facings: Renew the clutch driven plate.
6. Cracked pressure plate face: Renew the clutch pressure plate assembly.

TIP *Clutch shudder is usually most evident when reversing up an incline. As loose or damaged engine mountings are a cause for clutch shudder, thoroughly check the engine mounting rubbers and mounting hardware for damage or looseness before removing the clutch for inspection.*

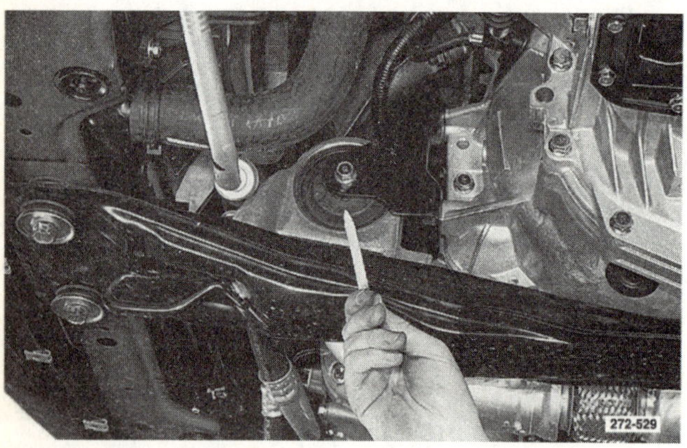

Loose or worn engine mountings can cause clutch shudder.

CLUTCH GRAB

1. Gummy driven plate facings: Renew the clutch driven plate.
2. Cracked pressure plate face: Renew the clutch pressure plate assembly.
3. Driven plate cushion springs broken or weak: Renew the driven plate.
4. Loose or damaged engine mountings: Check and renew the engine mountings as necessary.
5. Clutch slave cylinder piston sticking: Check and overhaul the slave cylinder.

i *Inspect the external operation of the release lever before removing the clutch for inspection.*

INSUFFICIENT CLUTCH RELEASE

1. Low fluid level in the master cylinder reservoir: Replenish the fluid in the master cylinder and bleed the system. Check the hydraulic system for leaks.
2. Air in the hydraulic system: Check the hydraulic system for leaks and bleed the system.

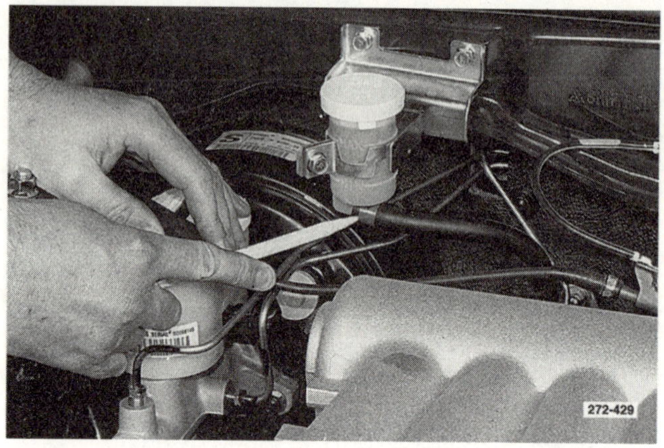

Check for fluid leaks in the clutch hydraulic system starting at the reservoir.

3. Clutch slave cylinder piston sticking: Check and overhaul the slave cylinder.
4. Excessive clutch pedal freeplay: Check and adjust the freeplay to Specifications.
5. Leaking master cylinder or slave cylinder piston seals: Overhaul the master cylinder or slave cylinder as necessary.
6. Broken release bearing clip: Check and repair the release mechanism.
7. Faulty clutch release mechanism: Check and repair the release mechanism.
8. Clutch driven plate distorted: Renew the clutch driven plate.
9. Driven plate hub tight on the input shaft splines: Check and renew the driven plate or the input shaft.

TIP *In most cases failure of the clutch to disengage is due to air in the hydraulic system or excessive clutch pedal freeplay. Check the floor covering behind the clutch pedal for signs of fluid leaks. Ensure that all adjustments are correct and bleed the hydraulic system before removing the clutch for inspection.*

Check that the release bearing clip is firmly attached to the diaphragm and that it is not damaged or worn.

RELEASE BEARING NOISE

1 Dry, worn or seized release bearing: Check and renew the release bearing and clip.
2 Damaged pressure plate diaphragm spring: Check and renew the clutch pressure plate assembly.
3 Cracked or damaged release bearing clip: Check and renew the clutch pressure plate assembly.

TIP *Depress the clutch pedal, with the engine running, to check for bearing noise. If the noise is still evident the release bearing is faulty. If the noise ceases, the gearbox input shaft bearing may be faulty. Refer to Manual Transaxle (chapter 10). If the release bearing is faulty and requires renewing, always check the other clutch components for serviceability.*

Clutch Unit and Release Bearing

To Install Clutch Unit — Clutch aligning tool

TO REMOVE AND DISMANTLE

Because of the pull type clutch fitted to this vehicle, it is necessary to disconnect the release bearing clip during transaxle removal. Always refer to the procedure in Manual Transaxle(chapter 10) to avoid damaging components.

1 Remove the transaxle as described in Manual Transaxle (chapter 10).
2 Suitably mark the pressure plate cover in relation to the flywheel to ensure replacement to the original position.
3 Progressively loosen and remove the pressure plate retaining bolts, working in a diagonal sequence.

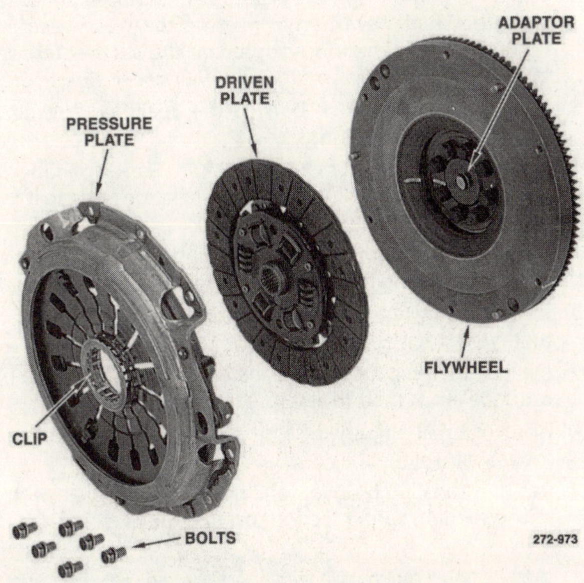

Clutch pressure plate, driven plate and flywheel. Note the direction of the driven plate.

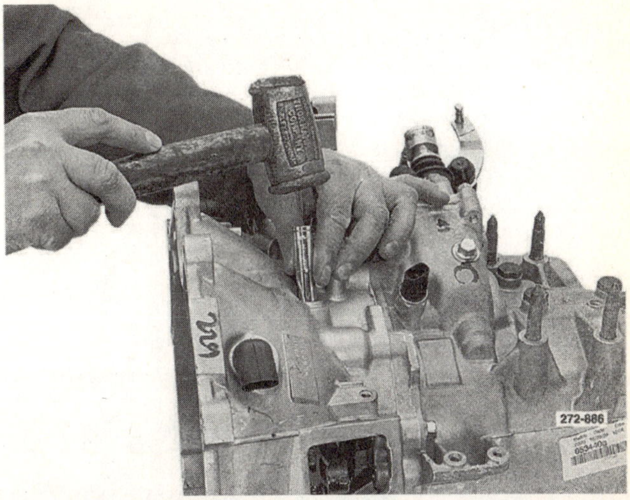

Tap the cover plug into the transaxle housing to allow removal of the release fork shaft.

4 Remove the pressure plate and driven plate assembly from the flywheel noting the installed direction of the driven plate hub.
5 Remove the rubber dust boot surrounding the clutch release lever.
6 Using an 11 mm socket or tube carefully tap the cover plug into the transaxle housing. Remove the cover plug from the housing and, if damaged, discard the plug.
7 Insert a flat bladed screwdriver between the release fork and the return springs and carefully release the springs.
8 Remove the lock bolt from the release fork shaft and noting the installed position of the return springs, withdraw the shaft from the transaxle housing.
9 Remove the springs, the fork and the bearing from the housing. Withdraw the felt pads from the fork.

TO CHECK AND INSPECT

TIP *If the clutch has done considerable work, it is recommended that a clutch kit is installed to ensure a long working life for the clutch mechanism. A clutch repair kit normally contains a new pressure plate and clip, driven plate and release bearing.*

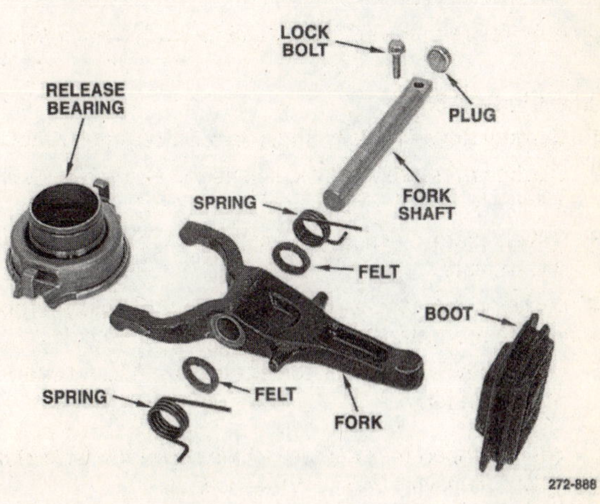

Clutch release components removed from the clutch housing.

Renew the driven plate if the friction material is worn to within 0.3mm of the rivets.

1 Clean the inside of the transaxle housing and the release fork using rags and a suitable solvent. Do not immerse the release bearing in solvent.
2 Check that the driven plate facings are not glazed or contaminated with oil.
3 If the driven plate facings are worn down to the rivets, check the flywheel and pressure plate facings for scoring.
4 Check the hub of the driven plate for looseness and the cushion springs for wear or breakage.
5 Check the pressure plate and flywheel for high spots, glazing, scoring and cracking. If the flywheel is scored enough so you can feel it with your fingernail, remove the flywheel as described in Engine (chapter 4) and take it to a brake/clutch specialist for machining. Light glazing can be removed with emery paper.
6 Check the diaphragm spring for bent, cracked or fatigued fingers. Check the release bearing clip for wear and deterioration. It should be possible to rotate the inner hub of the clip smoothly within the pressure plate.
7 Check the release bearing for noise or roughness. Do not immerse the release bearing in cleaning fluid. The bearing is lubricated when manufactured and requires no further lubrication in service.

Check the diaphragm spring fingers for cracks and distortion.

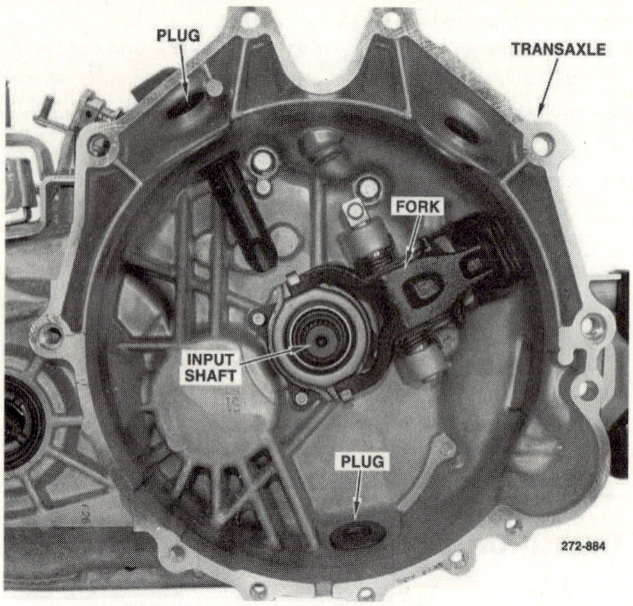

Clutch release components installed. Also shown is the inspection hole plugs used when removing the transaxle.

8 Renew all unserviceable parts.

TO ASSEMBLE AND INSTALL

Assembly and installation is a reversal of the removal procedure with attention to the following points:

1 Clean all components except the driven plate and release bearing in suitable solvent. Apply high melting point grease to the release shaft and to the bore of the fork.
2 Install the felt pads to the fork, retaining them with a dab of grease.
3 Engage the bearing in the fork and place the bearing and the fork over the slider.
4 While holding the springs in the position noted on removal, slide the shaft down, ensuring that the felt pads remain correctly seated and do not prevent the shaft entering the bore of the fork.
5 Tighten the shaft lock bolt to the specified torque and engage the return springs.
6 Apply a small amount of Permatex Auto Gasket or equivalent sealant to the cover plug prior to installation and tap it home until it is flush with the housing.
7 Install the dust boot to the clutch housing. Check the release lever and bearing for free smooth operation.
8 Clean the flywheel and pressure plate friction surfaces with methylated spirits.
9 If a new driven plate is to be installed, slide the driven plate onto the input shaft to test its ease of movement.
10 Apply a small amount of high melting point grease to the input shaft splines.
11 Position the clutch driven plate onto the flywheel with the raised cushion spring section towards the pressure plate.

Some clutch driven plates are marked "Flywheel Side" near the hub to assist in installation.

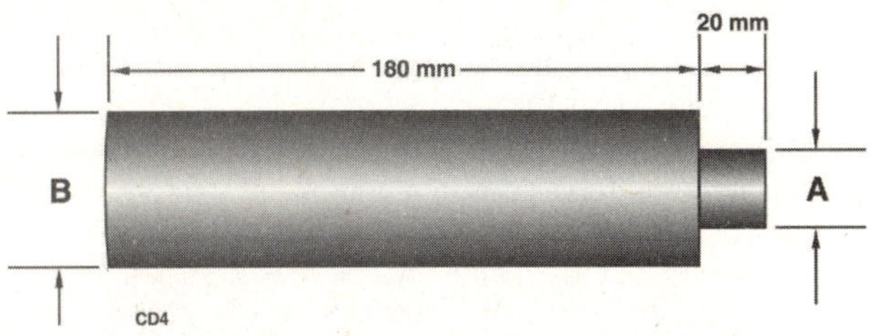

A clutch aligning tool can be fabricated from a length of round section of wood or metal. Dimension A is the diameter of the alignment hole in the flywheel adapter plate. Dimension B is the inside diameter of the driven plate hub.

12 Install the pressure plate to the flywheel. If the original pressure plate is being installed, ensure that the mating marks are aligned.

13 Using a clutch aligning tool, align the centre of the driven plate with the centre of the flywheel adapter plate.

14 Tighten the pressure plate retaining bolts evenly and progressively in a diagonal sequence to the specified torque. Withdraw the aligning tool.

15 Install the transaxle assembly as described in Manual Transaxle (chapter 10).

16 Check and adjust the clutch pedal height and free travel as described later in this chapter under the Clutch Pedal heading.

Master Cylinder

TO REMOVE AND INSTALL

1 Working inside the vehicle, remove the lower dash panel, as follows:

a Remove the screws retaining the bonnet lever to the driver side lower dash panel and position the lever clear of the work area. Remove the screw cover from the top of the panel.

b Remove the lower dash panel screws and disconnect the panel from each side of the steering column. Partially remove the panel and where fitted, rotate the footwell lamp bulb holder anticlockwise and separate it from the panel. Remove the driver side lower dash panel from the vehicle.

2 Remove the clip and clevis pin retaining the master cylinder pushrod to the clutch pedal. Be careful not to lose the washers.

3 Working in the engine compartment, place a cloth beneath the clutch pipe fitting and using a pipe spanner, loosen the hydraulic pipe from the master cylinder.

4 Remove the hydraulic pipe from the clutch master cylinder and plug the pipe to prevent the entry of dirt. Remove the master cylinder mounting nuts.

5 Remove the reservoir mounting bolts and withdraw the master cylinder and reservoir from the vehicle ensuring that the clutch fluid is not spilt on the vehicles paintwork.

If clutch fluid is spilt on the paintwork during removal of the master cylinder, wash the area immediately with clean water and allow to dry naturally. Do not wipe the area with a cloth.

Installation is a reversal of the removal procedure with attention to the following points:

1 Tighten the hydraulic pipe union and mounting nuts to the securely.

2 Bleed the clutch hydraulic system as described later in this chapter.

3 Road test the vehicle and check for correct clutch operation. If necessary, check and adjust the clutch pedal height and free travel as described later in this chapter under the Clutch Pedal heading.

TO DISMANTLE

If the master cylinder is worn or leaking it is often more economically viable to install a new or exchange cylinder rather than recondition the old one. Check the price difference before proceeding.

1 With the master cylinder removed as previously described, drain and discard the fluid from the master cylinder reservoir.

2 Release the hose clamp and disconnect the reservoir hose from the master cylinder.

3 Fold back the rubber boot and slide it along the master cylinder push rod.

4 Remove the retaining snap ring and withdraw the pushrod and washer.

5 Tap the end of the cylinder on a piece of timber and remove the piston assembly from the master cylinder bore. Alternatively, low air pressure may be carefully applied to the hydraulic pipe aperture, using a wad of cloth to cover the end of the bore in case the piston ejects with force.

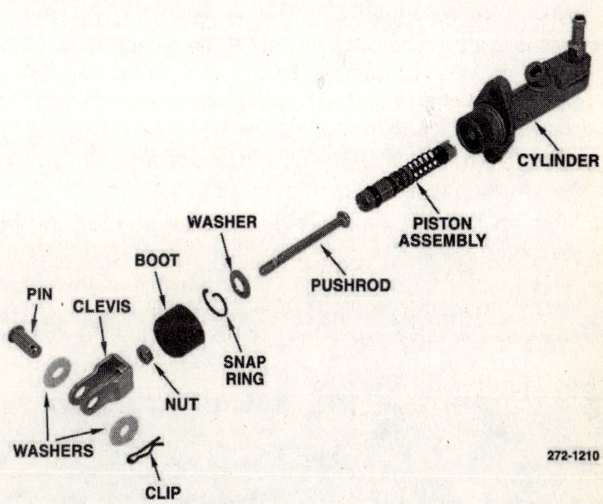

Master cylinder dismantled.

TO CLEAN AND INSPECT

1 Thoroughly clean the master cylinder components and the inside of the cylinder with methylated spirits. Do not use petrol or other mineral spirits. Dry the components with compressed air if available.

2 Inspect the cylinder bore for pitting, scoring, damage or wear. Renew the master cylinder assembly if necessary.

3 Inspect the piston for wear and the return spring for breakage and loss of tension. The manufacturer warns against dismantling the piston. If any part is worn or damaged, renew the complete piston assembly.

4 Check the pushrod for wear and damage. Clean any sediment which may have accumulated in the bottom of the reservoir or hose.
5 Inspect the hose for deterioration and renew as necessary.

TO ASSEMBLE

Assembly is a reversal of the dismantling procedure with attention to the following points:

1 Lightly smear the piston contact surfaces with rubber grease.
2 Install the piston to the cylinder bore ensuring that the lip of the seals are not turned back on installation.
3 Install the pushrod and washer. While depressing the pushrod, install the snap ring to the groove in the end of the cylinder bore. Seat the rubber boot onto the cylinder.
4 Ensure that the piston returns fully and contacts the snap ring. Check that the operation of the piston is smooth.
5 Connect the reservoir and hose to the master cylinder and retain with the hose clamp.
6 Pour a small quantity of clean brake fluid into the reservoir and pump the piston in the bore until fluid emerges from the hydraulic pipe aperture.
7 Install the master cylinder to the vehicle as previously described.

Slave Cylinder

TO REMOVE AND INSTALL

1 Disconnect the hydraulic pipe from the slave cylinder. Seal the pipe to prevent the entry of dirt and the loss of fluid.
2 Remove the bolts retaining the slave cylinder and damper and withdraw the slave cylinder from the vehicle. If necessary the damper can also be removed at this time after disconnecting the hydraulic pipe.

Installation is a reversal of the removal procedure with attention to the following points:

1 Tighten the hydraulic pipe union and the retaining bolts securely.
2 Bleed the hydraulic system as described later in this chapter.
3 Road test the vehicle and check for correct clutch operation.

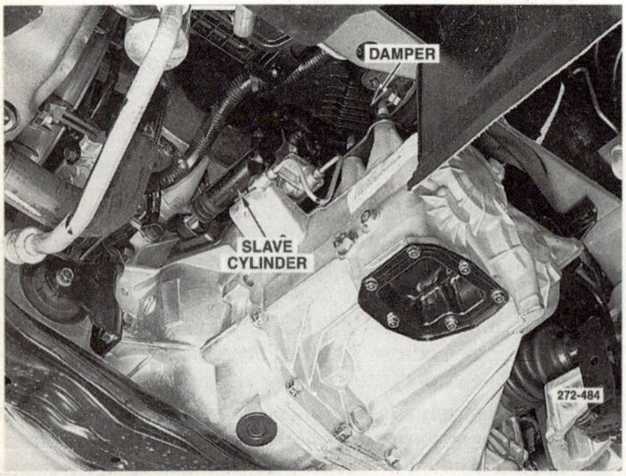

Slave cylinder and damper installed.

TO DISMANTLE

If the slave cylinder is worn or leaking it is often more economically viable to install a new or exchange cylinder rather than recondition the old one. Check the price difference before proceeding.

1 With the slave cylinder removed as previously described, remove the rubber dust boot and pushrod from the end of the slave cylinder.
2 Tap the end of the cylinder on a piece of timber and remove the piston and spring from the cylinder bore. Alternatively, low air pressure may be carefully applied to the hydraulic pipe aperture, using a wad of cloth to cover the end of the bore in case the piston ejects with force.
3 Remove the bleeder valve from the cylinder body.
4 Remove the banjo bolt and disconnect the fitting and sealing washers from the slave cylinder.
5 Tap the end of the cylinder with a piece of timber to dislodge the valve under the banjo bolt.

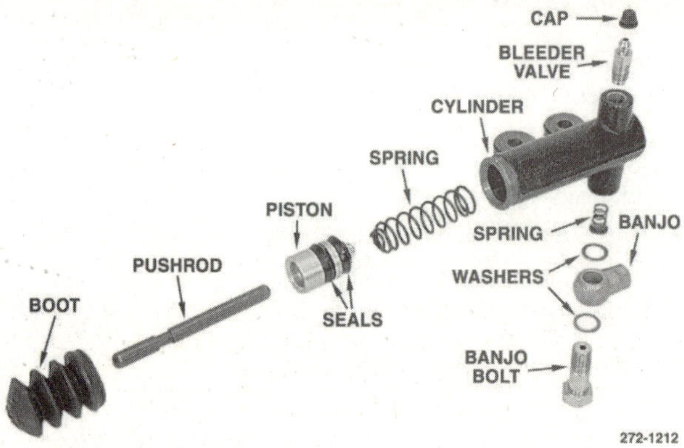

Slave cylinder dismantled.

TO CLEAN AND INSPECT

1 Thoroughly clean the slave cylinder components and the cylinder bore with methylated spirits. Do not use petrol or other mineral spirits.
2 Inspect the cylinder bore for pitting, scoring, damage or wear. Renew the cylinder if necessary.
3 Inspect the piston for wear and the spring for breakage and loss of tension. Renew as necessary.

TIP *If the condition of the seals or the piston is suspect, it is advisable to install a genuine repair kit which contains a preassembled piston and other components necessary to perform a thorough overhaul and ensure long service from the unit.*

4 Inspect the bleeder valve for blockage. Clear or renew as necessary.

TO ASSEMBLE

If the seal has been removed, ensure that the new seal is installed with the seal lip towards the pushrod end of the piston.

1 Lightly smear the piston and seal with rubber grease.
2 Install the spring to the cylinder bore with the larger end of the spring entering first.

3 Install the piston to the cylinder bore ensuring that the lip of the seals are not turned back on installation. Push the piston fully into the cylinder.
4 Install the pushrod to the rubber boot ensuring that the end of the boot correctly locates in the pushrod groove. The longer part of the rod goes inside the boot.
5 Install the boot and pushrod to the slave cylinder ensuring that the end of the boot correctly locates in the slave cylinder groove.
6 Install the bleeder valve to the cylinder body.
7 Install the valve, the banjo fitting and the banjo bolt using new sealing washers. Ensure that the rubber on the spring is towards the banjo bolt.
8 Install the slave cylinder to the vehicle as previously described and bleed the hydraulic system as described under the heading that follows.

Bleeding the Clutch System

TO BLEED

Bleeding the hydraulic system is necessary when some portion of the hydraulic system has been disconnected or fluid has been drained allowing air to enter the system.

1 Where applicable, drain the master cylinder reservoir of the old brake fluid using a syringe.
2 Refill the master cylinder reservoir with clean brake fluid and ensure that the reservoir is maintained at least three quarters full throughout the entire operation.
3 Remove the rubber cap from the slave cylinder bleeder valve and attach a transparent hose to the valve. Immerse the other end of the hose in a small glass container partially filled with clean brake fluid.

Bleeding the clutch hydraulic system.

TIP *If desired a one person bleeding kit may be used. These kits usually consist of a transparent tube with a one way ball valve. With a one person bleeding kit, it is not necessary to close the bleeder valve after each pedal stroke, however follow the kit manufacturers instructions.*

4 Have an assistant slowly pump the clutch pedal several times and then hold the pedal to the floor.
5 Open the bleeder valve and allow the fluid and air to flow into the container. Close the bleeder valve when all the fluid and air stops flowing from the hose.
6 Complete steps 4 and 5 until a constant stream of new brake fluid, without any air bubbles, is being discharged into the glass jar.
7 Remove the hose, install the rubber cap to the bleeder valve and check the operation of the clutch. If necessary, top up the reservoir with the specified brake fluid to the Maximum mark and install the reservoir cap.

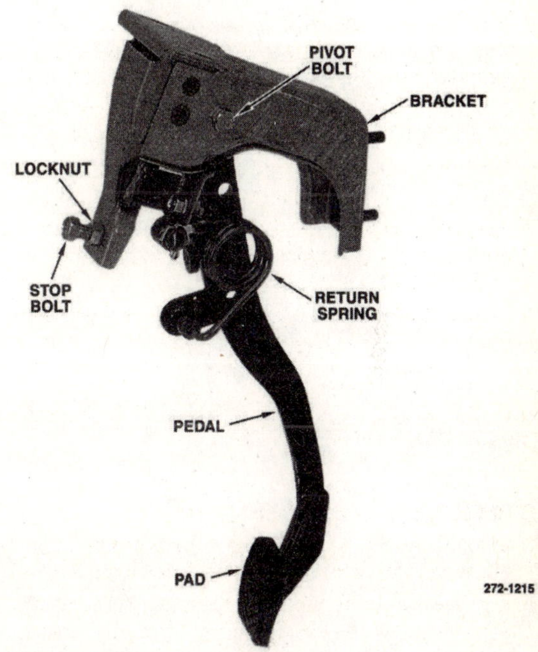

Clutch pedal assembly removed from the vehicle.

Clutch Pedal

If adjustment is required to the pedal height or freeplay, remove the lower dash panel as follows:

1 Remove the screws retaining the bonnet lever to the driver side lower dash panel and position the lever clear of the work area. Remove the screw cover from the top of the panel.
2 Remove the lower dash panel screws and disconnect the panel from each side of the steering column. Partially remove the panel and where fitted, rotate the footwell lamp bulb holder anticlockwise and separate it from the panel. Remove the driver side lower dash panel from the vehicle.

TO ADJUST HEIGHT

1 Fold back the carpet below the clutch pedal to reveal the insulation covering the floor panel.
2 Position a ruler to the side of the clutch pedal with the end resting on the insulation. Measure the distance from the centre of the pedal pad to the floor as shown in the illustration.
3 If the distance is not 168 – 173 mm, loosen the locknut and adjust the pedal stop bolt until the pedal height is correct, tighten the locknut. Check the pedal free travel as described under the following heading.

On models with cruise control, disconnect the wires from the cancel switch and adjust the switch. Ensure that the wires are securely connected when adjustment is complete.

TO ADJUST FREEPLAY

1 Position a ruler to the side of the clutch pedal with the end resting on the insulation.

2 Depress the pedal by hand until resistance is felt and measure the freeplay on the ruler as shown in the illustration.

3 If the distance is not 1 – 3 mm, loosen the pushrod locknut and adjust the pushrod as necessary until the specified freeplay is obtained. Tighten the locknut securely.

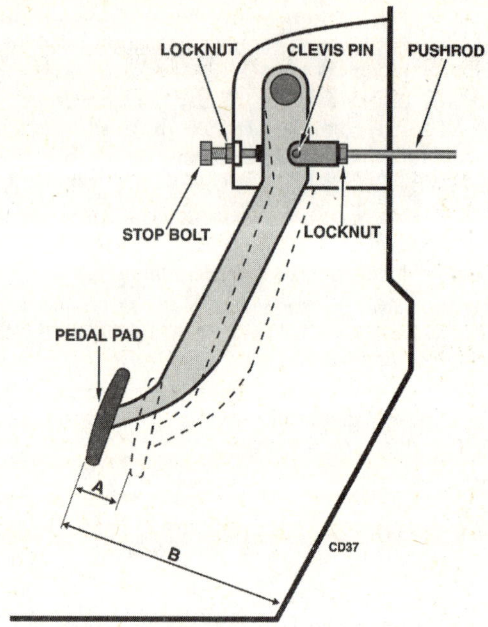

Clutch pedal measuring and adjusting points. Dimension A is clutch pedal freeplay. Dimension B is clutch pedal height.

Chapter 10
Manual Transaxle

Contents

Specifications

Type .... 5 speed synchromesh
Models F5M51-1-F5N
Lubricant:
- Type .... SAE 75W/85 GL 4 gear oil
- Capacity .... 2.8 litres

Gear ratios:
- First 3.583:1
- Second 2.105:1
- Third 1.407:1
- Fourth 1.031:1
- Fifth 0.804:1
- Reverse 3.416:1
- Final drive 3.736:1

Input shaft:
- [1] Adjustment method .... Spacers
- Front bearing —
  - Endfloat .... 0.05 – 0.17 mm
  - Spacer thickness .... 1.43 – 1.59 mm in increments of 0.08 mm
- Rear bearing —
  - Endfloat .... 0.01 – 0.12 mm
  - Spacer thickness .... 1.44 – 1.58 mm in increments of 0.07 mm
- Third gear —
  - Endfloat .... 0.01 – 0.09 mm
  - Spacer thickness .... 1.34 – 1.79 mm in increments of 0.09 mm
- Fifth gear —
  - Endfloat .... 0.01 – 0.09 mm
  - Spacer thickness .... 3.82 – 4.10 mm in increments of 0.04 mm

Output shaft:
- [1] Adjustment method .... Spacers
- Preload .... 0.13 – 0.18 mm
- Spacer thickness .... 1.36 – 1.68 mm in increments of 0.04 mm
- Bearing —
  - Endfloat .... 0.01 – 0.09 mm
  - Spacer thickness .... 0.86 – 1.46 mm in increments of 0.03 mm
- Third gear —
  - Endfloat .... 0.01 – 0.09 mm
  - Spacer thickness .... 2.81 – 3.09 mm in increments of 0.04 mm

Synchro ring to gear clearance limit .... 0.5 mm
Preload:
- Differential case .... 0.05 – 0.11 mm
- Output shaft .... 0.13 – 0.18 mm

Differential side gear backlash .... 0.025 – 0.150 mm

TORQUE WRENCH SETTINGS

Ball joint to steering knuckle nut .... 60 – 72 Nm
Centre member retaining bolts:
- Front .... 88 Nm
- Rear .... 52 Nm

Clutch bearing slider bolts .... 18 Nm
[3] Differential ring gear bolts .... 132 Nm
Drive shaft hub nut .... 196 – 255 Nm
Drive shaft support bearing bolts .... 45 Nm
Engine mountings:
- Left hand side —
  - Mounting nuts .... 69 – 93 Nm
  - [2] Through bolt .... 69 Nm
- Front and rear —
  - Bracket to transaxle bolts .... 69 Nm
  - [2] Through bolts .... 69 Nm
  - Retaining bolts .... 44 Nm

Engine pipe flange nuts .... 49 Nm
Filler and drain plugs .... 32 Nm
Interlock plate retaining bolt .... 30 Nm
Lower cover bolts .... 7 Nm
Reverse idler shaft lock bolt .... 48 Nm
Reverse lamp switch .... 32 Nm
Selector cable cover panel .... 12 Nm
Selector housing cover bolts .... 18 Nm
Selector lever assembly .... 12 Nm
Selector lever mounting bolt .... 18 Nm
Shift cable support bracket .... 18 Nm
Stabiliser bar link nuts .... 35 – 44 Nm
Starter motor retaining bolts .... 27 – 34 Nm
Tie rod end castellated nut .... 29 Nm
Transaxle case retaining bolts .... 44 Nm
Transaxle case to clutch housing bolts .... 44 Nm

[1] Spacer sets are available from authorised dealers.
[2] Tighten with the weight of the transaxle on the mounting.
[3] Apply Permatex Lock Bolt on threads.

Description

The transaxle fitted to the range of vehicles covered by this manual have five forward gears and one reverse gear. Full synchromesh is available on all gears, including reverse, which are helical cut and in constant mesh to provide smooth and quiet operation. In addition, the first, second and third gears are fitted with a double synchroniser to provide smoother gear change.

The first/second gear synchroniser assembly is splined to the output shaft and the third/fourth and fifth gear synchroniser assemblies are splined to the input shaft.

The synchroniser assemblies couple the selected gear to the shaft by way of the teeth on the inside of the synchroniser sleeve which mesh with the corresponding teeth on the gear.

Synchronisation is achieved when the synchroniser ring engages the cone on the side of the gear and brings the gear and the synchroniser assembly to the same speed, providing quiet engagement.

The gear change mechanism is held by spring loaded detent balls, which engage grooves in the shift rails. The load of the detent balls must be overcome before the rails can be moved to select or disengage a gear. The gearshift has an interlock mechanism, which prevents the selection of two gears simultaneously.

The transmission and differential assemblies are accommodated in the same casing (the transaxle). The same oil is used and the one oil level check covers both assemblies. Oil leakage is prevented by a sealing compound applied to the mating surfaces of each transaxle case component, and oil seals on the input shaft, drive shafts, speedometer drive and selector shaft.

The gear linkage consists of a remote, floor mounted gear lever assembly connected to the transaxle by two flexible selector cables. The cables are designed to keep noise and vibration to a minimum and the cable length is set during manufacture and are nonadjustable.

The drive from the transaxle is transmitted to the front wheels through drive shafts, which have a tripod joint on the transaxle end and a Birfield joint on the wheel end. The left hand side tripod joint is splined to the differential side gear and retained by an expanding snap ring. The right hand side tripod joint is splined to a relay shaft and also retained by an expanding snap ring.

The relay shaft rotates in a bearing, which is mounted to the cylinder block and reduces torsional vibrations and torque steer tendencies. Drive shaft servicing is covered in Front Axle and Suspension (chapter 13).

Extensive knowledge and a range of special tools are required to overhaul the manual transaxle making it not economically viable for the home mechanic to do so. Should a fault occur that cannot be rectified using the procedures described in this chapter, it is recommended that the vehicle be taken to an authorised dealer or manual transaxle specialist. If the transaxle must be removed for overhauled by a specialist or replacement with an exchange unit, the removal and installation procedure is covered in this chapter.

Routine Maintenance

TO CHECK OR RENEW TRANSAXLE OIL

1 Raise the front of the vehicle with a jack and support the vehicle with chassis stands under the recommended points. Refer to General Information (chapter 1) for the correct trolley jack and chassis stand positions to prevent damage or injury.

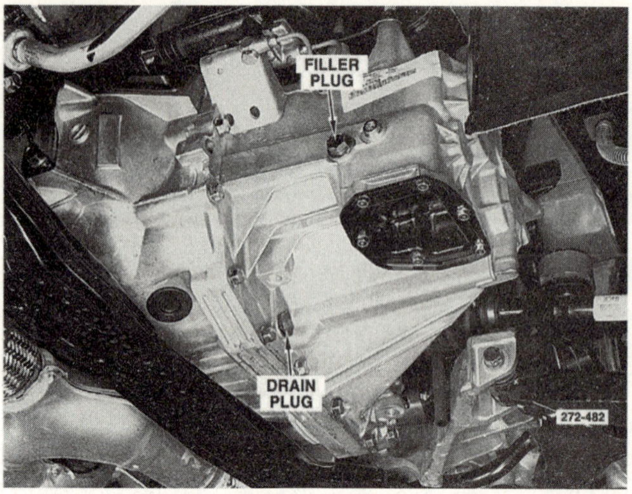

Manual transaxle drain and filler plugs.

2 Remove the bolts and clips retaining the front splashguard and withdraw the guard from the vehicle. Clean the area around the transaxle drain plug with a cloth.

It is important that the vehicle is level when checking the transaxle oil level.

3 Remove the filler plug from the transaxle and, using a finger or a bent piece of wire, check the oil level. The correct level is when the oil is level with or fractionally below the bottom of the plug hole. Overfilled units should be drained until the oil is level with the bottom of the plug hole.

4 If the oil is being renewed, place a suitable container beneath the drain plug and remove the plug from the bottom of the transaxle. Allow all of the oil to drain from the transmission and then install and tighten the drain plug to the specified torque.

5 If the oil level is low or is being renewed, use a clean length of garden hose and a suitable funnel to fill the transmission to the correct level, with the specified oil.

6 When satisfied that the level is correct, install the filler plug and tighten to the specified torque. Wipe around the plug/s with a cloth after installation.

DRIVE SHAFTS

Routine maintenance items relating to the drive shafts are described in Front Axle and Suspension (chapter 13).

Trouble Shooting

DIFFICULT GEAR CHANGE

1 Incorrect lubricant or low lubricant level: Drain the transaxle and refill with the correct quantity and grade of oil. Refer to the Routine Maintenance heading.

2 Faulty clutch or release mechanism: Check and overhaul the clutch or the release mechanism. Refer to Clutch (chapter 9).

3 Gear change selector cables sticking or stretched: Renew the selector cables.

Periodically lubricate the selector assembly bush.

4 Gear lever assembly worn or damaged: Check and renew the faulty components as necessary.

5 Worn or damaged transaxle internal components: Have the transaxle assembly inspected and if necessary overhauled by a specialist. Alternatively install an exchange unit.

Check the clutch for correct operation. If reverse gear can be selected without the engine stumbling or stalling, this is a good indication that the clutch is disengaging sufficiently.

SLIPPING OUT OF GEAR

1 Damaged or loose engine mountings: Renew or tighten the engine mountings as necessary.

2 Selector cables stretched or damaged: Check the selector cables and renew if necessary.

3 Worn or damaged transaxle internal components: Have the transaxle assembly inspected and if necessary overhauled by a specialist. Alternatively install an exchange unit.

Check the engine mountings for deterioration.

In most cases, slipping out of gear is caused by wear or damage to internal components of the transaxle. However, prior to removing the transaxle, check the engine mounting rubbers and the mounting hardware as vibrations caused by deteriorated mountings can cause the transaxle to slip out of gear.

GEARS CLASH ON CHANGING DOWN

1 Faulty clutch or release mechanism: Check or overhaul the clutch or the release mechanism. Refer to Clutch (chapter 9).

2 Lubricating oil too heavy: Drain the transaxle and fill with the specified type and quantity of lubricating oil.

3 Worn or damaged transaxle internal components: Have the transaxle assembly inspected and if necessary overhauled by a specialist. Alternatively install an exchange unit.

Check the clutch for correct operation as previously described under Difficult Gear Change before removing the transaxle for inspection. Also check that the correct type of transaxle oil is being used.

NOISY TRANSAXLE

1 Transaxle mounting bolts or engine mountings loose: Tighten the transaxle mounting bolts and check all the engine mountings.

2 Incorrect lubricant or low lubricant level: Fill to the correct level with the specified lubricant.

3 Worn or damaged gears or bearings in the transaxle: Have the transaxle assembly inspected and if necessary overhauled by a specialist. Alternatively install an exchange unit.

First check the transaxle lubricant level. If the level is low, check for signs of leakage and rectify as necessary. To isolate transaxle noise in neutral, run the engine, depress the clutch pedal and engage any gear. If the noise ceases with the clutch pedal depressed, it is an indication that the noise is in the transaxle.

Selector Lever

TO REMOVE AND INSTALL

1 Remove the centre console as described in Body (Chapter 17).

2 Remove the retaining clips and disconnect the selector cables from the selector levers.

3 Remove the spring clips and raise the selector cables clear of the assembly.

4 Remove the retaining bolts and manoeuvre the lever assembly from the vehicle.

5 If necessary, mark the installed position of the gear selector in relation to the base plate, remove the retaining bolts and separate the gear selector from the base plate.

Installation is a reversal of the removal procedure with attention to the following points:

1 Ensure that the insulator pad on the bottom of the base plate is in good condition.

2 If removed, install the gear selector to the base plate, ensuring that the marks made during removal are aligned. Tighten the retaining bolts securely.

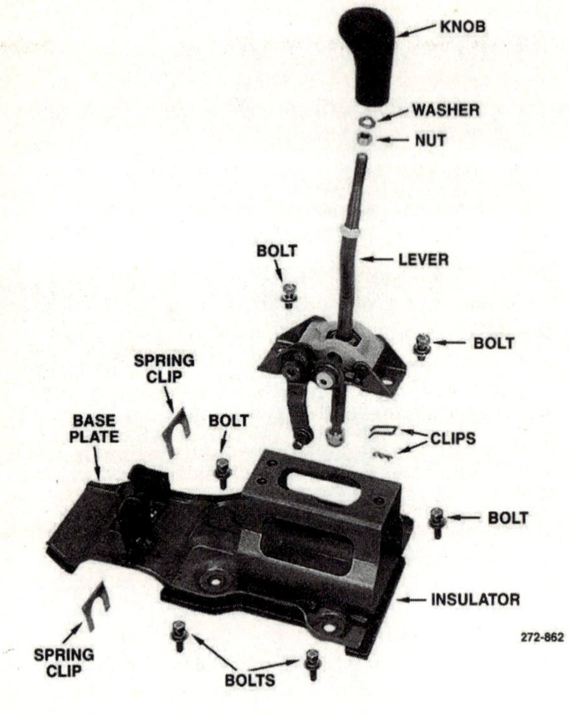

Gear lever components dismantled.

3 Apply multipurpose grease to the plastic bush on the right hand side selector cable.
4 Engage the cables to the levers and install the retaining clips.
5 Place the cables into the brackets and install the spring clips.

Selector Cables

TO REMOVE AND INSTALL

1 Loosen the air cleaner inlet hose clamps, disconnect the PCV valve hose and remove the air inlet hose. Refer to Fuel and Engine Management (chapter 7) if necessary.
2 Remove the clips retaining the selector cables to the transaxle selector levers. Noting their installed position, withdraw the washers and disconnect the cables from selector lever pins.

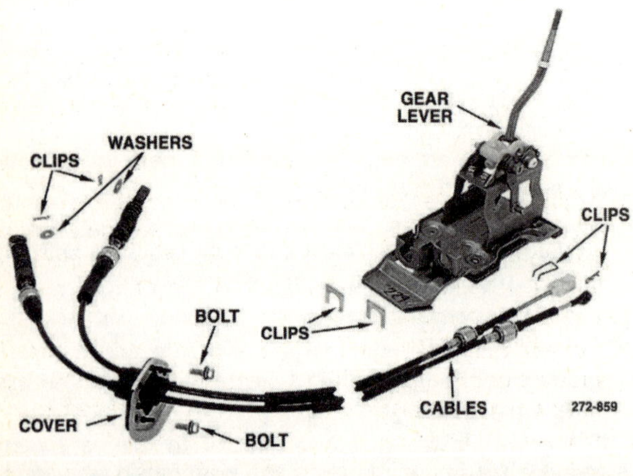

Selector cable components.

3 Using a screwdriver, prise the selector cables from the brackets.

Ensure that the screwdriver will not damage the cable retaining clips.

4 Remove the selector lever as previously described in this chapter.
5 Remove the bolts retaining the driver side footrest and withdraw it from the vehicle.
6 Remove the clip retaining the carpet to the bracket and fold the carpet down.
7 Remove the screws retaining the support bracket and withdraw the bracket.
8 Remove the bolts retaining the selector cable cover panel to the firewall.
9 Carefully manoeuvre the cables from the inside of the vehicle.

TIP *Wrap the ends of the cables in a clean cloth to prevent staining of the seats or carpet.*

Installation is a reversal of the removal procedure with attention to the following points:

1 Working inside the vehicle, guide the cables through the firewall, ensuring that the arrow on the cover panel is pointing upwards. Tighten the retaining bolts securely.
2 Ensure that the selector cables are not interfering with the immobiliser ECU wiring
3 Install the gear selector assembly and connect the selector cables to the levers.
4 Connect the selector cables to the selector arms in the engine compartment, ensuring that the washers are placed in the position noted on removal. Install the retaining clips.
5 Road test the vehicle and check for satisfactory gear selection.

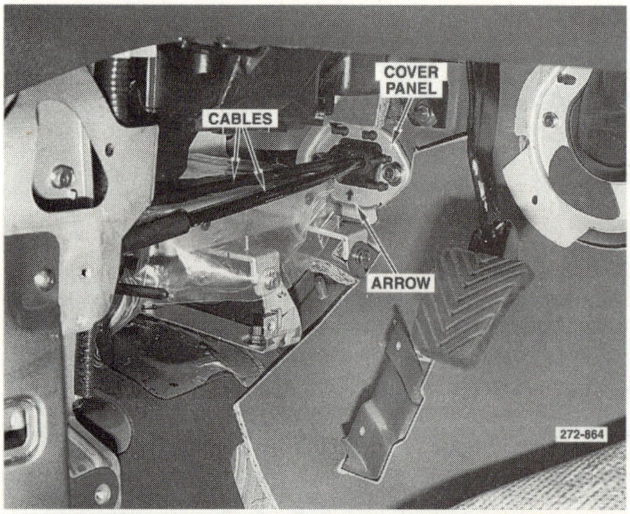

When installing the selector cables, ensure that the arrow on the cover panel is pointing upwards.

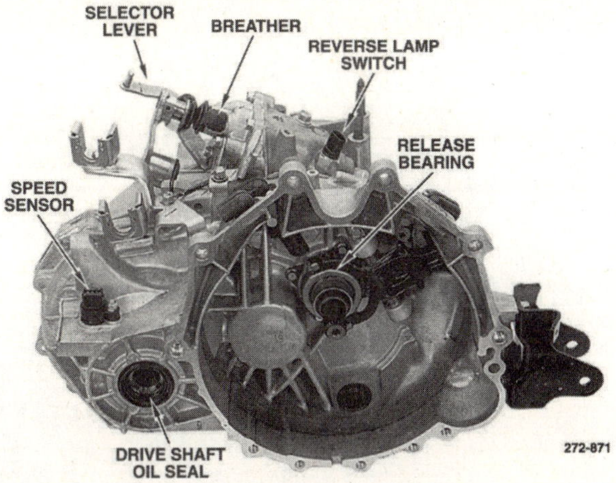

Transaxle removed, showing various components.

Oil Seals

Before renewing the oil seals ensure that the breather at the top of the transaxle housing is not blocked.

TO RENEW DRIVE SHAFT OIL SEAL

1 Raise the front of the vehicle and support on chassis stands under the recommended points. Refer to General Information (chapter 1) for the correct trolley jack and chassis stand positions to prevent damage or injury.

Remove the front wheel on the relevant side of the vehicle.

2 Remove the bolts and clips retaining the splashguards and withdraw the guards from the vehicle.

3 Remove the drain plug and drain the oil from the transaxle.

4 Working under the relevant side of the vehicle, remove the split pin and nut retaining the tie rod end to the steering knuckle. Discard the split pin. Disconnect the tie rods using a puller or alternatively by placing a suitable dolly or hammer against one side of the steering arm eye and striking the opposite side with a hammer.

5 Remove the brake hose retaining clip from the suspension unit.

6 On models with ABS, remove the bolt retaining the wheel sensor wiring to the suspension unit and then remove the sensor retaining bolt. Remove the sensor from the suspension unit and secure it clear of the work area.

7 Remove the stabiliser bar link nuts from the stabiliser bar and noting the installed position, withdraw the stabiliser bar link from the vehicle.

8 Separate the suspension ball joint from the steering knuckle, by placing a suitable dolly or hammer against one side of the steering knuckle eye and striking the opposite side with a hammer. If necessary, refer to Front Axle and Suspension (chapter 13).

TIP *Loosely install the ball joint retaining nut to prevent damage to the thread of the ball joint stud.*

9 To renew the driver side oil seal, proceed as follows:

a Remove the bolts retaining the drive shaft support bearing and separate the support bearing from the cylinder block.

Method of removing the passenger side drive shaft.

b With the aid of an assistant, withdraw the driver side drive shaft assembly from the transaxle using care not to damage the transaxle housing. Place suitable sized timber blocks between the suspension unit and the vehicle body to keep the drive shaft clear of the work area.

10 To renew the passenger side oil seal, proceed as follows:

a While an assistant holds the passenger side suspension unit, insert a lever at the bottom of the passenger side inner drive shaft joint and while supporting the top of the joint with an other bar, lever the inner joint from the transaxle housing. Use a piece of timber as a fulcrum for the lever to prevent damage to the transaxle housing.

Do not allow the assistant to pull the suspension unit as damage to the drive shaft inner joint may result.

b To provide ample working space, hold the suspension unit away from the transaxle, by placing a suitably sized timber block between the suspension unit and the inner mudguard. Secure the drive shaft clear of the work area being careful not to strain the drive shaft joints or damage the boots. Discard the snap ring from the end of the inner joint.

If difficulty is experienced removing the passenger side oil seal, the transaxle or the front suspension crossmember may have to be removed. Refer to the relevant chapter for the removal procedure of components.

11 Noting its installed depth, carefully prise the oil seal from the transaxle housing. Ensure that the transaxle housing is not damaged while removing the oil seal. Use a timber block, while levering the seal. Ensure that no dirt or foreign matter enters the transaxle.

12 Inspect the seal running surface of the drive shaft for grooves or damage. Polish out very fine imperfections using fine wet or dry paper. If badly grooved or damaged install a new drive shaft.

13 Using a lint free cloth, clean the seal housing. Check the housing for damage.

Installing a drive shaft oil seal using a large socket as a mandrel.

14 Drive the new oil seal, with the lip facing the transaxle, into the housing using a tube or mandrel that pushes on the outer circumference of the oil seal, to the depth noted during removal.

> *Ensure that the tube or mandrel does not damage the transaxle housing.*

15 Apply a small amount of multipurpose grease to the running surface of the oil seal and the drive shaft.

16 Install a new snap ring to the passenger side drive shaft inner joint, remove the timber block supporting the suspension unit and carefully guide the joint into the housing until the snap ring securely retains the shaft.

17 On the driver side, remove the timber block and guide the drive shaft into the transaxle housing, being careful not to damage the oil seal or the housing. Install the drive shaft support bearing to the cylinder block and tighten the retaining bolts to the specified torque.

18 Install the tie rod end and suspension ball joint retaining nuts and tighten to the specified torque. Install a new split pin on the tie rod end castellated nut.

19 Ensure that the brake hose is securely clipped to the suspension unit. Check carefully that the hose will not catch on anything.

20 Refill the transaxle as described earlier in this chapter.

TO RENEW INPUT SHAFT OIL SEAL

A leaking input shaft oil seal is indicated by transaxle oil dripping from the bottom of the clutch housing.

1 Remove the transaxle as described later in the chapter.

2 Remove the release bearing from the clutch housing, referring to Clutch (chapter 9).

3 Remove the bolts retaining the clutch bearing slider and withdraw the slider from the transaxle.

4 Using a screwdriver, carefully prise the oil seal out of the transaxle noting its installed depth. It may be necessary to drill a small hole in the seal, insert a self tapping screw and using multigrips on the screw head withdraw the seal. With both methods use care to avoid damaging the bearing, input shaft or transaxle housing.

5 Inspect the seal running surface of the input shaft for grooves or damage. Polish out very fine imperfections using fine wet or dry paper.

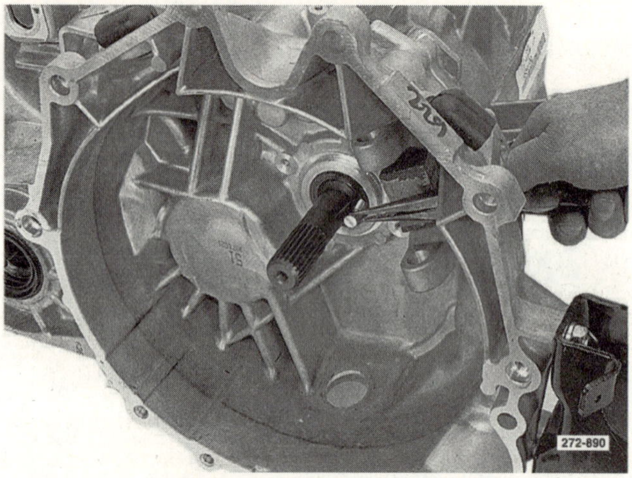

Removing the input shaft oil seal using a self tapping screw.

6 Using a lint free cloth, clean the seal housing. Check the housing for damage.

7 Ensuring that the oil seal is inserted squarely in the transaxle housing, tap the seal in to the housing using a suitable drift to the depth noted during removal.

8 Lubricate the lip of the oil seal with a multipurpose grease.

9 Install the clutch bearing slider and tighten the retaining bolts to the specified torque.

10 Install the release bearing, referring to Clutch (chapter 9).

11 Install the transaxle as described later in this chapter.

TO RENEW SELECTOR SHAFT OIL SEAL

Due to the lack of space around the selector shaft oil seal, it is recommended to remove the transaxle, as described later in this chapter.

1 Place suitable protection around the engine compartment to prevent damage to the vehicle body.

2 Working in the engine compartment, remove the air cleaner assembly as described in Fuel and Engine Management (chapter 7).

3 Working in the engine compartment, remove the clips retaining the selector cables to the selector levers. Noting their installed position, withdraw the washers, and remove the cables from selector lever pins.

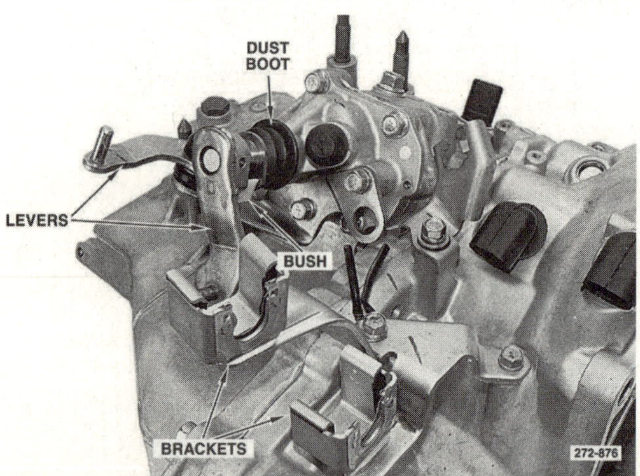

Selector assembly installed.

Removing the selector shaft oil seal.

4 Using a screwdriver, prise the selector cables from the brackets.

> *Ensure that the screwdriver will not damage the cable retaining clips.*

5 While an assistant supports the left hand side of the selector shaft with a socket or tube and a hammer, carefully tap the roll pin out of the shaft. Withdraw the selector arm from the shaft.

> *Removing the roll pin without supporting the shaft could damage the shaft and/or the housing.*

6 Slide the selector lever from the selector shaft, ensuring that the selector bush is not misplaced.

7 Ease the dust boot from the shaft. Check the condition of the boot and renew if necessary.

8 Using a chisel or similar, carefully remove the oil seal from the housing.

Installing the selector shaft oil seal.

9 Inspect the seal running surface of the selector shaft for grooves or damage. Polish out very fine imperfections using fine wet or dry paper.

10 Using a lint free cloth, clean the seal housing. Check the housing for damage.

11 Lubricate the lip of the oil seal with a multipurpose grease.

12 Insert the oil seal squarely in the housing, ensuring that the shoulder of the seal is seated against the housing.

13 Ensure that the rubber boot and selector shaft are installed as noted on dismantling, and that the selector bush is seated on the selector arm pin.

14 Check the fluid level and refill with the specified transaxle fluid, as previously described.

TO RENEW SPEED SENSOR O RING

1 Place suitable protection around the engine compartment to prevent damage to the vehicle body.

2 Loosen the air cleaner inlet hose clamps, disconnect the PCV valve hose and remove the air inlet hose. Refer to Fuel and Engine Management (chapter 7) if necessary.

3 Disconnect the speed sensor wiring connector.

> TIP *If difficulty experienced disconnecting the wiring connector, remove the speed sensor retaining bolt, withdraw the sensor and then disconnect the wiring connector.*

4 Clean around the speed sensor to prevent dirt entering the transaxle.

5 Remove the retaining bolt and withdraw the sensor from the transaxle.

6 Using a flat bladed screwdriver, carefully remove and discard the O ring from the outside of the sensor housing. Sparingly lubricate the O ring, using transaxle fluid and install the new O ring to the sensor.

7 Install the sensor, tighten the retaining bolt securely and connect the wiring.

8 Check the fluid level and refill with the specified transaxle fluid, as previously described.

Location of the speed sensor O ring.

Transaxle Assembly

The pull type clutch mechanism on this vehicle, makes it necessary to clip the clutch release lever bearing to the pressure plate. Trying to remove the transaxle without releasing the clip will result in damaged components. Follow the procedure for releasing the clip carefully.

TO REMOVE AND INSTALL

1 Raise the front of the vehicle with a jack and support the vehicle with chassis stands under the recommended points. Refer to General Information (chapter 1) for the correct trolley jack and chassis stand positions to prevent damage or injury.

2 Remove the air cleaner assembly as described in Fuel and Engine Management (chapter 7).

3 Disconnect the vehicle speed sensor and the reverse lamp wiring connectors.

4 Remove the clips retaining the selector cables to the transaxle selector levers. Withdraw the washers, noting their installed position, remove the cables from selector arm pins and place the cables clear of the work area.

5 Using a screwdriver, prise the selector cables from the brackets.

Ensure that the screwdriver will not damage the cable retaining clips.

6 Remove the spring clip retaining the clutch hydraulic hose to the bracket on the transaxle housing.

7 Remove the bolts retaining the clutch sleeve cylinder and the damper to the transaxle housing. Place the cylinder with the hydraulic hose still connected, clear of the work area.

Do not depress the clutch pedal while the slave cylinder is removed or the piston will eject from the slave cylinder.

8 Remove the bolts and clips retaining the front and the passenger side splashguards and withdraw the guards from the vehicle.

9 Working under the vehicle, remove the drain plug and drain the transaxle oil into a suitable container. Replace the drain plug and tighten to the specified torque.

10 Remove the nuts retaining the engine pipes to the exhaust manifolds, lower the pipes and support them using a piece of wire tied to the body of the vehicle. Discard the gaskets.

Do not allow the pipe to hang unsupported, as damage to the flex joint may occur.

11 Disconnect the wiring from the starter motor, remove the retaining bolts and withdraw the starter motor from the vehicle. Refer to Engine Electrical (chapter 6), if necessary.

12 Remove the stabiliser bar link nuts from the stabiliser bar and control arm. Noting the installed position, withdraw the stabiliser bar links from the vehicle.

13 Loosen the ball joint nut and separate the lower control arm from the suspension unit on both sides of the vehicle. If necessary, hold a hammer or suitable dolly against one side of the steering knuckle eye and strike the opposite side with a hammer.

TIP *Loosely install the ball joint retaining nut to prevent damage to the thread of the ball joint stud.*

14 Remove and discard the split pin and remove the castellated nut retaining the tie rod end to the steering knuckle.

15 Disconnect the tie rod end from the steering knuckle using a puller or alternatively by holding a hammer or suitable dolly against one side of the steering knuckle eye and striking the opposite side with a hammer.

16 Remove the brake hose retaining clip from the suspension unit and place the hose clear of the suspension unit.

17 On models with ABS, remove the bolt retaining the wheel sensor wiring to the suspension unit and then remove the sensor retaining bolt. Remove the sensor from the suspension unit and secure it clear of the work area.

18 Repeat steps 14 to 17 on the opposite side of the vehicle.

19 Remove the bolts retaining the driver side drive shaft support bearing and separate the support bearing from the cylinder block.

20 With the aid of an assistant, withdraw the driver side drive shaft assembly from the transaxle using care not to damage the drive shaft oil seal or the transaxle housing. Place suitable sized timber blocks between the suspension unit and the vehicle body to keep the drive shafts clear of the work area.

21 While an assistant holds the passenger side suspension unit, insert a lever at the bottom of the passenger side inner drive shaft joint and while supporting the top of the joint with another bar, lever the inner joint from the transaxle housing. Use a piece of timber as a fulcrum for the lever to prevent damage to the transaxle housing.

Do not allow the assistant to pull the suspension unit, as damage to the drive shaft inner joint may result.

22 To provide ample working space, hold the suspension unit away from the transaxle, by placing a suitably sized timber block between the suspension unit and the inner mudguard. Secure the drive shaft clear of the work area being careful not to strain the drive shaft joints or damage the boots. Discard the snap ring from the end of the inner joint.

Method of removing the passenger side drive shaft.

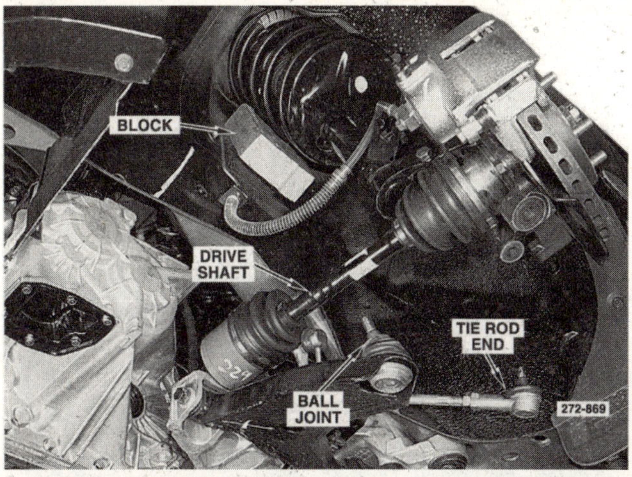

Securing the suspension unit away from the vehicle body using a timber block.

23 Remove the through bolts from the front and rear transmission mountings. Release the clip retaining the A/C pipe to the centre member. Remove the retaining bolts and withdraw the centre member from the vehicle.

24 Remove the bolts retaining the rear transaxle mounting bracket to the transaxle and remove the bracket from the vehicle.

25 Noting their installed position, remove the transaxle retaining bolts from the bottom of the transaxle.

26 With suitable protection on the jack head, support the engine under the sump.

Placing a piece of timber between the sump and jack will evenly distribute the weight across the sump.

27 Loosen the left hand side engine mounting nuts and remove the nut from the through bolt. Noting the installed direction of the through bolt, withdraw the bolt. Carefully lower the floor jack, until sufficient clearance is obtained to remove the mounting.

28 Disengage the clutch release bearing as follows:

a Suitably mark the installed position of the high tension leads in the distributor cap, and using a twisting motion, disconnect the leads from the cap.

b Remove the screws retaining the distributor cap and withdraw the cap from the distributor.

c Remove the inspection hole plugs, located at the bottom of the transaxle, and below the distributor.

d Have an assistant insert a long flat bladed screwdriver between the release bearing and the pressure plate clip, through the bottom inspection hole.

e Insert another screwdriver through the top inspection hole and place the tip of the screwdriver between the release bearing and the pressure plate clip.

f With both screwdrivers in position, gently pull the release lever towards the passenger side of the opening. Simultaneously turn both screwdrivers 90° and slowly release the lever.

g The release lever should travel all the way to the driver side of the opening, indicating the clip and the release bearing are separated. If the lever will not move to the driver side, repeat the procedure.

Removing the transaxle without disengaging the release bearing will cause damage to the pressure plate and/or release bearing.

29 Support the transaxle using a floor jack. Secure the transaxle to the jack using rope or chain.

Ensure that the weight of the transaxle does not hang on the clutch driven plate when the transaxle is being withdrawn as damage to the clutch assembly could result.

It is good practice to check the serviceability of the clutch components while the transaxle is removed. Refer to Clutch (chapter 9).

30 Remove the bolts retaining the transaxle clutch housing to the engine.

31 Lower the jack that is supporting the engine sufficiently to allow the transaxle to be withdrawn from the vehicle.

32 Ease the transaxle from the engine and lower the transaxle to the floor.

Installation is a reversal of the removal procedure with attention to the following points:

1 If the clutch plate has been removed, ensure that the clutch plate splines are aligned with the centre of the adaptor plate, as described in Clutch (chapter 9).

2 Sparingly apply high melting point grease to the input shaft splines.

3 Align the transaxle with the engine ensuring that clutch housing mating face is parallel with the rear of the engine.

4 Ensure that the weight of the transaxle does not hang on the clutch driven plate during installation.

5 Tighten all bolts to the specified torque.

6 Install a new snap ring to the left hand drive shaft inner joints.

7 Ensure that all wiring is securely connected.

8 Push the clutch bearing release lever towards the passenger side of the vehicle, to engage the release bearing to the pressure plate clip. Install the distributor cap and the high tension leads according to the marks made during removal.

9 Install the clutch slave cylinder and the damper. If the piston has been removed or the hydraulic pipe disconnected, bleed the hydraulic system as described in Clutch (chapter 9).

10 Fill the transaxle with the specified type and quantity of lubricant.

Method of disengaging the clutch release bearing. Pull the release lever towards the passenger side of the vehicle.

Chapter 11
Automatic Transaxle

Contents

Specifications

Models	
3.0 litre engines	F4A51 4-speed
3.5 litre engines	F4A51 4-speed or F5A51 5-speed
Type	4 or 5 speed with lockup torque converter
Lubricant:	
Type	
TE/TF/TH and KE/KF/KH models	Mitsubishi ELC4-SP2
TJ/TL/TW and KJ/KL/KW models	Mitsubishi ELC4-SP3
Capacity —	
Dry refill	8.5 litres
[1] Service refill	5.5 litres
Gear ratios (F4A51 only):	
First	2.842:1
Second	1.495:1
Third	1.000:1
Fourth	0.731:1
Reverse	2.720:1
Final drive —	
3.0 litre	3.735:1
3.5 litre	3.497:1
Overdrive solenoid resistance	11 - 15 ohms

TORQUE WRENCH SETTINGS

Ball joint to steering knuckle nut	60 - 72 Nm
Centre member bolts:	
Front	88 Nm
Rear	52 Nm
[2] Differential ring gear bolts	135 Nm
Drain plug	32 Nm
Drive shaft support bearing bolts	45 Nm
Drive shaft hub nut	206 Nm
Driveplate to converter bolts	33 - 38 Nm
Driveplate to crankshaft bolts	74 Nm
Engine mountings:	
Left hand side —	
Mounting nuts	69 - 93 Nm
[3] Through bolt	69 Nm
Front and rear —	
Bracket to transaxle bolts	69 Nm
[3] Through bolts	69 Nm
Retaining bolts	44 Nm
Inhibitor switch bolts	11 Nm
Oil filter	12 Nm
Selector lever assembly bolts	12 Nm
Selector shaft nut	22 Nm
Side cover bolts	11 Nm
Speed sensor bolt	5 Nm
Stabiliser bar link nuts	35 - 44 Nm
Starter motor bolts	27 - 34 Nm
Tie rod end castellated nut	10 Nm
Transaxle housing to engine bolts	74 Nm
Valve body bolts	11 Nm

[1] When servicing the transaxle, a further 5.5 litres of fluid is required for flushing.
[2] Apply Permatex Lock Bolt on threads.
[3] Tighten with the weight of the transaxle on the mounting.

Description

The vehicles covered in this manual are equipped with a new design, computer controlled 4 or 5 speed automatic transaxle. The computer system constantly analyses the driving style of the driver and adjust its operation accordingly. The "smart logic" system, using five sensors, takes into consideration throttle position, vehicle speed and brake usage to calculate the most powerful and, at the same time, economical gear shift pattern.

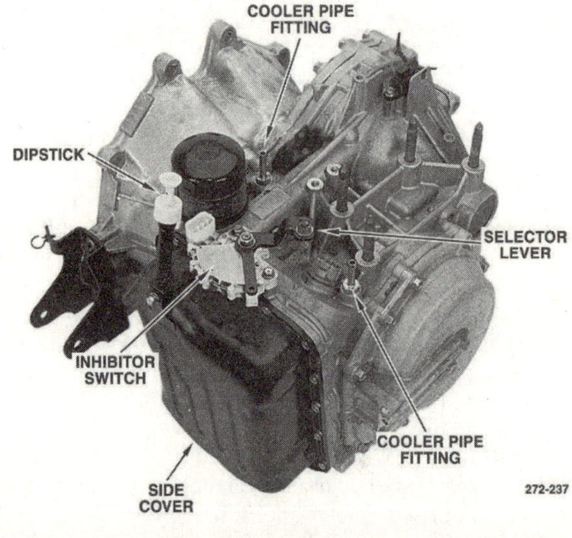

Front view of the transaxle showing various components.

Sports models are fitted with sequential type gear selector system. In sports mode, the driver has complete control of gear selection, in a similar manner to a manual transaxle vehicle.

The hydraulic system consists of a single front oil pump and a set of electronically controlled solenoid valves which direct fluid pressure to the various clutches and brake bands. The lockup torque converter provides direct drive when the vehicle speed is above approximately 70 km/h and the transaxle is in top gear.

Fourth gear (overdrive) is controlled by a solenoid located inside the transaxle and the coolant temperature sensor via the engine ECU. To reduce the load on the engine and the transaxle, the ECU prevents fourth gear being selected until the engine coolant reaches normal operating temperature.

The hydraulic fluid passes through an oil cooler which is incorporated in the bottom tank of the radiator. The fluid is circulated through a disposable, spin-on type oil filter.

The transmission and the differential are incorporated in one housing (the transaxle) and one fluid level check covers both components. The transaxle fluid level is checked using a dipstick, located near the battery.

The selector lever is floor mounted and connected to the transaxle by an adjustable cable. When the selector lever is in Park, the transaxle is locked by a pawl, preventing vehicle movement. Serious damage may occur to the transaxle if Park is selected while the vehicle is moving.

An inhibitor switch is installed to the transaxle to ensure that the engine can only be started in Park or Neutral. The reverse lamp switch is also incorporated in the inhibitor switch.

The Neutral indicator lamp in the instrument cluster operates as a transaxle malfunction indicator lamp, while the transaxle is in the Drive position. If the transaxle fluid overheats, the lamp will flash rapidly, approximately twice every second. A single flash every second, or an intermittent flash indicates an internal fault.

If the transaxle fluid is overheating, proceed as follows:

- **a** Park the vehicle on a safe place, move the selector to Park and leave the engine idling. After a few minutes, move the selector to Drive and check that the lamp has stopped flashing.
- **b** Check the level of the automatic transaxle fluid as described in Routine Maintenance.

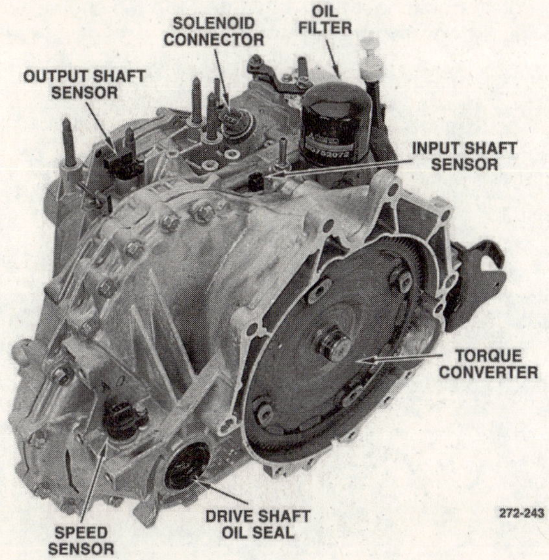

Rear view of the transaxle showing various components.

- **c** If towing a trailer or caravan, ensure that the total towed weight does not exceed the manufacturers specification. Refer to the Operators Manual.

If the lamp is flashing any other code, other than overheating, the cause of the problem should be investigated and rectified by an authorised dealer or an automatic transaxle specialist.

When tuning or testing the engine, ensure that the handbrake is firmly applied and the selector lever is in Park.

It is not possible to push start the vehicle. To avoid damage through lack of lubrication of internal parts, the vehicle should only be towed in Neutral for a distance not more than 25 km and at a maximum speed of 50 km/h. If the vehicle is to be towed more than 25 km, or if the transaxle is damaged, the vehicle should be towed with the front wheels raised from the ground.

Extensive knowledge and specialised equipment is required to test and overhaul the automatic transaxle making it not economically viable for the home mechanic to do so. Therefore, should a fault occur that cannot be rectified using the procedures described in this chapter, it is recommended that the vehicle be taken to an authorised dealer or an automatic transaxle specialist.

If the transaxle must be removed for overhaul by a specialist or replacement with an exchange unit, the removal and installation procedure is covered in this chapter.

Routine Maintenance

TRANSAXLE FLUID

Only the recommended automatic transaxle fluid should be used when topping up or changing the fluid in the system. The fluid level should be checked at the intervals recommended in Routine Maintenance (chapter 2). It is important that the transaxle fluid is maintained at the correct level using only the specified fluid or erratic transaxle operation may occur.

When working on the automatic transaxle, cleanliness is essential. Do not reuse transaxle fluid and do not allow foreign matter to enter the dipstick tube.

To Check and Top Up

It is difficult to obtain an accurate fluid level check when the vehicle has been operating at sustained high speed, in city traffic, during hot weather, or as a towing vehicle. In any of these

Location of the automatic transaxle dipstick.

Checking the automatic transaxle fluid level. Inset shows dipstick markings.

circumstances, allow the transaxle to cool for approximately 30 minutes before checking the fluid level.

The fluid level should be checked when the transaxle is at operating temperature.

1 Drive the vehicle for a minimum of 16 kilometres to bring the transaxle to operating temperature.

2 Park the vehicle on a level surface and firmly apply the handbrake. With the engine idling, move the selector lever through all the gear positions, momentarily stopping at each position, and select Neutral. Allow the engine to idle in Neutral during the checking procedure.

> *An inaccurate fluid level may be obtained if the transaxle is checked in Park or any other gear other than Neutral.*

3 Open the bonnet and clean around the top of the dipstick tube to ensure that no dirt or foreign matter enters the tube when the dipstick is withdrawn.

4 Withdraw the dipstick and wipe it clean with a lint free cloth. Replace the dipstick fully into the filler tube.

5 Withdraw the dipstick and check the fluid level. The fluid level should be at the Hot level mark on the dipstick. While checking the fluid level, check the fluid for contamination, discolouration or a burnt smell, indicating a mechanical fault within the transaxle which will require the services of an authorised dealer or an automatic transaxle specialist.

6 If the fluid level is low, add sufficient amount of the specified automatic transaxle fluid through the dipstick tube until the fluid is at the Hot level mark on the dipstick. To avoid a false reading, allow a few minutes for the fluid to drain completely from the filler tube before replacing the dipstick and checking the level.

> *Do not overfill the transaxle as foaming and unsatisfactory operation will result. If it is necessary to add a substantial amount of fluid to the transaxle, the cause of the fluid loss should be immediately investigated and rectified.*

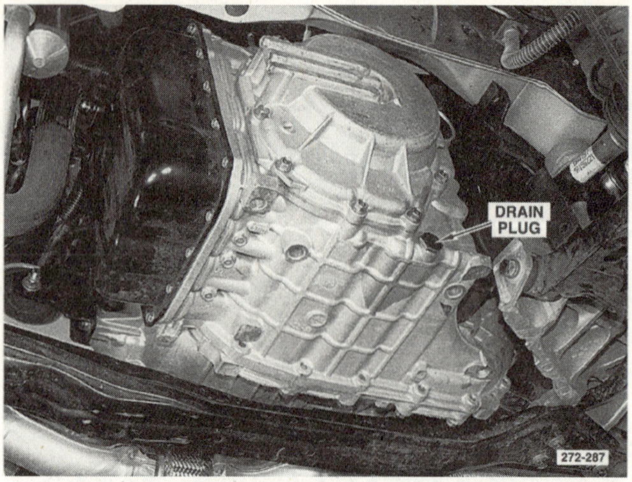

Location of the transaxle drain plug. Left hand splashguard removed.

To Renew Fluid and Filter

The automatic transmission filter is only fitted to TE/TF/TH/TJ and KE/KF/KH/KJ models.

1 Bring the transaxle to operating temperature, raise the front and rear of the vehicle and support on chassis stands, so that the transaxle is reasonably level. Refer to General Information (chapter 1) for the positions of the jacking points.

2 Remove the bolts and scrivets retaining the front and passenger side splashguards and withdraw the guards. Clean the area around the transaxle drain plug with a cloth.

3 Release the clip retaining the right hand side oil cooler hose and disconnect the hose from the radiator cooler pipe. Connect a suitably length of hose over the radiator cooler pipe and place the end of the hose into a container capable holding about 6 litres of hot transaxle fluid.

> *Use care while draining the fluid to avoid scalding as the fluid may be very hot.*

4 Place the transaxle in Neutral, start the engine and leave it to idle for no longer than 1 minute. Stop the engine, when the fluid stops discharging.

Removing the oil filter, using a filter removal tool. TE/TF/TH/TJ and KE/KF/KH/KJ models.

5 Place a container under the drain plug and remove the drain plug and drain the fluid. Check the condition of the drain plug sealing washer and renew as necessary.

6 On TE/TF/TH/TJ and KE/KF/KH/KJ models, renew the filter as follows:

a Remove the oil filter with a filter removal tool, turning it in an anticlockwise direction.

b Smear transaxle fluid onto the sealing gasket of the new filter and tighten the filter by hand three quarters of a turn after it contacts the filter housing.

TIP *Before installing the new filter ensure that the sealing gasket from the old filter has not adhered to the filter sealing seat on the transaxle housing.*

7 When the transaxle has completely drained, install and tighten the drain plug securely.

i *If the transaxle fluid is contaminated with metal or friction material particles, consult an automatic transmission specialist or an authorised dealer. If badly contaminated, the manufacturer recommends to overhaul the transaxle, renew the oil filter and to flush the transaxle and cooler, using clean automatic transaxle fluid.*

8 Add the specified new fluid through the transaxle dipstick tube until the level reaches halfway between the bottom edge of the dipstick and the Hot mark.

9 Repeat from step 3 again to flush the system. The oil filter does not require changing again.

10 When satisfied, that the fluid has drained again and it is clean, install the drain plug with a serviceable sealing washer, and tighten to the specified torque.

11 Add the specified new fluid through the dipstick tube, until the level reaches halfway between the bottom edge of the dipstick and the Hot mark.

12 Start the engine and check the fluid level when the transaxle has reached operating temperature, ensuring that the fluid level is at the Hot level mark.

Fluid Leakage

TIP *Most fluid leaks are more apparent when the vehicle is at operating temperature. Use care to prevent scalding from the hot fluid.*

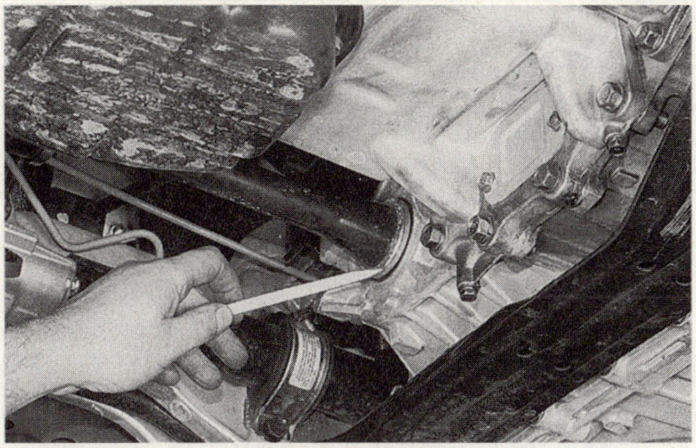

Check the drive shaft oil seals for leakage.

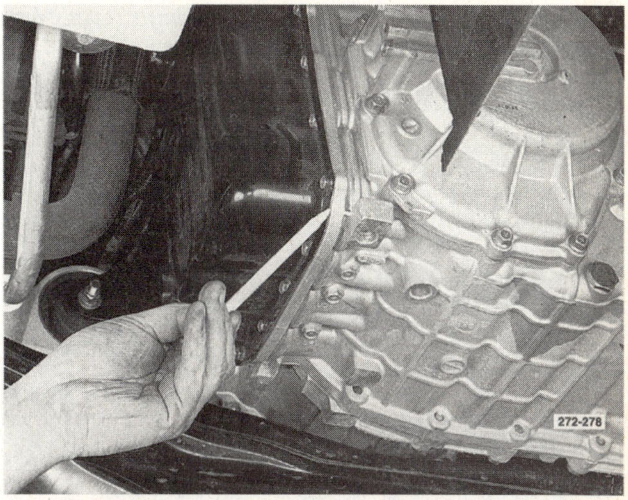

Check between the transaxle housing and the side cover for oil leaks.

1 Raise the front of the vehicle and support on chassis stands. Refer to General Information (chapter 1) for the positions of the jacking points. Remove the bolts and scrivets retaining the front and the passenger side splashguards and withdraw the guards.

2 Check the oil filter and the cooler pipe fitting washers for fluid seepage and leaks.

3 Check side cover and the oil seals for leaks.

4 Check the oil cooler hose connections at the lower radiator tank.

5 If the source of the leak cannot be easily identified, it may be necessary to clean the transaxle housing with suitable degreasing fluid and repeat the inspection a short time later.

SELECTOR CABLE

1 Check the selector cable for kinks, stiffness or damage.

2 Check the adjustment of the selector cable by confirming that the vehicle will only start in Park and Neutral.

3 Apply the foot and handbrake and attempt to start the vehicle in Reverse or in Drive. If the vehicle starts in any gear, check the adjustment of the selector cable and inhibitor switch.

4 Check the operation of the reverse lamps. Apply the handbrake, turn the ignition On and place the transaxle in

Check the oil cooler hose connections for fluid leaks.

Reverse. Ensure that the reverse lamps only operate when the vehicle is in Reverse.

5 Ensure that the gear selector can only be moved out of Park after pressing the button on the side of the selector handle.

DRIVE SHAFTS

Routine maintenance items relating to the drive shafts are described in Front Axle and Suspension (chapter 13).

Trouble Shooting

NO DRIVE IN D RANGE

1 Incorrect transaxle fluid level: Drain or top up the transaxle fluid to the correct level.
2 Incorrect selector cable adjustment: Check and adjust the selector cable.
3 Incorrect or contaminated transaxle fluid: Renew the fluid and filter using only the recommended type of fluid, closely following the procedure in the previous heading.
4 Worn or damaged transaxle internal components: Have the transaxle assembly inspected and if necessary, overhauled by a specialist. Alternatively install an exchange unit.
5 Broken drive shaft: Inspect the drive shafts and renew as necessary.

NO REVERSE IN R RANGE

1 Low fluid level in the transaxle: Check the fluid level in the transaxle and top up if necessary.
2 Incorrect selector cable adjustment: Check and adjust the selector cable.
3 Worn or damaged transaxle internal components: Have the transaxle assembly inspected and if necessary, overhauled by a specialist. Alternatively install an exchange unit.

NO HOLD IN PARK POSITION

1 Incorrectly adjusted selector cable: Check and adjust the selector cable.
2 Worn, loose or bent selector lever, selector cable fittings, bushes etc: Check the lever, cable and bushes. Repair or renew as necessary.
3 Selector lever loose on the selector shaft: Repair or renew the selector lever.

SLIPPING OR ROUGH IN UPSHIFT

1 Low fluid level in the transaxle: Check the fluid level and top up if necessary.
2 Incorrect selector cable adjustment: Check and adjust the selector cable.
3 Faulty output shaft speed sensor: Inspect the wiring connector and if necessary, renew the speed sensor.
4 Worn or damaged transaxle internal components: Have the transaxle assembly inspected and if necessary, overhauled by a specialist. Alternatively install an exchange unit.

ENGINE WILL NOT START IN P OR N POSITION OR WILL START IN ANY RANGE

1 Inhibitor switch faulty or incorrectly adjusted: Check and adjust or renew the inhibitor switch as necessary.
2 Incorrectly adjusted selector cable: Check and adjust the selector cable.

Selector Cable

TO REMOVE AND INSTALL

1 Working in the engine compartment, remove the air cleaner assembly as described in Fuel and Engine Management (chapter 7).
2 Remove the battery, referring to Engine electrical (chapter 6) if necessary.
3 Release the clips retaining the wiring harness to the rear of the battery shroud and remove the shroud from the vehicle.
4 Remove the clip retaining the inner selector cable to the selector lever and lower the inner cable from the pin. Remove the spring clip retaining the outer selector cable to the transaxle.
5 Working inside the vehicle, remove the screw from the centre console side trim panel. Pull the bottom of the panel away from the carpet to disengage the retaining clips and then manoeuvre it downward to separate it from the console. Repeat this on the other side trim panel.
6 Remove the centre console from the vehicle. Refer to Body (chapter 17), if necessary.

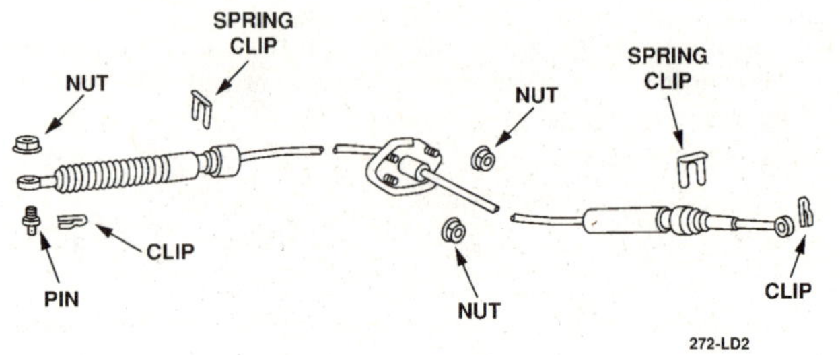

Selector cable components.

7 Release the Velcro bands, retaining the carpet and fold the carpet down, away from the work area.
8 Disconnect the wiring connectors from the engine ECU.
9 Remove the bolts retaining the ECU and withdraw it from the vehicle.
10 Remove the clip and disconnect the inner selector cable from the selector lever.
11 Remove the spring clip retaining the outer selector cable to the selector lever assembly.
12 Remove the nuts retaining the outer cover and rubber seal to the firewall.
13 Working in the engine compartment, withdraw the selector cable from the vehicle.

Installation is a reversal of the removal procedure with attention to the following points:

1 Guide the new cable through the aperture on the firewall, ensuring that the cable is correctly seated, without interfering with other components.
2 Ensure that the rubber seal between the selector cable and the firewall is not damaged and correctly seated.
3 Ensure that the selector cable retaining clips are fully seated in their grooves.
4 If necessary, adjust the selector cable using the procedure that follows.

TO ADJUST

1 Working in the engine compartment, remove the air cleaner assembly as described in Fuel and Engine Management (chapter 7).

2 Firmly apply the handbrake and move the gear selector lever inside the vehicle to Neutral. In this position it should be possible to insert a 6 mm drill bit through the shift lever and into the boss on the inhibitor switch. If the drill bit will not enter the inhibitor switch, proceed as follows:

a Remove the selector cable locknut and disconnect the inner cable.

b Move the transaxle selector lever clockwise 2 clicks to the Park position and then bring it back two clicks to Neutral. The drill bit should slide into the selector lever and the boss on the inhibitor switch. If necessary, loosen the inhibitor switch bolts and rotate the switch until the drill bit is a neat sliding fit. Tighten the switch retaining bolts without moving the switch.

Checking the inhibitor switch adjustment in Neutral using a 6mm drill shank.

3 Confirm that the selector lever inside the vehicle is still in the Neutral and have an assistant inside the vehicle, lightly push the selector lever towards Reverse without depressing the selector lever button.

4 Connect the cable and tighten the selector cable locknut securely. Be careful not to disturb the transaxle selector lever.

5 Ensure that the selector lever operation is smooth and the quadrant indicator corresponds with the correct range as it is selected.

6 Check that the engine will start only when the selector lever is in Park and Neutral. If necessary, check the inhibitor switch as described later in this chapter.

7 Check the operation of the reverse lamps. If necessary, check the inhibitor switch as described later in this chapter.

Selector Lever — Standard

TO REMOVE AND INSTALL

1 Remove the centre console referring to Body (chapter 17), if necessary.

2 Disconnect the wiring connector. Release the clip retaining the wiring and the connector to the selector lever mounting bracket.

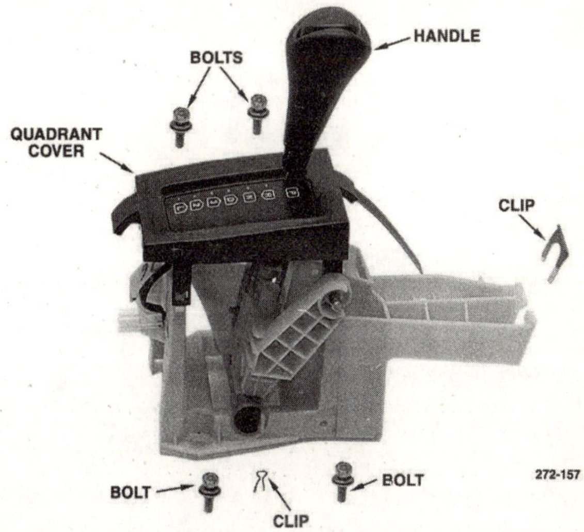

Selector lever assembly removed, except Sports models.

3 Remove the clip retaining the inner selector cable to the selector lever pivot shaft.

4 Remove the spring clip retaining the outer selector cable to the selector lever assembly.

5 Remove the bolts retaining the selector lever assembly to the floor panel and manoeuvre the assembly from the vehicle.

6 If necessary, remove the quadrant cover, as follows:

a Carefully prise out the button on the top of the handle and withdraw the button and the spring. Noting the installed position of the spring seat, withdraw it from the handle.

b Using a suitable sized deep socket, remove the retaining nut and withdraw the handle from the selector.

c Remove the bolts retaining the quadrant cover to the selector assembly and raise the cover.

d Rotate the illumination bulb holder anticlockwise and withdraw the holder from the quadrant cover. Withdraw the quadrant cover from the lever assembly.

Installation is a reversal of the removal procedure with attention to the following points:

1 If removed, install the quadrant cover and tighten the retaining bolts securely. Install the bulb holder.

2 If removed, install the selector handle and tighten the retaining nut securely. Install the spring seat in the position noted during removal and press the button into the handle, until the clips engage the handle.

3 Install the selector lever assembly and tighten the retaining bolts to the specified torque.

4 Connect the inner and outer selector cable to the assembly, ensuring that the retainers are securely engaging their grooves.

5 Install the centre console, referring to Body (chapter 17), if necessary.

6 Check the operation of the selector lever. If necessary, adjust the selector cable as previously described.

TO DISMANTLE

As the only parts available at the time of publication are the selector handle and the quadrant cover, the selector should not be dismantled further.

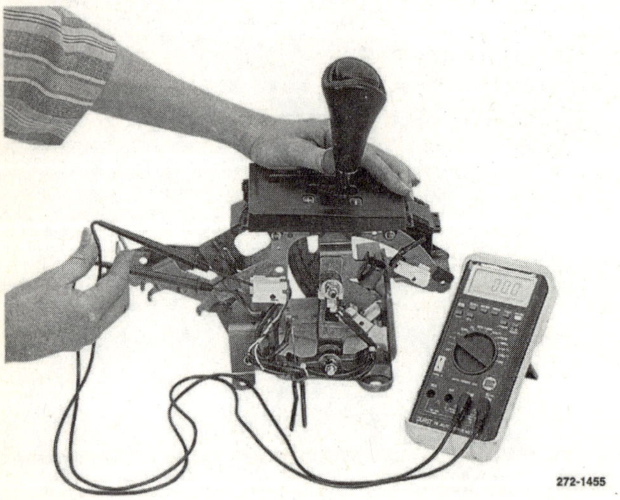

Testing the selector switches on a Sports model using an ohmmeter.

Selector Lever — Sports

TO TEST SELECTOR SWITCHES

1 Remove the centre console referring to Body (chapter 17) if necessary.
2 Disconnect the selector switch wiring connector. Release the clip retaining the wiring and the connector to the selector lever mounting bracket.
3 Trace the wiring from the appropriate switch to the wiring connector and insert a ohmmeter, ensuring that the probes of the meter are making good contact with the spade terminals in the connector.
4 Move the selector lever to activate the switch.
5 The meter should show continuity, while the switch is activated. There should be no continuity once the switch is released. If necessary, remove the selector assembly and renew the switch.
6 If the operation of the switch is satisfactory and a fault is still present, refer the vehicle to an authorised dealer or an automatic transmission specialist.

TO REMOVE AND INSTALL

The removal and installation procedure of a sport selector lever is similar to the standard model selector lever, with attention to the following points:

1 Disconnect the selector switch wiring connector.
2 Using long nose pliers, remove the clip retaining the inner selector cable to the selector lever pivot shaft.
3 Remove the spring clip retaining the outer selector cable to the selector lever assembly.
4 Remove the bolts retaining the selector lever assembly to the floor panel and manoeuvre the assembly from the vehicle.
5 If necessary, remove the selector switch from the selector lever as follows:

a Trace the wiring of the switch to the wiring connector and suitably mark the installed location of the wires in the connector.

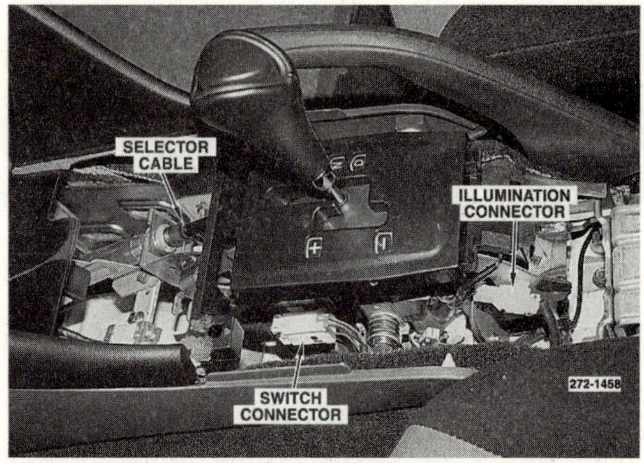

Sports selector lever with the centre console removed.

As there are no factory made markings on the wiring connector, ensure that the markings are easily recognisable to assist in the correct replacement of the wires.

b Insert a suitable thin probe through the front of the connector, depress the tab retaining the spade of the appropriate wire in the connector and pull the wire from the rear of the connector.
c Remove the screws securing the switch retaining bracket and withdraw the bracket as an assembly.
d Remove the retaining screw and separate the switch from the bracket.

Installation is a reversal of the removal procedure with attention to the following points:

1 If removed, install the switch retaining bracket to the selector lever assembly, tightening the retaining screws securely.
2 Ensure that the wires are located in the wiring connector according to the marks made on removal.
3 Using a suitable probe push the wire into the back of the connector ensuring that the tab in the wiring connector is securing the wire in position.
4 Ensure that the wiring connections are secure and are making good contact.
5 Install the selector lever as described under the previous heading.

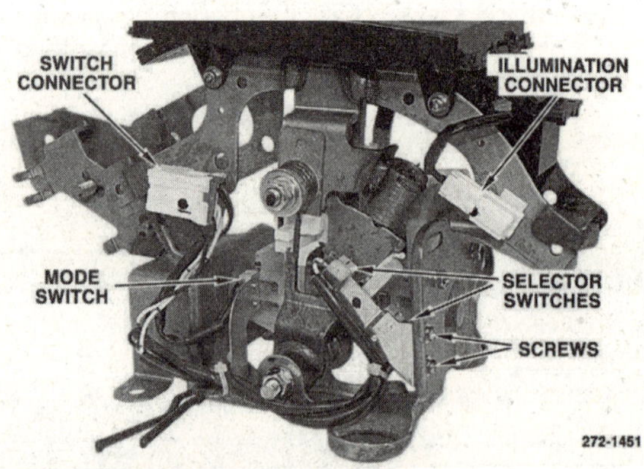

Location of the various switches and wiring connectors on a Sports model selector assembly.

Inhibitor Switch

Prior to checking and adjusting the inhibitor switch, ensure that the selector cable is correctly adjusted as previously described.

TO CHECK AND ADJUST

1 Working in the engine compartment, remove the air cleaner assembly as described in Fuel and Engine Management (chapter 7).

2 Ensure that the selector cable is correctly adjusted, as previously described.

3 Disconnect the wiring connector from the inhibitor switch.

4 Connect an ohmmeter between the inhibitor switch terminals 9 and 10 and check that continuity exists only when the selector lever is in Neutral or Park.

5 If necessary, select Neutral, loosen the retaining bolts and move the inhibitor switch until a 6 mm drill bit can be inserted into the selector lever hole and into the inhibitor switch boss. If the adjustment does not rectify the fault, renew the inhibitor switch.

6 Move the selector lever to Reverse and check that continuity exists between the inhibitor switch terminals 7 and 8. Check that continuity fails when the selector lever is not in Reverse. If a fault is indicated, renew the inhibitor switch.

TO REMOVE AND INSTALL

1 Working in the engine compartment, remove the air cleaner assembly as described in Fuel and Engine Management (chapter 7).

2 Remove the spring clip and washer retaining the inner selector cable to the transaxle selector lever and lower the inner cable.

3 Remove the nut and washer from the selector shaft and withdraw the selector lever from the transaxle.

4 Mark the installed position of the inhibitor switch to aid assembly and remove the retaining bolts. Withdraw the inhibitor switch from the transaxle.

Installation is a reversal of the removal procedure with attention to the following points:

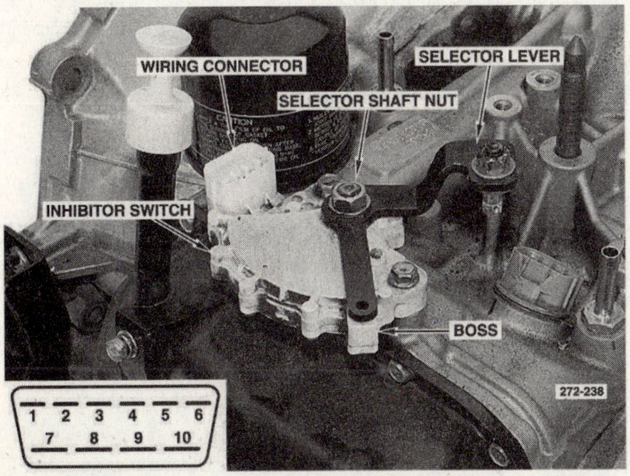

Location of the inhibitor switch. Inset shows terminal identification.

1 Install the inhibitor switch and while holding the switch in the position noted on removal, tighten the retaining bolts to the specified torque.

2 Install the selector lever, the washer and the nut. Tighten the selector shaft nut to the specified torque.

3 Check the operation of the inhibitor switch and if necessary, adjust as previously described.

Oil Seals

TO RENEW DRIVE SHAFT OIL SEAL

1 Raise the front of the vehicle and support it on chassis stands. Refer to General Information (chapter 1), if necessary. Remove the front wheel on the relevant side of the vehicle.

2 Remove the retaining bolts and scrivets and withdraw the relevant splashguard from the vehicle.

3 Allow the fluid to cool, if necessary and remove the drain plug and drain the fluid from the transaxle.

4 Working under the vehicle, remove the split pin and castellated nut retaining the tie rod end to the steering knuckle. Discard the split pin. Disconnect the tie rod using a puller or alternatively by placing a suitable dolly or hammer against one side of the steering arm eye and striking the opposite side with a hammer.

5 Remove the brake hose retaining clip from the suspension unit.

6 On models with ABS, remove the bolt retaining the wheel sensor wiring to the suspension unit and then remove the sensor retaining bolt. Remove the sensor from the suspension unit and secure it clear of the work area.

7 Remove the stabiliser bar link nuts from the stabiliser bar and noting the installed position, withdraw the stabiliser bar link from the vehicle.

8 Separate the suspension ball joint from the steering knuckle, by placing a suitable dolly or hammer against one side of the steering knuckle eye and striking the opposite side with a hammer. If necessary, refer to Front Axle and Suspension (chapter 13).

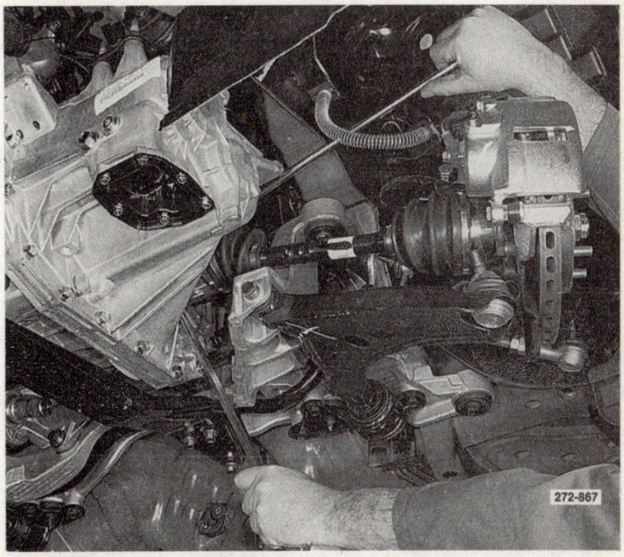

Method of removing the left hand drive shaft. Manual transaxle model shown, automatic models are similar.

TIP *Loosely install the ball joint retaining nut to prevent damage to the thread of the ball joint stud.*

9 When renewing the driver side oil seal, proceed as follows:

a Remove the bolts retaining the drive shaft support bearing and separate the support bearing from the cylinder block.

b With the aid of an assistant, withdraw the driver side drive shaft assembly from the transaxle using care not to damage the transaxle housing. Place suitable sized timber blocks between the suspension unit and the vehicle body to keep the drive shaft clear of the work area.

10 When renewing the passenger side oil seal, proceed as follows:

a While an assistant holds the passenger side suspension unit, insert a lever at the bottom of the passenger side inner drive shaft joint and while supporting the top of the joint with an other bar, lever the inner joint from the transaxle housing. Use a piece of timber as a fulcrum for the lever to prevent damage to the transaxle housing.

TIP *Do not allow the assistant to pull the suspension unit, as damage to the drive shaft inner joint may result.*

b To provide ample working space, hold the suspension unit away from the transaxle, by placing a suitably sized timber block between the suspension unit and the inner mudguard. Secure the drive shaft clear of the work area being careful not to strain the drive shaft joints or damage the boots. Discard the snap ring from the end of the inner joint.

If difficulty experienced removing the passenger side oil seal, the transaxle or the front suspension crossmember may have to be removed. Refer to the relevant chapter for the removal procedure of components.

11 Carefully prise the oil seal from the transaxle housing. Ensure that the transaxle housing is not damaged while removing the oil seal. Use a timber block, while levering the seal. Ensure that no dirt or foreign matter enters the transaxle.

12 Inspect the seal running surface of the drive shaft for grooves or damage. Polish out very fine imperfections using fine wet or dry paper. If badly grooved or damaged, install a new drive shaft.

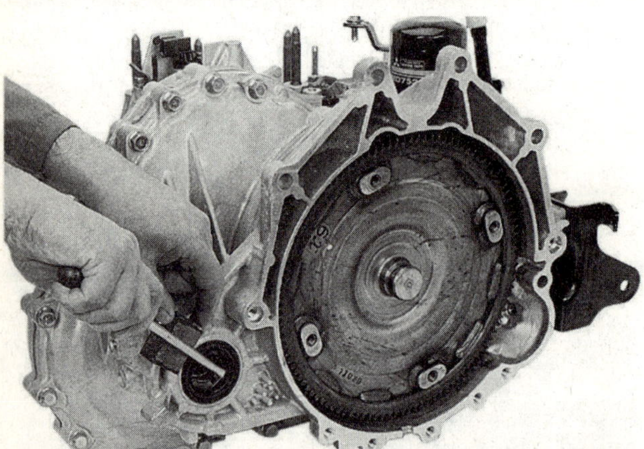

272-246

Using a screwdriver and a wood block to remove the drive shaft oil seal.

272-248

Installing the drive shaft oil seal.

13 Using a lint free cloth, clean the seal housing. Check the housing for damage.

14 Drive the new oil seal, with the lip facing the transaxle, into the housing using a tube or mandrel that pushes on the outer circumference of the oil seal, until the outer edge of the seal is flush with the transaxle housing.

Ensure that the tube or mandrel does not damage the transaxle housing.

15 Apply multipurpose grease to the running surface of the oil seal and the drive shaft.

16 Install a new snap ring to the passenger side drive shaft inner joint, remove the timber block supporting the suspension unit and carefully guide the joint into the housing, until the snap ring securely retains the shaft.

17 On the driver side, remove the timber block and guide the drive shaft into the transaxle housing, being careful not to damage the oil seal or the housing. Install the drive shaft support bearing to the cylinder block and tighten the retaining bolts to the specified torque.

18 Install the tie rod end and suspension ball joint retaining nuts and tighten to the specified torque. Install a new split pin on the tie rod end castellated nut.

19 Refill the transaxle as described earlier in this chapter.

TO RENEW TORQUE CONVERTER OIL SEAL

1 Remove the transaxle from the vehicle using the procedure described later in this chapter.

2 Place a straight edge across the mounting face of the transaxle housing and while pushing the converter towards the rear of the transaxle, measure the distance between the straight edge and the centre mounting lug on the converter. Note the measurement.

3 Place absorbent paper or old, absorbent rags under the transaxle. Carefully remove the converter from the housing.

In most cases a large quantity of fluid will leak from the converter and the transaxle when the converter is removed.

Removing the converter oil seal, using a self tapping screw.

4 Carefully prise the oil seal from the housing. If necessary, insert a small self tapping screw into the body of the seal. Using pliers, withdraw the seal by gripping the self tapping screw. Ensure that the transaxle housing is not damaged while inserting the screw.

TIP *Use a small piece of timber as a fulcrum for the pliers to prevent damage to the transaxle housing.*

5 Inspect the seal running surface of the torque converter for grooves or damage. Polish out very fine imperfections using fine wet or dry paper.

TIP *If the grooves are excessive, consult an automatic transaxle specialist, or install an exchange converter.*

6 Check the oil pump for oil leaks. If the oil pump is leaking, it is recommended that the transaxle be taken to an authorised dealer or an automatic transaxle specialist to dismantle the transaxle and renew the oil pump seal.

Check the seal running surface of the torque converter for damage.

While the torque converter is removed, check the oil pump for leaks.

7 Using a lint free cloth, clean the seal housing. Check the housing for damage.

8 Drive the oil seal, with the lip facing the transaxle, into the housing using a tube or mandrel that pushes on the outer circumference of the oil seal. The face of the new oil seal must be flush with the front of the housing.

i *Ensure that the tube or mandrel does not damage the input shaft.*

9 Apply a few drops of clean transaxle fluid to the running surface of the oil seal and the converter.

Installing the torque converter oil seal using a large socket as a mandrel.

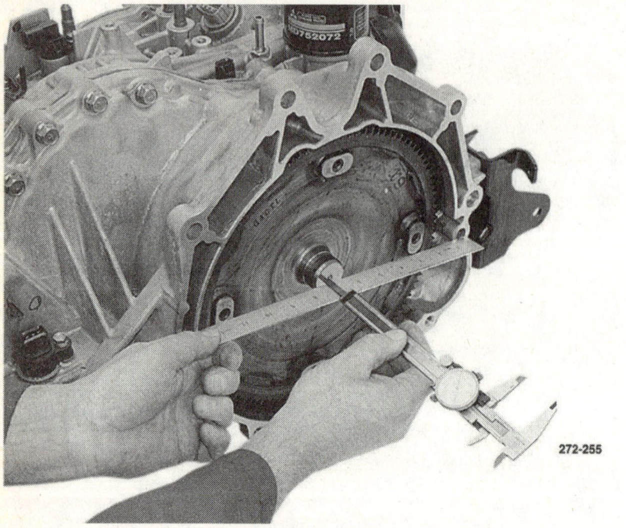

Checking that the installed depth of the torque converter is at least 6.5mm.

10 Carefully install the converter to the transaxle, ensuring that the drive tangs of the converter do not damage the seal. Place a straight edge across the housing. Ensure that the distance between the converter lug and straight edge is as noted on removal. The distance should not be less than 6.5 mm.

11 If the distance is less than specified, push the converter towards the transaxle while carefully rotating it, until the drive tang is correctly seated.

12 Install the transaxle as described previously in this chapter.

13 Check the fluid level and refill with the specified transaxle fluid, as previously described.

TO RENEW SELECTOR SHAFT OIL SEAL

If the selector shaft oil seal is leaking, it is recommended that the transaxle be taken to an authorised dealer or an automatic transaxle specialist to dismantle the transaxle and renew the seal.

TO RENEW TRANSAXLE SENSOR O RINGS

The procedure to renew the input and output shaft and speed sensor O rings is similar and can be done using the following procedure. However, if the solenoid connector O ring is leaking, it is recommended that the transaxle be taken to an authorised dealer or an automatic transaxle specialist to dismantle the transaxle and renew the O ring.

1 Place suitable protection around the engine compartment to prevent damage to the vehicle body.

2 Remove the air cleaner assembly as described in Fuel and Engine Management (chapter 7).

3 Disconnect the relevant sensor wiring connector.

> TIP *If difficulty is experienced disconnecting the speed sensor wiring connector, remove the speed sensor first as described in operations 4 and 5 and then disconnect the wiring connector.*

4 Clean around the relevant sensor to prevent dirt entering the transaxle.

5 Remove the retaining bolt and withdraw the sensor from the transaxle.

Location of the speed sensor o ring.

6 Using a flat bladed screwdriver, carefully remove the O ring from the outside of the sensor housing. Sparingly lubricate the new O ring, using transaxle fluid and install the O ring to the sensor.

7 Install the sensor and tighten the retaining bolt securely. Connect the wiring ensuring that it is making good contact.

8 Check the fluid level and refill with the specified transaxle fluid, as previously described.

Transaxle Assembly

> S *To Remove and Install Transaxle – Suitable lifting tackle, floor jack and long chassis stands*

To remove the transaxle using the following procedure, the vehicle must be raised so that there is at least 550 mm distance between the front bumper and the ground plus the height of the floor jack.

TO REMOVE AND INSTALL

1 Remove the air cleaner assembly as described in Fuel and Engine Management (chapter 7).

2 Remove the battery, referring to Engine Electrical (chapter 6), if necessary.

3 Release the clips retaining the wiring harness to the rear of the battery shroud and remove the shroud from the vehicle.

4 Disengage the clips retaining the wiring to the bracket beneath the battery, remove the retaining bolts and withdraw the bracket.

5 If applicable, remove the engine cover retaining bolts and withdraw the cover.

6 Remove the bolt retaining the wiring harness to the transaxle side cover and remove the bolt retaining the earth cable to the vehicle body. Position the harness toward the cooling fans, clear of the work area.

7 Disconnect the wiring connectors from the transaxle and also from the distributor, radio suppressor, TPS, ISC valve, coolant temperature sensor, temperature gauge sender and the large injector connector. Place the harness clear of the work area.

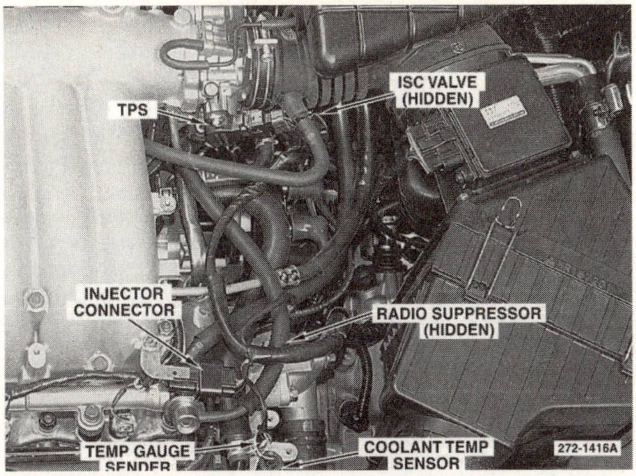

Disconnect the connectors from the components shown.

TIP *If difficulty is experienced disconnecting the speed sensor wiring connector, remove the sensor first as previously described and disconnect the wiring connector. Reinstall the sensor to the transaxle to prevent the entry of dirt.*

8 Remove the spring clip and washer retaining the inner selector cable to the transaxle selector lever and lower the inner cable, away from the selector lever. Remove the bolts retaining the outer selector cable and place the cable clear of the work area.

9 Raise the front and rear of the vehicle with a jack and support the vehicle with chassis stands under the recommended points. Refer to General Information (chapter 1) for the correct floor jack and chassis stand positions to prevent damage or injury. Remove the front wheels.

10 Remove the bolts and scrivets retaining the passenger side splashguard and withdraw the guard. Position the guard located around the drive shaft clear of the shaft.

11 Remove the drain plug and drain the transaxle fluid into suitable container.

12 Remove the nuts retaining the front and rear engine pipes to the exhaust manifolds, lower the pipes and support them using a piece of wire tied to the body of the vehicle.

i *Do not allow the pipe to hang unsupported, as damage to the flexible joint may occur.*

13 Remove the starter motor as follows:

a Remove the nut and bolts retaining the starter motor heat shield.

b Raise the rubber cover and remove the nut retaining the battery cable to the solenoid. Release the wiring connector from the solenoid spade terminal.

c Remove the retaining bolts and withdraw the starter motor from the vehicle.

TIP *The top bolt retaining the starter motor is accessible from the transaxle side.*

14 Working on the passenger side of the vehicle, remove and discard the split pin and remove the castellated nut retaining the tie rod end to the steering knuckle.

15 Disconnect the tie rod end from the steering knuckle using a puller or alternatively by holding a hammer or suitable dolly against one side of the steering knuckle eye and striking the opposite side with a hammer.

16 Remove the brake hose retaining clip from the suspension unit.

17 On models with ABS, remove the bolt retaining the wheel sensor wiring to the suspension unit and then remove the sensor retaining bolt. Remove the sensor from the suspension unit and secure it clear of the work area.

18 Remove the stabiliser bar link nuts from the stabiliser bar and, noting the installed position withdraw the stabiliser bar links from the vehicle.

19 Loosen the ball joint nut and separate the lower control arm from the suspension unit. If necessary, hold a hammer or suitable dolly against one side of the steering knuckle eye and strike the opposite side with a hammer.

TIP *Loosely install the ball joint retaining nut to prevent damage to the thread of the ball joint stud.*

20 Repeat steps 14 to 19 on the driver side of the vehicle.

21 Remove the bolts retaining the driver side drive shaft support bearing and separate the support bearing from the cylinder block.

22 With the aid of an assistant, withdraw the driver side drive shaft assembly from the transaxle using care not to damage the drive shaft oil seal or the transaxle housing. Place suitable sized timber blocks between the suspension unit and the vehicle body to keep the drive shaft clear of the work area.

23 Remove the passenger side drive shaft from the transaxle as follows:

a While an assistant holds the passenger side suspension unit, insert a lever at the bottom of the passenger side inner drive shaft joint and while supporting the top of the joint with another bar, lever the inner joint from the transaxle housing. Use a piece of timber as a fulcrum for the lever to prevent damage to the transaxle housing.

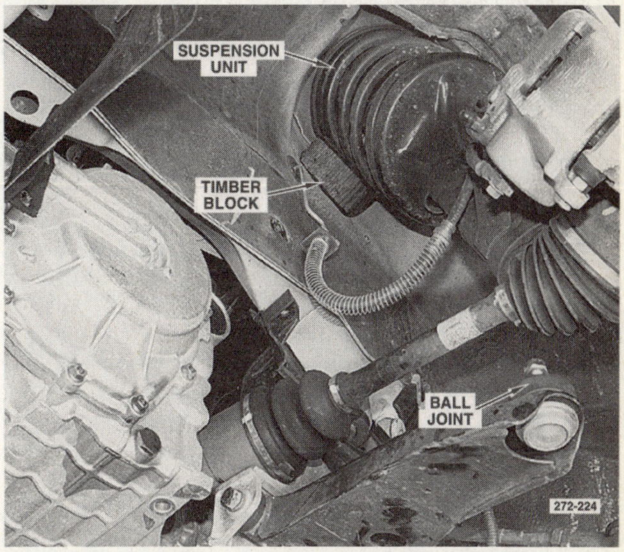

Securing the suspension unit away from the vehicle body to provide more working space.

TIP *Do not allow the assistant to pull the suspension unit, as damage to the drive shaft inner joint may result.*

b To provide ample working space, hold the suspension unit away from the transaxle, by placing a suitably sized timber block between the suspension unit and the inner mudguard, as shown in the illustration. Secure the drive shaft clear of the work area being careful not to strain the drive shaft joints or damage the boots. Discard the snap ring from the end of the inner joint.

24 Remove the bolts and the plate covering the torque converter to driveplate retaining bolts.

25 Using a large screwdriver, rotate the torque converter through the starter motor aperture, levering on the ring gear until one of the retaining bolts is exposed through the cover plate hole.

26 Remove the bolt and rotate the torque converter to expose the next bolt. Repeat this until all bolts are removed. Prior to removing the last bolt, place a matchmark between the driveplate and torque converter to ensure that it is returned to the same position. Push the torque converter away from the driveplate.

27 Remove the nuts and washers and, noting their installed position, withdraw the through bolts from the mountings.

28 If fitted, release the clip retaining the air conditioning pipe to the centre member. Remove the retaining bolts and withdraw the centre member from the vehicle.

29 Remove the bolts retaining the rear transaxle mounting bracket to the transaxle and remove the bracket from the vehicle.

30 Remove the transaxle retaining bolts from the bottom of the transaxle.

31 With suitable protection on the floor jack head, support the engine under the sump.

Placing a piece of timber between the sump and jack will evenly distribute the weight across the sump.

32 Loosen the transaxle mounting nuts and remove the nut from the through bolt. Noting the installed direction of the through bolt, withdraw the bolt and the damper. Carefully lower the floor jack, until sufficient clearance is obtained to remove the mounting and withdraw the mounting from the engine compartment.

33 Attach suitable length of chain to the two transaxle mounting studs closest to the torque converter and raise the lifting tackle until it has taken the weight of the transaxle.

34 Remove the remaining transaxle retaining bolts. Carefully separate the transaxle from the cylinder block dowel pins, ensuring that the converter remains attached to the transaxle.

35 Lower the transaxle and with the aid of an assistant manoeuvre the transaxle from beneath the vehicle.

TIP *Place a creeper board beneath the transaxle to make moving the transaxle easier.*

Installation is a reversal of the removal procedure with attention to the following points:

1 If a new or exchange transaxle is being installed, flush the oil cooler with some clean transaxle fluid to prevent dirt etc. entering the new transaxle, closely following the procedure described under the Routine Maintenance heading. Ensure that a new oil filter is installed to the transaxle.

2 Ensure that the torque converter is fully engaged in the front of the transaxle by measuring the distance between the mounting lug in the centre of the torque converter and the mounting face of the transaxle housing. Ensure that the distance is at least 6.5 mm before installing the transaxle to the vehicle. If the distance is less than 6.5 mm, push the converter towards the transaxle while carefully rotating it, until the drive tangs are seated correctly.

3 Ensure that the transaxle housing and the cylinder block mating faces are clean and free of any burrs.

4 Apply a small amount of multipurpose grease to the outer surface of the locating lug.

5 Align the marks on the driveplate and converter made during removal.

6 Install the bolts retaining the transaxle housing to the cylinder block. Tighten the bolts to the specified torque.

7 Install the bolts securing the converter to the driveplate and tighten evenly to the specified torque, ensuring that the marks made on removal align.

8 Install a new snap ring to the passenger side drive shaft inner joint, and install the drive shaft.

9 Install the driver side drive shaft and tighten the drive shaft support bearing retaining bolts to the specified torque.

10 Ensure that all wiring is securely connected to the transaxle.

11 Fill the transaxle with the specified lubricant as previously described.

12 Check and if necessary, adjust the selector linkage as previously described.

13 Check the transaxle for leaks and rectify as necessary.

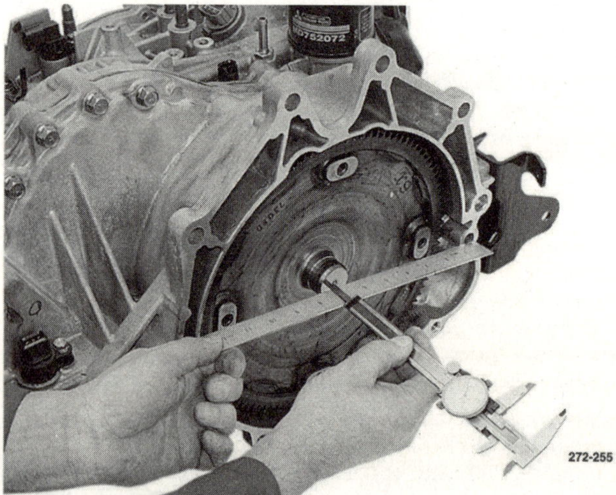

272-255

Checking the installed depth of the torque converter is at least 6.5mm.

Chapter 12
Steering

Contents

Airbag Warning

Use extreme caution when working around the airbag system. When carrying a live airbag assembly, always carry it with the trim cover facing away from you.

When storing the assembly on a bench, always store it with the trim cover facing up and avoid placing anything near or on the assembly.

Specifications

Steering gear:	
Make	TRW
Type	Constant ratio power assisted rack and pinion
Steering wheel freeplay (maximum):	
Engine stopped	10 mm
Engine running	30 mm
Turns lock to lock	3.08
Steering pump type	Variable flow vane impeller
Fluid type	Mitsubishi power steering fluid
Steering gear preload (measured at pinion)	0.8 – 1.6 Nm
Tie rod end rotating torque	0.5 – 2.5 Nm

TORQUE WRENCH SETTINGS

Intermediate shaft pinch bolts	18 Nm
Pressure banjo fitting to pump	24 Nm
Pump retaining nut and bolt	39 Nm
Rack pad locknut	59 Nm
Steering column retaining bolts	10 Nm
Steering gear mounting bolts	69 Nm
Steering wheel nut	30 Nm
Tie rod ball housing to rack	88 Nm
Tie rod end stud castellated nut	29 Nm
Tie rod end locknut	54 Nm

Description

The power steering gear is a rack and pinion type with integral power cylinder. A rotary valve is incorporated in the pinion assembly to direct hydraulic pressure to the appropriate side of the piston in the power cylinder.

This hydraulic pressure assists the mechanical components, which operate in a similar manner to a manual rack and pinion steering system.

The steering gear is mounted on the lower section of the cross-member by rubber mountings and a bracket.

Hydraulic pressure is supplied by a pump which is operated by an engine drive belt from the crankshaft pulley. The fluid reservoir is remote from the pump and is attached to the inner guard. A pressure switch in the power steering pump monitors the system pressure and signals the engine ECU to adjust the engine idle speed by allowing additional air to bypass the throttle valve when pump pressure reaches a certain amount.

Power steering fluid is drawn from the reservoir to the power steering pump. From the pump, fluid is supplied under pressure to the power steering gear. After leaving the power steering gear, fluid runs through the return line which is also a tube type oil cooler, located between the front bumper and the radiator. The return line has a long section running along the drivers side inner guard in the engine compartment to help keep the power steering fluid cool.

No lubrication of the steering gear or the steering pump is required in service and in the event of a loss of power assistance, the steering will continue to operate but will require substantially increased effort. However, as the steering gear and the pump are lubricated by the power steering fluid, the cause of the loss of power assistance or the fluid loss must be determined and promptly repaired to prevent damage to the steering pump and to a lesser extent, the steering gear.

Tie rods are connected to each end of the rack by ball housings which are retained by staked metal tabs. Sealed ball joint tie rod ends connect the tie rods to the steering knuckles.

The steering gear assembly is protected against the entry of dust and water by a seal on the top of the pinion housing and by corrugated rubber boots attached to the tie rods and the ends of the steering gear housing.

The energy absorbing steering column is designed to compress in the event of a severe front end collision. The energy absorbing components are the column tubes, the steering shaft and the mounting bracket breakaway capsules. All these components should be handled with care when removing or installing the steering column assembly.

Some models are equipped with airbags. The drivers airbag is fitted behind the horn pad and the passenger airbag is behind a panel in the dashboard above the glove box. Refer to the Airbag Warning heading in this chapter, prior to doing any work on the steering assembly.

Due to the need for specialised equipment, it is not practical to overhaul the steering gear assembly or the steering pump. If these components are faulty it is recommended that these units be replaced with exchange units, through an authorised dealer or a steering specialist. The removal and installation procedure is included in the text.

Where fitted, disable the airbags before carrying out repairs on the steering assembly. Refer to Body (chapter17).

Routine Maintenance

PUMP DRIVE BELT

In most cases a loose or worn drive belt can be heard squealing when a load is placed on the pump, such as when the steering wheel is turned or as the engine is accelerated.

Inspect the drive belt for breakage, glazing, wear and renew if necessary. If the belt is still serviceable but loose, correctly tension as described in Engine (chapter 4).

TO CHECK FOR FLUID LEAKS

Using a suitable solvent, clean around all power steering components and hoses or pipe fittings where fluid leakage might prevail and then start the engine. Turn the steering wheel from one steering lock to the other several times and stop the engine. Check for leakage. Where leakage is found at pipe or hose fittings, tighten and recheck. Renew the fitting O rings if necessary.

Where leakage is found in items such as the steering gear assembly or pump assembly, specialist attention is necessary.

HYDRAULIC LINES AND HOSES

Check the hydraulic lines and hoses for signs of deterioration especially chafing. Ensure that the fittings at the pump and steering gear are secure and all retaining brackets installed.

POWER STEERING FLUID

Check the fluid level in the power steering reservoir using the dipstick incorporated in the cap.

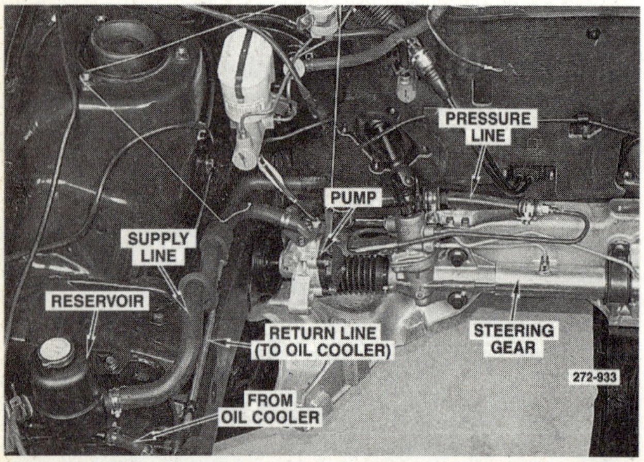

Power steering components, viewed with the engine removed.

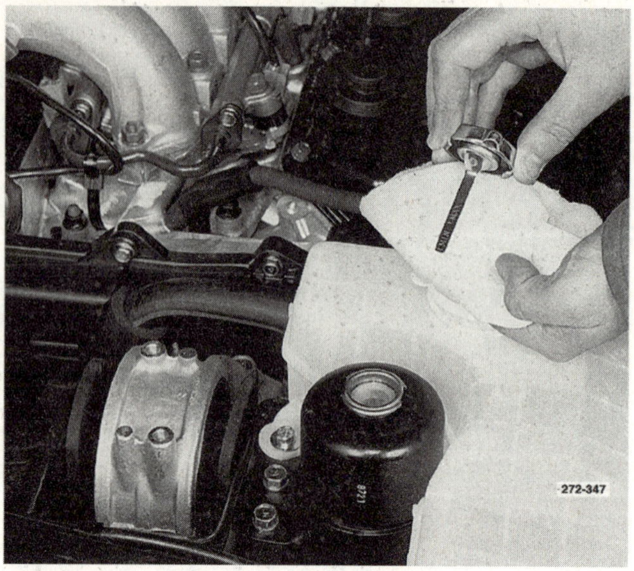

Checking the power steering fluid level.

If necessary, add fluid until the level reaches the MAX mark on the dipstick. Do not overfill the reservoir.

STEERING GEAR AND TIE ROD ENDS

1. Raise the front of the vehicle and support on chassis stands. Refer to General Information (chapter 1) for the positions of the vehicle jacking points.
2. Inspect the steering gear mountings and ensure that the bolts are secure. Also check that the rubber bushes are not deteriorated.
3. Check the rubber boots at each end of the steering gear assembly for tears or deterioration.
4. Check that the tie rod ends are securely connected to each steering knuckle.
5. Lower the vehicle to the ground and have an assistant rotate (rock) the steering wheel back and forth approximately 40 mm either side from the straight ahead position.
6. Check for any looseness, or freeplay, indicating worn components at all of the steering components while an assistant rocks the steering wheel as previously described.

Checking the steering gear for loose and deteriorated mountings.

Checking the steering gear rubber boots for deterioration and lubricant leakage.

TIP *Looseness will be visible by movement between components as the steering wheel is rotated. Common places of wear are tie rods, tie rod ends and the intermediate shaft universal joints. If the tie rods are suspected, remove the dust boot from the steering gear and peel it along the tie rod until the ball socket on the tie rod is visible. Then check the freeplay as previously described.*

Trouble Shooting

EXCESSIVE PLAY OR LOOSENESS IN STEERING GEAR

1. Tie rod end loose or worn: Tighten or renew the tie rod end.
2. Steering gear assembly loose on the crossmember: Inspect for damage and tighten the steering gear mounting bolts to the specified torque.
3. Worn or loose intermediate shaft: Renew the shaft or tighten the intermediate shaft clamps.
4. Steering wheel retaining nut loose: Remove the airbag or horn pad and tighten the steering wheel, as described later in this chapter.
5. Steering gear worn: Overhaul or renew the steering gear.

Check the tie rod ends for wear.

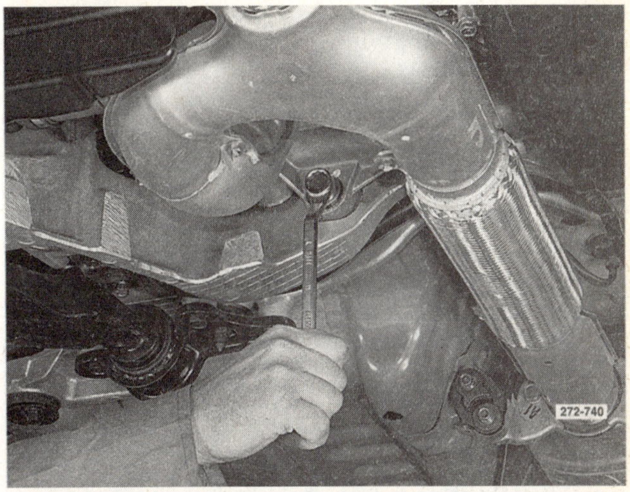

Ensure the steering gear mounting bolts are secure.

Looseness or freeplay in the steering gear is best assessed with the weight of the vehicle on the front wheels.

Start the engine and have an assistant turn the steering wheel from left to right while the inspection is being made.

HEAVY STEERING

1. Low or uneven tyre pressures: Check and inflate the tyres to the recommended pressure.
2. Low fluid level in the power steering reservoir: Check for leakage and repair or renew any faulty components as necessary. Top up the reservoir fluid.
3. Suspension worn or out of alignment: Check the front suspension for wear, renew any faulty components as necessary and realign the front wheels.
4. Binding in the intermediate shaft or steering shaft joints: Check the joints and renew the intermediate shaft or the steering column if necessary.
5. Insufficient steering pump pressure: Have a specialist check the pump pressure and renew the pump assembly if necessary.
6. Faulty or worn steering gear: Overhaul or renew the steering gear.

Ensure that both front tyres have ample depth of tread and are inflated to the correct pressure. Check the front suspension components for wear as described in Front Axle and Suspension (chapter 13) before having the alignment checked.

STEERING PULLS TO ONE SIDE

1. Uneven tyre wear or pressures: Check the condition of the tyres and inflate to the recommended pressure.
2. Incorrect front wheel alignment: Check and align the front wheels.
3. Dragging brakes: Check each wheel for dragging brakes and rectify as necessary. Refer to the Trouble Shooting heading in Brakes (chapter 15).
4. Damaged suspension: Inspect and renew the damaged components as necessary.
5. Faulty steering gear: Overhaul or renew the steering gear.

FRONT WHEEL WOBBLE OR SHIMMY

1 Uneven tyre wear or pressures: Check the condition of the tyres and inflate to the recommended pressure.
2 Tyre or wheel imbalance: Check the wheels for missing balance weights. Balance the wheels as necessary.
3 Worn or loose hub bearings: Check the condition of the front hub bearings and renew the hub bearings if necessary.
4 Faulty or worn suspension units: Check and renew the suspension units as described in Front Axle and Suspension (chapter 13).

Check for excessive play or looseness in the steering gear as previously described. Raise the front of the vehicle and check the hub bearings by manipulating the front wheel in and out at the top and bottom. No perceptible movement in the hub bearings should be felt.

STEERING ERRATIC OR WANDERING

1 Smooth tyres: Check and renew the tyres as necessary.
2 Incorrect wheel alignment: Check and adjust the wheel alignment.
3 Steering gear loose on the crossmember: Inspect the bushes for damage and renew if necessary. Tighten the steering gear mounting and bolts to the specified torque.
4 Worn tie rod ends: Renew the worn tie rod ends.
5 Worn or loose hub bearings: Renew the hub bearings.
6 Wear in the steering gear: Overhaul or renew the steering gear.

FAILURE OF POWER ASSISTANCE

1 Loose or broken drive belt: Renew the drive belt as described in Engine (chapter 4).
2 Low fluid level in the pump reservoir: Check for leakage and repair or renew the faulty components as necessary. Top up the reservoir fluid.
3 Insufficient pump pressure: Check the drive belt condition and the reservoir fluid level. Remove and overhaul the power steering gear assembly.

If the power assistance fails, the vehicle should only be driven the minimum distance necessary for repairs to be performed. If the power steering fluid has leaked, it is strongly recommended that the drive belt is removed to prevent pump damage.

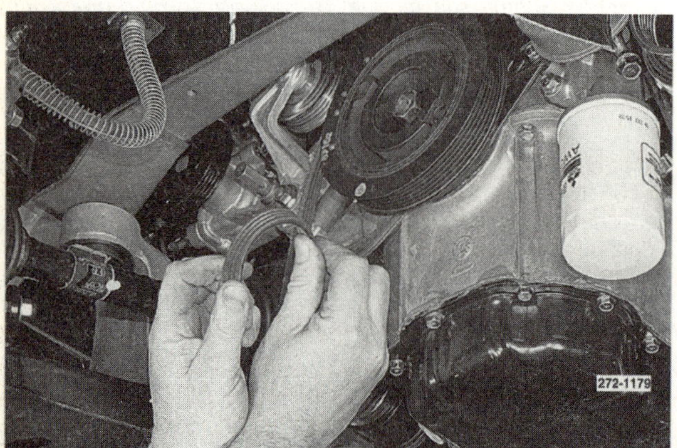

Checking the power steering pump drive belt for cracks or deterioration.

NOISE IN POWER STEERING

1 Loose or defective drive belt: Check and, if necessary, adjust or renew the drive belt as described in Engine (chapter 4).
2 Air in the steering hydraulic system: Bleed the air from the system.
3 Loose or incorrectly installed power steering hoses or pipes: Tighten and/or correctly install the pipes or hoses.
4 Loose or misaligned pump mounting brackets: Tighten and/or align the mounting brackets.

Restricted pipes or hoses can also cause noisy power steering. Check for kinks or damage to the pipes and hoses.

Hydraulic System

TO CHECK TURNING EFFORT

With the tyre pressures and the front wheel alignment correctly adjusted and the vehicle placed on a level surface, apply the handbrake and start the engine.

With the engine at the correct operating temperature and idling speed, turn the steering wheel from lock to lock several times to warm the power steering fluid and then centre the steering wheel.

Attach a spring scale to one of the steering wheel spokes as close as possible to the outer edge and measure the effort to turn the steering wheel in both directions. The turning effort should be approximately equal in both directions and should be approximatley 27 Newtons. If the required effort is excessive, refer to Trouble Shooting (Heavy Steering) in this chapter.

Checking the steering wheel turning effort using a spring scale.

TO BLEED THE SYSTEM

1 Check that the pump reservoir is filled to the Maximum mark on the dipstick. Top up if necessary.
2 Start the engine and turn the steering wheel slowly from lock to lock 3 or 4 times. Check the level in the reservoir again and top up if necessary.

It is normal for the power steering pump to be noisy while there is air present in the system.

3 Run the engine below 1 000 rpm and turn the steering wheel to full left or right lock for 2 – 3 seconds. Turn the wheel to the other lock and keep it there for 2 – 3 seconds.

4 If necessary, top up the reservoir to the correct level and repeat step 3. Check that the fluid in the reservoir is not cloudy or foamy.

It is normal for the fluid to become foamy if air has been in the system. For example, if a power steering line has been removed from the pump or steering gear. If the fluid has become foamy, let the vehicle sit with the engine Off until the fluid has a chance to settle and then proceed with step 5.

5 With the engine idling, check the fluid level in the reservoir.

TO RENEW HYDRAULIC LINES

Pressure Line

The pressure line is made up of two steel pipes joined by a flared nut fitting above the centre of the steering gear assembly. The line running from the pump is a flexible hose with a solid pipe with hose fittings crimped on each end. The second pressure line runs from the banjo fitting to the steering gear. This is a solid steel pipe. The two pipes can be renewed separately.

1 Raise the front of the vehicle and support it on chassis stands. Refer to General Information (chapter 1) for the positions of the jacking points.

2 Position a container under the steering gear.

3 Remove the bolts retaining the pressure fluid line brackets to the steering gear and the crossmember. Using a suitable pipe spanner, loosen the high pressure fluid line fitting at the steering gear and allow the fluid to drain into the container. The high pressure line is the pipe that is looped around the rotary valve and the fitting is beneath the return line fitting.

TIP *The use of a normal open ended spanner may distort the pipe fitting. A pipe spanner should be used where possible.*

4 Remove the union nut from the banjo pipe fitting at the power steering pump and disconnect the pipe from the pump. Discard the O rings.

Installation is a reversal of the removal procedure with attention to the following points:

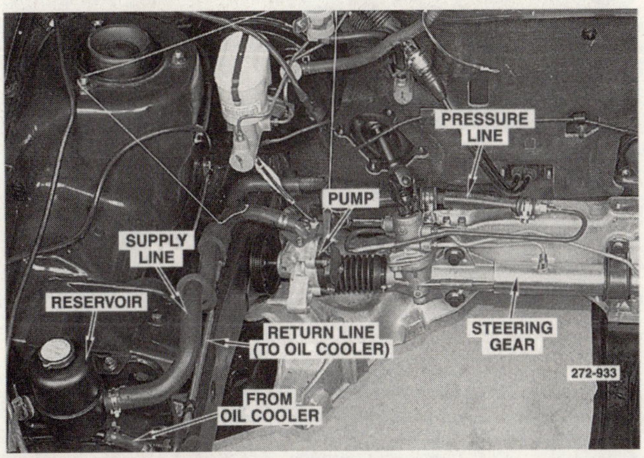

Power steering components, viewed with the engine removed.

1 Ensure that new O ring is used on the fitting at the steering gear and that the fitting is fitted squarely into the steering gear valve housing.

2 Install new O rings to the union stud at the pump.

3 Ensure that the lines are routed correctly.

4 Refill and bleed the power steering system and check for leaks as previously described in this chapter.

Return Line

The return line consists of a steel pipe which loops in front of the radiator to act as the oil cooler. Rubber hoses, retained by hose clamps join the return line to the steering gear and the power steering reservoir.

The cooler pipe and the two hoses can be renewed separately.

1 To renew the rubber hoses, proceed as follows:

a Drain the reservoir by disconnecting the return line at the reservoir and allowing the fluid to drain into a container.

b Using pliers, crimp the hose clamp tangs together and manoeuvre the clamp along the hose.

c Twist and remove the hose ends from the fitting and remove the hose from the vehicle.

d When installing the new hose, lubricate the inside of the hose with power steering fluid to ease installation.

2 To renew the steel cooler pipe, proceed as follows:

a Drain the reservoir, as described above.

b Remove the front bumper and grille. Refer to the appropriate headings in Body (Chapter 17).

c Remove the bolts retaining the windscreen washer bottle to the vehicle. Partially remove the bottle and disconnect the hose and wiring connector from the pump. Remove the windscreen washer bottle from the vehicle.

d Remove the rubber hose connecting the damaged steel pipe, as previously described.

e Support the weight of the engine and remove the right hand side engine mounting as described in Engine (chapter 4).

f Remove the bolts retaining the clamps and rubber grommets. Manoeuvre the steel pipe up from beside the engine and feed it back into the engine compartment and then from the vehicle.

g Install the new steel return/cooler pipe and the rubber hoses.

h Position the rubber grommets of the retaining clamps carefully, ensuring there is no metal to metal contact between the steel pipe and the retaining clamp, and the pipe is not touching the body or the subframe of the vehicle. Tighten the retaining bolts securely.

Power steering oil cooler, viewed with the front bumper removed.

If any of the plastic coated clamps have the plastic worn away they should be renewed to prevent the pipe rubbing against the metal of the clamp.

i Install the front bumper and grille. Refer to Body (Chapter 17).

j Install the windscreen washer bottle, ensuring that the hose and wiring are connected.

3 Refill and bleed the power steering system and check for leaks as previously described in this chapter.

Steering Wheel

To Remove Tight Steering Wheels— Steering wheel puller
To Remove Airbag — T30 Torx socket.

TO REMOVE AND INSTALL

1. Where fitted, disable and remove the airbag as described in Body (chapter 17).
2. To remove the horn pad from a vehicle not equipped with an airbag, remove the screws from either side of the horn pad. Partially remove the horn pad, disconnect the horn wiring connector and withdraw the horn pad from the steering wheel.

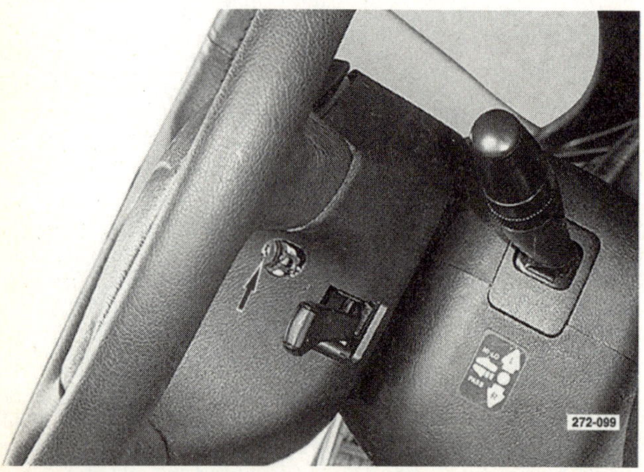

On models with an airbag, the retaining screws have Torx heads. On models without an airbag, normal screws are fitted.

If dismantling of the horn pad is necessary to fix a problem with the horn switch, refer to Body Electrical (chapter 16).

3. Loosen the steering wheel retaining nut but do not completely remove it at this stage. Mark the relationship of the steering wheel to the steering shaft to aid in the correct positioning of the wheel on installation.
4. Hold the top and bottom of the steering wheel and rock the wheel back and forth at the same time moving your hands around the steering wheel. Continue to do this until the steering wheel separates from the steering shaft.

   If the steering wheel fails to separate using the above method, install a steering wheel puller to the steering wheel and tighten the centre bolt of the puller until the wheel is loose.

Airbag removed showing the steering wheel retaining nut.

Do not strike the end of the puller or the steering shaft as sharp blows can cause irreparable damage to the collapsible steering shaft.

5. Remove the retaining nut and remove the steering wheel from the shaft.

Installation is a reversal of the removal procedure with attention to the following points:

1. Ensure that the marks on the steering wheel and steering shaft are aligned.
2. Ensure that the horn wire is correctly routed.
3. On models with an airbag, install and enable the airbag as described in Body (chapter 17).

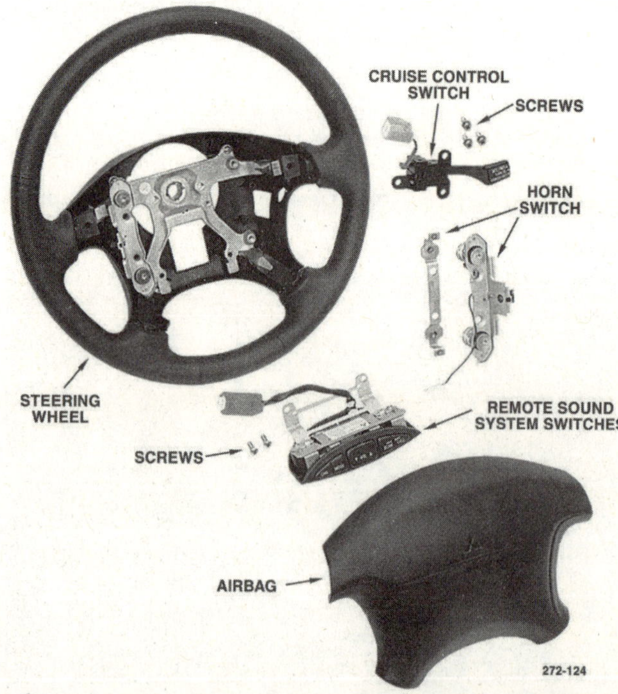

Steering wheel components. Verada Xi model shown.

Clock Spring

The clock spring consists of a ribbon wire that can rotate with the steering wheel and transmit power to and from the various switches on the steering wheel as well as supplying power to trigger the drivers airbag. The ribbon wire within the clock spring can be easily broken by rotating the clock spring past its normal operating limits. If the clock spring is removed without the front wheels in the straight ahead or if unsure that it is centralised, it is very important that the procedure printed on the clock spring is followed on installation to prevent damage. The steering must be set in the straight ahead position prior to installing the clock spring.

To Remove and Install

1. Set the front wheels to the straight ahead position and disconnect the negative battery terminal.
2. Disable and remove the driver airbag. Refer to Body (chapter 17).
3. Remove the steering wheel as previously described.

> ⚠ *Avoid turning the steering column or clock spring to prevent damage to the clock spring when it is reinstalled.*

4. Remove the screws retaining the steering column covers and remove the covers from the column.
5. Disconnect the horn wiring connector from beneath the steering column. Unclip the clock spring wiring from along the steering column. Remove the retaining screws and remove the clock spring. Do not rotate the clock spring.

Installation is a reversal of the removal procedure with attention to the following points:

> ⚠ *If the clock spring has been moved, hold the inner hub and rotate the outer hub in a clockwise direction until resistance is felt. Rotate the outer hub anti clockwise approximately 3 and 4/5 turns and then align the two alignment marks. Install the clock spring in this position.*

1. Ensure that the front wheels are still in the straight ahead position.
2. Install the clock spring and tighten the retaining screws.

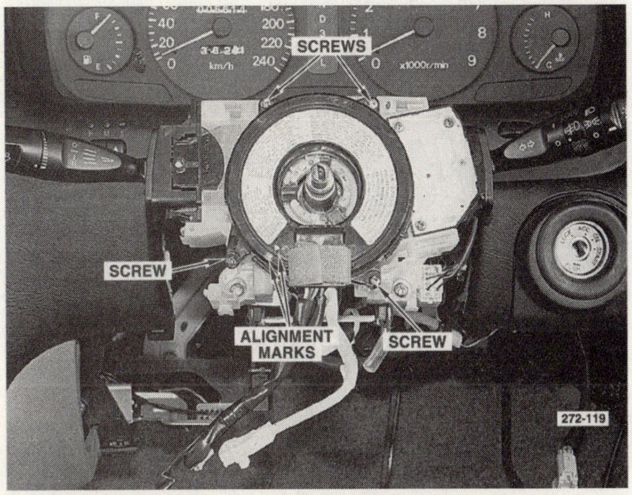

Steering wheel and covers removed showing the clock spring retaining screws and the two alignment arrows.

3. Secure the wiring to the base of the steering column and connect the horn wiring connector. Do not connect the airbag wiring connector beneath the column at this stage.
4. Install the steering column covers, ensuring that the dowels on the top cover locates into the top of the combination switch base plate. Install the lower cover and tighten the retaining screws.
5. Install the steering wheel as previously described.
6. Install and enable the airbag as described in Body (chapter 17).

Ignition Switch and Lock Cylinder

The removal and installation procedures for the ignition switch and lock cylinder are fully covered in Body Electrical (chapter16). The lock assembly renewal is covered in this chapter under the Steering Column Assembly heading.

Steering Column Assembly

TO REMOVE AND INSTALL

1. Disconnect the negative battery terminal. Set the front wheels in the straight ahead position.
2. If applicable, disable the airbag as described in Body (chapter 17).
3. Remove the steering wheel as previously described.
4. On models not fitted with an airbag, remove the lower dash panel as follows:

   a. Remove the screws retaining the bonnet lever to the lower dash panel and position the lever clear of the work area.
   b. Remove the screw cover from the top of the drivers side lower dash panel. Remove the remaining screws from the lower dash panel, disconnect the panel from the top left side of the dashboard and remove the panel from the vehicle.

5. Remove the screws from the bottom of the ignition lock surround panel. Remove the shield from beneath the lock cylinder and then manoeuvre the surround panel from the dashboard. Disconnect the key reader wiring connector and where fitted, remove the illumination bulb holder by rotating the bulb holder anticlockwise after partially removing the key reader from around the lock cylinder.
6. Remove the retaining screws and manoeuvre the upper and lower steering column covers from the column.
7. Where fitted, remove the clock spring as previously described.
8. Separate the various switch wiring connectors.
9. Mark the relation of the intermediate shaft coupling to steering shaft to aid assembly. Remove the pinch bolt.
10. Remove the steering column retaining bolts and manoeuvre the steering column from the vehicle.
11. Inspect the column for signs of damage and check that the steering shaft rotates freely in the bearings.
12. If the ignition lock is damaged and needs to be renewed proceed as follows:

    a. Remove the shear bolts securing the lock cylinder housing to the steering column. To assist in the removal of these bolts, use a sharp centre punch to loosen the bolts. Remove the lock assembly from the steering column. Discard the bolts.

Removing one of the shear bolts from the ignition lock assembly.

> **TIP** *In difficult cases, it may be necessary to drill the head of the shear bolts and use a screw extractor to remove the bolts.*

b Install new shear bolts on assembly. Ensure that the lock assembly operates correctly before tightening the bolts so that the heads shear off.

Installation is a reversal of the removal procedure with attention to the following points:

1. Align the matchmarks made on the steering shaft and the intermediate shaft.
2. Install the steering column, ensuring that the wiring is correctly routed and not trapped between the column and the body.
3. Tighten the steering column retaining bolts to Specifications.
4. Tighten the intermediate shaft pinch bolts to Specifications.
5. Connect the wiring to the key reader and where fitted the illumination bulb holder. Install the shield and the ignition lock surround panel and tighten the retaining screws securely.
6. Connect the wiring to the combination switch and then install the upper and lower steering column covers.
7. On models without an airbag, install the lower dash panel beneath the steering column.
8. On models with an airbag, enable the airbag as described in Body (chapter 17).

Intermediate Shaft Assembly

TO REMOVE AND INSTALL

1. Raise the front of the vehicle and support it on chassis stands. Refer to General information (chapter 1) for the position of the jacking points. Place the front wheels in the straight ahead position and remove the key from the ignition.

> *It is important that the key is removed on models with an airbag to prevent the steering wheel rotating past its normal lock to lock movement. If the steering wheel is rotated more than three turns from the straight ahead position, the clock spring will be destroyed.*

2. Working under the vehicle, mark the installed position of the lower intermediate shaft universal joint to the pinion on the steering gear, to speed up alignment on assembly. Remove the pinch bolt.

> **TIP** *Because of the pinch bolt flat, the intermediate shaft will only fit one way in the universal joint. Marking it will make installation easier when working in this tight area.*

3. Working inside the vehicle, remove the bolts retaining the cover plate gasket and retainer to the firewall.

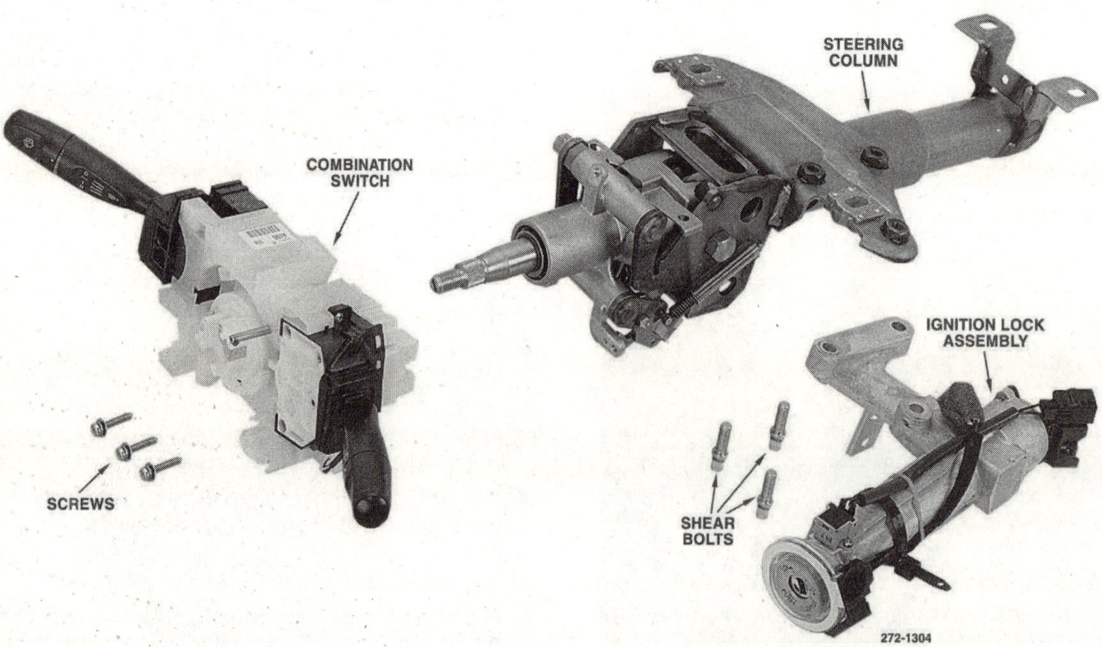

Steering column and associated components.

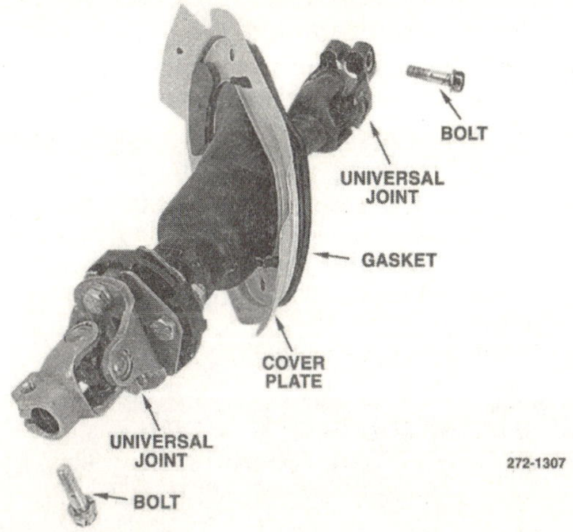

Intermediate shaft assembly.

4 Mark the installed position of the intermediate shaft in relation to the steering shaft and remove the upper pinch bolt from the intermediate shaft. Move the shaft up, towards the steering column to disengage the intermediate shaft from the steering gear and then disengage the top of the intermediate shaft from the steering column.

5 Manoeuvre the intermediate shaft from the vehicle.

> *If difficulty is encountered it may be necessary to remove the steering column in order to remove the intermediate shaft. The procedure to remove the steering column is described earlier in this chapter.*

6 Inspect the universal joints for wear and damage. The intermediate shaft, if faulty must be renewed as an assembly.

7 Inspect the cover plate gasket and retainer. Also check the condition of the bearing on the intermediate shaft. If damaged, the intermediate shaft will have to be renewed as an assembly.

Installation is a reversal of the removal procedure with attention to the following points:

1 Install the intermediate shaft, ensuring that the marks made on removal align.

2 Tighten the pinch bolts to Specifications.

3 If the steering wheel was not locked, or it is suspected that the clock spring is not centralised, centralise the clock spring as previously described.

Steering Pump

As specialist knowledge and equipment are required to overhaul the power steering pump, it is not a worthwhile repair proposition for the home mechanic. Therefore it is recommended that a faulty pump be replaced by an exchange unit obtained through an authorised dealer or a steering specialist.

TO REMOVE AND INSTALL

1 Clean the area around the pump using degreaser etc. to prevent dirt entering the pump or hydraulic system.

2 Working in the engine compartment, disconnect the supply hose and allow the fluid to drain into the container.

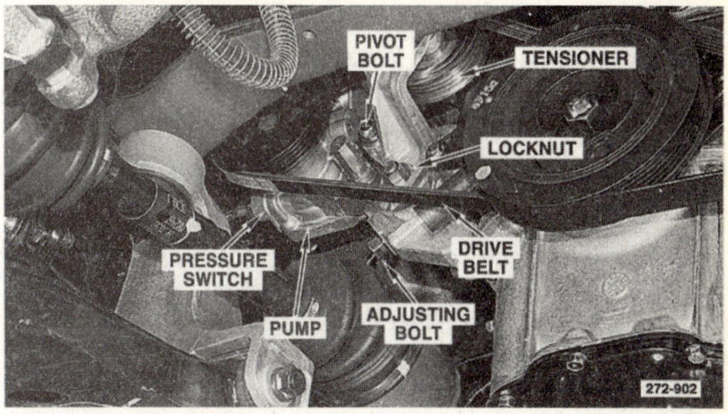

Power steering pump location, viewed from beneath the vehicle.

3 Loosen and remove the banjo nut and disconnect the pressure line from the pump. Seal the lines and the pump to prevent the entry of dirt.

4 Loosen the power steering pump belt tensioner pivot bolt and locknut. Loosen the adjuster and remove the drive belt.

5 Disconnect the wiring connector from the pressure switch.

6 Remove the upper retaining nut and the lower mounting bolt and then withdraw the pump from beneath the vehicle.

7 Remove and discard the O rings from the pressure line banjo stud.

Installation is a reversal of the removal procedure with attention to the following points:

1 Wipe the exterior of the pump with cloth and suitable solvent. Be careful not to allow the solvent to enter the pressure and return line apertures.

2 Inspect the pump for damage or cracks and renew the pump if necessary.

3 Ensure that all pipes and pump connections are clean and unobstructed. Install new O rings to the pressure line banjo stud and lubricate with power steering fluid.

4 Manoeuvre the power steering pump into position on the engine and tighten the upper retaining nut and the lower mounting bolt.

5 Install and tighten the pressure line banjo nut. Install the return line securely, ensuring that the lines are not fouling or chafing and then connect the wiring to the pressure switch.

6 Install the pump drive belt and tension as described in Engine (chapter 4).

7 Fill the system with the recommended fluid, and bleed the system as previously described.

8 Check for leaks and repair as necessary.

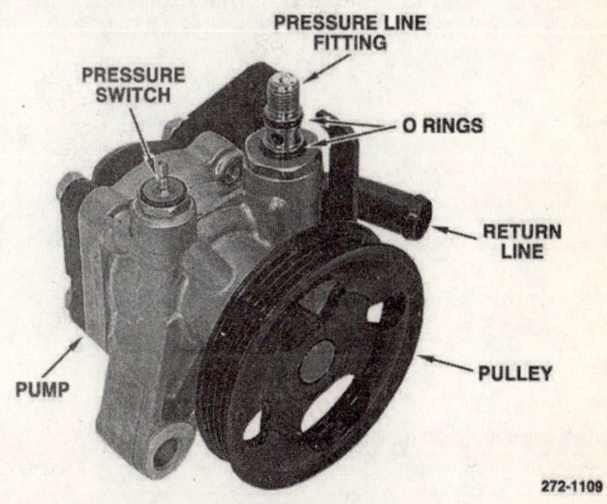

Power steering pump removed from the vehicle.

Steering Gear Assembly

To overhaul the steering gear, extensive knowledge and a range of special tools are required, therefore it is not a worthwhile proposition for the home mechanic.

If the steering gear has a fault such as leaking seals or a problem with the rack and pinion, it is recommended that an exchange unit is installed. Exchange units should be available through an authorised dealer or a steering specialist. A detailed removal and installation procedure is included for this purpose.

The procedures to renew the rubber boots, tie rod ends and tie rod assemblies are described later in this chapter.

If removing the steering gear assembly from the vehicle, it is necessary to remove the ignition key and engage the steering lock so that the steering column cannot be moved past the normal lock to lock position once the steering gear assembly has been removed. Damage to the clock spring will result if the above procedure is not carried out.

TO REMOVE AND INSTALL

1 If applicable, disable the airbag as described in Body (chapter 17).

2 Raise the front of the vehicle and support it on chassis stands. Refer to General information (chapter 1) for the position of the jacking points. Remove the front wheels.

3 Remove the front crossmember. Refer to Front Axle and Suspension (chapter 13).

4 Remove the mounting bolts and remove the steering gear assembly from the crossmember.

5 Thoroughly clean the steering gear assembly except the rubber components in a suitable solvent. Be careful not to allow the solvent to enter the pressure and return pipe apertures.

6 Check the steering gear assembly housing for cracks and damage and renew as necessary.

7 Check the pipes and connections for cracks, damage or obstructions and renew as necessary.

8 Check the mounting bushes for wear and damage. Renew as necessary.

Installation is a reversal of the removal procedure with attention to the following points:

1 Install the steering gear to the crossmember and with the aid of an assistant securing the crossmember, install the steering gear mounting bolts and tighten to the specified torque.

2 Install the crossmember to the vehicle as described in Front Axle and Suspension (chapter 13).

3 Fill the power steering reservoir to the correct level and bleed the system as previously described.

4 Check all power steering pipe connections for leaks and repair as necessary.

5 Check and if necessary, adjust the front wheel toe in as described in Front Axle and Suspension (chapter 13). After adjustment, ensure that the rubber boots are not distorted or twisted and the rubber boot clamps and clips are facing downward.

If the steering was unlocked and rotated with the steering gear removed, the clock spring should be centralised as described under the Clock Spring heading before the steering column is rotated.

6 Road test the vehicle and check for correct steering operation and that the steering wheel is correctly centralised.

TO RENEW TIE ROD END

1 If applicable, disable the airbag as described in Body (chapter 17).

2 Raise the front of the vehicle and support it on chassis stands. Refer to General information (chapter 1) for the position of the jacking points. Remove the front wheels.

3 Mark the installed position of the tie rod end and the locknut in relation to the tie rod and loosen the locknut.

4 Remove the split pin and remove the castellated nut retaining the tie rod end to the steering knuckle.

5 Disconnect the tie rod end from the steering knuckle using a ball joint puller or alternatively by holding a hammer or suitable dolly against one side of the steering knuckle eye and striking the opposite side with a hammer.

6 Remove the tie rod end from the tie rod, counting the number of turns required to separate the two components.

7 Screw the new tie rod end onto the tie rod using the same number of turns required to remove the original tie rod end and align the marks made on removal. This will ensure that toe in is set closely to what it was originally, however it will only be accurate if the new tie rod is identical to the old tie rod.

8 Connect the tie rod end to the steering knuckle. Install and tighten the castellated nut to the specified torque. Install a new split pin to the castellated nut.

9 Tighten the tie rod end locknut securely, aligning the marks made on removal.

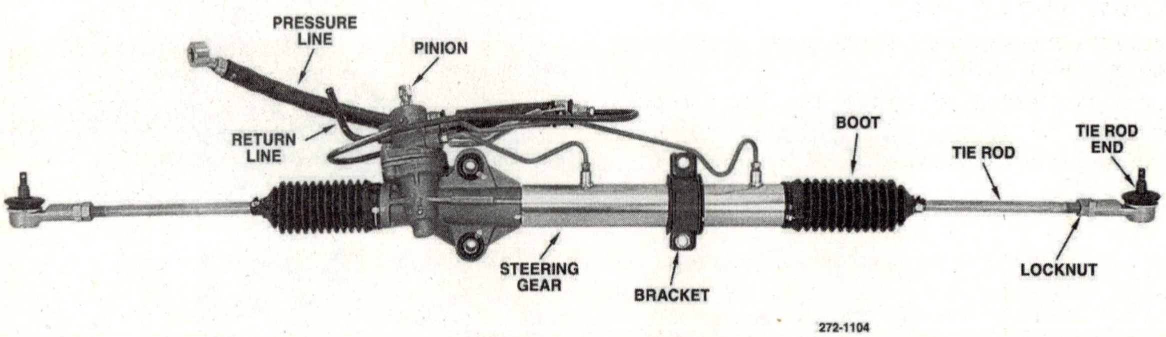

Steering gear removed from the vehicle.

10 Install the front wheel and lower the vehicle to the ground.

11 Have the front wheel toe in checked and if necessary adjusted by a tyre specialist as soon as possible. For more information on wheel alignment refer to Front Axle and Suspension (chapter 13).

12 Where applicable enable the airbag as described in Body (chapter 17).

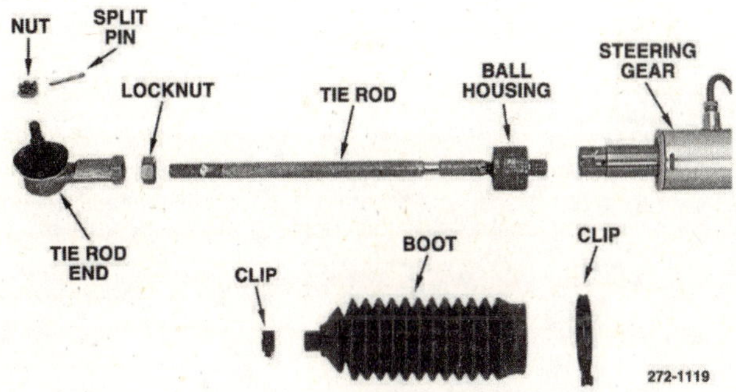

Tie rod, tie rod end and rubber boot.

TO RENEW STEERING GEAR RUBBER BOOT

1 Remove the tie rod end as previously described and remove the tie rod end locknut.

2 Remove the outer clip and cable tie retaining the boot ends to the tie rod and the steering gear assembly. Slide the boot off the tie rod.

TIP *Wrap insulating tape over the threads to prevent damage to the boot.*

3 Fill the corrugations inside the new boot with molybdenum disulphide grease or equivalent and carefully slide the boot over the tie rod.

4 Engage the end of the new boot with the groove in the steering gear assembly.

5 Install the small retaining clip, tie rod end and locknut as previously described.

6 Position the rubber boot, so the end of the rubber boot is in the groove provided in the tie rod. Install the clip and a new cable tie to the boot ensuring that the boot is not twisted or dented.

7 Install the tie rod end as previously described.

TO RENEW TIE ROD ASSEMBLY

1 Remove the steering gear assembly from the vehicle as described later in this chapter.

2 Remove the steering gear rubber boots as previously described.

3 Secure the tie rod shaft in a vice.

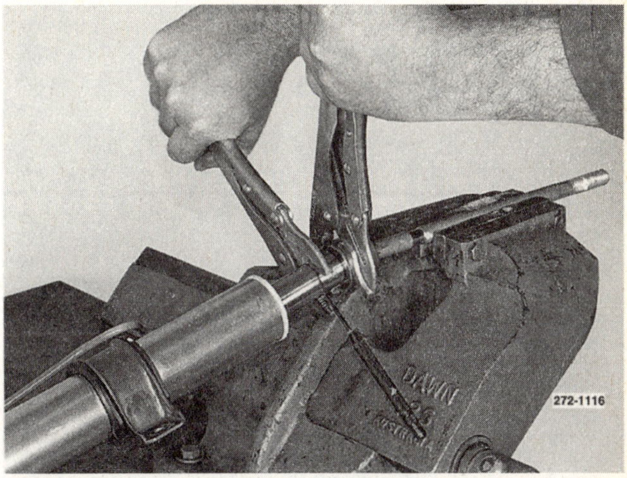

Removing the tie rod from the steering gear.

4 Insert a pin punch into the hole in the side of the steering gear rack shaft and secure it using vice grips, as shown in the illustration.

5 Install a second pair of vice grips onto the tie rod ball housing and unscrew the ball housing from the steering gear rack shaft.

6 Lubricate the ball housing of the new tie rod with a molybdenum disulphide grease or equivalent.

7 Install the new tie rod to the end of the rack and tighten the tie rod ball housing holding the tie rod and steering gear as previously described.

8 Stake the ball housing against the rack using a pin punch.

9 Install the steering gear rubber boots and tie rod ends as previously described.

10 Install the steering gear assembly as previously described in this chapter.

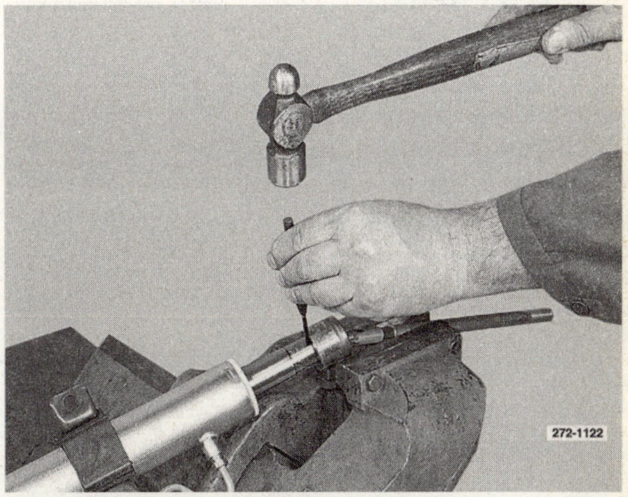

Staking the steering gear to the tie rod.

Chapter 13
Front Axle and Suspension

Contents

Specifications

Type ... Independent MacPherson struts with lower control arms and stabiliser bar
Shock absorber ... Double acting
Hub bearing end float (maximum) ... 0.05 mm
Lower ball joint starting torque ... 10 – 22 Nm
Stabiliser bar link ball joint starting torque ... 1.7 – 3.2 Nm
Front wheel alignment:
Toe in ... 0 ± 3 mm
Camber ... 0 deg ± 30 min
Caster ... 3 deg ± 30 min
Toe out on turns with inner wheel at 20 deg ... 22 deg

TORQUE WRENCH SETTINGS

Ball joint to steering knuckle nut ... 60 – 72 Nm
Brake caliper anchor plate bolts ... 90 Nm
Centre member:
  Front bolts ... 88 Nm
  Rear bolts ... 52 Nm
Lower control arm:
  [1]Front pivot bolt nut ... 100 – 120 Nm
  [1]Rear pivot nut ... 100 Nm
Rear pivot bush to crossmember bolts ... 70 – 93 Nm
Crossmember to body nut and bolts ... 88 Nm
Drive shaft nut ... 196 – 255 Nm
Drive shaft support bearing bolts ... 45 Nm
Stabiliser bar:
Bracket bolts ... 35 – 44 Nm
Link nuts ... 35 – 44 Nm
Steering gear to intermediate shaft pinch bolt ... 18 Nm
Suspension unit:
Upper mounting nuts ... 40 – 50 Nm
To steering knuckle nuts and bolts ... 95 – 105 Nm
Piston rod nut ... 60 Nm
Tie rod end castellated nut ... 29 Nm

[1]The vehicle should be lowered to the ground and bounced up and down several times before tightening these nuts.

Description

The front suspension is an independent type comprising two MacPherson strut suspension units mounted vertically on each side of the vehicle. The lower end of the suspension unit is bolted to the steering knuckle, which in turn houses the front hub bearings. The steering knuckle pivots on the lower control arm by means of a ball joint.

The lower control arm pivots at its inner ends on rubber bushes, secured to an aluminium crossmember that is bolted to the vehicle underbody.

Each front suspension unit assembly comprises a tubular hydraulic, shock absorber type suspension unit, surrounded at the upper end by a coil spring. On top of the coil spring is the upper mounting which attaches to the underside of the front wheel housing. The piston rod of the suspension unit is attached to the centre of the upper mounting by a rubber mounted bearing.

When a suspension unit is defective, it is recommended that both suspension units be renewed as a pair.

The camber, caster and steering axis inclination are set in production and cannot be adjusted. Any variation in these angles will be caused by worn or damaged components.

Front suspension components.

The drive from the transaxle is transmitted to the front wheels through drive shafts that have a tripod constant velocity joint on the transaxle end of the shaft and a Birfield constant velocity joint splined to the front hub and retained by a nut. The right hand side drive shaft has a support bearing to reduce torsional vibration and torque steer tendencies.

The front wheel bearings are of the double row ball type and they have no provision for adjustment or greasing in service.

Routine Maintenance

1 Check the shock absorber condition by bouncing the front of the vehicle up and down (one side at a time) and observe the vehicle. If the shock absorbers are serviceable the vehicle should come to rest in a single movement, if it bounces two or three times before stopping, the suspension units should be dismantled and the shock absorber portion renewed.

Leaking fluid is a sure sign that the suspension units require renewal.

2 Check the front suspension components for wear by raising the front of the vehicle and supporting on chassis stands under the jacking points, leaving the front wheels and suspension unsupported. Refer to General Information (chapter 1) for more information.

With an assistant pushing and pulling the front wheel in and out at the top and then the bottom, check for excessive looseness in the hub bearings, suspension ball joints and pivot bushes.

The suspension ball joint is spring loaded in its socket to compensate for normal wear, therefore if any movement can be detected, the ball joint should be renewed.

3 Check the lower control arm pivot bushes by having an assistant firmly push the wheel towards the rear and then the front of the vehicle.

4 Inspect the stabiliser bar mounting rubbers and the link ball joints for wear, damage and deterioration.

5 Inspect the vehicle ride height from side to side, if it is lower on one side than the other, remove the coil spring and check its free length against a new spring. If one spring is unserviceable, it is a good practice to install two new springs as a matching pair.

Suspension units that are leaking fluid must be renewed.

Check the lower control arm bushes for wear.

Ensure that the tyre pressures are correct and the vehicle is unladen before checking the ride height.

6 Inspect the condition of the wheel bearings as described under Steering Knuckle and Hub Assembly, later in this chapter.

7 Check the inner and outer drive shaft boots for cracks, tears or other damage.

Trouble Shooting

FRONT END NOISE

1 Loose or worn stabiliser bar link ball joints or bushes: Check, tighten or renew the bushes.

2 Loose or worn lower control arm bushes: Check, tighten or renew the bushes.

3 Noise in the suspension unit: Check, overhaul or renew the suspension units on both sides.

4 Loose suspension unit assembly upper mounting or piston rod nut: Tighten the upper mounting or piston rod nuts.

5 Loose or worn suspension ball joint: Renew the ball joint.

6 Loose or worn steering gear: Tighten the steering gear mounting nuts and bolts or overhaul the steering gear.

Check the stabiliser bar link ball joints for wear.

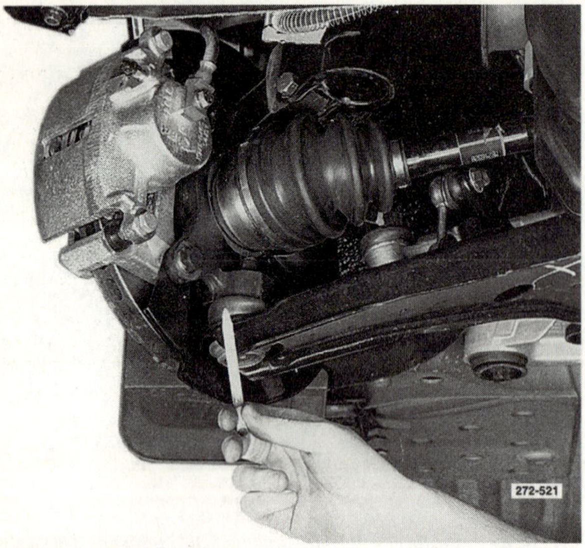

Check the ball joints for wear.

7 Loose drive shaft hub nut or worn front hub bearings: Tighten the hub nut or renew the front hub bearings.

8 Loose crossmember nuts and bolts: Check all components thoroughly, renew any doubtful components and tighten the nuts and bolts securely.

> *To check the front suspension components for wear, refer to Routine Maintenance in this chapter.*

POOR OR ERRATIC ROAD HOLDING ABILITY

1 Low or uneven tyre pressures: Check and inflate the tyres to the recommended pressure.

2 Defective suspension unit: Check, overhaul or renew the suspension units on both sides.

3 Incorrect front or rear wheel alignment: Check and adjust the front and/or rear wheel alignment, or renew components as necessary.

4 Weak or broken front coil spring: Renew the front coil springs as a matching pair.

5 Loose or worn lower control arm bushes, ball joints, upper suspension mounting or piston rod nut: Tighten or renew the faulty components.

6 Faulty rear suspension components: Check the rear suspension components as described in Rear Axle and Suspension (chapter 14).

7 Defective stabiliser bar link ball joints or loose mounting nuts: Check and tighten or renew the links.

> TIP *For a quick test of suspension unit condition, refer to Routine Maintenance in this chapter.*

NOISE OR VIBRATION IN DRIVE SHAFTS

1 Lack of lubricant due to damaged drive shaft joint boots: Check the drive shaft joints for wear, repack the drive shaft joints and renew the drive shaft boots.

Rotate the drive shaft and carefully check the boots for splits and damage.

2 Worn or damaged inner or outer drive shaft joint: Check and renew the worn components as necessary.

3 Loose drive shaft hub nut: Tighten the hub nut to the specified torque.

4 Faulty or worn support bearing: Check the support bearing and renew if necessary.

5 Bent or twisted drive shafts: Check and renew the drive shafts as necessary.

> TIP *Noise or vibration in the front end can also be caused by excessive tyre or wheel imbalance. Drive shaft joint noise is usually more pronounced when moving slowly with the steering on full lock.*

Suspension Unit

> *To Dismantle and Assemble — Spring compressor*

TO REMOVE AND INSTALL

1 Raise the front of the vehicle, support it on chassis stands and remove the front wheel. Refer to General Information (chapter 1) for the positions of the jacking points.

2 If the suspension unit is to be dismantled, working in the engine compartment, remove the dust cap and loosen the piston rod nut. Do not remove it at this stage, as serious injury may result.

3 On models with ABS, remove the bolt retaining the wiring clamp to the lower part of the suspension unit and secure the wire away from the work area.

4 Remove the bolts and nuts retaining the suspension unit to the steering knuckle. Disconnect the steering knuckle from the suspension unit.

5 Mark the outside of the suspension unit upper mounting to the inner mudguard panel and remove the retaining nuts.

> TIP *There is a small rubber protrusion on the upper mounting on most models that indicates the outside.*

Where applicable, the small rubber protrusion on the upper mounting indicates the outside.

6 Manoeuvre the suspension unit from the vehicle using care not to damage the drive shaft boot.

7 Secure the lower end of the suspension unit assembly in a vice and install a spring compressor. Compress the coil spring and remove the piston rod nut and washer. Discard the piston rod nut.

The coil spring is partilly compressed and has a large amount of potential energy. Removal of the piston rod nut without a spring compressor installed may cause serious injury to the operator.

8 Mark the position of the coil spring in relation to the upper and lower spring seats and to the insulators.

9 Remove the upper mounting, upper spring seat, insulator, spring, bump stop/dust cover and lower insulator from the suspension unit.

10 With the suspension unit vertical, move the piston rod up and down several times. A firm resistance with no slack or tight spots should be evident in both directions and there should be no fluid leakage.

If the suspension unit hydraulic operation is not satisfactory, it is recommended that the suspension units on both sides be renewed.

11 Check the upper mounting assembly for damage and fatigue. Check the bearing for wear. Check all rubber components for wear, damage and deterioration. Renew worn or damaged components.

12 Check the coil spring for fatigue by comparison with a new spring for the same model vehicle. Also, check the spring for cracks and damage. If the coil spring is defective, it is recommended that the springs on both sides be renewed.

TO ASSEMBLE AND INSTALL

Assembly and installation is a reversal of the removal and dismantling procedure with attention to the following points:

1 Install the lower insulator to the suspension unit ensuring that it is correctly positioned on the lower spring seat.

2 Install the dust cover/bump stop and compressed coil spring ensuring that the markings made on removal correspond with the marking on the lower spring seat.

3 Install the upper insulator and upper spring seat and mounting, ensuring that marks made on removal or the protrusion on the upper mounting, will face the outside of the vehicle when installed.

4 Install the washer and a new piston rod nut to the suspension unit. Tighten the nut securely while holding the upper mounting by hand to stop it turning.

Suspension unit with spring compressors installed.

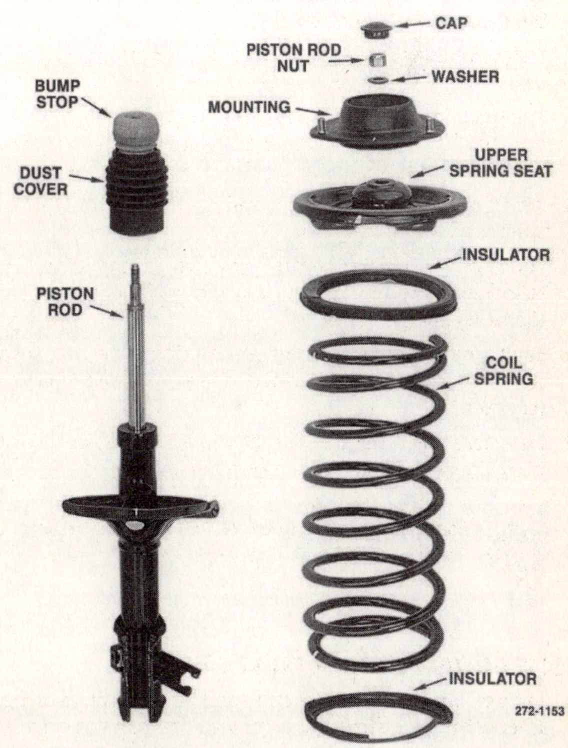

Suspension unit dismantled.

5 Slowly release the coil spring compressor from the suspension unit assembly ensuring that the spring components are seating correctly. Tighten the piston rod nut to the specified torque.

TIP *If difficulty is encountered in holding the suspension unit when tightening the piston rod nut, install the suspension unit to the vehicle then tighten the piston rod nut to Specifications.*

6 Pack the recess on top of the upper mounting bearing with multipurpose grease and install the dust cap. Ensure that the grease does not contaminate the rubber components of the upper mounting.

7 Install the suspension unit to the upper mounting and tighten the nuts finger tight. Install the suspension unit to steering knuckle bolts and tighten the retaining nuts to the specified torque.

8 Install the ABS wiring bracket to the suspension unit and tighten the bolt securely.

9 Install the front wheel and lower the vehicle to the ground. Tighten the upper mounting retaining nuts to the specified torque.

If not previously done, tighten the piston rod nut to the specified torque and install the dust cap. Refer to step 6.

10 Check and if necessary, adjust the front wheel alignment.

Steering Knuckle and Hub Assembly

To Remove and Install Hub Bearing — Press and press plates, bearing separator
To Check Hub End Float — Dial gauge

TO INSPECT

1 Raise the front of the vehicle and support it on chassis stands. Refer to General Information (chapter 1) for the postions of the jacking points.

2 Spin the wheel and listen for rumbling noises, which will indicate worn front wheel bearings. Grasp the wheel at the top and bottom and feel for excessive play in the wheel bearing. If the hub assembly is not serviceable, it should be overhauled.

3 If in doubt about the serviceability of the hub assembly, remove the front wheel and check the hub end float as follows:

a Remove the brake caliper anchor plate retaining bolts. Slide the caliper assembly from the disc and suspend the caliper from the coil spring using a piece of wire. Withdraw the brake disc from the hub.

Do not apply the brake pedal while the caliper is removed.

b Secure a dial gauge to the suspension unit with its plunger against the hub face. Zero the gauge.

c Check the hub end float by pulling and pushing the hub in an axial direction and noting the reading on the dial gauge.

Checking the hub end float using a dial gauge.

If the hub end float is outside Specifications, ensure that the drive shaft nut is tightened to the specified torque.

d If the drive shaft nut is tight but excessive end float still exists, the hub assembly must be dismantled and worn components renewed.

TO REMOVE AND DISMANTLE

If the hub assembly is dismantled for any reason, the bearings and seals must be renewed. These components will be damaged during the dismantling operation. Because of the tools required and the difficulty level involved, it is recommended that an exchange steering knuckle is installed or that a front end specialist or authorised dealer performs the hub overhaul. Check the price difference between an overhauled unit before proceeding to dismantle the front hub.

1 Raise the front of the vehicle, support it on chassis stands and remove the front wheel. Refer to General Information (chapter 1) for the postions of the jacking points.

2 Remove the split pin from the end of the drive shaft. While an assistant applies the brakes, loosen the drive shaft nut approximately one turn. A long breaker bar will be necessary, as the drive shaft nut is very tight.

The hub bearing may be damaged if the drive shaft nut is removed with the vehicle weight resting on the suspension.

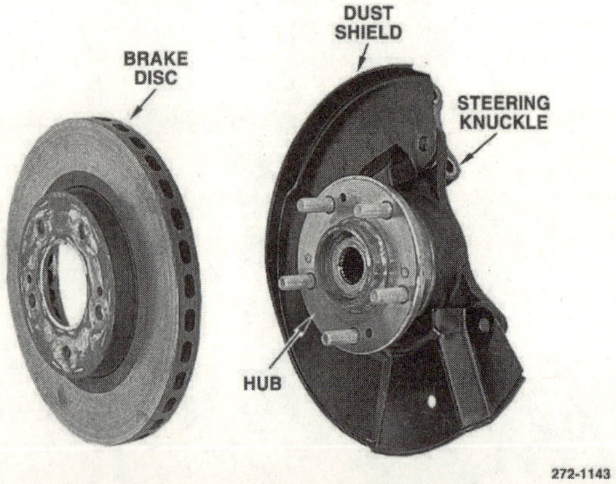

Steering knuckle, hub and brake disc.

3 Remove the brake caliper anchor plate retaining bolts. Slide the caliper assembly from the disc and suspend the caliper from the coil spring using a piece of wire. Withdraw the brake disc from the hub.

Do not apply the brake pedal while the caliper is removed.

4 On models with ABS, remove the sensor and wiring retaining bolts and place the sensor clear of the work area.

5 Loosen, but do not remove the nut retaining the ball joint to the steering knuckle. Disconnect the ball joint from the steering knuckle using a ball joint separator or alternatively by placing a hammer or a suitable dolly against one side of the steering knuckle eye and striking the opposite side with a hammer. Remove the retaining nut and discard.

6 Remove the split pin and castellated nut from the tie rod end and disconnect the tie rod from the steering knuckle using a similar procedure to the ball joint.

Ensure that the steering is in the straight ahead position and remains so while removing the drive shaft.

7 Using a suitable puller, press the drive shaft through the hub assembly, being careful not to damage the drive shaft joint boot. As the drive shaft is pushed from the hub, it will be necessary to pull the lower part of the suspension unit outwards to provide clearance.

If a puller is not available, a blow from a heavy copper/hide faced hammer should dislodge the drive shaft. To prevent damage to the threads, ensure that the drive shaft nut is installed level with the end of the shaft and the castellations on the nut are facing the hub, before using the soft faced hammer.

8 Leave the drive shaft installed in the transaxle and secure it away from the working area using wire or rope. Use care not to place excessive strain on the joints or damage the rubber boots.

9 Remove the bolts and nuts retaining the suspension unit to the steering knuckle and hub assembly, remove the assembly from the vehicle.

10 To dismantle the steering knuckle and hub assembly, proceed as follows

a Prise the oil seal from the inner side of the steering knuckle using a screwdriver and discard.

b Using internal snap ring pliers, withdraw the snap ring from the steering knuckle.

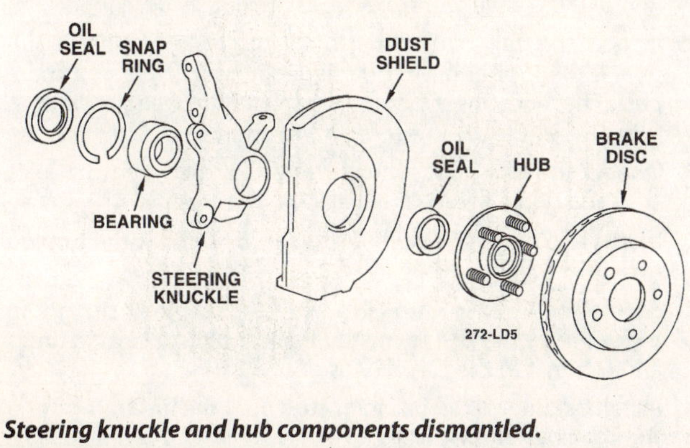

Steering knuckle and hub components dismantled.

c Remove the retaining bolt and manoeuvre the dust shield from the steering knuckle.

d Using a press and suitable press plates, push the hub through the steering knuckle. The outer bearing race will remain on the hub.

TIP: An alternative method to remove the hub from the steering knuckle is to mount the steering knuckle in a vice and attach a suitable slide hammer to the wheel studs.

e Using a suitable bearing separator remove the bearing race from the hub and discard.

f Remove the outer seal from the steering knuckle using a suitable lever and discard the oil seal.

g Using a suitable puller remove the outer race from the hub and discard.

h Thoroughly clean the steering knuckle and hub in a suitable solvent and inspect for wear, cracks and damage. Once the hub is dismantled, renewal of the bearings and seals is essential, renew other components as necessary.

TO ASSEMBLE AND INSTALL

Assembly and installation are a reversal of the removal and dismantling procedure with attention to the following points:

When grease is referred to in the assembly procedure, use high melting point multipurpose grease.

1 If the steering knuckle and hub were dismantled, install new bearings as follows:

a Using a suitable metal tube or mandrel that pushes on the outer race only, press the bearing assembly into the steering knuckle.

TIP: If inner races or ball carriers separate from the bearing assembly, they must be kept clean and installed to the side from which they came. The bearing comes prepacked with grease.

b Apply grease to the outer oil seal and install the seal into the steering knuckle.

c Support the hub on press plates with the wheel mounting flange down.

d Position the steering knuckle squarely onto the hub. Using a suitable diameter steel tube or mandrel that abuts the inner race of the bearing, press the bearing and steering knuckle assembly fully onto the hub.

Take care not to damage the bearing when assembling the steering knuckle to the hub.

e Install the snap ring to the steering knuckle using internal snap ring pliers.

f Apply grease to the inner oil seal and install the seal into the steering knuckle.

g Ensure that the hub rotates smoothly in the bearings. Check the assembly procedure if necessary.

h If removed, install the dust shield to the steering knuckle.

2 Install the steering knuckle to the suspension unit assembly lower mounting and tighten the nuts and bolts to the specified torque.

3 Install the drive shaft into the hub and install the nut on the shaft.

4 Install the ball joint to the steering knuckle. Install a new self locking nut and tighten to the specified torque.

5 Install the tie rod to the steering knuckle, tighten the castellated nut to the specified torque and install a new split pin.

6 Install the brake caliper assembly and tighten the anchor plate bolts to the specified torque.

7 With the aid of an assistant applying the brakes, tighten the drive shaft nut to the specified torque. Align the hole in the drive shaft and install a new split pin.

8 On models with ABS, install the sensor and wire retainer, secure with the retaining bolts.

9 Install the front wheel and lower the vehicle to the ground.

Lower Control Arm

TO REMOVE AND INSTALL

1 Raise the front of the vehicle, support it on chassis stands and remove the front wheels. Refer to General Information (chapter 1) for the location of the jacking points.

2 Remove the stabiliser link nut from the lower control arm.

3 Loosen, but do not remove the nut retaining the ball joint to the steering knuckle. Disconnect the ball joint from the steering knuckle using a ball joint separator or alternatively by placing a hammer or a suitable dolly against one side of the steering knuckle eye and striking the opposite side with a hammer. Remove the retaining nut and discard.

4 Remove the bolts retaining the rear pivot bush to the cross-member.

5 While supporting the lower control arm, remove the front pivot bolt nut and washers and withdraw the lower control arm from the vehicle.

6 Inspect the lower control arm for cracks, dents and distortion and renew if necessary.

7 Inspect the lower control arm bushes and bolts for wear and distortion. If the front pivot bush is worn, renew the lower control arm as a complete assembly.

8 If the rear pivot bush is worn on models with a nut retaining the rear pivot bush, remove the nut and remove the bush assembly.

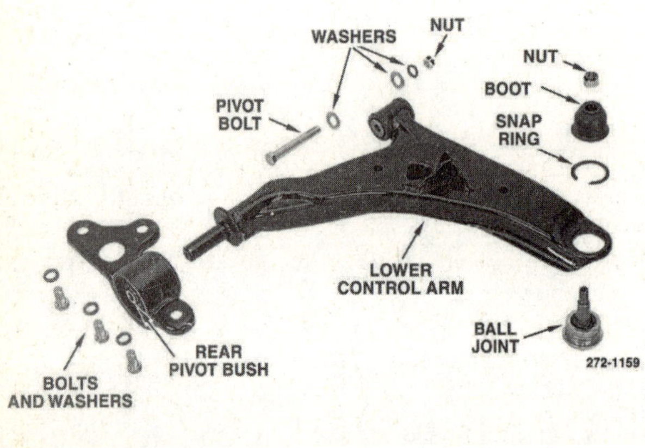

Lower control arm components. Models without a nut retaining the rear pivot bush.

9 If the rear pivot bush is worn on models with a pressed on rear pivot bush, proceed as follows:

a Remove the bush from the lower control arm using a bearing separator and a press.

b Lubricate the inside of the new bush with soapy water. Position the new bush over the shaft so that the metal bracket is parallel with the control arm.

c Using a press and a suitable tube or mandrel that pushes on the inner tube of the bush only, press the bush fully onto the control arm.

If the metal bracket of the bush is not parallel with the control arm, it can be slightly twisted once installed to the control arm.

10 Inspect the ball joint for wear and damage as described under the heading that follows and renew if necessary.

Installation is a reversal of the removal procedure with attention to the following points:

1 Install the lower control arm to the vehicle and tighten the rear pivot bush to crossmember bolts to the specified torque. Tighten all other fasteners finger tight at this stage. Ensure that the rear pivot shaft is correctly seated in the rear bush assembly.

2 Install the stabiliser bar link and tighten the retaining nut to the specified torque.

3 Install the ball joint stud to the steering knuckle. Install a new self locking nut and tighten the nut to the specified torque.

4 Install the front wheels and lower the vehicle to the ground. Bounce the front of the vehicle up and down several times to stabilise the suspension and settle the bushes.

5 Hold the front pivot bolt and tighten the nut to the specified torque with the weight of the vehicle resting on the suspension.

6 Where fitted, tighten the rear pivot nut to the specified torque.

Ball Joint

To Renew Ball Joint — Press and suitable tubes

TO CHECK

1 Remove the lower control arm, as previously described.

2 Inspect the ball joint rubber boot for cracks and deterioration. If the rubber boot is damaged it is available as a spare part. The boot must be installed carefully using a smooth tube that abuts the lower retaining ring.

3 Inspect the ball joint for wear and play between the joint stud and body. Renew the ball joint if necessary.

If unsure of the condition of the ball joint check the turning torque as follows:

a Secure the lower control arm, with the ball joint upright in a vice, move the ball joint stud back and forth several times and return it to the central position.

b Install the nut on the ball joint stud until the nut bottoms on the shoulder of the thread.

Using a press to install the ball joint to the lower control arm.

c Using a low range torque wrench, measure the torque required to start the ball joint stud turning in the body. Renew the ball joint if the starting torque is not within 10 – 22 Nm.

If the ball joint requires renewal proceed as follows:

4 Prise the rubber boot from the ball joint and discard.

5 Remove the retaining snap ring from the ball joint.

6 Support the lower control arm below the ball joint using a suitable tube that is large enough to allow the joint to pass through and press the ball joint from the lower control arm.

7 Reverse the removal procedure and press the new ball joint into the lower control arm.

> *When pressing the new ball joint into the lower control arm, ensure that reasonable pressure is required, especially when the joint is almost home. If the ball joint is not tight in the lower control arm, the lower control arm must be renewed.*

8 Install a new retaining snap ring ensuring that it seats correctly in the groove provided.

9 Apply some multipurpose grease to the inside of a new rubber boot and install the boot over the ball joint stud and seat carefully using a smooth tube that abuts the lower retaining ring.

10 Install the lower control arm to the vehicle as previously described.

Stabiliser Bar

TO RENEW LINKS

1 Raise the front of the vehicle, support it on chassis stands and remove the front wheels. Refer to General Information (chapter 1) for the postions of the jacking points.

2 Remove the nuts retaining the links to the lower control arm and the stabiliser bar and withdraw the links from the vehicle.

3 Inspect the link ball joints for wear and play between the joint stud and link body. Renew the link as an assembly if necessary. If unsure of the condition of the ball joints check the starting torque as follows:

a Secure the link with the ball joints upright in a vice, move each ball joint stud back and forth several times and return it to the central position.

Checking the starting torque of the stabiliser bar link ball joints.

b Install a nut on each ball joint stud until the nut bottoms on the shoulder of the thread.

c Using a low range torque wrench, measure the torque required to start the ball joint stud turning in the link. Renew the link assembly if the starting torque of either ball joint is not within 1.7 – 3.2 Nm.

4 If the rubber boots require renewal, release the spring wire clip and withdraw the old boot. Tape the hexagon portion of the stud with some electrical tape or similar.

Apply some multipurpose grease to the inside of a new rubber boot and install the boot to the ball joint. Remove the tape and secure the boot with the wire clip.

TO RENEW BUSHES

1 Raise the front of the vehicle, support it on chassis stands and remove the front wheels. Refer to General Information (chapter 1) for the postions of the jacking points.

2 Remove the nuts retaining the links to the lower control arms and disconnect the links.

3 Remove the bolts retaining each lower control arm rear pivot bush to the crossmember. Remove the lower control arm front pivot bolt nuts and washers and allow the lower control arms to hang down approximately 150 mm. Use wire to hold the lower control arms in this position to prevent damage to the ball joints.

4 If the manufacturers paint marks are missing, suitably mark the installed position of the rubber bushes on the stabiliser bar.

5 Remove the stabiliser bar bracket bolts and ease the stabiliser down to allow removal of the bushes.

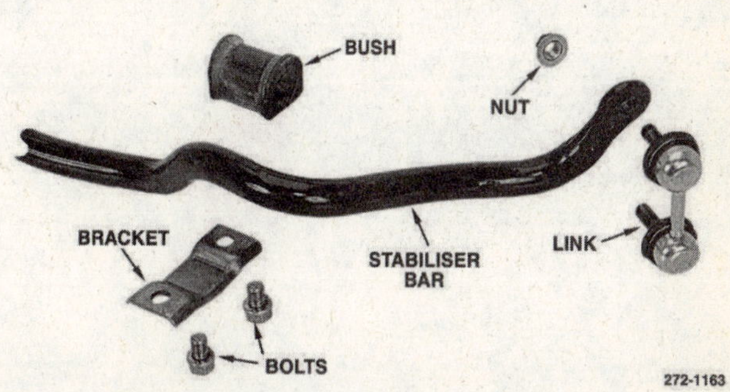

Stabiliser bar components.

Installation is a reversal of the removal procedure with attention to the following points:

1 Ensure that the stabiliser bar mounting bushes are located according to the paint marks noted on removal.

2 Install the stabiliser bar brackets and tighten the retaining bolts securely.

3 Push the inner ends of the lower control arms back into position. Tighten the rear pivot bush bolts and loosely install all other fasteners.

4 Install the front wheels and lower the vehicle to the ground. Bounce the front of the vehicle up and down several times to stabilise the suspension and settle the bushes.

5 Tighten the front lower control arm pivot bolt and where fitted, the rear pivot bush nut to Specifications.

TO REMOVE AND INSTALL

1 Raise the front of the vehicle, support it on chassis stands and remove the front wheels. Refer to General Information (chapter 1) for the location of the jacking points.

2 Remove the stabiliser bar to link nuts, and leave the links connected to the lower control arms.

3 Remove the bolts retaining each lower control arm rear pivot bush to the crossmember. Remove the lower control arm front pivot bolt nuts and washers and allow the lower control arms to hang down approximately 150 mm. Use wire to hold the lower control arms in this position to prevent damage to the ball joints.

4 Remove the retaining bolts and scrivets and withdraw the splashguard from the vehicle.

5 Remove the engine pipe from the vehicle as described in Engine (chapter 4). Use care not to damage the oxygen sensor.

6 While supporting the middle of the centre member remove the bolts retaining the rear of the centre member to the crossmember. Lower the rear of the centre member enough to allow removal of the stabiliser bar.

7 If the manufactures paint marks are missing, suitably mark the installed position of the rubber bushes on the stabiliser bar and remove the bolts retaining the stabiliser bar brackets. Withdraw the brackets.

8 Manoeuvre the stabiliser bar between the centre member and withdraw the bar from the vehicle.

9 Examine the stabiliser bar and brackets for damage, cracks and bends.

10 Inspect the stabiliser bar rubber bushes and links for wear and deterioration. Normally it is standard practice to renew the rubber bushes as a matter of course and renew other parts as necessary.

11 Check the stabiliser bar ball joint links as previously described.

Installation is a reversal of the removal procedure with attention to the following points:

1 Ensure that the stabiliser bar mounting bushes are located according to the paint marks noted on removal.

2 Install the stabiliser bar brackets and tighten the retaining bolts securely.

3 Push the inner ends of the lower control arms back into position. Tighten the rear pivot bush bolts and loosely install all other fasteners.

4 Install the front wheels and lower the vehicle to the ground. Bounce the front of the vehicle up and down several times to stabilise the suspension and settle the bushes.

5 Tighten the front lower control arm pivot bolts and where fitted, the rear pivot bush nut to Specifications.

Crossmember

TO REMOVE AND INSTALL

1 Disconnect the negative battery terminal.

2 On models with a driver airbag, to prevent damage to the clock spring behind the steering wheel, set the front wheels of the vehicle to the straight ahead position and remove the ignition key. Disable the airbag. Refer to Body (chapter 17) for the correct procedure.

Crossmember viewed with lower control arms and engine pipe removed. The retaining nuts and bolts are indicated by the large arrows.

3 Raise the front of the vehicle, support it on chassis stands and remove the front wheels. Refer to General Information (chapter 1) for the postions of the jacking points.

4 Remove the stabiliser bar to link nuts, and leave the links connected to the lower control arms.

5 Remove the bolts retaining the lower control arm rear pivot bushes to the crossmember. Remove the control arm front pivot bolts, nuts and washers and allow the lower control arms to hang down approximately 150 mm. Use wire to hold the control arms in this position to prevent damage to the ball joints.

TIP *If necessary, the lower control arms can pivot forwards to give more space around the work area.*

6 Disconnect the steering gear tie rod ends from the steering knuckles on each side of the vehicle. Refer to Steering (chapter 12) for the procedure to disconnect the tie rods.

7 Remove the retaining bolts and scrivets and withdraw the splashguard from underneath the front of the vehicle.

8 Remove the front and rear engine mounting through bolts. Refer to Engine (chapter 4) if necessary.

9 Remove the retaining bolts and manoeuvre the centre member from underneath the vehicle.

10 Place a suitable container beneath the vehicle and disconnect the power steering return line from the pipe on the inner guard. Disconnect the pressure line from the power steering pump at the banjo nut or bolt. Matchmark the steering gear pinion shaft to the intermediate shaft using quick drying paint or similar. Remove the steering gear to intermediate shaft pinch bolt .

11 Remove the front engine pipe from the vehicle as described in Engine (chapter 4).

12 With a floor jack supporting the crossmember, remove the nuts and bolts retaining the crossmember to the vehicle underbody. Lower the jack and manoeuvre the assembly from under the vehicle.

TIP *The aid of an assistant will ease removal of the crossmember and prevent damage to the steering gear boots.*

13 Carefully inspect the crossmember for cracks, distortion or other damage and renew if necessary.

Installation is a reversal of the removal procedure with attention to the following points:

1 Check the crossmember rubber bushes and renew if necessary.

2 Ensure that the stabiliser bar rubber bushes are serviceable, renew if necessary. Install the bracket bolts and tighten securely.

3 When installing the crossmember ensure that the steering intermediate shaft to steering gear pinion matchmark is aligned. Use care to avoid damaging the steering gear boots.

4 Tighten the intermediate shaft pinch bolt to the specified torque.

5 Do not fully tighten the lower control arm to crossmember fasteners until the vehicle is back on its wheels and has been bounced up and down several times to stabilise the suspension and settle the bushes.

6 On models with a driver airbag enable the airbag as described in Body (chapter 17).

7 Ensure that the steering lines are connected using new sealing washers. Top up the steering reservoir and bleed the system as described in Steering (chapter 12).

Front Wheel Alignment

Specialised equipment is required to accurately measure the front wheel alignment, except for front wheel toe in. Therefore it is not a worthwhile proposition for the home mechanic to perform a complete wheel alignment check. The vehicle should be taken to a wheel alignment specialist.

The front wheel camber and caster are not adjustable and any variation from specifications will be caused by worn or damaged components.

Prior to performing a wheel alignment, thoroughly inspect the steering gear, tie rods, ball joints, lower control arm bushes, coil springs and suspension units. Renew or repair as necessary.

The tread of the front tyres should be examined for excessive or uneven wear as certain conditions of tyre wear are indicative of damaged or worn components in the suspension, steering linkage and/or wheels and bearings. Refer to Tyre Wear Trouble Shooting in Wheels and Tyres (chapter 19). If the tyres are found to be defective, renew with serviceable tyres.

The vehicle should be unladen, except for the normal amount of fuel. The tyres should be inflated to the normal pressure.

TO CHECK AND ADJUST TOE IN

The following procedures may be used as a temporary measure to check and adjust the toe in. The vehicle should be taken to a wheel alignment specialist at the earliest opportunity.

1 With the vehicle on a level floor, raise the front of the vehicle and support it on chassis stands.

2 Spin each front wheel in turn and using a piece of chalk, mark a thin line around the periphery of each tyre as near to the centre as possible.

3 Lower the front of the vehicle to the ground and bounce the vehicle up and down several times to stabilise the suspension. Set the front wheels in the straight ahead position.

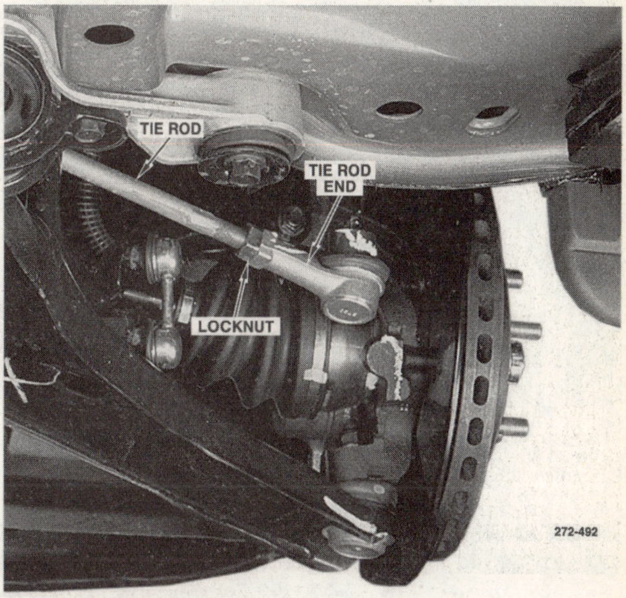

Toe in is adjusted at the tie rod ends.

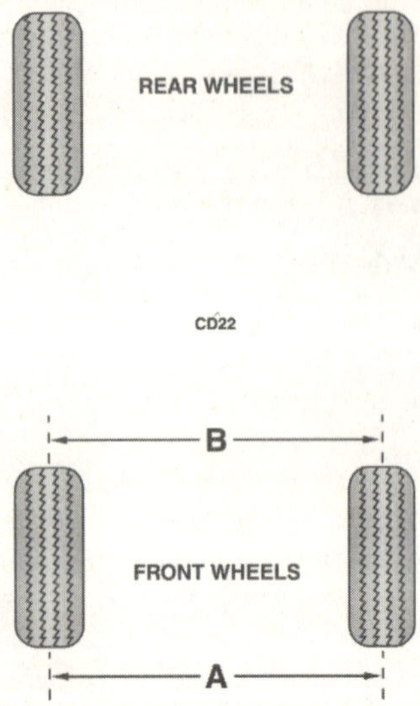

Front wheel toe in/toe out. If dimension A is less that B, this termed toe in. If dimension A is more than B, this is termed toe out. Refer to Specifications.

4 Mark the centre chalk line at the front of both tyres at approximately 20 – 25 cm above the ground.

5 Using a telescopic gauge or tape measure, measure and record the distance between the chalk marks on the centres of the tyres.

6 Maintaining the wheels in a straight ahead position, roll the vehicle forward until the chalk marks are the same distance above the ground, but at the rear of the front tyres.

7 Measure and record the distance between the centres of the tyres.

8 The difference between the front and rear measurements will give the toe in/toe out reading. If the larger measurement is at the front, the wheels are in a toe out position. If the larger measurement is at the rear, the wheels are in a toe in position. Compare the reading obtained with Specifications.

9 If the toe in is not to Specifications, adjust it as follows:

a Using pliers, expand the clips retaining the steering gear boots to the tie rods and move the clips onto the tie rods.

b Loosen the locknuts at both tie rod ends.

c Rotate the tie rods in small, equal amounts to bring the toe in to Specifications.

It is important to make equal adjustments to each tie rod to maintain the central position of the steering gear.

d Tighten the tie rod end locknuts securely, ensuring that the tie rods do not move from the adjusted position.

e Locate the steering gear boots in the groove provided on the tie rods, ensuring that they are not twisted or deformed, and install the retaining clips.

f Check the position of the steering wheel with the front wheels in the straight ahead position. If necessary, adjust each tie rod an equal amount to centralise the steering wheel. Do not remove and reposition the steering wheel, especially on models with an airbag.

g It is advisable to check the toe in measurement after the adjustment is completed and the tie rod end locknuts are tightened. Have the wheel alignment checked by a specialist at the earliest opportunity.

Drive Shafts

If the drive shafts are worn or damaged, it is usually more economically viable to install an exchange assembly from a drive shaft specialist. If the rubber boots require renewal, the procedure is covered in this chapter.

TO REMOVE AND INSTALL

1 Remove the wheel cover, if fitted and loosen the front wheel nuts.

2 Remove the split pin from the end of the drive shaft nut. While an assistant applies the brakes, loosen the drive shaft nut approximately one turn. A long breaker bar will be necessary, as the drive shaft nut is very tight.

The hub bearing may be damaged if the drive shaft hub nut is removed with the vehicle weight resting on the suspension.

3 Raise the front of the vehicle, support it on chassis stands and remove the front wheels. Refer to General Information (chapter 1) for the postions of the jacking points. Remove the drive shaft nut completely.

4 Where fitted, remove the ABS speed sensor retaining bolt and place the speed sensor clear of the work area. Remove the drive shaft nut.

5 Position a container under the transaxle. Remove the drain plug and drain the transaxle oil. Ensure that the drain plug sealing washer is serviceable and install the drain plug.

TIP *If necessary, refer to the relevant transaxle chapter for the location of the drain plug.*

6 Remove the split pin and loosen the castellated nut retaining the tie rod end to the steering knuckle.

7 Disconnect the tie rod end from the steering knuckle using a ball joint separator or alternatively by placing a hammer or a suitable dolly against one side of the steering knuckle eye and striking the opposite side with a hammer. Remove the retaining nut.

8 Remove the nut retaining the ball joint to the steering knuckle. Disconnect the ball joint using a similar procedure to the tie rod. Discard the ball joint self locking nut.

9 Place the hub assembly in the straight ahead position and ensure that it remains so while removing the drive shaft.

10 Using a suitable puller, push the outer end of the drive shaft through the hub taking care not to damage the hub inner oil seal or ABS sensor rotor, where fitted.

In the absence of a puller, position the drive shaft retaining nut so that the flat face of the nut is level with the end of the shaft. The castellations on the nut should face the hub.

Strike the end of the shaft with a heavy copper/hide hammer to loosen the splines, remove the nut and using a suitable drift, drive the shaft through the hub.

TIP *It may be necessary to temporarily install the ball joint to make the assembly more stable.*

Method of removing the left hand drive shaft from the transaxle. The bar or lever at the top is used to prevent the inner joint binding in the transaxle.

11 If removing the left hand side drive shaft, release the drive shaft from the transaxle, by placing a suitable bar or lever at the top to prevent the inner joint cocking and binding in the transaxle and levering at the bottom with a tyre lever or large flat bladed screwdriver. The drive shaft snap ring can be very tight on some vehicles.

12 To provide ample working space, hold the suspension unit away from the transaxle by placing a suitable sized wooden block between the suspension unit and the inner mudguard.

Take care not to damage the drive shaft oil seal and avoid pulling on the drive shaft if it is to be reinstalled.

13 If removing the right hand side drive shaft, it is necessary to remove the engine pipe as described in Engine (chapter 4). Remove the support bearing retaining bolts and guide the drive shaft from the transaxle.

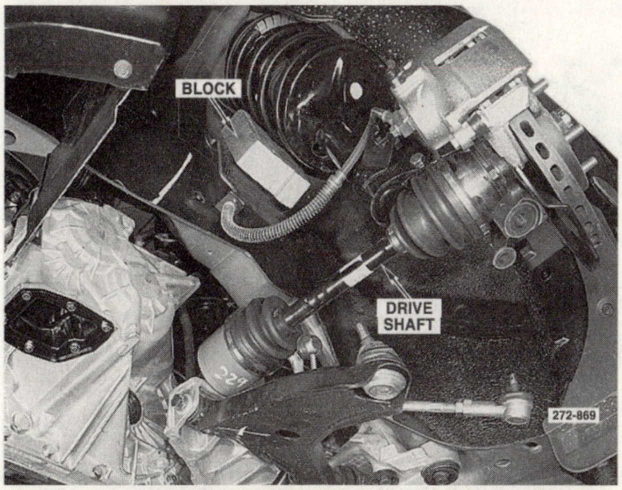

Using a wooden block to hold the suspension unit away from the transaxle will give more clearance to allow removal of the drive shaft from the hub.

14 Withdraw the drive shaft from the vehicle and plug the hole in the transaxle to prevent the entry of dirt. Remove and discard the drive shaft snap ring

Installation is a reversal of the removal procedure with attention to the following points:

1 Check that the drive shaft joints and boots are serviceable. On the right hand side check that the support bearing turns freely without any roughness or noise.

2 Install a new snap ring to the left hand side drive shaft prior to installation into the transaxle.

3 Check the transaxle oil seal for wear and deterioration and if necessary, renew as described in Manual Transaxle (chapter 10).

4 Apply a smear of multipurpose grease to the lip of the steering knuckle inner oil seal and to the drive shaft oil seal.

5 Install the drive shaft hub nut and while an assistant applies the brakes, tighten the nut to the specified torque. Install a new split pin to the nut.

6 On the right hand drive shaft install and tighten the support bearing retaining bolts to the specified torque.

7 Fill the transaxle with the correct quantity and grade of lubricating oil. Refer to the relevant transaxle chapter if necessary.

TO RENEW SUPPORT BEARING

1 Remove the right hand drive shaft as previously described.

2 Release the tripod joint rubber boot retaining clip (large) and while holding the rubber clear of the work area, remove the internal snap ring. Slide the outer housing from the tripod and rollers assembly. Tape the rollers to the trunnion assembly to prevent them falling off. Note the direction of the trunnion and rollers for assembly purposes.

3 Support the tripod outer housing on press plates and pressing on the welch plug inside the housing, press the relay shaft from the outer housing. Discard the welch plug and the relay shaft snap ring.

4 Using press plates positioned under the support bearing housing and pressing on the end of the shaft, push the shaft through the bearing.

Pressing the welch plug and relay shaft from the tripod housing.

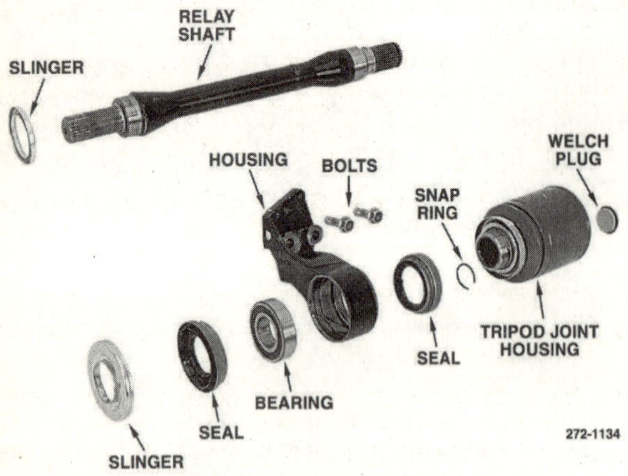

Relay shaft and support bearing components.

5 Note the installed position and using a suitable lever , remove the dust seals from the bearing housing.

6 Using a press and a suitable tube or mandrel, press the bearing from the housing.

7 Press the new bearing into the bearing housing using a suitable tube or mandrel the pushes on the outer race only. Continue pressing until the bearing contacts the shoulder of the bore in the bearing housing.

8 Apply some multipurpose grease to the inside of the new dust seals and install seals to the bearing housing.

9 Using a press and suitable attachments. Press the shaft through the bearing ensuring that the seal lip does not turn back as the shaft passes through.

10 Install a new snap ring to the end of the relay shaft, and using suitable attachments press the tripod joint housing onto the relay shaft until the snap ring seats.

11 Install a new welch plug to the inner bore of the tripod joint housing.

12 Further assembly and boot renewal for the right hand side drive shaft follows the procedure described below.

TO RENEW BOOTS

The manufacturer uses different types of grease for the inner tripod joint and the outer constant velocity joint. To ensure the specified grease is used on the components, always use the grease supplied in the repair kit.

It is a good practice to renew both drive shaft boots at the same time.

1 Remove the relevant drive shaft, as previously described.

2 Remove the retaining clips and pull the rubber boot back from the drive shaft inner joint.

3 Mark the installed position of the tripod joint housing in relation to the shaft and withdraw the large internal snap ring from the housing. Slide the trunnion and drive shaft assembly from the joint housing, holding onto the rollers to prevent them dropping from the trunnion.

Do not mark the tripod joint housing with a centre punch or chisel.

4 Remove the excess grease and remove the small retaining snap ring from the end of the drive shaft.

5 Mark the installed position of the trunnion assembly and drive shaft and withdraw the trunnion assembly from the drive shaft.

It may be necessary to use a hammer and brass drift to remove the trunnion assembly from the drive shaft. Do not hit the rollers or damage may result.

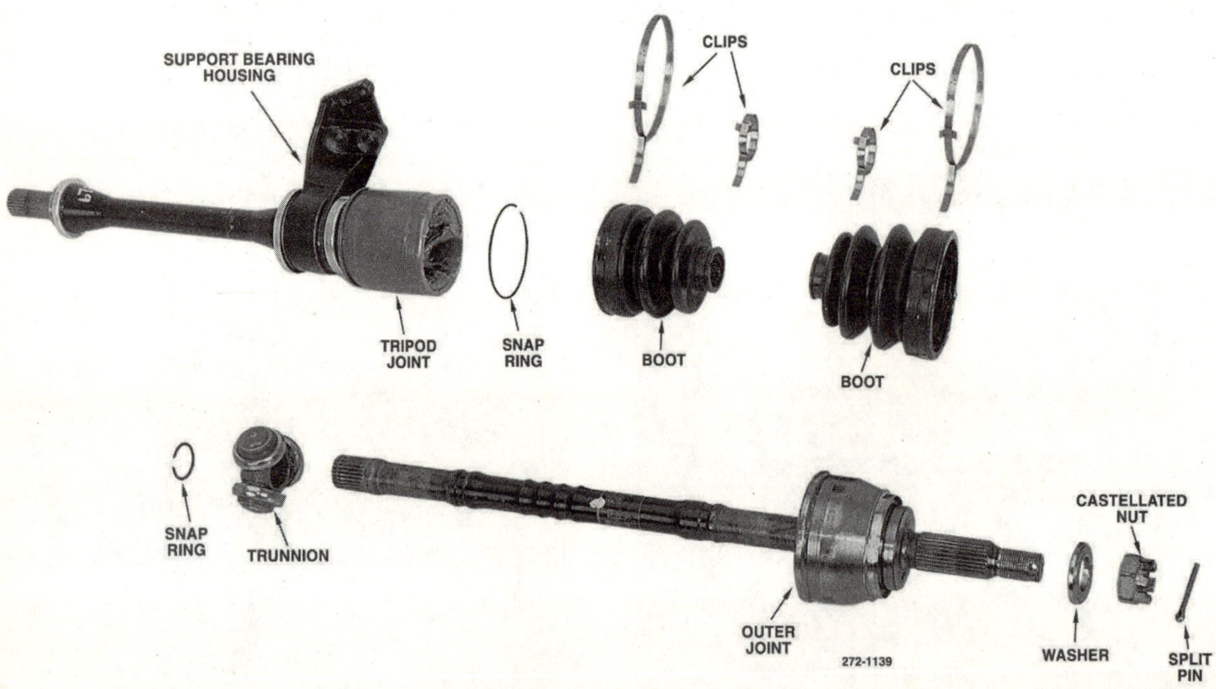

Right hand drive shaft dismantled.

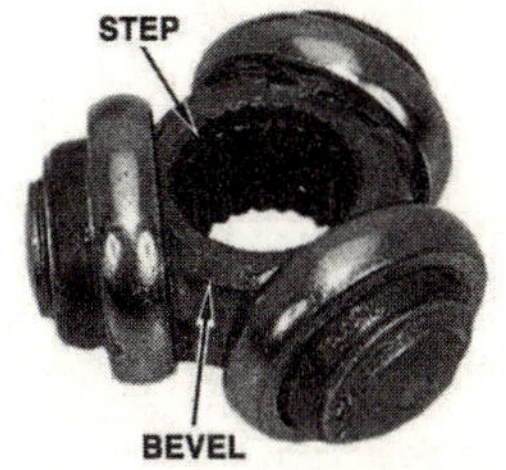

Install the trunnion with the bevel and step towards the outer joint.

6 Remove the rubber boot and the retaining clip.

7 Clean the tripod joint housing, trunnion assembly and drive shaft splines as much as possible without using solvent. Do not dismantle the trunnion assembly.

8 Remove the boot retaining clips from the outer joint.

9 Withdraw the rubber boot from the drive shaft, noting the installed positions to aid assembly.

10 Remove as much of the old grease as possible from the outer drive shaft joint without using solvent. Note the amount of old grease removed to aid assembly.

11 Cover the splines on the inner end of the drive shaft with electrical tape to protect the rubber boots during installation.

12 Work the special grease supplied with the outer boot repair kit into the outer joint and apply the remainder to the inside of the outer drive shaft boot.

13 Slide the rubber boot onto the shaft, outer end first, in the positions noted on dismantling. Install and tighten the new small retaining clip.

14 Install the small retaining clip and the new inner boot to the drive shaft.

15 Install the trunnion assembly onto the drive shaft in the positions marked on dismantling and install the small retaining snap ring in the groove on the end of the drive shaft.

TIP *The marking on the trunnion assembly must face towards the outer joint.*

16 Pack the special grease supplied in the repair kit into the tripod joint housing.

17 Slide the tripod joint housing onto the trunnion assembly to the position noted on removal. If no marks were made refer to the illustration. Ensure that the trunnion assembly is pushed fully into the tripod joint housing.

18 Pack the excess grease forced out of the joint assembly into the rubber boot and into the cavity in the housing behind the installed trunnion assembly. Install the large internal snap ring.

19 Slide the rubber boot onto the tripod joint housing ensuring that the ends of the rubber boot are correctly positioned as noted during dismantling. Securely tighten the small retaining clip. Slide the trunnion to the mid position in the tripod joint housing.

20 Release the excess air from the boots, by carefully inserting a small screwdriver between the joint and the boot and slightly raising the boot.

21 Install new large boot retaining clips to inner and outer rubber boots. Tighten the clips securely.

22 Install the drive shaft as previously described.

Chapter 14
Rear Axle and Suspension

Part 1. Sedan

Contents

Specifications

Type .... Multi link independent with struts
Shock absorber .... Hydraulic or gas/hydraulic
Hub bearing:
- End float (maximum) .... 0.05 mm
- Turning resistance .... Less than18 N

Rear wheel alignment:
- Toe in (at centre of tread) .... 3 ± 3 mm
- Camber:
  - With 205/65R 15 tyres .... Minus 40 min ± 30 min
  - With 215/60R 16 tyres .... Minus 50 min ± 30 min

TORQUE WRENCH SETTINGS

ABS speed sensor retaining bolt .... 7.8 Nm
Brake caliper anchor plate bolts .... 49 – 59 Nm
[1] Control arm pivot bolt nuts .... 118 Nm
Crossmember to body nuts .... 88 Nm
Hub assembly to carrier bolts .... 74 – 88 Nm
[1] Lower control arm pivot bolt nuts .... 98 Nm
Stabiliser bar:
- Bracket to body bolts .... 19 Nm
- Link nuts .... 44 Nm

Suspension unit:
- Upper mounting to body nuts .... 44 Nm
- Piston rod nut .... 20 – 25 Nm
- Lower mounting eye to carrier bolt .... 98 Nm

[1] Toe control arm:
- Cam bolt nut .... 69 – 78 Nm
- Ball joint nut .... 28 Nm

[1] Trailing arm:
- Front pivot to body nut .... 137 – 157 Nm
- Rear pivot to carrier nut .... 118 – 137 Nm

Upper control arm:
- Mounting bracket to body bolts .... 39 Nm
- [1] Inner pivot nuts .... 57 Nm
- [1] Outer pivot to carrier nut .... 98 Nm

[1] Tighten with the weight of the vehicle on the suspension.

Description

The independent multi link rear suspension consists of a suspension unit, a rear axle carrier bracket, upper and lower control arms, a trailing arm and a toe control arm. Rubber bushes are installed at all pivot points except the outer toe control arm and lower carrier bracket to give smooth and quiet operation. The toe control arm uses a ball joint at the outer end and the lower carrier bracket uses a spherical joint.

The suspension unit consists of a sealed shock absorber enclosed at the upper end by a coil spring on which an upper mounting assembly is located and attached to the underside of the rear wheel housing.

The shock absorber piston rod is attached to the centre of the suspension unit upper mounting assembly.

The lower end of the suspension unit is attached the rear axle carrier to which the brake caliper and hub assembly is mounted.

Rear suspension components with brake disc and anchor plate removed.

To maintain the rear axle carrier in correct alignment, a trailing arm is mounted to the body forward of the suspension, and control arms are mounted front and rear of the suspension cross-member. The rear toe control arm has an eccentric inner pivot bolt to allow corrections to the rear wheel alignment.

A stabiliser bar is fitted to Sports models to reduce body roll when cornering. The stabiliser bar is mounted to the underbody via rubber bushes and attached at the outer ends to the axle carriers via ball joint links.

TIP *At the time of writing, a worn or damaged rear suspension bush necessitated renewing the complete component i.e. carrier bracket. Renewal of these bushes follows standard workshop practice and it is possible that the bushes may become available as separate components. Check with your spare parts supplier regarding price and availability before renewing a complete component for a worn bush.*

Routine Maintenance

1 Check the shock absorber condition by bouncing the rear of the vehicle up and down (one side at a time) and observe the vehicle. If the shock absorbers are serviceable the vehicle should come to rest in a single movement. If it bounces two or three times before stopping, the suspension units should be dismantled and the shock absorbers renewed.

2 Check the rear suspension components for wear by raising the rear of the vehicle and supporting on chassis stands, leaving the wheels and suspension unsupported. Refer to General Information (chapter 1) for the correct vehicle jacking points.

Check the suspension units for fluid leakage which would indicate failure of the shock absorber shaft seals.

With an assistant pushing and pulling the rear wheel in and out, check for excessive looseness and wear in the wheel bearings, the suspension arm bushes and the toe control arm ball joint.

3 On Sports models, inspect the stabiliser bar mounting and link joints and the suspension arm bushes for wear, damage and deterioration. Ensure that all mounting nuts and bolts are tight.

Suspension units that are leaking fluid will not be operating correctly and must be renewed.

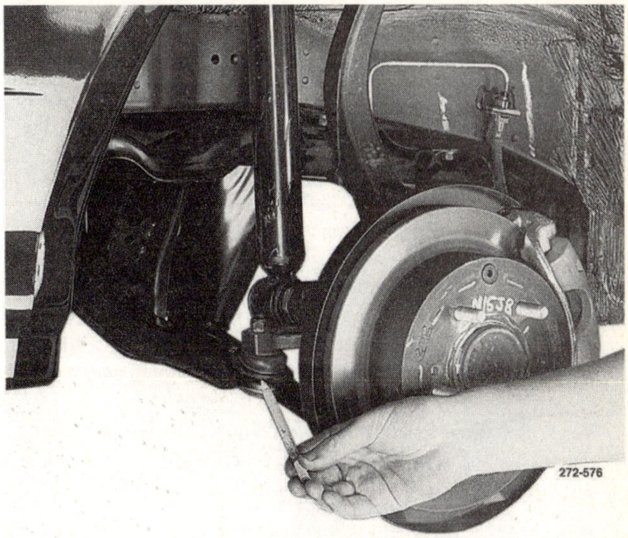

Check the toe control arm ball joint for wear.

4 Inspect the vehicle ride height. If it is lower on one side, remove the coil spring from the suspension unit and check its free length against a new spring. If the spring is found to be unserviceable, it is a good practice to install two springs as a matching pair.

Trouble Shooting

REAR END NOISE

1 Defective suspension unit or mounting: Renew faulty components.

2 Worn rear hub bearings: Check and renew the hub bearings as necessary.

3 Loose or worn control arm bushes or pivot bolts: Check and tighten or renew worn components.

4 Worn toe control arm ball joint: Check the ball joint for wear and renew the toe control arm if necessary.

5 Worn carrier bracket spherical joint: Check for wear in the joint and renew the carrier bracket if necessary.

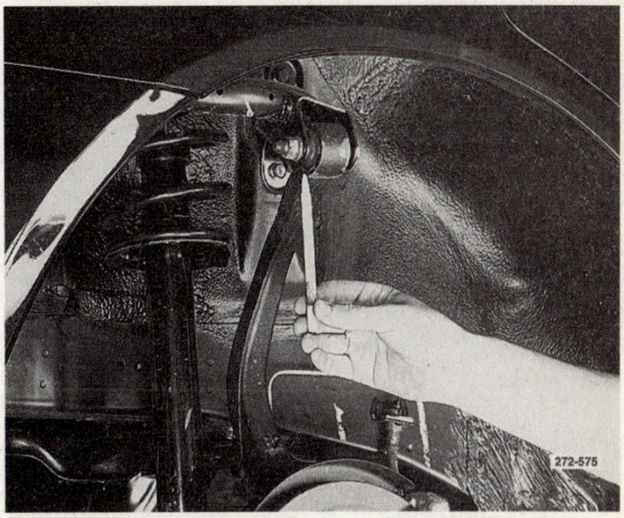

Check all suspension bushes for wear and loose pivot bolts.

6 Broken coil spring: Renew the coil spring, preferably in matching pairs.

7 Loose or worn trailing arm bushes or pivot bolts: Check and tighten or renew worn components.

Refer to the Routine Maintenance heading for a quick guide to suspension unit condition.

To check control and trailing arm bushes or pivot bolts, insert a lever between the suspect unit and its mounting and lever the unit back and forth checking for excessive movement.

Rear hub bearing noise can be diagnosed by raising and supporting the rear of the vehicle, spinning one wheel at a time and listening for a rumbling noise.

Inspect the coil springs for breaks. If the spring is found to be unserviceable, it is good practice to install two new coil springs as a matching pair.

POOR OR ERRATIC ROAD HOLDING ABILITY

1 Low or uneven tyre pressures: Inflate to the recommended pressures.

2 Incorrect front or rear wheel alignment: Check and adjust the wheel alignment as necessary.

3 Defective suspension unit: Renew the faulty suspension unit, preferably in pairs.

4 Loose or broken stabiliser bar or worn links — Sports models: Check and tighten or renew faulty components.

5 Loose or worn control arm bushes or pivot bolts: Check and tighten or renew worn components.

6 Worn toe control arm ball joint: Check the ball joint for wear and renew the toe control arm if necessary

7 Loose or worn trailing arm bushes or pivot bolts: Check and tighten or renew worn components.

8 Loose rear hub bearings: Check and renew the rear hub bearings as necessary.

Check the serviceability of the suspension units, the trailing and control arm bushes and pivot bolts as previously described.

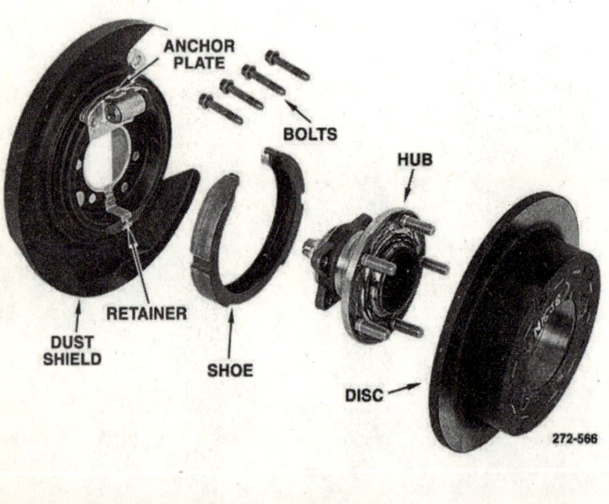

Rear hub, disc and handbrake components.

If incorrect rear wheel alignment is suspected, it is recommended that the vehicle be taken to a wheel alignment specialist or an authorised dealer to be checked and if necessary, adjusted.

Rear Hub

The rear hub is a sealed unit and should not be dismantled. Models with ABS brakes use a different hub to those without, so ensure that the correct hub is used if renewal is necessary.

To Check Hub Bearing End Float — Dial gauge

TO CHECK, REMOVE AND INSTALL

1 Raise the rear of the vehicle and support it on chassis stands. Refer to General Information (chapter 1) for the position of vehicle jacking points.

2 Remove the rear wheel and release the handbrake. Remove the brake caliper anchor plate bolts and slide the brake caliper from the rear disc. Support the caliper from the upper control arm using wire or similar.

Do not allow the caliper to hang from the brake hose.

3 Withdraw the brake disc from the hub. If necessary, refer to Brakes (chapter 15) for more Information.

4 On models with ABS, remove the ABS sensor from the carrier.

5 Spin the hub and check for roughness and play. If roughness and play exist checking end float with a dial gauge is unnecessary, the hub should be renewed.

6 Secure a dial gauge to the suspension unit or the caliper anchor plate mounting with its plunger set at zero and abutting the outer edge of the hub flange.

Check the hub end float by pulling and pushing the hub in and out. Note the reading on the dial gauge.

If hub end float is outside Specifications, install a new or exchange hub on installation.

7 Remove the hub assembly retaining bolts and remove the hub assembly from the axle carrier bracket. Secure the anchor plate to the carrier bracket to prevent the assembly falling and getting damaged.

8 Check that no grease has weeped from the hub assembly. Renew the hub assembly if necessary.

9 On models with ABS, ensure that the pulse ring is not damaged. If necessary the pulse ring can be removed using a bearing separator and a press and a new pulse ring pressed on.

10 Install the hub assembly to the carrier bracket with the flat on the hub at the bottom. Install and tighten the retaining bolts to Specifications.

11 On models with ABS, install the ABS sensor to the carrier.

12 Install the rear brake disc to the hub assembly.

13 If the brake hose was disconnected for any reason, bleed the brakes as described in Brakes (chapter 15).

14 Install the rear wheel and lower the vehicle to the ground.

Suspension Unit

To Remove and Install Coil Spring — Coil spring compressor

TO REMOVE AND DISMANTLE

1 Remove the rear seat as follows:

a Locate the two retaining clips at the front of the rear seat cushion. While pulling each clip towards the front of the vehicle, lift up the front of the cushion in the area of the clips and manoeuvre it from the vehicle.

b Remove the bolts retaining the rear seat backrest to the vehicle body and ease the side backrest upwards off the hooked brackets at the top and withdraw it from the vehicle.

2 Prise out the clips retaining the front of the parcel shelf to the body, raise the front of the parcel shelf just enough to gain access to the upper mounting nuts and support the shelf in this position using a block of timber or similar.

3 Remove the dust cover from the suspension unit upper mounting and remove the upper mounting nuts.

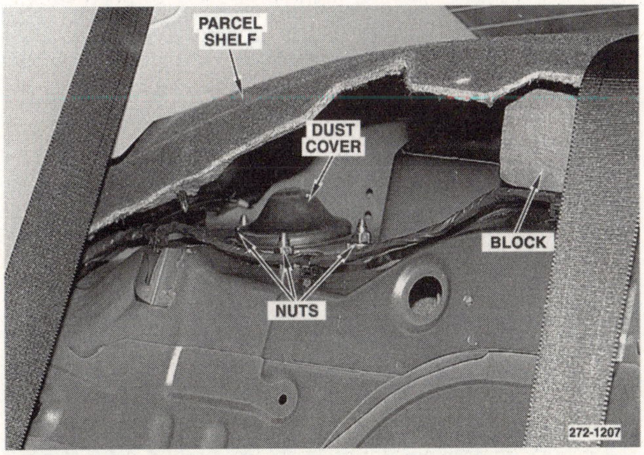

Suspension unit upper mounting. Using a timber block to support the parcel shelf as shown avoids the trouble of removing it.

4 If the suspension unit is to be dismantled, loosen the piston rod nut, but do not remove it at this stage.

The coil spring is partially compressed and has a large amount of potential energy. Removal of the piston rod nut may cause serious injury to the operator.

5 Raise the rear of the vehicle and support it on chassis stands. Remove the rear wheel. Refer to General Information (chapter 1) for the correct vehicle jacking points.

6 Remove the lower suspension unit to carrier retaining bolt and washers. Manoeuvre the suspension unit from the vehicle.

7 Thoroughly clean the suspension unit and secure it in a vice equipped with soft jaws.

8 Using spring compressors, compress the coil spring. Remove and discard the piston rod nut.

To avoid injury, do not remove the piston rod nut without a spring compressor installed.

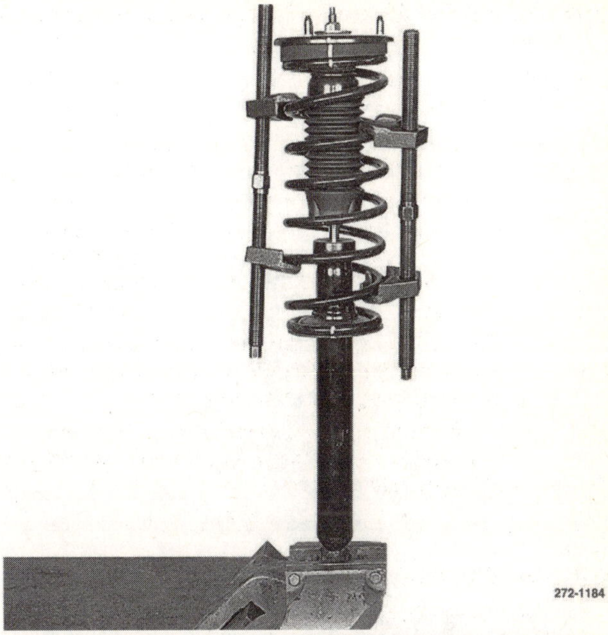

Spring compressors installed. Note the alignment marks on the upper and lower mountings, insulator and spring.

9 Mark the installed position of the spring, upper mounting and insulator. Withdraw the washer, bush, upper mounting and insulator, the coil spring, sleeve, bush dust cover and the bump stop and from the piston rod.

10 Inspect the coil spring for cracks and damage. If one coil spring is found to be unserviceable, it is good practice to install two new rear springs as a matching set.

11 Inspect all rubber components for damage and deterioration and renew as necessary. If the rubber bush in the lower

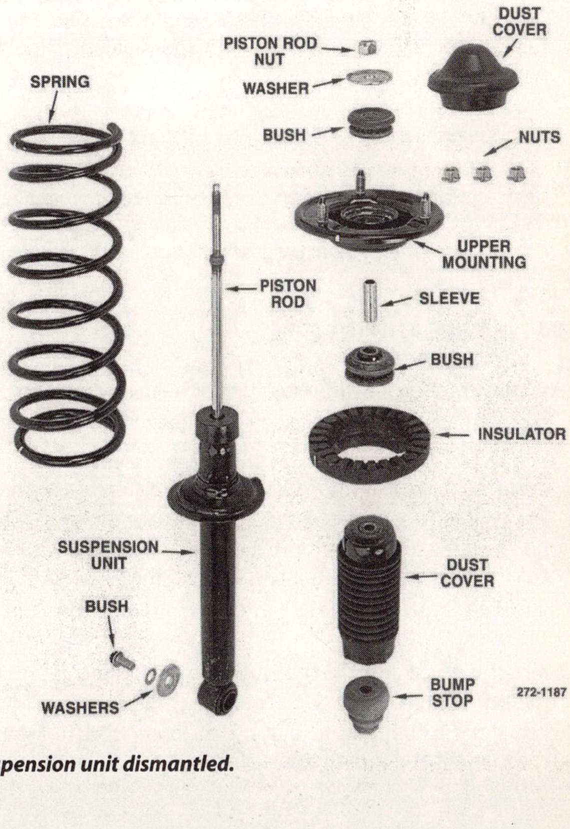

Suspension unit dismantled.

Check for defective shock absorber operation by working the piston rod up and down.

end of the suspension unit is worn, renew the suspension unit.

12 Check the operation of the shock absorber as follows:

With the suspension unit vertical, move the piston rod up and down several times. A firm pressure with no slack spots should be evident over the entire operating range of the suspension unit.

Check the shock absorber for fluid leakage. If the leakage is excessive, i.e. more than just a very slight weep around the seal, and/or no pressure is felt, the suspension unit should be renewed, preferably in pairs.

A gas shock absorber strut should return at a steady speed without assistance.

> *If a shock absorber strut is gas pressurised, the word GAS is usually stamped on the body of the strut. Do not incinerate or puncture gas filled struts.*

TO ASSEMBLE AND INSTALL

Assembly and installation is a reversal of the removal and dismantling procedure with attention to the following points:

1 Secure the suspension unit in a soft jawed vice.

2 Install the bump stop and the cover to the piston rod ensuring that the bump stop sits on the piston rod shoulder.

3 Using spring compressors, compress the coil spring and position it on the suspension unit. Ensure that the lower end of the coil spring is abuts the stop pressed in to the suspension unit.

4 Install the bush with the lip facing the top. Install the collar to the bush.

5 Install the insulator to the spring seat, aligning the marks made on removal.

6 Install the spring seat to the suspension unit, ensuring that the marks made during removal are correctly aligned.

> *When installing the upper mounting, ensure that the D shape in the upper mounting is correctly aligned with the piston rod.*

7 Install a new piston rod nut and tighten securely. Final tightening will be done when the suspension unit is installed.

8 Carefully remove the spring compressors from the suspension unit.

9 Position the suspension unit in the rear wheel housing and with the aid of an assistant, install the suspension unit upper mounting assembly retaining nuts and tighten them to the specified torque.

10 Supporting the weight of the carrier, install the suspension unit to the carrier, install the retaining bolts and nuts and tighten to the specified torque.

11 On sports models, install the stabiliser bar link to the suspension unit, install and tighten the retaining nut to Specifications.

12 If applicable, install the bolt retaining the ABS speed sensor wire to the suspension unit.

13 Lower the vehicle to the ground and tighten the piston rod nut to the specified torque. Install the dust cap and the parcel shelf retaining clips.

Trailing Arm

TO REMOVE AND INSTALL

1 Raise the rear of the vehicle and support it on chassis stands. Remove the rear wheels. Refer to General Information (chapter 1) for the position of the vehicle jacking points.

2 Remove the rear trailing arm pivot bolt, nut and washer from the carrier bracket.

3 Prise out the plastic cap and remove the front trailing arm pivot bolt, nut and washer from the body. Remove the trailing arm from the vehicle, collect the rubber stoppers from each side of the trailing arm bush.

4 Inspect the trailing arm bush for wear and deterioration. If the bush is worn, renew the trailing arm as an assembly.

Inspect the trailing arm for cracks and damage. Inspect the pivot bolts and nuts for wear on the shaft or threads. Inspect the stoppers for wear and deterioration. Renew components as necessary.

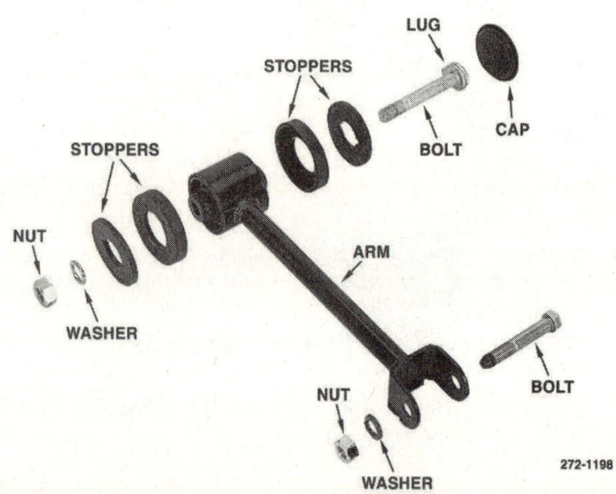

Trailing arm components.

Installation is a reversal of the removal procedure with attention to the following points:

1 Install the trailing arm to the front mounting ensuring that the stoppers are not displaced. Install the pivot bolt, washer and nut, ensuring that the anti-spin lug on the pivot bolt head is located correctly on the bracket. Do not tighten the nut at this stage.
2 Install the trailing arm to the carrier bracket mounting and install the pivot bolt and nut. Do not tighten the pivot bolt nut at this stage.
3 Lower the vehicle to the ground. With the weight of the vehicle on the wheels bounce the rear of the vehicle several times. Tighten the trailing arm pivot bolt nuts to the specified torque.
4 If a new trailing arm has been installed, it will be necessary to have the rear wheel alignment checked and adjusted as necessary. Refer to the Rear Wheel Alignment heading in this chapter.

Upper Control Arm

TO REMOVE AND INSTALL

1 Raise the rear of the vehicle and support it on chassis stands. Remove the rear wheels. Refer to General Information (chapter 1) for the position of the vehicle jacking points.
2 Remove the upper control arm to carrier bracket pivot bolt, nut and washer. Separate the two components.
3 Remove the bolts retaining the upper control arm mounting brackets to the vehicle body and manoeuvre the control arm from the vehicle.
4 Inspect the control arms bushes for wear and deterioration. Inspect the control arm for cracks and damage. If necessary, remove the pivot bolts and nuts and renew the upper control arm. The bushes are not available separately. If dismantled, examine the bolts and nuts for wear and damage. Renew components as necessary.
5 Inspect the upper carrier bracket bush for wear and deterioration and renew the carrier bracket if necessary.

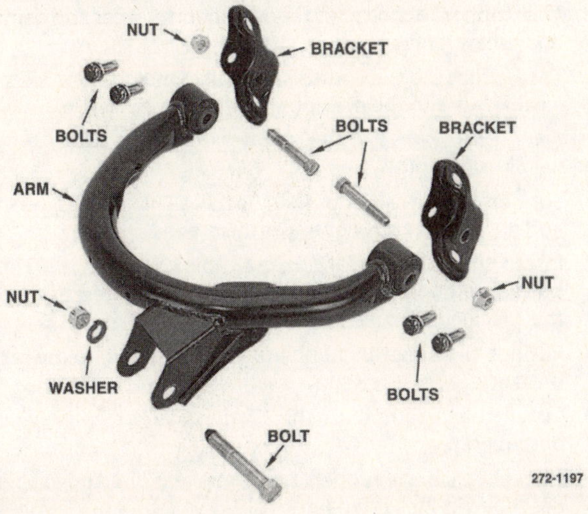

Upper control arm dismantled.

Installation is a reversal of the removal procedure with attention to the following points:

1 If the mounting brackets have been disturbed, install but do not tighten the pivot bolts and nuts.
2 Install the upper control arm to the vehicle and tighten the mounting bracket bolts securely.
3 Connect the outer end of the control arm to the carrier bracket and install the pivot bolt nut and washer. Do not tighten the pivot bolt nut at this stage. Install the rear wheel.
4 Lower the vehicle to the ground and bounce the rear up and down several times. With the weight of the vehicle on the wheels, tighten the pivot bolt nuts to the specified torque.

Lower Control Arm

TO REMOVE AND INSTALL

1 Raise the rear of the vehicle and support it on chassis stands. Remove the rear wheels. Refer to General Information (chapter 1) for the position of the vehicle jacking points.
2 Remove the nuts and remove the pivot bolts from the vehicle. Withdraw the lower control arm from the vehicle.
3 Inspect the control arms bush for wear and deterioration. Inspect the control arm for cracks and damage. Examine the bolts and nuts for wear and damage. Renew components as necessary.
4 Inspect the lower carrier bracket spherical joint for wear and deterioration and renew the carrier bracket if necessary.

Installation is a reversal of the removal procedure with attention to the following points:

1 Install the lower control arm to the vehicle, with the protective plate towards the front of the vehicle and insert the pivot bolts. Do not tighten the pivot bolt nuts at this stage. Install the rear wheel.
2 Lower the vehicle to the ground and bounce the rear up and down several times. With the weight of the vehicle on the wheels, tighten the pivot bolt nuts to the specified torque.

Toe Control Arm

To Remove Toe Control Arm —
Ball joint separator

1 Raise the rear of the vehicle and support it on chassis stands. Remove the rear wheels. Refer to General Information (chapter 1) for the position of the vehicle jacking points.
2 Using quick drying paint, suitably mark the position of the eccentric cam bolt and plate to preserve the rear toe in setting on installation.
3 Remove the toe control arm ball joint nut and discard. Using a ball joint separator disconnect the ball joint from the carrier bracket.
4 Hold the cam bolt using a socket and suitable extension and remove the nut. Collect the nut, washer and cam plate. Remove the cam bolt noting the position of the cam plate.
5 Withdraw the toe control arm from the vehicle.

Before removing the toe control arm, mark the position of the eccentric cam bolt and plate to preserve the rear toe in setting.

6 Inspect the toe control arm bush for wear and deterioration. Inspect the arm for cracks and damage. The ball joint stud should move smoothly with a small amount of drag, the stud must not be loose in the arm. Renew the toe control arm if necessary.

TIP *If the ball joint boot is worn or damaged the boot is available as a spare part. Apply multipurpose grease to the ball joint and the inside of the boot when fitting a new boot and push the boot into place using a suitable smooth tube or mandrel.*

7 Inspect the pivot bolt eccentrics and nut for wear and damage and renew as necessary.

Installation is a reversal of the removal procedure with attention to the following points:

1 Install the toe control arm to the crossmember and retain with the cam bolt. Ensure that the cam bolt and plate is located as marked on removal. Install the nut and washer but do not tighten the nut at this stage.

2 Insert the ball joint stud through the carrier bracket and retain with a new nut. Do not fully tighten the nut at this stage.

3 Install the rear wheel and lower the vehicle to the ground. Bounce the rear up and down several times.

4 With the weight of the vehicle on the wheels, tighten the ball joint nut to the specified torque.

5 Rotate the cam bolt nut until the marks placed on the cams during removal align. Tighten the cam bolt nut to the specified torque.

Carrier Bracket

To Disconnect Toe Control Arm — Ball joint separator

TO REMOVE AND INSTALL

1 Remove the rear hub as previously described.

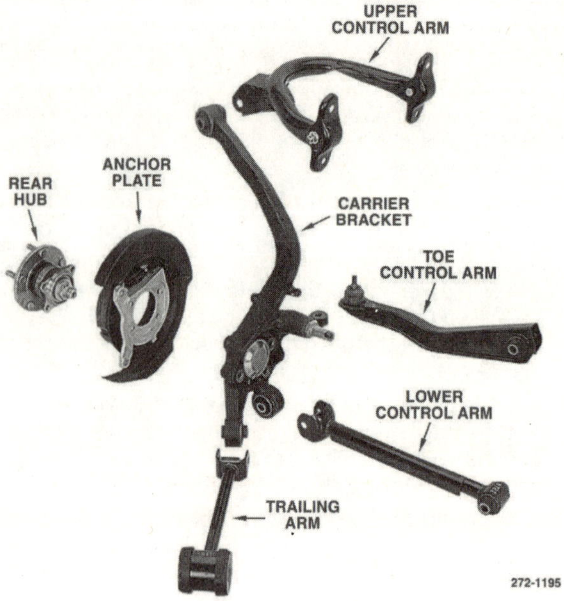

Sedan rear suspension arms and carrier bracket.

2 Separate the anchor plate from the carrier bracket and support the assembly clear of the work area using wire or similar.

3 Remove the toe control arm ball joint nut and discard. Using a ball joint separator disconnect the ball joint from the carrier bracket.

4 Remove the bolts, nuts and washers retaining the control arm, suspension unit eye bolt and trailing arm to the lower part of the carrier bracket and separate these components.

5 Supporting the carrier bracket, remove the upper control arm to carrier bracket pivot bolt, nut and washer. Remove the carrier bracket from the vehicle.

6 Wipe clean the carrier bracket. Ensure that the carrier bracket bushes and the spherical joint are not immersed in solvent otherwise damage will occur.

7 Check the carrier bracket for wear, cracks and damage. Pay particular attention to the rubber bushes and the spherical joint.

8 Check the pivot bolts nuts and washers for wear and damage and renew if necessary.

9 Ensure that the toe control arm ball joint and boot are serviceable, refer to the previous heading, if necessary.

Installation is a reversal of the removal procedure with attention to the following points:

1 Supporting the carrier bracket, install all pivot bolts, washers and nuts but do not tighten at this stage.

2 Install the toe control arm ball joint stud to the carrier bracket and install a new retaining nut. Tighten the nut to the specified torque.

3 Position the brake anchor plate and dust shield to the carrier bracket.

4 Install the rear hub to the carrier bracket as previously described.

5 On rear disc brake models, install the rear disc and caliper as described in Brakes (chapter 15).

6 If applicable, install the bolt retaining the ABS speed sensor wire to the suspension unit.

7 Lower the vehicle to the ground and bounce the rear up and down several times. With the weight of the vehicle on the wheels, tighten the pivot bolt nuts to the specified torque.

8 If a new carrier bracket was installed it will be necessary to check the rear wheel alignment as described under the appropriate heading in this chapter.

Stabiliser Bar — Sports models

TO REMOVE AND INSTALL

1 Raise the rear of the vehicle and support it on chassis stands. Refer to General Information (chapter 1) for the positions of the vehicle jacking points.

2 Remove the retaining nuts and disconnect the stabiliser links from the lower control arms and the ends of the stabiliser bar. Remove the links from the vehicle.

3 If the manufacturers marks are missing, apply paint marks on the stabiliser bar adjacent to the mounting rubbers to help with alignment on installation.

4 Remove the retaining bolts and remove the stabiliser bar mounting brackets from the crossmember.

5 Manoeuvre the stabiliser bar from the vehicle.

6 Check the stabiliser bar for bends, cracks and damage. Renew as necessary.

7 Check the mounting rubbers for wear and deterioration and the mounting brackets for bends, cracks and damage. Renew parts as necessary.

8 Check the stabiliser links for bends, wear and damage. Check the condition of each ball joint by moving the stud and checking for excessive freeplay and wear. Renew as necessary.

Installation is a reversal of the removal procedure with attention to the following points:

1 Ensure that the mounting rubbers are installed adjacent to the manufacturers marks or the paint marks made during the removal procedure. The splits in the mounting rubbers should be to the top rear when correctly installed.

2 Tighten all the nuts and bolts securely or to the specified torque .

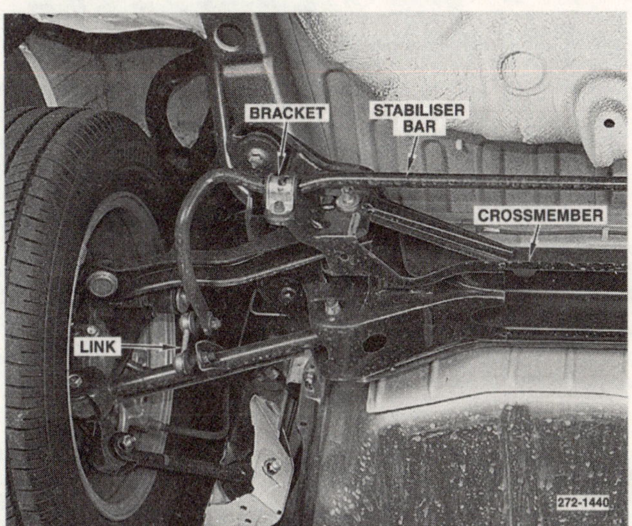

Stabiliser bar installed. Sports models.

Crossmember

TO REMOVE AND INSTALL

1 Raise the rear of the vehicle and support it on chassis stands. Remove the rear wheels. Refer to General Information (chapter 1) for the position of the vehicle jacking points.

2 On Sports models, remove the stabiliser bar as previously described.

3 Remove the lower control arm and the toe control arm from each side of the vehicle as previously described in this chapter.

4 Support the crossmember under the centre using a jack. Remove the mounting nuts and lower the crossmember from the vehicle.

5 Check the crossmember for cracks and damage. Check the rubber mounting bushes for deterioration. Renew components as necessary.

Installation is a reversal of the removal procedure with attention to the following points:

1 Install the crossmember to the vehicle and tighten the mounting nuts to the specified torque.

2 Install the lower control arm and toe control arm to the vehicle as previously described.

Rear Wheel Alignment

ADJUSTMENT

Specialised equipment is required to measure the rear wheel alignment, except for the rear wheel toe in. Therefore, as it is not a worthwhile proposition for the home mechanic to perform a complete wheel alignment, the vehicle should be taken to a wheel alignment specialist.

The rear wheel camber is not adjustable and any variation from Specifications will be caused by worn or damaged suspension components.

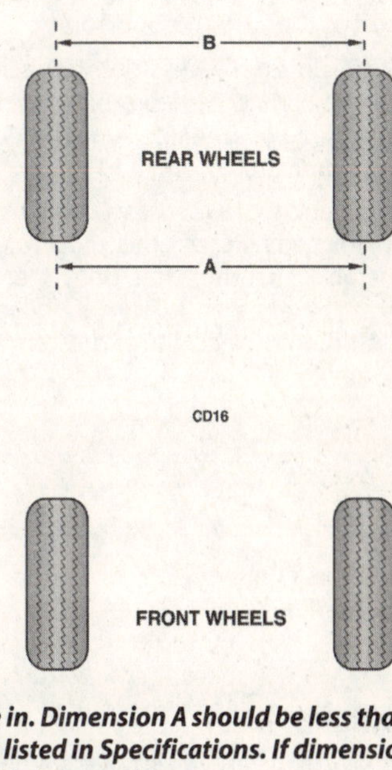

Rear wheel toe in. Dimension A should be less than dimension B by the amount listed in Specifications. If dimension A is more than dimension B, this is termed toe out.

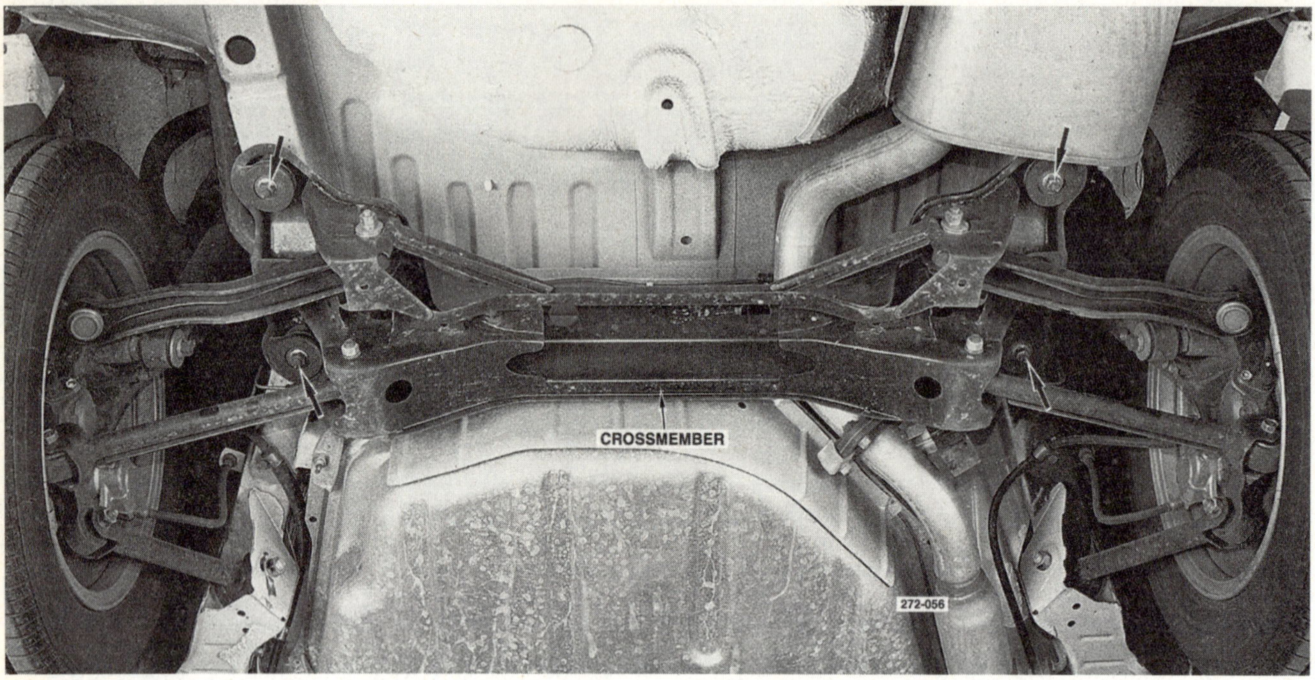

Rear crossmember installed. Arrows indicate the mounting nuts.

Prior to performing a rear wheel alignment, perform a thorough inspection of the springs and suspension components and renew as necessary.

The tread of the rear tyres should be examined for excessive or uneven wear as certain tyre conditions are indicative of damaged or worn components in the suspension and/or wheels and hub bearings. Refer to Wheels and Tyres (chapter 19). If the tyres are found to be defective, replace with serviceable tyres.

The vehicle should be unladen except for the normal amount of fuel and with the tyres inflated to the normal pressures.

TO CHECK AND ADJUST TOE IN

The following procedure may be used as a temporary measure to check and adjust toe in. The vehicle should be taken to a wheel alignment specialist at the earliest opportunity.

1. Position the vehicle on a level floor, bounce the vehicle a few times and check that the vehicle is level. If the vehicle is not level or is sagging on one side, worn or loose suspension components are indicated.
2. Measure the distance between the centre of the rear toe control arm pivot bolts. Repeat this procedure on the other side of the vehicle ensuring that both measurements are equally balanced to less than 1.0 mm by loosening the pivot bolt nuts and turning the eccentric adjusters as necessary.
3. Set the front wheels in the straight ahead position.
4. Mark a vertical chalk line on the centre of the rear tyres parallel with the imaginary centre of the axle shaft and behind the rear suspension.
5. Using a tape measure, measure and record the distance between the centres of the tyres.
6. Maintaining the wheels in a straight ahead position, roll the vehicle forward until the chalk marks are the same distance above the floor but to the front of the rear suspension.
7. Measure and record the distance between the tyres. When the lowest figure is to the front, the difference between the two figures is the toe in, which should be within Specifications. If adjustment of the toe in is necessary, loosen the pivot bolt nuts and turn the eccentric pivot bolts until the correct toe in is obtained.

The adjustment must be made equally to both sides to retain the correct left/right balance.

8. Tighten the pivot bolt nuts to the specified torque, ensuring that the pivot bolt does not move.

Part 2. Station Wagon

Contents

Specifications

Type .... Five link tubular axle with coil springs
Shock absorber .... Double acting telescopic
Hub bearing:
End float .... 0.05 mm
Turning resistance .... 1.8 Nm
Wheel alignment:
Toe in .... 3 ± 3 mm
Camber .... 30 min ± 30 min

TORQUE WRENCH SETTINGS

[1] Lower control arm to underbody bolts .... 230 Nm
[1] Lower control arm to axle bolts .... 230 Nm
[1] Upper control arm bolts .... 230 Nm
[1] Shock absorber bolt (upper) .... 45 Nm
[1] Shock absorber nut (lower) .... 45 Nm
[1] Panhard rod to body retaining nut .... 110 Nm
[1] Panhard rod to axle retaining nut .... 150 Nm
Brake caliper anchor plate bolts .... 49 – 59 Nm
Hub bearing retaining nut .... 200 – 260 Nm

[1] The vehicle should be lowered to the ground before these bolts are tightened to the specified torque.

Description

The tubular rear axle is located by four control arms, a panhard rod and two telescopic shock absorbers.

The lower control arms hold the axle square to the vehicle and provide the lower mountings for the coil springs.

Lateral location of the axle is provided by the panhard rod.

Mounts on the axle accommodate the rear of the lower control arms, the lower shock absorber mounting points and the rear of the upper control arms. Rubber bushes are installed at all pivot points to give smooth and quiet operation.

To prevent excessive noise being transferred inside the vehicle, an insulator is used on top of the coil spring. The lower end of the spring is located in the lower control arm with a small rubber insulator.

The bump stop rubber is located on the subframe directly behind the spring and over the axle on each side.

Station wagon rear suspension components.

Routine Maintenance

1 Check the shock absorber condition by bouncing the rear of the vehicle up and down (one side at a time) and observe the vehicle. If the shock absorbers are serviceable the vehicle should come to rest in a single movement. If it bounces two or three times before stopping, the shock absorbers should be renewed.

2 Check the rear suspension components for wear by raising the rear of the vehicle and supporting on chassis stands, leaving the wheels and suspension unsupported. Refer to General Information (chapter 1) for the correct vehicle jacking points.

Check the shock absorbers for fluid leakage which would indicate failure of the shock absorber shaft seals.

With an assistant pushing and pulling the rear wheel in and out, check for excessive looseness and wear in the wheel bearings and suspension arm bushes.

3 Inspect the suspension arm bushes for wear, damage and deterioration. Ensure that all mounting nuts and bolts are tight.

4 With the vehicle unladen and the tyres correctly inflated, inspect the vehicle ride height. If it is lower on one side, remove the coil spring and check its free length against a new spring. If the spring is found to be unserviceable, it is a good practice to install two springs as a matching pair.

Trouble Shooting

NOISE IN SUSPENSION

1 Defective shock absorbers or mountings: Renew faulty components as necessary. Renew the shock absorbers as a matching pair.

2 Loose or defective control arm bushes: Check, tighten or renew as necessary.

3 Worn coil spring insulator: Renew as necessary.

4 Broken coil spring: Renew the coil spring, preferably as a matching pair.

Check the shock absorber bushes for wear and deterioration.

5 Worn or damaged hub bearings: Renew as necessary.

6 Loose stabiliser bar bolts: Check the bolts and bar for damage and renew or tighten as necessary.

As a quick guide to shock absorber condition, bounce the vehicle up and down (one side at a time) and observe if the vehicle comes to rest in a single movement. If the vehicle bounces two or three times before coming to rest, the shock absorber should be removed and tested.

To check the control arm bushes and bolts, insert a lever between the suspect unit and its mounting and lever the unit back and forth, checking for excessive movement. Check the rear spring coils for breaks. Also check the insulators for damage or distortion.

If a spring or shock absorber is found to be unserviceable, it is advisable to install two new springs or shock absorbers as a matching pair.

REAR WHEELS NOT IN ALIGNMENT WITH FRONT WHEELS

1 Spring badly sagged on one side: Renew the defective coil springs, preferably as a pair.

2 Distorted or bent axle: Renew the axle assembly.

3 Bent control arm: Renew the control arm.

4 Control arm mounting bolt missing or broken: Renew the bolt.

5 Distortion of body or front crossmember: Check and rectify as necessary.

To check this condition, measure from the centre of the front wheels to the centre of the rear wheels on both sides. Compare the measurements which must be equal. Before measuring, ensure that the front wheels are in the straight ahead position.

POOR OR ERRATIC ROAD HOLDING ABILITY

1 Low or uneven tyre pressures: Inflate the tyres to the recommended pressures.

Check that all rear suspension nuts and bolts are secure.

2 Incorrect wheel alignment: Check and correct the rear wheel alignment.

3 Defective shock absorber: Renew the faulty shock absorber, preferably in pairs.

4 Loose or worn control arm bushes or mounting bolts: Check and tighten or renew worn components.

5 Loose rear hub bearings: Check and renew the rear hub.

Check the serviceability of the shock absorbers, bushes and mounting bolts as previously described. If incorrect wheel alignment is suspected, it is recommended that the vehicle be taken to a specialised workshop or an authorised dealer to be checked and if necessary repaired.

Rear Hub

TO CHECK, REMOVE AND INSTALL

1 Raise the rear of the vehicle and support it on chassis stands. Remove the rear wheels. Refer to General Information (chapter 1) for the position of the vehicle jacking points.

2 Remove the rear wheel and release the handbrake. Remove the brake caliper anchor plate bolts and slide the brake caliper from the rear disc. Support the caliper from the upper control arm using wire or similar. Do not allow the caliper to hang from the brake hose.

3 Withdraw the brake disc from the hub. If necessary, refer to Brakes (chapter 15) for more Information.

4 On models with ABS, remove the ABS sensor from the anchor plate.

5 Spin the hub and check for roughness and play. If roughness and play exist checking end float with a dial gauge is unnecessary, the hub should be renewed.

6 Secure a dial gauge to the lower control arm or the caliper anchor plate mounting with its plunger set at zero and abutting the outer edge of the hub flange.

Check the hub end float by pulling and pushing the hub in and out. Note the reading on the dial gauge.

If hub end float is outside Specifications, install a new or exchange hub on installation.

Checking the rear hub end float using a dial gauge.

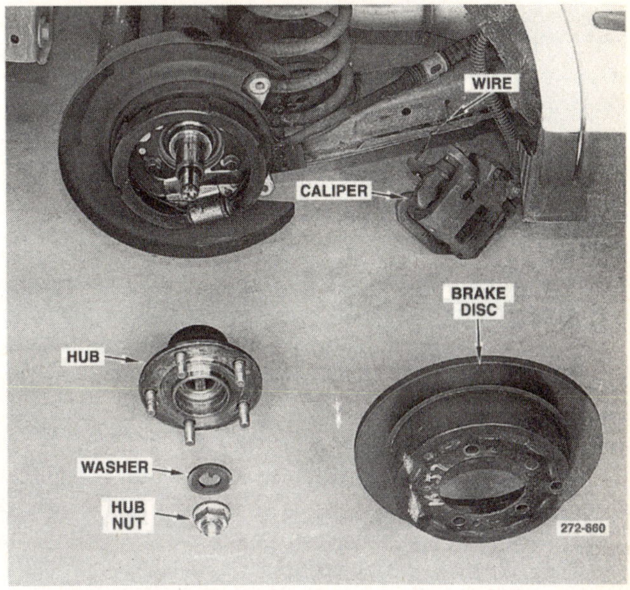

Rear hub and associated components.

7 Remove the hub assembly dust cap using a suitable thin coal chisel and hammer or by prising it off using a thin screwdriver.

8 Release the staking of the hub nut using a small, thin coal chisel or punch and remove the hub nut using a socket and a breaker bar. Discard the hub nut.

9 Withdraw the hub and thrust washer from the stub axle.

10 On models with ABS, ensure that the pulse ring is not damaged. If necessary, the pulse ring can be removed using a bearing separator and a press and a new pulse ring pressed on.

Installation is a reversal of the removal procedure with attention to the following points:

1 Check that the stub axle is not damaged, cracked or worn. If necessary renew the rear axle assembly.

2 Install the hub and thrust washer. Install a new hub nut and tighten to the specified torque.

3 Check that the hub rotates freely and without end play, before staking the hub nut to the indentation in the stub axle using a pin punch and a hammer.

4 Install the dust cap securely.

Shock Absorber

TO REMOVE AND INSTALL

1 Raise the rear of the vehicle and support it on chassis stands placed beneath the rear axle as close as possible to the rear wheels.

2 Working under the vehicle, remove the bolt and washer from the upper shock absorber mounting and the nut and washers from the lower mounting. Remove the shock absorber from the vehicle.

3 With the shock absorber removed, examine the mounting rubbers for wear and test the operation of the shock absorber as described under the following side heading. Renew as necessary.

Installation is a reversal of the removal procedure with attention to the following points:

1 Check the rubber bushes for wear and deterioration and renew the shock absorber as necessary.

2 Ensure that the mounting nuts and bolts are tightened securely when the weight of the vehicle is on the rear wheels.

TO CHECK

The extent to which a shock absorber can be tested without special equipment is limited to the following checks.

1 Check the body of the shock absorber for dents or damage and for fluid leakage. The shock absorber cannot be repaired in service and should be renewed if defective.

2 Mount the shock absorber upright in a vice by the lower mounting eye.

3 Grasp the upper half of the shock absorber, pull up to the fully extended position and slowly push down until the shock absorber is fully compressed. Renew the shock absorber if the operation is noisy.

4 Hydraulic shock absorbers should be able to be compressed and released with a firm resistance over the entire range without any slack spots. It is advisable to compress and release the shock absorber several times to remove any air bubbles if the shock absorber has been lying on its side prior to this test. Renew the shock absorber if necessary.

5 Gas shock absorbers should be able to be compressed fully and quietly. When released check that the upper part of the shock absorber returns at a constant speed when released. Renew the shock absorber if necessary.

If a shock absorber is gas pressurised, the word GAS is usually stamped on the body of the shock absorber. Do not incinerate or puncture gas filled shock absorbers.

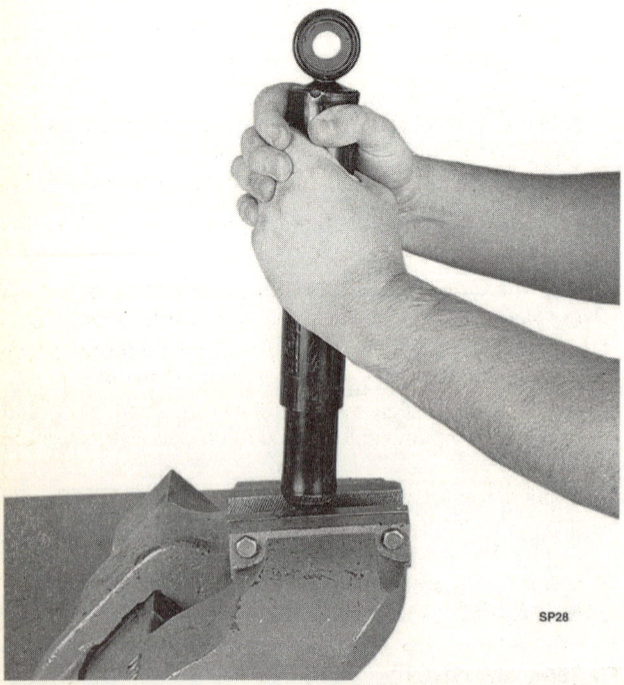

Checking for defective shock absorber operation.

Coil Spring

TO REMOVE AND INSTALL

1 Raise the rear of the vehicle and support it on chassis stands. Remove the rear wheels. Refer to General Information (chapter 1) for the position of the vehicle jacking points.

Do not fully lower the jack at this stage.

2 Remove the nut and washers and disconnect the shock absorber from its lower mounting.

3 Release the handbrake cable from the lower control arm.

4 Lower the jack until the spring is clear of the subframe.

5 Remove the spring from the vehicle and withdraw the insulators.

6 Check the spring for cracks and damage. Compare the free length with that of a new spring. If necessary, renew both springs as a matching pair.

Installation is a reversal of the removal procedure with attention to the following points:

1 Ensure that the upper and lower insulators are correctly positioned.

2 The shock absorber lower mounting nuts must be tightened after the weight of the vehicle is on the rear wheels.

Always install the spring to the side from which it was removed.

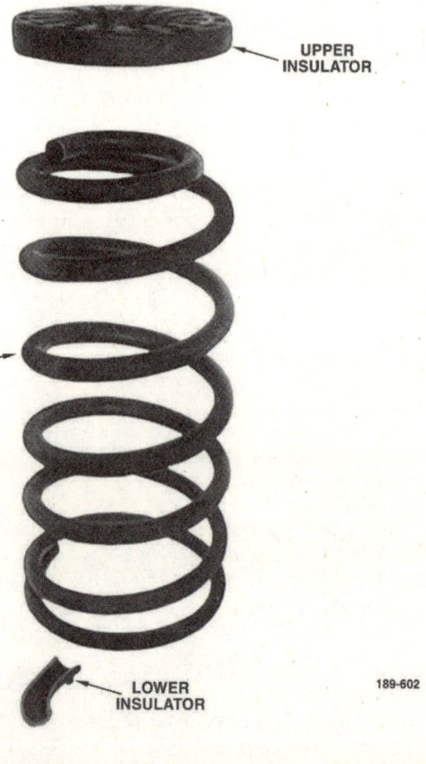

Rear spring and insulators.

Upper Control Arm

To Remove and Install Bushes — Press and press plates, suitable spacer

TO REMOVE AND INSTALL

1 Raise the rear of the vehicle and support it on chassis stands. Remove the rear wheels. Refer to General Information (chapter 1) for the position of the vehicle jacking points.

Leave the jack in the raised position underneath the rear axle.

2 Remove the coil spring as previously described.
3 Raise the jack until the axle is approximately in its normal position.
4 Remove the upper control arm mounting bolts, nuts and washers.
5 Remove the upper control arm.
6 Check the bushes and mounting bolts for wear or damage.

Installation is a reversal of the removal procedure with attention to the following points:

1 Coat the threads of the control arm mounting bolts with Permatex Lock Nut or equivalent prior to installation.
2 With the control arm in position, install the front mounting bolt, washer and nut finger tight.
3 Set the jack to align the rear mounting hole and install the bolt, nut and washer finger tight.
4 Install the coil spring as previously described, lower the vehicle to the ground and tighten the upper control arm mounting bolts to the specified torque.

TO RENEW BUSHES

1 Remove the upper control arm as previously described.
2 Using a suitable press and press plates or pipe, support the upper control arm around the outer diameter of the eye of the control arm in which the bush is housed. Ensure that there is no obstruction to prevent the bush being pressed out.
3 Using a suitable spacer of a diameter slightly smaller than the diameter of the bush, press the spacer and bush out of the upper control arm. Ensure that the control arm is not being distorted while pressing out the bush.

Installation is a reversal of the removal procedure with attention to the following points:

1 Using a suitable press and press plates, support the upper control arm and position the new bush above the recess of the control arm eye. Position the bush so the semi-circular cut outs in the bush are in line with the upper control arm beam.
2 Using a suitable spacer, press the bush into the eye of the upper control arm until the bush projects an equal distance each side of the arm. Ensure that the control arm is not being distorted while pressing in the new bush.
3 Install the upper control arm as previously described.

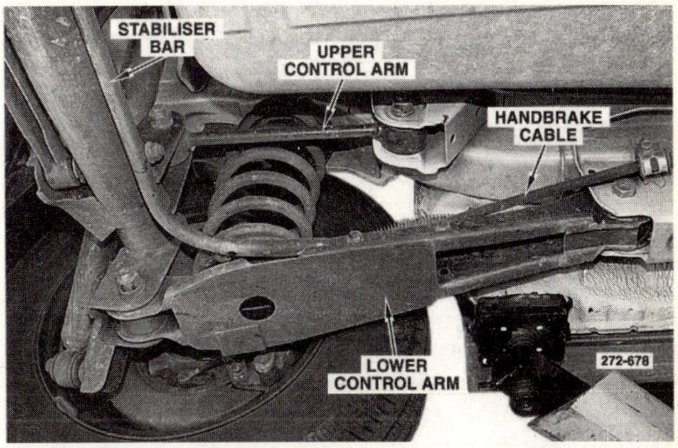

Upper and lower control arms.

Lower Control Arm

To Remove and Install Bushes — Press and press plates, suitable spacer

TO REMOVE AND INSTALL

1 Raise the rear of the vehicle and support it on chassis stands. Remove the rear wheels. Refer to General Information (chapter 1) for the position of the vehicle jacking points.

Leave the jack in position under the rear axle at this stage.

2 Disconnect the shock absorber from its lower mounting.
3 Remove the handbrake cable from the lower control arm.
4 Lower the jack fully and withdraw the coil spring and insulators.
5 Raise the jack until the rear axle is approximately in its normal position.
6 Remove the stabiliser bar to control arm retaining bolts.
7 Remove the lower control arm mounting bolts, nuts and washers and remove the lower control arm.
8 Check the bushes and mounting bolts for wear and damage.

Installation is a reversal of the removal procedure with attention to the following points:

1 Coat the threads of the control arm mounting bolts with Permatex Lock Nut or equivalent prior to installation.
2 With the control arm in position, install the front control arm mounting bolt, washer and nut. Finger tighten only.
3 Set the jack to align the control arm rear mounting and install the mounting bolts, washers and nuts. Finger tighten only.
4 Install the spring ensuring that the upper and lower insulators are correctly positioned and the spring ends sit in the mating step.
5 Lower the vehicle to the ground and tighten the control arm mounting bolts to the specified torque.

TO RENEW BUSHES

1 Remove the lower control arm as previously described.

2 Using a suitable press and press plates or pipe, support the lower control arm securely and squarely around the bush housing. Ensure that there is no obstruction to prevent the bush being pressed out.
3 Using a spacer of a diameter slightly smaller than the outer diameter of the bush, press the spacer and bush out of the lower control arm.

When pressing out the bush ensure that the lower control arm does not become distorted, especially in the area directly around the bush housing.

Installation is a reversal of the removal procedure with attention to the following points:

1 Using a suitable press and press plates or pipe, support the lower control arm securely and squarely.
2 When installing the new bush to the front of the lower control arm, ensure that the bush is installed from the chamfered side of the bush recess of the control arm. Install the bush so that the largest of the two semi-circular cut outs is to the rear of the lower control arm.
3 To install the rear bush to the lower control arm it will be necessary to lubricate the bush and bush recess of the control arm with a soapy solution. To aid in the installation of the bush, the exposed rubber of the bush which is to be pressed into the control arm should be compressed to a smaller diameter to prevent the rubber being damaged during installation.
4 When installing the front or the rear lower control arm bushes, press them into position until each side of the bush projects an equal distance from each side of the arm.
5 Install the lower control arm as previously described.

Panhard Rod

To Remove and Install Bushes — Press and press plates, suitable spacer

TO REMOVE AND INSTALL

1 Raise the rear of the vehicle and support it on chassis stands. Remove the rear wheel. Refer to General Information (chapter 1) for the correct vehicle jacking points.

Leave the jack in position under the rear axle assembly at this stage.

2 Remove the panhard rod mounting bolts and remove the panhard rod from the vehicle.
3 Check the bushes and mounting bolts for wear and damage.

Installation is a reversal of the removal procedure with attention to the following points:

1 Coat the threads of the panhard rod mounting bolts with Permatex Lock Nut or equivalent prior to installation.
2 Tighten the mounting bolts to the specified torque with the vehicle still supported on the chassis stands.

TO RENEW BUSHES

1 Remove the panhard rod from the vehicle as previously described.
2 Using a suitable press and press plates or pipe, securely support the panhard rod around the eye of the rod in which the bush is installed.

Ensure that there is clearance for the bush to be pressed through the eye of the rod.

3 Using a suitable spacer of a diameter slightly smaller than the diameter of the bush, press the spacer and bush out of the panhard rod. Ensure that the panhard rod is not being distorted while pressing out the bush.

Installation is a reversal of the removal procedure with attention to the following points:

1 Ensure that the panhard rod is securely and squarely supported when pressing in the new bush.
2 Lubricate the bush and bush housing with a soapy solution prior to installation.
3 The rubber lip of the bush should be compressed to a smaller diameter to prevent the bush being damaged during installation.
4 When installed, ensure that the bush projects an equal distance each side of the rod.
5 Install the panhard rod to the vehicle as previously described.

Stabiliser Bar

TO REMOVE AND INSTALL

1 Raise the rear of the vehicle and support it on chassis stands. Remove the rear wheel. Refer to General Information (chapter 1) for the correct vehicle jacking points.
2 Remove the bolts retaining the stabiliser bar to the lower control arms and manoeuvre the stabiliser bar from the vehicle.

Installation is a reversal of the removal procedure with attention to the following points:

1 Inspect the stabiliser bar and bolts for cracks and damage and renew as necessary.
2 Tighten the stabiliser bar bolts securely.

Axle Assembly

TO REMOVE AND INSTALL

1 Raise the rear of the vehicle and support it on chassis stands. Remove the rear wheel. Refer to General Information (chapter 1) for the correct vehicle jacking points. Remove the rear wheels.
2 Remove the rear brake calipers, brake discs and anchor plates as described in Brakes (chapter15).
3 Release the handbrake cables from the lower control arms.
4 Release the brake hose mounting brackets from the axle assembly.
5 Using a jack, raise the axle assembly to its normal position.
6 Separate the panhard rod from the axle assembly as previously described.
7 Remove each shock absorber from its lower mounting and lower the jack. Remove the coil springs from the vehicle.
8 Raise the axle to its normal position.

Station wagon rear suspension components.

9 Remove the bolts securing the upper and lower control arms to the axle.

10 Lower the jack and withdraw the axle assembly from the vehicle.

11 Clean the axle assembly with a suitable cleaning solvent and check for cracks, distortion and damage. Renew as necessary.

Installation is a reversal of the removal procedure with attention to the following points:

1 Using a jack, support the axle assembly in position under the vehicle.

2 Coat the threads of the control arm mounting bolts with Permatex Lock Nut or equivalent, install the bolts to the upper and lower control arm mountings. Do not fully tighten these bolts at this stage.

3 Install the coil springs ensuring that the upper and lower insulators are correctly positioned.

4 Coat the threads of the shock absorber mounting bolts with Permatex Lock Nut or equivalent, raise the jack and connect the shock absorbers to the mounts.

5 Install the rear brake anchor plates, brake discs and brake calipers to the axle assembly as described in Brakes (chapter 15).

6 Install the brake hose mounting brackets to the axle assembly.

7 Install the handbrake cables to the lower control arms.

8 Bleed the brake hydraulic system. Refer to Brakes (chapter 15).

9 Lower the vehicle to the ground and tighten the upper and lower control arms and shock absorber mounting bolts and nuts to the specified torque.

Chapter 15
Brakes

Contents

Specifications

Type:
- Front ... Single piston caliper with ventilated disc
- Rear ... Single piston caliper incorporating drum type handbrake with solid disc

Operation:
- Footbrake ... Vacuum assisted, hydraulic dual circuit with optional ABS
- Handbrake ... Mechanical on rear wheels

Master cylinder:
- Type ... Tandem
- Bore diameter ... 25.4 mm
- Proportioning valve split point –
  - Sedan ... 1.72 – 2.7 Mpa
  - Wagon ... 3.68 – 4.67 Mpa

Brake fluid type ... Dot 4

Front discs:
- Diameter ... 276 mm
- Standard pad thickness ... 7.9 mm
- Minimum pad thickness ... 1 mm
- Brake disc thickness —
  - Standard ... 24 mm
  - Minimum ... 22.4 mm
  - Runout limit ... 0.05 mm

Rear discs:
- Diameter ... 258 mm
- Standard pad thickness ... 8.8 mm
- Minimum pad thickness ... 1 mm
- Brake disc thickness —
  - Standard ... 10.4 mm
  - Minimum ... 8.4 mm
- Runout limit ... 0.06 mm
- Handbrake drum diameter limit ... 169 mm

Handbrake shoe lining thickness:
- Standard ... 2.8 mm
- Minimum ... 1.5 mm

Brake booster pushrod setting ... 9.2 – 9.4 mm

Brake pedal:
- Height ... 168 – 173 mm
- Freeplay ... 3 – 8 mm

Handbrake lever travel ... 3 – 5 notches at 200 N pull

ABS sensor resistance ... 1.4 – 1.8 k/ohms

TORQUE WRENCH SETTINGS

ABS:
- Front and rear sensor bolts ... 25 Nm
- Rear sensor retaining bracket bolt ... 25 Nm

Brake booster retaining nuts ... 14 Nm

Brake pedal pivot bolt and nut ... 30 Nm

Front brakes:
- Anchor plate bolts ... 90 Nm
- Brake hose union bolt ... 29 Nm
- Caliper lock bolts ... 32 Nm

Master cylinder retaining nuts ... 10 Nm

Rear brakes:
- Anchor plate bolts ... 49 – 59 Nm
- Brake hose union bolt ... 29 Nm
- Caliper lock bolts ... 32 Nm
- Hub bolts sedan ... 74 – 88 Nm
- Hub nut wagon ... 200 – 260 Nm

Wheel nuts ... 90 – 108 Nm

Description

The brakes are applied by hydraulic pressure in two independent circuits by means of a tandem master cylinder.

The circuits are split diagonally, the left hand front and the right hand rear brakes are connected in one circuit while the right hand front and the left hand rear brakes are connected in the other circuit.

Should a malfunction occur in one circuit, the remaining circuit is capable of stopping the vehicle safely.

The proportioning valve is installed in front of the firewall on the passenger side of the vehicle. The proportioning valve regulates the hydraulic pressure to the rear brakes to prevent premature locking of the rear wheels during severe braking. There are different proportioning valve split points for sedan and station wagon models due to the heavier loads a station wagon is expected to carry. If a proportioning valve is to be renewed, ensure that the correct one is fitted.

To reduce the pedal effort during brake application, a vacuum brake booster is installed between the master cylinder and the vehicle firewall. The valve and rod assembly of the booster is on the same axis as the brake pedal pushrod and in the case of a vacuum failure, will act as a single pushrod without assistance. Vacuum to the booster is supplied by a hose connected to the inlet manifold. A check valve is installed in the booster to prevent loss of vacuum.

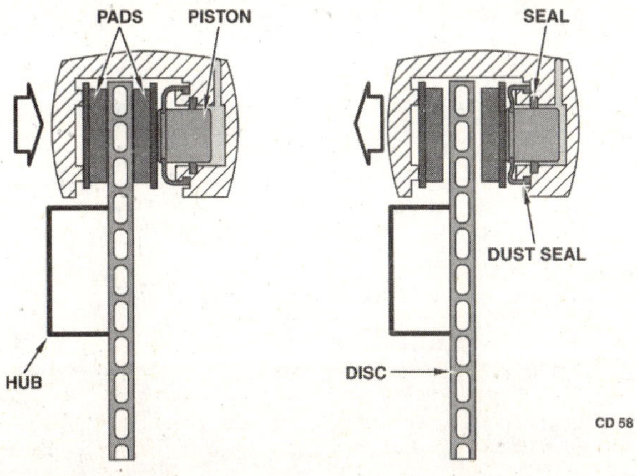

Single acting piston type disc brake system.

The front disc brakes consist of a ventilated discs and self adjusting, single piston floating calipers which slide on guide pins, that are attached to the anchor plate.

As pad wear takes place, the caliper piston is allowed to slide outward through the seal to take up a new position in the bore of the caliper housing. Elastic deformation of the piston seal takes place when the brakes are applied. This deformation retracts the piston when the brakes are released, maintaining clearance between the pads and the disc.

The rear disc brakes operate using the same principals as the front disc brakes, except a solid disc is used and a small drum is cast into the inner part of the disc for the handbrake mechanism.

The handbrake is operated by a floor mounted lever which is connected to the rear brakes through flexible cables. The cables operate an expanding actuator which forces the one piece brake shoes against the brake drum which is an integral part of the brake disc.

Some vehicles are equipped with an antilock braking system (ABS) to prevent the wheels locking in an emergency braking situation. Some ABS equipped models also incorporate traction/slip control.

The ABS system uses sensors at each wheel which signal the ABS ECU when a wheel is locking. The ABS ECU then directs the operation of the hydraulic modulator to reduce the hydraulic pressure to the particular wheel to maintain minimum rotation. The pressure to all wheels are controlled individually.

Resetting the round portions of the reservoir cap seal after topping up the brake fluid.

The ABS ECU is located behind the passenger side of the dashboard, while the modulator is located in front of the firewall in the left hand side of the engine compartment.

If the ABS braking system malfunctions, indicated by illumination of the dashboard warning lamp, check the basics as listed in Trouble Shooting and if it fails to cure the problem, refer the vehicle to an authorised dealer for repair.

In the event of an ABS malfunction, the brakes revert to a normal braking system.

Routine Maintenance

TO CHECK BRAKE FLUID

The brake fluid should be checked regularly ensuring that the level is between the Min and Max marks on the side of the reservoir. If the level is low, clean any dirt from around the cap. Remove the cap and top up the reservoir with the specified type of brake fluid. Check that the two round portions of the seal are sitting flush with the rest of the seal, reposition as necessary. Also check the cap and seal for wear and renew if necessary.

> *If the brake fluid level is quite low, check the condition of the brake pads, as described later in this chapter. It is normal for the brake fluid level to drop slightly as the brake pads wear.*
>
> *Also, check for any signs of brake fluid leakage from the brake calipers, pipes and fittings. Repair any leaks referring to the appropriate headings in this chapter.*

TO RENEW BRAKE FLUID

The brake fluid should be renewed at intervals specified in Routine Maintenance (chapter 2). Renewing the brake fluid follows the same basic principal as bleeding the brake fluid, so refer to the relevant heading, later in this chapter, with attention to the following points:

1 Withdraw as much old fluid as possible from the brake master cylinder using a syringe.

2 Continue bleeding until a constant stream of new fluid, free from air bubbles is being discharged from each caliper/ wheel cylinder.

TO CHECK FRONT AND REAR BRAKES

The brakes should be examined at the intervals specified in Routine Maintenance (chapter 2) or whenever the condition of the brakes is suspect.

1 Raise the front and rear of the vehicle and support it on chassis stands. Refer to General Information (chapter 1) for the position of the jacking and support points.

2 Remove the wheels and check the brake pad lining thickness through the inspection hole in the caliper body. If the lining thickness is less than Specifications, renew the pads as described later in this chapter.

3 Check the machined faces on the brake disc for scoring, cracks, wear and signs of overheating. If the disc is heavily scored, refer it to a brake specialist for machining. If it is excessively worn, it should be renewed.

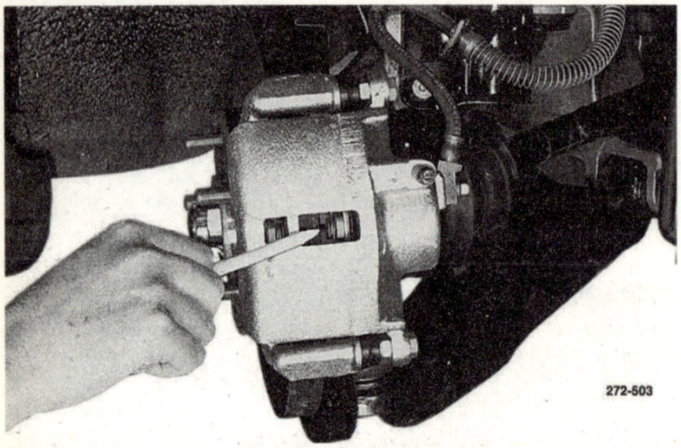

Check the brake pad thickness through the caliper inspection hole.

4 Periodically remove the rear calipers and brake discs as described under the heading To Remove and Install Brake Disc and check the handbrake shoes for wear or contamination. Check that the handbrake shoe contact surface has no hot spots or scoring. Repair or renew components as necessary.

TO CHECK HYDRAULIC COMPONENTS

Periodically, all hydraulic pipes, hoses and unions should be checked for cracks or splits, chafing, leaks and corrosion. Any component which is damaged or suspect should be renewed immediately.

The ideal time for examining the hydraulic components is when the wheels are removed for brake pad checking.

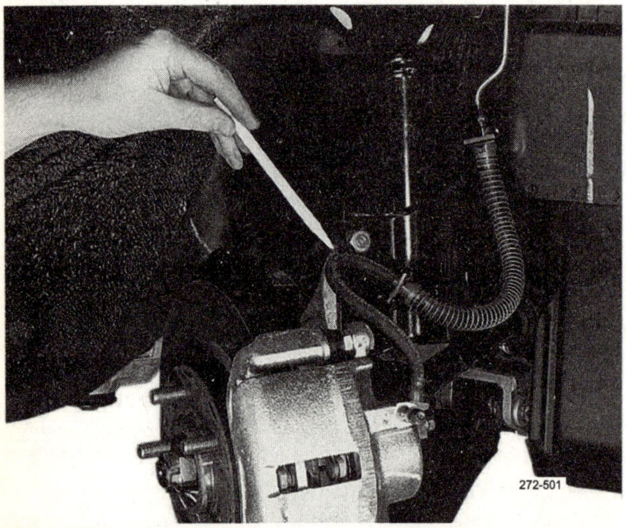

Check all brake hoses for chafing or deterioration.

PEDAL PADS

The brake pedal pad should be regularly inspected for splitting and other signs of wear. The pedal pad should be renewed if split or the tread is worn.

TO ADJUST HANDBRAKE

Apply the handbrake and check its effectiveness on a slope. If the handbrake is correctly adjusted it should hold the vehicle securely when the lever is applied firmly (with 200 N force) and its movement must be between 3 – 5 notches.

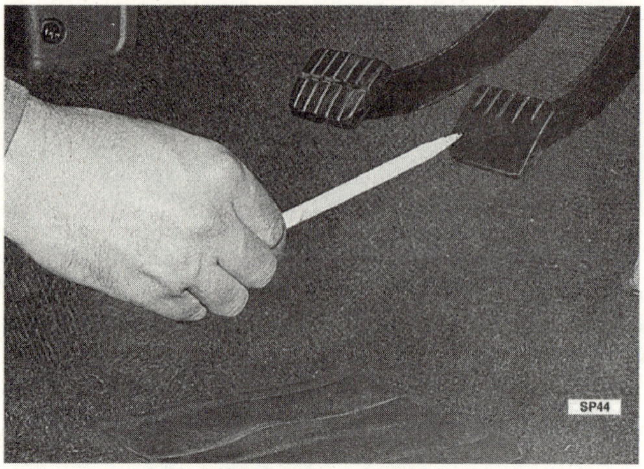

Worn or split pedal pads should be renewed.

1 Open the centre console and remove the centre compartment. With the handbrake released, check that the handbrake cables are slack. If necessary, loosen the handbrake cable adjusting nut until there is no tension on the cables.

2 Raise the rear of the vehicle and support on chassis stands under the rear jacking points. Refer to General Information (chapter 1) for the correct procedure. Remove the rear wheels.

3 Remove the adjustment hole plug in the disc and align the hole with the adjuster nut. The nut is at the top for sedans and at the bottom for station wagons.

4 Using a flat bladed screwdriver, turn the adjuster nut in increments until the disc will not rotate. Turn the adjuster in the opposite direction 5 notches, ensuring that the disc rotates without binding.

5 Apply the handbrake to centralise the shoes and check the adjustment. If necessary, turn the cable adjusting nut until the handbrake lever travel is 3 – 5 notches.

6 Check that with the handbrake released both rear wheels are free to rotate.

7 Check the handbrake operation with the vehicle parked on an incline. If satisfactory handbrake adjustment cannot be achieved, remove the rear disc as described later in this chapter and check the handbrake shoes and actuators.

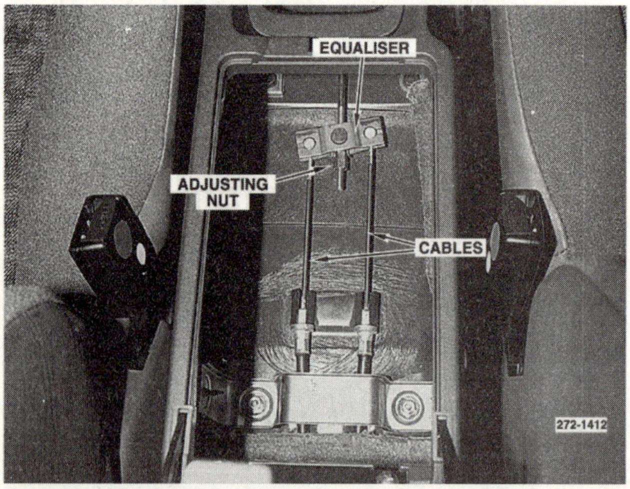

Centre compartment removed from console showing the handbrake cable adjusting nut.

Adjusting the handbrake shoe using a small flat bladed screwdriver.

Check the hydraulic pipe connections for leakage.

Trouble Shooting

BRAKE PEDAL HARD

1. Brake booster inoperative: Check that the vacuum hose is not restricted, leaking or disconnected. Check the operation of the brake booster as described later and renew as necessary.
2. Incorrect brake pads installed: Check and replace with the recommended type.
3. Seized caliper pistons or guide pins: Check and overhaul the calipers.
4. Restricted brake line from the master cylinder: Check the brake lines and remove the restriction or renew the line.
5. Seized pedal pivot: Rectify or renew the pivot shaft and bushes.

BRAKE DRAG

1. Seized caliper pistons or caliper guide pins: Check and overhaul the calipers.
2. Incorrectly adjusted handbrake shoe: Check the handbrake shoe adjustment.
3. Seized handbrake cables: Free up or renew the cable(s).
4. Incorrectly adjusted handbrake cable: Adjust the handbrake cable.
5. Blocked or restricted master cylinder ports, brake lines or hoses: Check and clean the master cylinder, the fluid reservoir and the brake lines.

TIP *To check brake drag, raise the vehicle and spin the wheels, one at a time, to check for binding. If the wheels are not binding, have an assistant apply and release the brakes. Check if the brakes release immediately. A blocked master cylinder port will cause binding on the two wheels fed by that particular circuit from the master cylinder. Open the bleeder valve on the suspect brake assembly to check if pressure build up is the cause of the binding.*

A seized handbrake cable will usually cause binding on a rear wheel. To check this condition, disconnect the handbrake cable and check if the wheel will then turn freely.

LOW SPONGY BRAKE PEDAL

1. Insufficient fluid in the hydraulic system: Check for leaks, replenish the fluid and bleed the hydraulic system.
2. Air in the hydraulic system: Bleed the hydraulic system.
3. Faulty master cylinder: Overhaul the master cylinder.
4. Incorrect or worn brake pads: Renew the brake pads.

TIP *A spongy brake pedal in most cases is caused by air in the hydraulic system. For air to enter the system one or more of the sealing rubbers or pipes must be sucking in air and is usually indicated by fluid leakage. Always rectify the cause of the trouble before bleeding the brake system.*

BRAKES LOCK ON APPLICATION

1. Gummy brake pads due to oil or fluid contamination: Rectify the source of the fluid leak and renew the brake pads.
2. Distorted brake disc: Machine or renew the faulty disc.
3. Incorrect brake pads installed: Check and renew the pads as a set with the correct type.

If this condition arises, remove all the wheels and check the friction material for oil contamination and excessive wear. A distorted brake disc will be indicated by pulsating brake pedal when the brakes are lightly applied.

BRAKE PEDAL PULSATES

1. Distorted brake disc: Check and machine or renew the brake disc as necessary.
2. Loose or worn hub bearings: Renew the bearings.

On vehicles with ABS, some pulsation of the brake pedal is normal during heavy braking.

Check and if necessary, renew the hub bearings as outlined in the relevant suspension chapter. Brake discs that are distorted must be machined. This job is best entrusted to a brake specialist who will also be able to determine if a new brake disc needs to be installed.

Check the brake disc for scoring.

BRAKE FADE

1. Brake pads saturated with hydraulic fluid: Rectify the leak and renew the brake pads as a complete set.
2. Incorrect brake pads installed: Check and install the correct pads in sets.
3. Brakes overheating: Refer to the Brakes Overheating heading.

TIP *In most cases brake fade is caused by overuse of the footbrake, which in turn causes a build up of heat at the friction material and discs. Once this excessive build up of heat is allowed to dissipate, the brakes should again function normally.*

BRAKES OVERHEATING

1. Brakes dragging: Refer to the Brake Drag heading.
2. Incorrect or worn brake pads: Renew the brake pads as a set.
3. Overuse of the footbrake: Revise driving habits.

BRAKE FAILURE

1. Faulty master cylinder: Overhaul or renew the master cylinder.
2. Loss of brake fluid due to a leaking caliper: Overhaul or renew the caliper assembly and bleed the hydraulic system.
3. Loss of brake fluid due to a fractured hose, pipe or faulty union: Renew the faulty components as necessary and bleed the hydraulic system.
4. Air in the hydraulic system: Locate the source of the air leak and rectify. Bleed the hydraulic system.
5. Water in the hydraulic fluid: Drain, flush, refill and bleed the hydraulic system.

To locate the source of a brake fluid leak, refill the master cylinder reservoir with brake fluid and have an assistant pump the brake pedal. Check for obvious signs of external leakage prior to dismantling and overhauling the brake system.

BRAKE NOISE

1. Brakes squeal on application: Glazed friction material. Remove and inspect the brake pads and deglaze the friction material with emery paper.

Front ABS sensor and pulse ring.

2. Grinding noise on application: Friction material worn away. Inspect the friction material on the brake pads. Renew the brakes pads as necessary and machine or renew the brake disc.

It is normal for brakes to emit an occasional squeal due to conditions such as moisture, dirt, salt and temperature variances.

In most cases excessive brake squeal is caused by the friction material on the brake pads becoming glazed. Unsuitable friction material or the omission of anti-squeal shims or anti-rattle springs may also cause brake noise. If the brake pads require renewal always renew the brake pads as a complete set.

If badly scored, the brake disc should be machined or renewed to restore full braking efficiency.

ABS WARNING LAMP STAYS ON

The ABS warning lamp, when illuminated for longer than 3 seconds after starting the engine, indicates a fault within the antilock braking (ABS) system. While specialised equipment is required to diagnose faults within this system, the following procedures outline basic checks that can assist in locating a fault prior to referring the vehicle to an authorised dealer.

Rear ABS sensor. Sedan model shown.

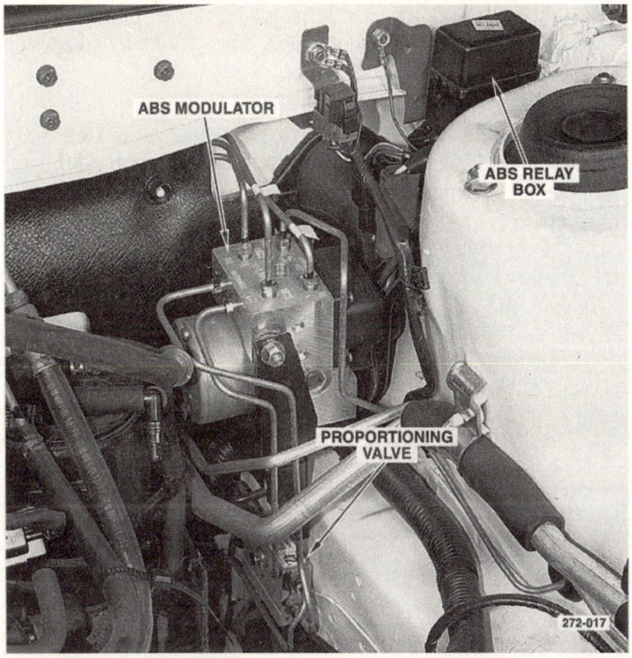

The ABS modulator is located on the passenger side of the engine compartment.

1 Front and rear hub bearings damaged or worn: Check the hub bearings as described in the relevant Axle and Suspension chapter and renew if necessary.
2 Damaged ABS wheel sensor or pulse ring: Check the wheel sensors and pulse rings for any signs of damage and renew components as necessary.
3 Faulty stop lamp switch: Ensure that the stop lamps are working. Refer to Body Electrical (chapter 16), if necessary.
4 Damaged or corroded wiring connectors: Disconnect and check each wheel sensor connector and the ABS modulator wiring connector for corrosion or backed out terminals. Repair or renew components as necessary.
5 Faulty ABS fuses: Refer to the wiring diagrams and check all fuses relevant to the ABS system.
6 Faulty ABS modulator: Refer the vehicle to an authorised dealer who has the necessary equipment to check the ABS modulator.

Master Cylinder

TO REMOVE AND INSTALL

1 Raise the bonnet and place a protective cover over the driver side mudguard to prevent damage to the vehicle paintwork.
2 Remove the master cylinder reservoir cap and, using a syringe, withdraw the brake fluid from the reservoir into a container.
3 Disconnect the wiring from the brake fluid level warning switch located on the side of the master cylinder reservoir.
4 Disconnect the brake pipes from the master cylinder. Plug the pipes to prevent entry of dirt.
5 Remove the nuts retaining the master cylinder to the brake booster. Carefully push the brake pipes away from the master cylinder.
6 Remove the master cylinder from the vehicle.

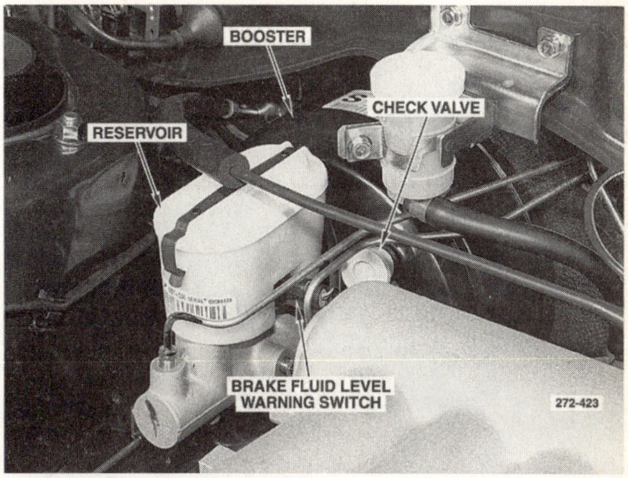

Brake master cylinder installed.

Care should be exercised when removing or installing the master cylinder assembly to ensure that brake fluid is not spilt on the vehicle paintwork. Brake fluid, if spilt, should be immediately washed away with water and allowed to dry naturally.

Do not depress the brake pedal or disturb the pushrod after the master cylinder has been removed as damage to the brake booster could result.

Installation is a reversal of the removal procedure with attention to the following points:

1 Secure the master cylinder in vice equipped with soft jaws and pour a small amount of clean brake fluid into the reservoir. Pump the master cylinder piston with a blunt rod until fluid begins to emerge from the outlets. If necessary, block the one outlet at a time to speed up the procedure.
2 Install the master cylinder to the brake booster and loosely install the retaining nuts. The design of the master cylinder will make it sit out approximately 2 – 3 mm away from the booster until the nuts are fully tightened. Connect the brake pipes to the master cylinder outlets using care not to crossthread the fittings. Do not fully tighten the pipes at this stage.
3 Tighten the master cylinder retaining nuts securely.
4 Securely tighten the brake pipes and connect the brake fluid level warning switch wiring connector.
5 Fill the master cylinder reservoir with clean brake fluid and bleed the brakes as described later in this chapter.

TO DISMANTLE

If the master cylinder is worn or leaking, it is often more economically viable to install a new or exchange cylinder rather than recondition the old one. Check the price difference before proceeding.

1 With the master cylinder removed, as previously described, drain the remaining brake fluid from the reservoir.
2 Mount the master cylinder securely in a vice equipped with soft jaws and remove the reservoir retaining bolt.
3 Pull the reservoir from the master cylinder and withdraw the grommets.
4 On models without ABS, depress the primary piston into the master cylinder, remove the piston stop bolt and O ring. Discard the O ring.

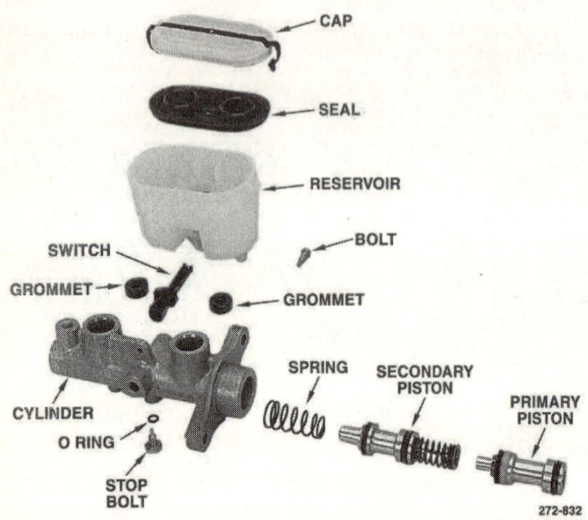

Brake master cylinder components. Models without ABS.

5 On models with ABS, depress the primary piston into the master cylinder and remove the stop pin from the master cylinder front reservoir grommet hole. A pair of long nose pliers may be necessary to remove the pin.

6 Lightly tap the master cylinder against a soft piece of timber until the pistons and spring can be withdrawn.

TO CLEAN AND INSPECT

1 Wash the master cylinder body, reservoir and cap in methylated spirits. Wash the remaining parts in clean brake fluid. Do not use petrol, kerosene or other cleaning solvents.

2 Check the cylinder bore for wear, scoring or pitting.

Do not hone the master cylinder bore. If the bore is pitted or worn, renew the master cylinder as an assembly.

3 Ensure that all ports between the reservoir and the cylinder bore are free of any obstruction.

4 The manufacturer recommends that the pistons and seals are not to be dismantled.

Where possible, use a genuine major repair kit when overhauling the master cylinder. The use of a major kit will ensure a thorough overhaul and long service from the unit.

TO ASSEMBLE

1 Liberally coat the cylinder bore with clean brake fluid. Install the spring to the secondary piston, ensuring that the small diameter end of the spring faces the piston assembly.

2 Insert the secondary piston assembly, spring end first, into the cylinder bore. Ensure that the seal lips are not turned back or damaged during this operation. Also make sure that the slot in the piston on ABS models will align with the stop pin on installation.

3 Insert the new primary piston assembly, spring end first, into the cylinder bore. Ensure that the seal lips are not turned back or damaged during this operation.

4 Using a blunt rod, push the piston assemblies into the cylinder bore and install the stop bolt with new O ring (non ABS) or the stop pin (ABS).

5 Ensure that both pistons move freely in the cylinder bore.

6 Lightly smear the reservoir and grommets with clean brake fluid and install the grommets into the master cylinder.

7 Push the reservoir into the grommets, align the retaining bolt hole and install the bolt. Tighten the retaining bolt securely.

8 Fill the reservoir with new brake fluid. Using a blunt rod, pump the pistons until fluid begins to emerge from the outlets. If necessary, block one port at a time to speed up this procedure.

9 Install the master cylinder to the vehicle as previously described.

Front Disc Brakes

To Check Disc Runout — Dial gauge
To Check Disc Thickness — Micrometer

TO RENEW BRAKE PADS

1 Raise the front of the vehicle and support on chassis stands. Refer to General Information (chapter 1) for the position of the vehicle jacking and support points. Remove the wheels.

2 Drain approximately two thirds of the brake fluid from the master cylinder reservoir using a syringe. Discard this fluid.

The fluid is drained from the reservoir to prevent overflow when the caliper piston is pushed back into its bore to facilitate pad renewal.

Complete work on one caliper at a time and renew brake pads as sets of four.

3 Hold the hexagon part of the lower guide pin with a spanner and remove the caliper lock bolt.

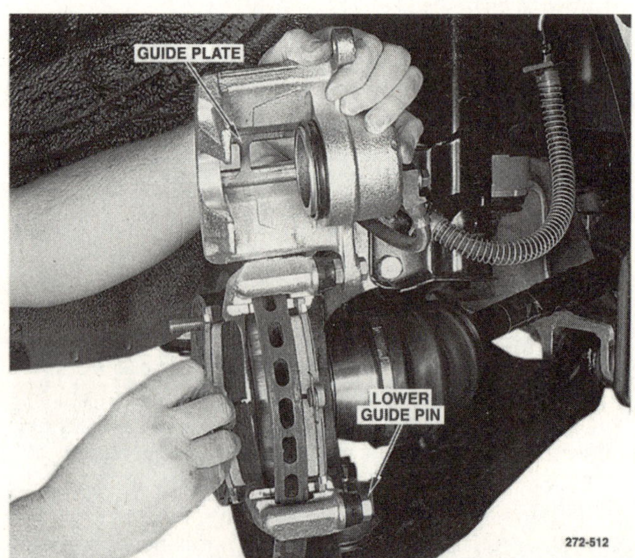

Renewing the front brake pads. Note the position of the H shaped guide plate.

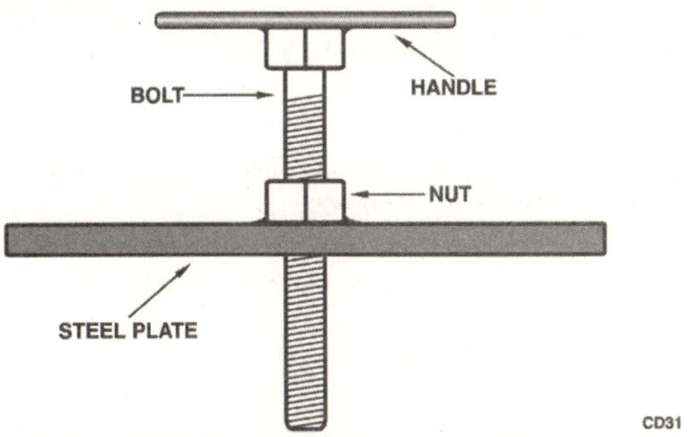

Illustration of a tool that can be fabricated to push the piston into the caliper bore.

4 Rotate the caliper assembly upwards, exposing the brake pads and anchor plate. Using wire or cord, support the caliper in this upright position.

5 Note the installed position of the H shaped guide plate in the caliper body. Withdraw the brake pads and shims from the anchor plate.

6 Ensure that the guide pins are moving freely in the caliper and that the dust boots are in good condition. If necessary, remove, clean and lubricate each guide pin with rubber grease.

7 Using a hammer handle, push the piston squarely into the caliper bore. If necessary, a G clamp or fabricated tool can be used for this operation. If leaks are evident or the dust seal is deteriorated, the caliper should be overhauled.

8 Check the condition of the H shaped guide plate. Renew if necessary, installing the plate in the position noted on removal.

9 Install the shims to the brakes pads and install the pads to the anchor plate ensuring that the pads are correctly seated.

10 Hold the brake pads together and rotate the caliper assembly until it is correctly positioned over the brake pads and the anchor plate.

11 Clean the threads of the caliper lock bolt and apply a drop of Permatex Lock Bolt or equivalent. Install the caliper lock bolt and tighten to the specified torque.

Pushing the piston into the caliper using a fabricated tool and an old brake pad.

12 Pump the brake pedal several times to bring the pads to their normal position alongside the disc and top up the brake master cylinder reservoir with the recommended brake fluid.

13 Install the wheels and lower the vehicle to the ground.

14 Road test the vehicle and check for correct brake operation.

TO OVERHAUL CALIPER

1 Raise the front of the vehicle and support it on chassis stands. Refer to General Information (chapter 1) for the position of the vehicle jacking and support points. Remove the front wheel.

2 Remove the union bolt connecting the flexible brake hose to the caliper body, disconnect the brake hose and plug the hose to prevent the loss of fluid and the entry of dirt. Discard the sealing washers.

3 Hold the hexagon portion of each guide pin with a spanner and remove the caliper lock bolts.

4 If necessary, remove the anchor plate retaining bolts and slide the anchor plate and brake pads off the brake disc.

5 Ease back the piston dust boot from the caliper body.

6 Place a piece of soft timber, approximately the thickness of a brake pad between the piston and the outer brake pad support. Using low pressure compressed air applied to the brake hose aperture, carefully force the piston from the caliper bore.

> *If the piston cannot be dislodged or compressed air is not available, connect the brake hose and slowly depress the brake pedal until the piston is forced from the caliper. If removing the piston in this way, only have one caliper removed from the disc at this time.*

7 Remove the dust seal from the piston.

8 Using a thin, blunt probe, preferably made from wood or plastic, lift and remove the piston seal from the groove in the caliper bore using care not to score the caliper bore.

9 Remove the cap and the bleeder valve from the caliper body and check that the valve is clear.

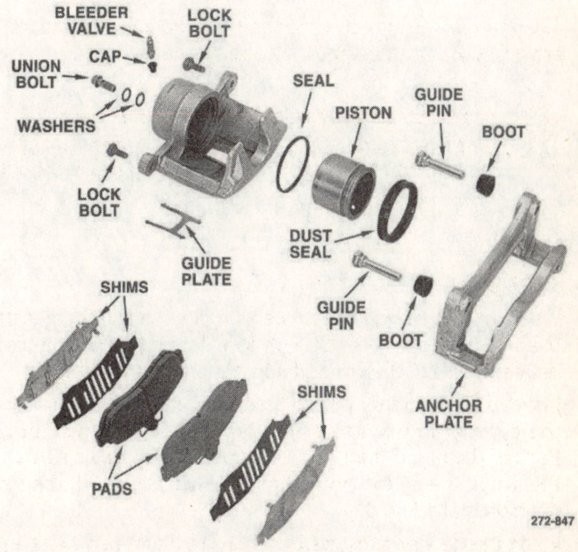

Front brake caliper components.

10 Secure the anchor plate in a vice fitted with soft jaws and withdraw the guide pins and dust boots from the anchor plate.

11 Clean all components in methylated spirits and arrange the components in order on a clean surface to aid assembly.

12 Inspect all caliper components for wear and corrosion. Pay particular attention to the pins, the piston and caliper bore. Renew worn or corroded components as necessary.

13 Ensure that the guide pins are moving freely in the anchor plate.

When overhauling a caliper, it is recommended that a genuine repair kit containing the rubber parts and, if supplied, the special grease is used. Use rubber grease if the repair kit does not contain the special grease.

14 Lubricate the new piston seal with brake fluid.

15 Install the piston seal into the groove in the caliper bore ensuring that the seal is not twisted and is correctly seated in the groove.

16 Lubricate the piston and the caliper bore with clean brake fluid.

17 Using a hammer handle, gently push the piston into the cylinder bore until the dust seal groove in the piston is just above the top of the caliper bore. Do not tilt or use force to install the piston.

18 Apply the special grease, or rubber grease to the inside of the piston dust seal, especially around the sealing lips. Install the lip of the dust seal in the piston groove.

19 Push the piston fully into the caliper bore and ease the dust seal lip into the groove in the caliper body ensuring that the seal is not twisted or damaged.

20 Apply the special grease or rubber grease to the guide pin bores in the anchor plate and inside the dust boots and guide pins.

21 If removed, slide the anchor plate over the brake disc and tighten the bolts to the specified torque.

If the brake pads have been removed from the anchor plate, install the brake pads as previously described.

22 Slide the caliper onto the anchor plate until it is correctly positioned over the brake pads.

23 Clean the threads of the caliper lock bolts and apply a drop of Permatex Lock Bolt or equivalent.

24 Install the guide pin lock bolts and tighten them to the specified torque.

25 Install the brake hose ensuring that new sealing washers are installed to each side of the brake hose connection. Tighten the union bolt to the specified torque.

26 Bleed the brakes as described later in this chapter.

BRAKE DISC

To Check

1 Raise the front of the vehicle and support it on chassis stands. Refer to General Information (chapter 1) for the position of the vehicle jacking and support points.

2 Raise the caliper and remove the brake pads as described under the previous heading, To renew Brake Pads.

3 Reverse the wheel nuts and tighten the disc against the hub.

4 Mount a dial gauge to a suspension component and position the plunger against the machined face of the disc approximately 10 mm from the outer edge of the disc.

5 Rotate the disc and measure the runout.

6 If the runout exceeds Specifications, check the front hub bearing end float as described in Front Axle and Suspension (chapter 13).

If the end float of the front hub bearing is within Specifications, proceed as follows:

a Remove the retaining bolts, withdraw the anchor plate and withdraw the brake disc.

b Change the position of the disc in relation to the hub. If this fails to correct the runout, machine or renew the disc as necessary.

7 Using a micrometer, check the thickness of the disc at several points. Renew the disc, if the thickness is less than specified.

To Remove and Install

1 Raise the front of the vehicle and support it on chassis stands. Refer to General Information (chapter 1) for the position of the vehicle jacking and support points. Remove the front wheel.

Checking the front disc brake runout using a dial gauge.

Using a micrometer to check the thickness of the disc. Check at several points.

2 Remove the anchor plate retaining bolts and slide the anchor plate, brake pads and caliper off the brake disc. Using wire or cord, suspend the caliper assembly, so that the weight of the caliper is not taken by the brake hose. Ensure that the shims remain attached to the back of the pads.

3 If the disc is to be reused, suitably mark the location of the disc in relation to the hub to aid assembly and withdraw the brake disc from the front hub.

TIP *Do not use a hammer to remove the disc. If the disc is difficult to remove, install the correct sized bolts in the threaded holes in the disc mounting flange and tighten the bolts alternately until the disc can be removed from the hub.*

4 Inspect the machined faces on the sides of the disc for scores, cracks, wear and signs of overheating. If the disc is badly scored it must be machined.

i *Disc machining is best entrusted to a reliable brake specialist who can advise on disc serviceability.*

5 Using a micrometer, measure the thickness of the disc at several positions around the machined faces on the side of the disc. Renew the disc if the thickness is below Specifications.

Installation is a reversal of the removal procedure with attention to the following points:

1 Ensure that the mating faces of the disc and the hub are clean and free from burrs and that the marks made on removal are aligned.

TIP *If a new disc is installed, remove the protective coating from the disc using solvent.*

2 Inspect the brake pads and renew as required.

3 Tighten the anchor plate bolts to the specified torque.

Rear Disc Brakes and Handbrake

To Check Disc Runout — Dial gauge
To Check Disc Thickness — Micrometer

TO RENEW BRAKE PADS

1 Raise the rear of the vehicle and support on chassis stands. Remove the rear wheels. Refer to General Information (chapter 1) for the position of the jacking and support points.

2 Drain approximately two thirds of the brake fluid from the master cylinder reservoir using a syringe. Discard this fluid. The fluid is drained from the reservoir to prevent overflow when the caliper piston is pushed back into its bore to facilitate pad renewal. Complete work on one caliper at a time and renew the brake pads as sets of four.

3 Hold the hexagon part of the lower guide pin with a spanner and remove the caliper lock bolt.

Removing the rear brake pads from the anchor plate.

4 Rotate the caliper assembly upwards, exposing the brake pads and anchor plate. Using wire or cord, support the caliper in this upright position.

5 Withdraw the brake pads and shims from the anchor plate.

6 Ensure that the guide pins are moving freely in the caliper and that the dust boots are in good condition. If necessary, remove, clean and lubricate each guide pin with rubber grease.

7 Using a hammer handle, push the pistons squarely into the caliper bores. If necessary, a G clamp or fabricated tool can be used for this operation. If leaks are evident or the dust seal is deteriorated, the caliper should be overhauled.

TIP *An illustration of the piston depressing tool is shown under the Front Disc Brake heading.*

8 Install the shims to the brakes pads and install the pads to the anchor plate ensuring that the pads are correctly seated.

9 Hold the brake pads together and rotate the caliper assembly until it is correctly positioned over the brake pads and the anchor plate.

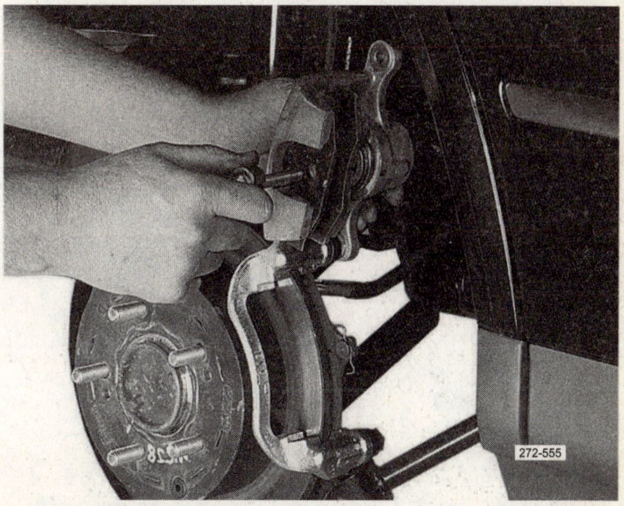

Pushing the piston into the caliper using a fabricated tool and an old brake pad.

10 Clean the threads of the caliper lock bolt and apply a drop of Permatex Lock Bolt or equivalent. Install the caliper lock bolt and tighten to the specified torque.

11 Pump the brake pedal several times to bring the pads to their normal position alongside the disc and top up the brake master cylinder reservoir with the recommended brake fluid.

12 Install the wheels and lower the vehicle to the ground.

13 Road test the vehicle and check for correct brake operation.

TO OVERHAUL CALIPER

1 Raise the rear of the vehicle and support on chassis stands. Remove the rear wheel. Refer to General Information (chapter 1) for the position of the vehicle jacking and support points.

2 Remove the union bolt connecting the flexible brake hose to the caliper body and plug the hose to prevent the loss of fluid and the entry of dirt. Discard the sealing washers.

3 Hold the hexagon portion of each guide pin with a spanner and remove the caliper lock bolts.

4 If necessary, remove the anchor plate retaining bolts and slide the anchor plate and brake pads off the brake disc.

5 Ease back the piston dust boot from the caliper body.

6 Place a piece of soft timber, approximately the thickness of a brake pad between the piston and the outer brake pad support. Using low pressure compressed air applied to the brake hose aperture, carefully force the piston from the caliper bore.

If the piston cannot be dislodged or compressed air is not available, connect the brake hose and slowly depress the brake pedal until the piston is forced from the caliper. If removing the piston in this way, only have one caliper removed from the disc at this time.

7 Remove the dust seal from the piston.

8 Using a thin, blunt probe, preferably made from wood or plastic, lift and remove the piston seal from the groove in the caliper bore using care not to score the caliper bore.

9 Remove the cap and the bleeder valve from the caliper body and check that the valve is clear.

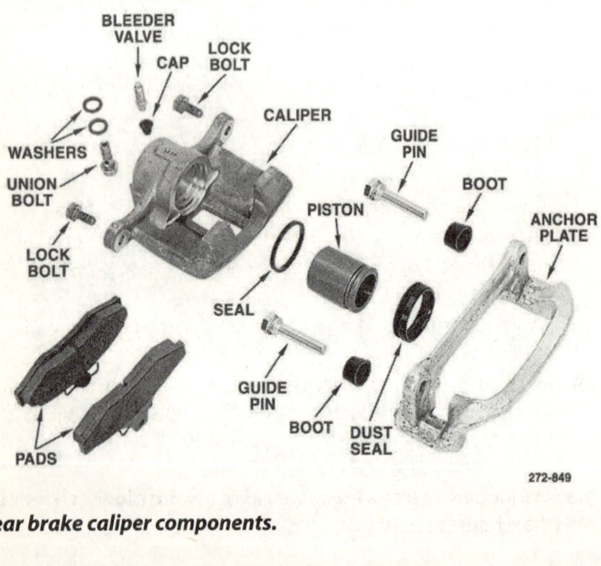

Rear brake caliper components.

10 Secure the anchor plate in a vice fitted with soft jaws and withdraw the guide pins and dust boots from the anchor plate.

11 Clean all components in methylated spirits and arrange the components in order on a clean surface to aid assembly.

12 Inspect all caliper components for wear and corrosion. Pay particular attention to the pins, the piston and caliper bore. Renew worn or corroded components as necessary.

13 Ensure that the guide pins are moving freely in the anchor plate.

When overhauling a caliper, it is recommended that a genuine repair kit containing the rubber parts and, if supplied, the special grease is used. Use rubber grease if the repair kit does not contain the special grease.

14 Lubricate the new piston seal with brake fluid.

15 Install the piston seal into the groove in the caliper bore ensuring that the seal is not twisted and is correctly seated in the groove.

16 Lubricate the piston and the caliper bore with clean brake fluid.

17 Using a hammer handle, gently push the piston into the cylinder bore until the dust seal groove in the piston is just above the top of the caliper bore. Do not tilt or use force to install the piston.

18 Apply the special grease, or rubber grease to the inside of the piston dust seal, especially around the sealing lips. Install the lip of the dust seal in the piston groove.

19 Push the piston fully into the caliper bore and ease the dust seal lip into the groove in the caliper body ensuring that the seal is not twisted or damaged.

20 Apply the special grease or rubber grease to the guide pin bores in the anchor plate and inside the dust boots and guide pins.

21 If removed, slide the anchor plate over the brake disc and tighten the bolts to the specified torque.

If the brake pads have been removed from the anchor plate, install the brake pads as previously described.

22 Slide the caliper onto the anchor plate until it is correctly positioned over the brake pads.

23 Clean the threads of the caliper lock bolts and apply a drop of Permatex Lock Bolt or equivalent.

24 Install the guide pin lock bolts and tighten them to the specified torque.

25 Install the brake hose ensuring that new sealing washers are installed to each side of the connection. Tighten the union bolt to the specified torque.

26 Bleed the brakes as described later in this chapter.

BRAKE DISC

To Check, Remove and Install

Checking, removal and installation of the rear brake disc follows the same basic procedure as the front brake disc with attention to the following points:

1 When removing the rear disc ensure that the handbrake is released and the disc is kept square as it is removed to avoid distorting the handbrake shoe. If necessary, remove the plug and loosen off the handbrake shoe adjuster nut to assist removal.

When the rear brake is removed, check the handbrake drum surface for scoring or excessive wear. Renew the rear discs as a matched pair if necessary.

2 If the disc is tight on the hub use suitable bolts in the holes provided in the disc to force the disc from the hub.

3 If a new pair of rear discs have been installed, burnish the handbrake shoes as described under the following heading.

TO CHECK AND OVERHAUL HANDBRAKE ASSEMBLY

1 Raise the rear of the vehicle and support on chassis stands. Remove the rear wheels. Refer to General Information (chapter 1) for the position of the vehicle jacking and support points.

2 Remove the brake disc assembly as previously described.

3 Check the linings for excessive wear, oil saturation or gumminess. If necessary, renew the shoe and linings as pairs with the corresponding lining and shoe assembly on the other wheel using the following operations.

4 Remove the rear hub as described in the relevant part of Rear Axle and Suspension (chapter 14). On sedan models, temporarily install a suitable nut and bolt to hold the anchor plate to the rear axle carrier.

5 Slide the brake shoe assembly downwards on sedans or upwards on station wagons and remove it from the vehicle.

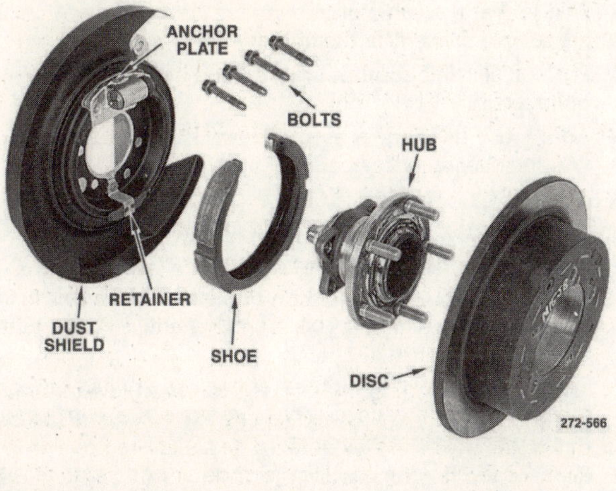

Rear hub, disc and handbrake components. Sedan models.

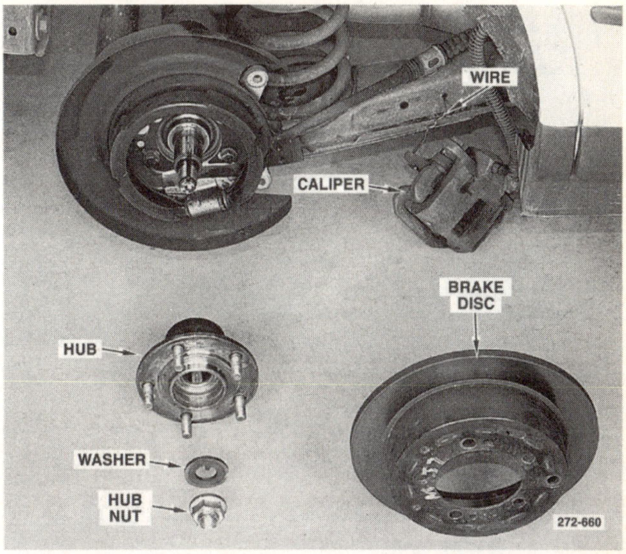

Rear hub, disc and handbrake components. Station wagon models.

If the actuator is not to be dismantled, retain the components using a rubber band.

6 If necessary, remove the anchor plate and dismantle the actuator as follows:

a On models with ABS, remove the wheel sensor retaining bolt, slide the sensor from the anchor plate and place it clear of the work area.

b On sedan models, remove the temporary nut and bolt retaining the anchor plate, disconnect the handbrake cable and withdraw the anchor plate from the rear axle carrier.

c On station wagon models, remove the bolts retaining the anchor plate to the axle assembly, disconnect the handbrake cable and withdraw the anchor plate from the rear axle assembly.

d Remove the adjuster screw and nut from the actuator. Check the adjuster pawl and the teeth on the adjuster nut for damage.

e Remove the plunger and pushrod from the actuator.

f If it is necessary to renew the dust boot on the rear of the anchor plate, drill out the retaining rivets using a suitable drill bit. Renew the rivets on assembly.

g Check the actuator housing and components for signs of corrosion or damage, renew components where necessary.

h Clean the anchor plate and the actuator components. Apply high melting point grease to the plunger, actuator bore and the inside of the lever.

i Assemble the actuator reversing the dismantling procedure.

7 Screw the adjuster fully into the nut and then screw it out approximately one quarter turn until the slot is parallel with the anchor plate.

8 Align the slot on the plunger until it is parallel with the anchor plate.

9 Install the anchor plate and dust shield to the vehicle, reversing the removal procedure. To prevent difficulties later on, ensure that the handbrake cable is connected as the assembly is being installed.

10 Manoeuvre the shoe onto the adjuster and plunger, ensuring that the shoe engages into the slots.

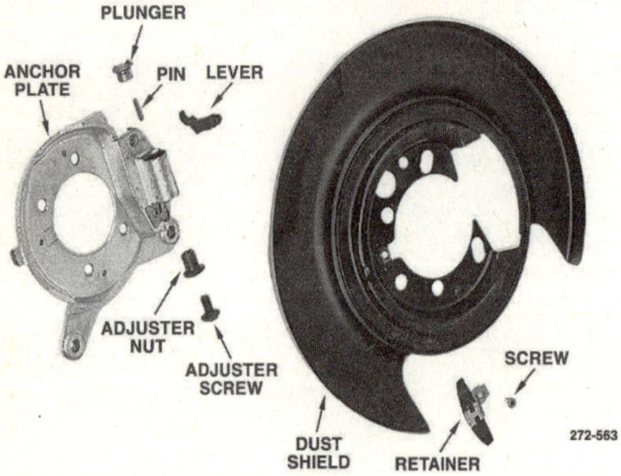

Handbrake anchor plate dismantled.

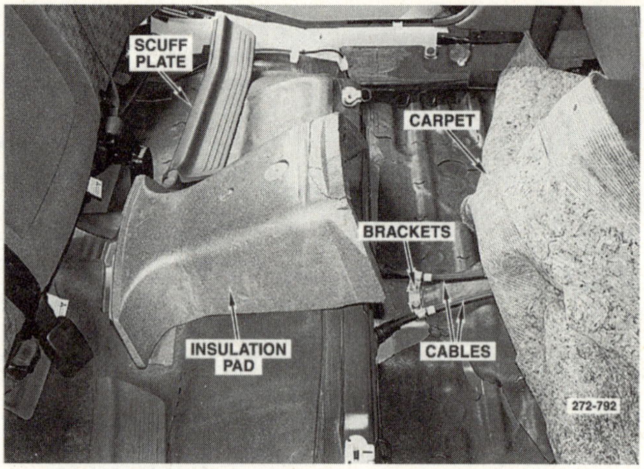

To remove the handbrake cables it is necessary to gain access to the area in front of the back seat.

11 Install the rear hub as described in the relevant part of Rear Axle and Suspension (chapter 14).

12 Adjust the handbrake as described under the Routine Maintenance heading in this chapter.

13 If new rear brake shoes have been installed, the manufacturer recommends burnishing the brake shoes using the following procedure:

a Find an area clear of traffic to prevent accidents and drive the vehicle forwards at 30 km/h. Apply the handbrake with a moderate force (50 N) and continue to drive for 50 metres. Repeat this step 5 times allowing a reasonable period in between for the rear shoes to cool.

b Drive the vehicle at 5 km/h in reverse and apply the handbrake using the same force as in step a. Continue to drive for 50 metres. Repeat this step twice.

14 To confirm that the handbrake is operating correctly, park the vehicle in a clear area, firmly apply the handbrake lever and attempt to drive off in first gear. On manual transaxle models the engine should stall and on automatics the vehicle should not move. Complete this test within 3 seconds to prevent clutch or transaxle damage.

If necessary, readjust the handbrake.

Handbrake Lever and Cables

TO REMOVE AND INSTALL REAR HANDBRAKE CABLES

1 Open the back doors and locate one of the seat cushion retaining clips. Pull the clip towards the front of the vehicle and lift the seat cushion upwards. Repeat this procedure on the other side of the cushion and manoeuvre the cushion from the vehicle.

2 Remove the skuff plates from each rear door opening by prising out using a large screwdriver or similar. Release the carpet from along the door openings and ease up in the centre section, releasing the retaining clip, if fitted.

3 Remove the console box from the centre console and loosen the bolts retaining the console to the floor.

4 While lifting the rear of the console as far as possible, ease the foam insulation pad from the vehicle to expose the handbrake cables and retaining brackets. Remove the bolts from the brackets.

5 Loosen the handbrake cable adjusting nut sufficiently to allow the handbrake cables to be disconnected from the equaliser and guides. Push the cables through the floor grommets but leave them in the vehicle at this stage.

6 Raise the rear of the vehicle and support it on chassis stands. Remove the rear wheels. Refer to General Information (chapter 1) for the position of the vehicle jacking and support points.

7 Remove the bolts retaining the handbrake cable support brackets to the underbody of the vehicle.

8 On sedan models, remove the rear hub as described in Rear Axle and Suspension (chapter 14). And disconnect the cable by manoeuvring the anchor plate around until the cable can be disconnected.

9 On station wagon models, remove the bolts retaining the cable to the lower control arm and disconnect the cable from the actuator lever.

10 Manoeuvre the cables from the vehicle.

11 Check that the handbrake cables are not damaged, frayed or kinked. Ensure that the inner cable slides smoothly.

12 Check the handbrake lever ratchet and pawl for damage and correct operation. Renew the handbrake lever, if necessary.

Installation is a reversal of the removal procedure with attention to the following points:

1 Install the cables to the actuator lever on the anchor plate. On sedan models install the rear hub as described in Rear Axle and Suspension (chapter 14).

2 Ensure that the cable grommets are in good condition and are sealing properly in the body.

3 Position all cable support brackets and tighten the retaining bolts securely.

4 Adjust the handbrake as previously described under the Routine Maintenance heading.

TO REMOVE AND INSTALL HANDBRAKE LEVER

1 Remove the rear section of the centre console from the vehicle as described in Body (chapter 17). Loosen and remove the handbrake cable adjusting nut. Slide the equaliser from the threaded rod.

2 Disconnect the wiring for the warning lamp switch. On some models, it may be necessary to remove the rear support frame bolts to allow clearance.

3 Remove the handbrake lever retaining bolts and withdraw the lever assembly from the vehicle.

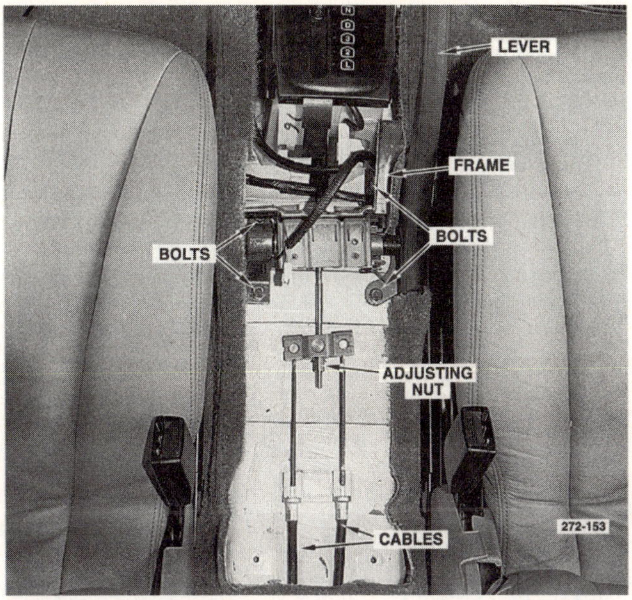

Centre console removed showing handbrake lever and associated components.

4 If necessary, separate the cables from the equaliser. Check the bushes for wear and renew if necessary.

5 Check the handbrake lever ratchet and pawl for damage and correct operation. Renew components as necessary.

Installation is a reversal of the removal procedure with attention to the following points:

1 Lubricate all contact surfaces and pivot points with multipurpose grease.

2 Adjust the handbrake as described under the Routine Maintenance heading.

Bleeding the Brake System

Bleeding the hydraulic system is necessary when some portion of the hydraulic system has been disconnected, the fluid has been drained allowing air to enter the system, or the brake fluid is being renewed.

There are four points in the system where bleeder valves are installed, one on each caliper. The procedure to bleed the brakes is the same for all models.

1 Raise the front and rear of the vehicle and support on chassis stands. Refer to General Information (chapter 1) for the position of the vehicle jacking and support points.

2 Fill the fluid reservoir with clean hydraulic brake fluid and maintain at least one third full throughout the entire operation.

3 Attach a transparent hose to the left hand rear bleeder valve and immerse the other end of the hose in a small amount of clean brake fluid contained in a clean glass jar held higher than the bleeder valve.

> **TIP** *If desired a one person bleeding kit may be used. These kits usually consist on a transparent tube with a one way ball valve. With a one person bleeding kit, it is not necessary to close the bleeder valve after each pedal stroke, however follow the kit manufacturers instructions.*

4 Unscrew the bleeder valve one complete turn.

Bleeding the brake system.

5 Have an assistant depress the brake pedal slowly to the full extent of its travel. Close the bleeder valve and allow the brake pedal to return without assistance.

6 Repeat steps 3 and 4 until a constant stream of new brake fluid, without any air bubbles, is being discharged into the glass jar. Hold the pedal down and tighten the bleeder valve.

> **i** *Do not allow the fluid in the reservoir to fall below one third full at any time during the operation or air will enter the system and the procedure will have to be repeated. Always use new brake fluid for topping up the reservoir.*

7 Perform the bleeding operation on the right front, right rear and left front bleeder valves, in that order.

8 Top up the reservoir with the specified brake fluid to the indicated level and install the reservoir cap.

9 If the brake pedal is spongy or travels to the floor, air has entered the system. Repeat the bleeding procedure ensuring that all the air is removed from the brake hydraulic system.

Brake Booster

TO CHECK OPERATION

1 With the engine switched Off, apply the footbrake several times to exhaust all vacuum from the system.

2 Firmly depress the brake pedal and start the engine. If the brake booster is functioning satisfactorily, a distinct downward movement of the brake pedal should be noticed.

Should the pedal fail to fall away when the engine has been started, the vacuum system can be considered inoperative.

> **i** *If the pedal continues to fall away, there is a fault in the hydraulic system.*

3 Ensure that the brake pedal is fully released, start the engine and run it at medium speed. Stop the engine.

Let the vehicle stand for 1 - 2 minutes, press the brake pedal two or three times and check its operation. If there is no vacuum assistance, the vacuum system has developed a leak or the check valve is defective.

Checking for vacuum at the brake booster check valve with the engine running.

Before removing the brake booster from the vehicle for inspection, disconnect the hose and check valve from the booster, start the engine and check that manifold vacuum is reaching the booster. Also test the check valve as described below.

TO TEST CHECK VALVE

1 With the engine switched off, apply the footbrake several times to exhaust all vacuum from the system.
2 Release the hose clamp and disconnect the vacuum hose from the inlet manifold.
3 Blow into the hose and check that air does not flow. Suck on the vacuum hose and check that air does flow. Renew the check valve if a fault is indicated.

TO RENEW CHECK VALVE

1 Depress the brake pedal several times to exhaust all vacuum from the brake system.
2 Release the clamp and disconnect the vacuum hose from the inlet manifold.
3 Pull the check valve and hose from the booster grommet. If necessary, remove the sealing grommet.

Installation is a reversal of the removal procedure with attention to the following points:

1 Check the condition of the vacuum hose and the sealing grommet. Renew as necessary.
2 Ensure that the hose clamps are secure and correctly located.

TO REMOVE AND INSTALL BOOSTER

1 Remove the lower dash panel from below the steering column, as follows:
a Remove the screws retaining the bonnet lever to the driver side lower dash panel and position the lever clear of the work area. Remove the screw cover from the top of the panel.
b Remove the lower dash panel screws and disconnect the panel from each side of the steering column. Partially remove the panel and where fitted, rotate the footwell lamp bulb holder anticlockwise and separate it from the panel. Remove the driver side lower dash panel from the vehicle.

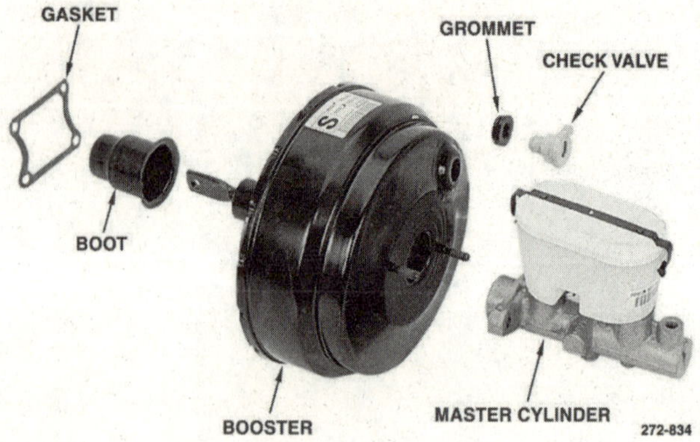

Brake booster and associated components.

2 Remove the retaining clip or split pin and withdraw the washer and clevis pin from the brake pedal.
3 Remove the brake booster retaining nuts.
4 Working in the engine compartment, pull the check valve and hose from the brake booster.
5 Remove the plenum chamber and support bracket from the engine. Refer to Engine (chapter 4) for the correct procedure.
6 On manual transaxle models, remove the clutch master cylinder reservoir and position clear of the work area ensuring that no fluid spills onto the vehicle paint.
7 Remove the master cylinder as previously described. Carefully position the brake pipes to one side and manoeuvre the brake booster from the engine compartment.

Excessive force exerted on the pipes may cause damage.

8 If damaged, remove the gasket located between the brake booster and the firewall.

Installation is a reversal of the removal procedure with attention to the following points:

1 Ensure that the vacuum hose is not leaking or collapsed and the connections are secure. Renew the check valve sealing grommet, if necessary.
2 If removed, install a new gasket to the brake booster, prior to installation.
3 Install the clevis pin and install the clip or split pin. On models with a split pin ensure that it is renewed.
4 If disturbed, ensure that the brake pedal return spring is correctly installed.
5 If a new booster has been installed, pushrod setting should be factory set. However if the pushrod setting has been disturbed it must be checked as described below.
6 Install the master cylinder and bleed the brakes as described previously in this chapter.

TO CHECK AND ADJUST PUSHROD SETTING

The pushrod is factory set and normally should not require adjustment unless the setting has been disturbed or a problem exists such as dragging brakes or excessive pedal travel. The manufacturer states that the master cylinder must be installed and bled, the engine started and the brake pedal applied twice, before attempting to adjust the pushrod setting.

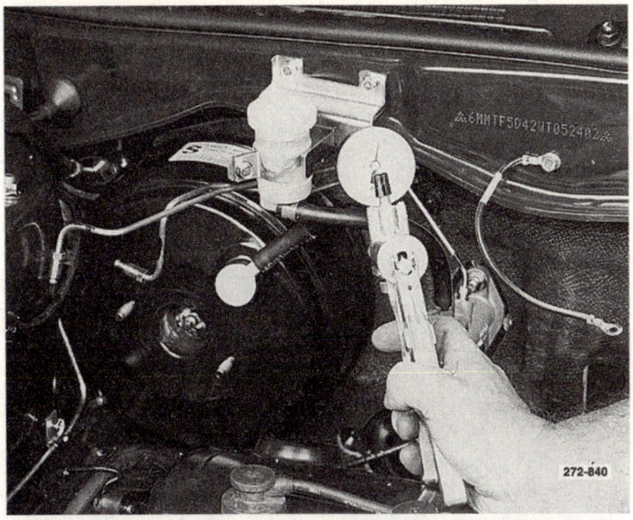

If using a hand operated vacuum pump to check the pushrod setting, ensure that 68 ± 5 kpa is applied to simulate engine vacuum.

Due to the difficultly in performing this procedure, it is recommended that it is left to an authorised dealer or a brake specialist.

> *The brake booster must be functioning correctly before attempting to adjust the pushrod setting as correct setting is reliant upon having engine vacuum (68 ± 5 kPa) in the booster while the check is performed. Alternatively a hand vacuum pump can be used to apply vacuum to the booster.*

1 Run the engine, apply the brake pedal twice then allow 10 seconds for the system to stabilise before switching off the engine.

2 Remove the master cylinder as previously described. Do not touch the brake pedal while the master cylinder is removed.

3 Using a straight edge and a pair of vernier calipers, measure the distance from the end of the master cylinder mounting of the booster to the centre of the pushrod. Allow for the thickness of the straight edge.

4 If the distance is not 9.2 – 9.4 mm, hold the knurled section of the pushrod with pliers and adjust the pushrod length until the correct distance is obtained.

5 Install the master cylinder and bleed the brakes as previously described.

Brake Pedal

TO CHECK PEDAL HEIGHT AND FREEPLAY

The dimensions for the pedal height are measured at right angles from the floor to the centre of the pedal pad, with the carpet folded down.

1 Fold down the top of the carpet, below the steering column.

2 Measure the distance from the centre of the pedal pad to the sound insulation sheet on the floor panel. If the brake pedal height is not 168 – 173 mm, proceed as follows:

a Remove the screws retaining the bonnet lever to the driver side lower dash panel and position the lever clear of the work area. Remove the screw cover from the top of the panel.

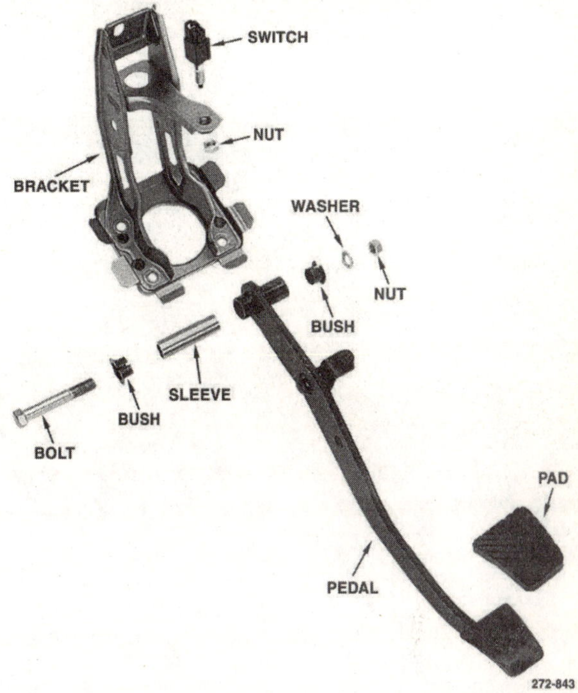

Brake pedal and associated components.

b Remove the lower dash panel screws and disconnect the panel from each side of the steering column. Partially remove the panel and where fitted, rotate the footwell lamp bulb holder anti clockwise and separate it from the panel. Remove the driver side lower dash panel from the vehicle.

d Disconnect the stop lamp switch wiring connector, loosen the locknut and unscrew the switch until it is clear of the pedal. Recheck brake pedal height, if it is still not correct, worn or damaged components exist such as the brake pedal or booster. Renew components as necessary.

3 Depress the brake pedal several times to exhaust all vacuum from the system and lightly press the brake pedal by hand until resistance is felt. Measure the distance and if the brake pedal freeplay is not 3 – 8 mm, check the brake pedal shaft bushes for wear and renew if necessary.

4 Screw the stop lamp switch in until the plunger is fully depressed, but the brake pedal does not move, and then back off the switch Þ to 1 turn. Tighten the locknut.

5 Connect the wiring and check that the stop lamps illuminate only when the brake pedal is depressed.

6 Install the lower dashboard panel.

Antilock Braking System (ABS)

DESCRIPTION

The antilock braking system (ABS) consists of a modulator located in front of the firewall in the leftr hand side of the engine compartment, an electronic control unit (ECU) located behind the instrument cluster, sensors mounted to the front steering knuckles, sensors mounted to the rear carrier brackets and pulse rings mounted to the drive shafts and to the inside of the rear hub assemblies.

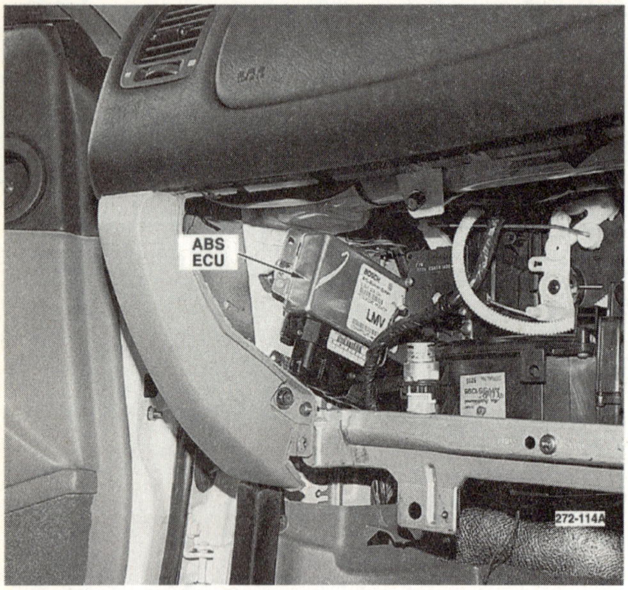

ABS ECU viewed with the glove compartment removed.

The slots in the pulse ring generate pulses in the wheel sensor which are transmitted to the ECU which then calculates wheel speed and deceleration rates.

The ECU signals the modulator assembly to increase or decrease hydraulic pressure to each wheel individually, to maintain minimum wheel rotation without locking.

Illumination of the dashboard warning lamp longer than 3 seconds after the engine is started, indicates a problem within the ABS system. Below are a several minor checks that can assist in tracing some of the more common faults within the ABS system. If these checks fail to find the fault, refer the vehicle to an authorised dealer or brake specialist with the equipment and knowledge needed to diagnose and repair the ABS system.

ABS SYSTEM BASIC CHECKS

1. Raise the front and rear of the vehicle and support on chassis stands. Refer to General Information (chapter 1) for the position of the jacking and support points.
2. Check the front and rear wheel bearings, ensuring that there is no excessive freeplay or wear. Refer to the relevant Axle and Suspension chapter, if necessary.
3. Inspect the front and rear ABS wheel sensors and front pulse rings for any signs of damage or being loose. Tighten or renew the ABS wheel sensors where necessary, or renew the drive shaft or the rear hub assembly.

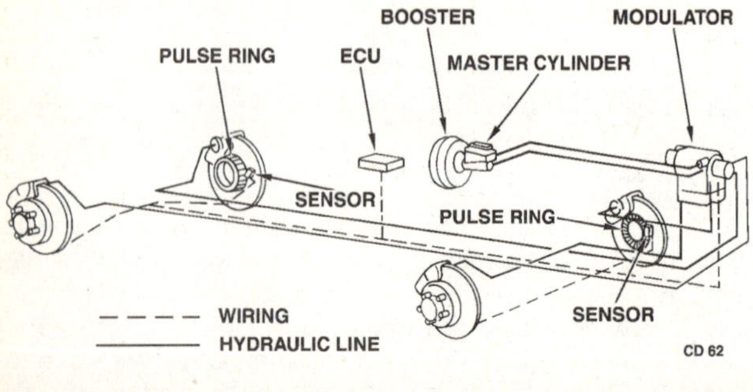

Typical ABS components.

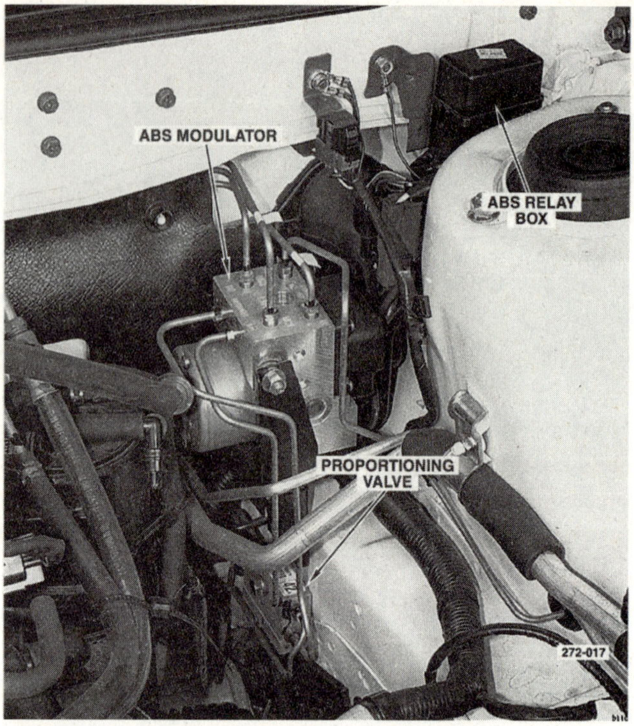

The ABS modulator is located in front of the firewall on the passenger side of the engine compartment.

The wheel sensors and pulse rings can be damaged from objects picked up by the wheels. Shake each sensor to ensure it is not loose or cracked.

4. Disconnect the wiring connectors from each wheel sensor. Check that the terminals are not corroded or backed out of the rear of the connector. Using an ohmmeter, check that the resistance between the sensor terminals is 1.4 – 1.8 k/ohms. Repair any damaged terminals or renew the wheel sensors where necessary.
5. Working in the engine compartment, check that the ABS modulator wiring connector is making good contact and the terminals are free of corrosion. Repair where necessary.

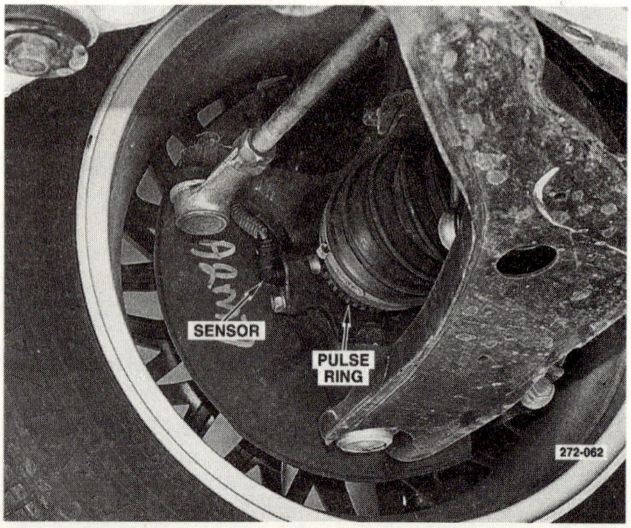

Front ABS sensor and pulse ring.

6 If no fault can be found, it is recommended that the vehicle be referred to an authorised dealer or brake specialist with the specialised equipment to isolate the fault.

TO RENEW WHEEL SENSORS

1 Remove the relevant wheel/s from the vehicle.
2 Disconnect the wiring connector and manoeuvre the sensor wiring from any brackets.
3 Remove the bolt retaining the sensor to the steering knuckle, on the front brakes, or the rear axle/rear axle carrier on the rear brakes, and remove the sensor from the vehicle.

Installation is a reversal of the removal procedure with attention to the following points:

1 Check that the pulse ring is not damaged or distorted prior to installing the wheel sensor.
2 Ensure that the wiring connector is not corroded or damaged and that is making good electrical contact.

TO RENEW PULSE RINGS

Rear Pulse Ring

The rear ABS wheel sensor pulse ring can be renewed separately and is pressed onto the rear hub assembly. If damaged the pulse ring can be renewed using a bearing separator and a press. Care must be taken when pressing the new pulse ring on to ensure

Rear ABS sensor. Sedan model shown.

that it is square. Refer to the Rear Axle and Suspension (chapter 14) for the correct procedure to remove and install the rear hub.

Front Pulse Ring

At the time of publication the front pulse ring was only available as part of the drive shaft assembly and if damaged, must be renewed as a complete unit. Consult your parts supplier prior to purchase.

Chapter 16
Body Electrical

Contents

Routine Maintenance

LAMPS

Check the park lamps, stop lamps, reverse lamps, indicators, hazard lamps, fog lamps and, where fitted, fog lamps and driving lamps for correct operation. Ensure that each lamp is illuminated clearly and the lens and lens reflector is not damaged. Renew components as necessary.

Check the bulb for a broken filament, blacked out bulb or corroded base on any lamps that are not working. If necessary, renew the bulb with one of the same voltage and wattage, which is marked on the side of the base. Clean the bulb base with methylated spirits or similar if corroded.

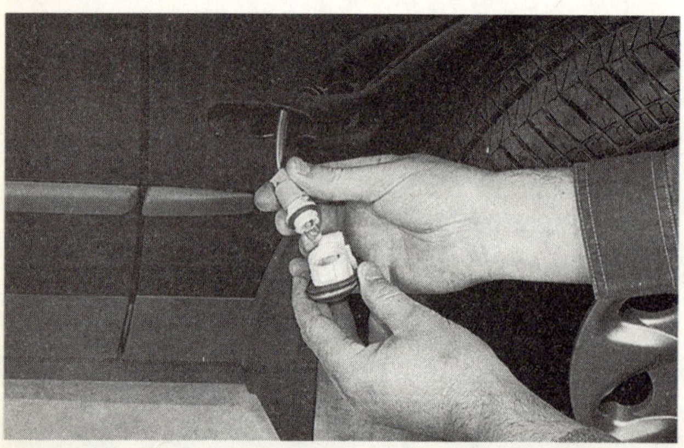

Removing the bulb holder from the side turn signal lamp.

Check and adjust the headlamp aim as described under the Lamps heading in this chapter.

TIP

Refer to the Lamps heading in this chapter for procedures to check bulbs and remove and install lamps.

It is important that bulbs are replaced only with one of the same voltage and wattage. Using higher wattage headlamp bulbs than the manufacturer recommends could result in excessive loads on the headlamp circuit.

HORN

All models are fitted with one low note horn fitted behind the front bumper. Some models have a second high note horn.

If the horn does not work, refer to the Horn heading in this chapter for procedures to check components.

Front bumper removed showing the horn location. On some models a second horn is located in the same position on the driver side of the vehicle.

WIPERS

Check the water level in the windscreen washer bottle and top up if necessary.

Ensure that nothing is caught between the wiper blade rubbers and the windscreen. Check that the wiper arms are making secure contact with the glass.

Renew any wiper blades that appear perished, torn or which are not cleaning the screen properly. It is normal to renew the wiper blades regularly. Refer to the Windscreen Wipers heading in this chapter if necessary.

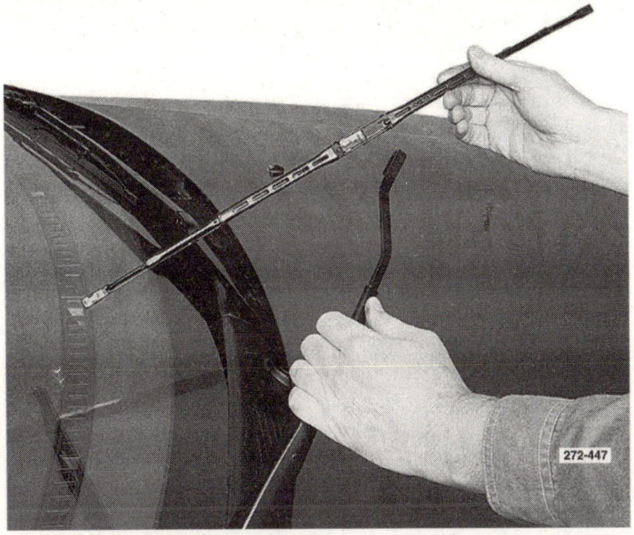

Check and renew the wiper blades at regular intervals.

Windscreen washer bottle location. On station wagons, the tailgate washer is also supplied by this washer bottle.

BATTERY

Routine maintenance on the battery is covered in Engine Electrical (chapter 6).

Lighting System Trouble Shooting

LAMP OR LAMPS FAIL TO OPERATE

1 Faulty bulb: Check and renew the bulb.

2 Fuse or fusible link blown: Rectify the fault and renew as necessary.

3 Open circuit in the wiring or connections: Check and rectify the fault.

4 Faulty lamp switch: Check and renew the switch.

5 Relay faulty: If possible, substitute a known good relay of the same type to check the circuit. Renew the relay if faulty.

Switch On the lamp concerned and, using a test lamp, check for power in the lamp circuits. This is best done by starting at the lamp wiring connector and working back to the power source.

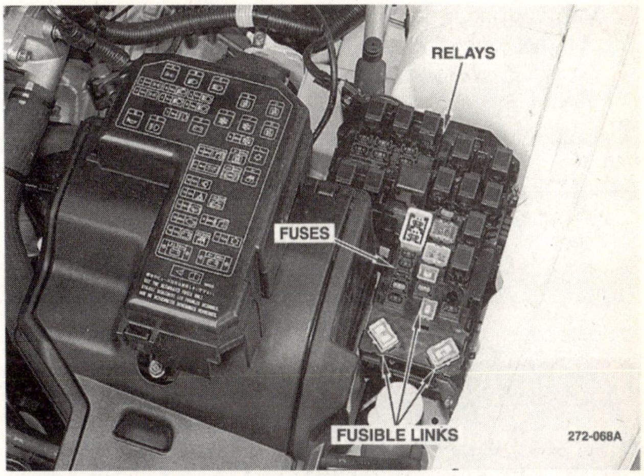

Location of the engine compartment fuses, fusible links and relays.

Checking for power at the right hand headlamp.

LAMP OR LAMPS INCORRECTLY ILLUMINATED

1 Lamp or lamps incorrectly earthed: Check the lamp earth circuit for loose or poor contact, either at the lamp body or the wiring, and repair as necessary.

2 Incorrect bulbs installed: Check the bulb wattage and voltage and renew with the correct type if necessary.

3 Dirty or damaged lamp reflector or lens: Clean or renew the lamp as necessary.

4 Faulty bulb: Check with a known serviceable bulb and renew if necessary.

TIP *The most common cause for this condition is incorrect lamp earthing. Check the lamps at their earthing points.*

LAMPS FLARE WITH ENGINE SPEED INCREASE

1 Poor earth connection between the battery and the engine: Check the battery earth lead and the earth between the battery and the body.

2 Faulty battery: Check and renew the battery.

3 Battery in a low state of charge: Recharge the battery and check the charging system.

4 High resistance or faulty connections between the battery and the alternator: Check the circuit and rectify the fault.
5 Faulty alternator voltage regulator: Check the alternator output.
6 Faulty alternator: Check and repair or renew as necessary.

The most common cause for this condition is dirty terminals on the earth leads. Check all earth leads at their earthing points.

Turn Signal Lamp Trouble Shooting

TURN SIGNAL LAMPS DO NOT OPERATE AND NO SOUND FROM FLASHER UNIT

1 Fuse blown: Rectify the fault and renew the fuse.
2 Faulty flasher unit: Renew the flasher unit.
3 Faulty turn signal switch: Renew the switch.
4 Fault in the wiring circuit: Check and repair the fault.

TURN SIGNAL WARNING LAMP DOES NOT FLASH BUT AUDIBLE CLICKING FROM FLASHER UNIT

1 Warning lamp bulb blown: Check and renew the bulb.
2 Fault in the warning lamp wiring: Check and repair the fault.

When renewing bulbs, ensure that a new bulb of the correct voltage and wattage is used.

TURN SIGNAL LAMPS FLASH AT GREATER THAN NORMAL SPEED

1 Bulb blown in the front or rear turn signal lamps: Check and renew the bulb.
2 Faulty flasher unit: Check and renew the flasher unit.

HAZARD WARNING LAMPS DO NOT OPERATE

1 Fuse blown: Rectify the fault and renew as necessary.
2 Faulty flasher unit: Renew the flasher unit.
3 Faulty hazard warning switch: Renew the switch.
4 Fault in the wiring circuit: Check and repair the fault.

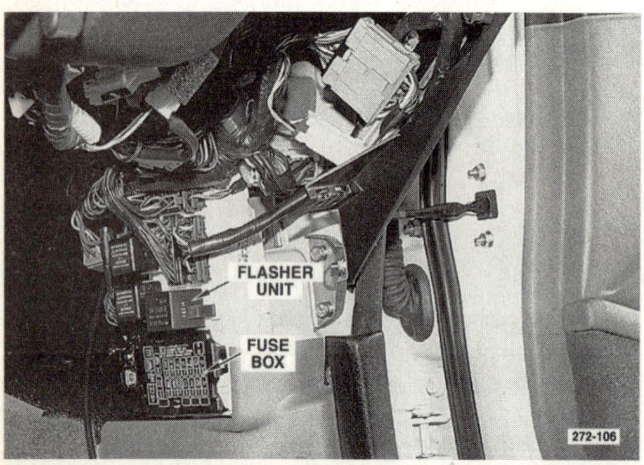

Lower dash panel removed showing the interior fuse box and the flasher unit.

Fuses, Fusible Links and Circuit Testing

Fuses protecting the various vehicle electrical circuits are located in a fuse box with access through the drivers side kick panel. Several fuses are also installed in the engine compartment fuse and relay box.

If a fuse is faulty or burnt out, use a replacement fuse of the specified amperage only. The specified amperage is marked on the fuse. The use of a higher amperage fuse than that specified could cause damage to the vehicle wiring harness or components.

TO CHECK AND RENEW A FUSE

1 Remove the fuse cover.
2 Using the information printed on the fuse box cover, locate the fuse protecting the circuit at fault. Pull the fuse from the fuse box and inspect it for serviceability.

A tool to remove the fuses is inside the interior fuse and relay box.

3 Should the fuse appear to be in a serviceable condition, test its continuity with an ohmmeter or test lamp.
4 Should the fuse appear to be faulty or burnt out, install a fuse of the same amperage.
5 If the circuit is still inoperative or the fuse burns out again, check the circuit as described under this heading or refer the problem to an automotive electrician or an authorised dealer.

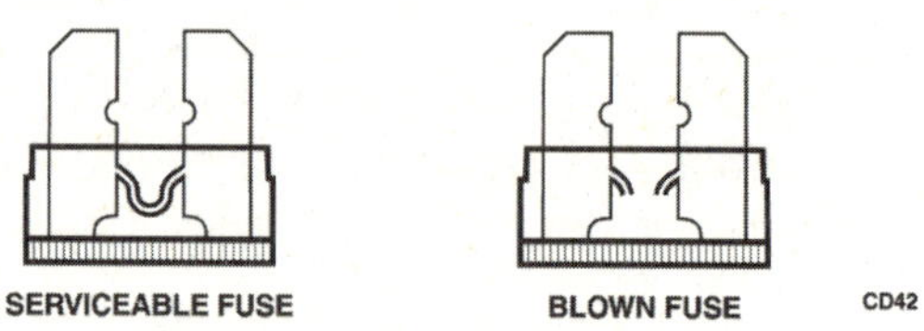

TO CHECK AND RENEW FUSIBLE LINKS

The fusible links are located in the engine compartment fuse and relay box adjacent to the battery.

If a fusible link is faulty or burnt out, use a genuine replacement fusible link of the specified amperage only. The specified amperage is marked on the fusible link. The use of a higher amperage fusible link than that specified could cause damage to the vehicle wiring harness.

A new fusible link should not be installed until the fault that caused the original link to fail has been repaired.

1 Working in the engine compartment adjacent to the battery, release the retaining clips and open the fusible link cover.
2 Inspect the condition of the fusible link through the window in the top.
3 The fusible links can be removed by pulling them from the housing, except for the No 23 (120 amp) and No 24 (60 amp) fusible links which are retained by a bolt on each terminal of the link.
4 Should the fusible link appear to be in a serviceable condition, test its continuity using an ohmmeter or test lamp.

5 If the fusible link is faulty, install a new fusible link and check the circuit for correct operation.

TO TEST CIRCUITS

1 Turn the circuit switch On.

2 Using a test lamp, check for power at one of the circuit connections. A fuse is a good starting point.

If power is available, reconnect the connector and continue checking towards the motor or bulb end of the circuit.

If power is not available, continue checking towards the battery or switch end of the circuit.

Ignition Switch and Lock Cylinder

The ignition switch is part of the steering lock and is renewed as described in Steering (chapter 12).

To Remove and Install — Lock Cylinder

1 Disconnect the negative battery terminal.

2 Remove the screws retaining the bonnet lever to the lower dash panel and position the lever clear of the work area.

3 Remove the screw cover from the top of the drivers side lower dash panel. Remove the remaining screws from the lower dash panel, disconnect the panel from the top left side of the dashboard and partially remove the panel. Where fitted, rotate the footwell lamp bulb holder anti clockwise and remove the panel from the vehicle.

4 Remove the screws from the bottom of the ignition lock surround panel. Remove the shield from beneath the lock cylinder and then manoeuvre the surround panel from the dashboard.

5 Disconnect the key reader wiring connector, loosen the cable tie and remove the screw from the base of the key reader. Where fitted remove the illumination bulb holder by rotating the bulb holder anticlockwise after partially removing the key reader from around the lock cylinder. Remove the key reader from the vehicle.

6 Insert the ignition key into the lock cylinder and turn the key to the ACC position.

7 Insert a small screwdriver or pin punch into the hole in the ignition lock housing and depress the locking latch. Withdraw the lock cylinder from the lock housing.

Installation is a reversal of the removal procedure with attention to the following points:

Ignition lock surround removed showing the key reader screw.

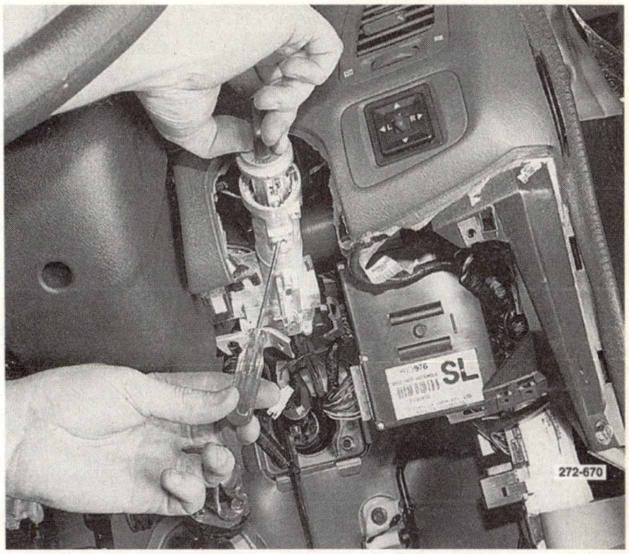

Removing the lock cylinder using a small screwdriver.

1 Turn the ignition key to the ACC position and install the lock cylinder in the lock housing. Depress the lock cylinder until it clicks into position.

2 Remove the key from the ignition.

3 Install the key reader, connecting the illumination lamp where fitted and the wiring connector and tighten the retaining screw.

4 Install the ignition lock assembly surround panel and shield. Tighten the retaining screws.

5 Install the lower dash panel and the bonnet lever.

Switches and Relays

TO TEST SWITCHES

Most switches have more than two wires entering the switch. One wire will be from the power source (battery or ignition), the second wire takes power from the switch to the component, or the component relay. A third wire may be used to power a small globe in the switch for illumination, in which case there will also be a fourth wire to earth the globe.

1 Access the switch by removing it from the switch holder, dashboard or component. Refer to the appropriate heading for a removal procedure if necessary.

2 Referring to the wiring diagrams, locate the colours of the wire going from the power source to the switch and the wire going from the switch to the component or the component relay.

3 Disconnect the wiring connector from the switch and using an ohmmeter, check the resistance between the appropriate terminals of the switch.

There should be no resistance with the switch in the On position and open circuit with the switch in the Off position. Renew the switch if necessary.

RELAYS

A relay can be described basically as a remotely operated switch. It can carry a greater current from the power source to the components than a small switch could. Having a relay in a circuit

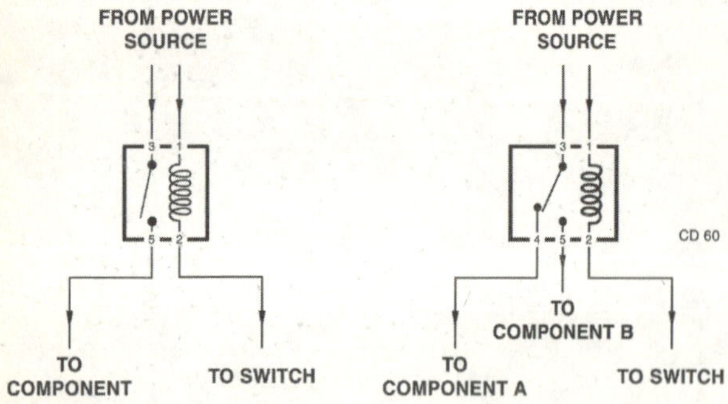

Typical four and five pin relays.

eliminates the need for heavy duty switches and thick wiring running to the switches and instrument cluster assembly. When the component switch is placed On, a small solenoid closes two contacts in the relay allowing power to run from the power source to the component.

In these illustrations, power is supplied to the solenoid from the power source. The switch, when On, completes the path to earth, energising the solenoid and closing the contacts allowing power to flow from the power source to the component.

On the five pin relay, the extra terminal allows power to flow from the power source to a component, such as a low beam headlamp bulb. When the solenoid is energised, the contact moves across allowing power to flow to the second component, such as a high beam headlamp bulb.

This is just one example of how a relay can be wired into a circuit. There are numerous applications and wiring combinations.

The various relays employed in the vehicle electrical system are located adjacent to the battery in the engine compartment and inside the vehicle behind the driver side kick panel.

Not all vehicles are equipped with all relays because of different equipment levels.

To renew a relay, pull the relay from the relay box. The small relays have an adapter bracket that sits between the relay and the relay box. To remove these relays, the clips have to be seperated from the lock tabs on the relay box using a small screwdriver prior to removing them. The adapter bracket can then be removed from relay, making testing the relay easier.

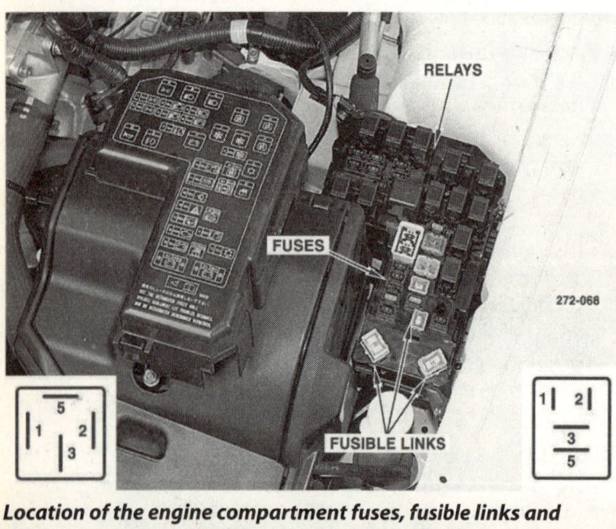

Location of the engine compartment fuses, fusible links and relays. Inset shows the relay terminal identification.

When installing a relay, ensure that the terminals are making good electrical contact. Many relays are of the same type. This can be confirmed by the part number printed on them. This is useful when trouble shooting as a known good relay can be substituted for a faulty one to check the circuit.

Refer to Fuel and Engine Management (chapter 7) for a procedure to check the fuel pump and the engine control relays as they are slightly different to relays located within the engine compartment fuse and relay box and the driver side fuse and relay box.

To Check

1. Remove the relay from the housing.
2. Using an ohmmeter, check that continuity exists between terminals 1 and 2, and not between 3 and 5.
3. Using jumper leads, connect terminals 1 and 2 to the battery terminals.

   Continuity should exist between terminals 3 and 5.

COMBINATION SWITCH

To Remove and Install

1. Set the front wheels to the straight ahead position and disconnect the negative battery terminal.
2. Where fitted, disable and remove the driver airbag. Refer to Body (chapter 17).
3. On models without an airbag, proceed as follows:

   a. Remove the screws retaining the bonnet lever to the lower dash panel and position the lever clear of the work area.

   b. Remove the screw cover from the top of the drivers side lower dash panel. Remove the remaining screws from the lower dash panel, disconnect the panel from the top left side of the dashboard and remove the panel from the vehicle.

   c. Remove the horn pad screws from each side of the steering wheel and partially remove the horn pad.
4. Disconnect the wiring from the horn switch and where fitted, the cruise control and sound system remote control assembly.
5. Unlock the steering column tilt mechanism and position the steering column to its lowest position. Lock the lever in that position.
6. Ensure that the steering wheel is in the straight ahead position and loosen the steering wheel retaining nut. Do not completely remove the nut.

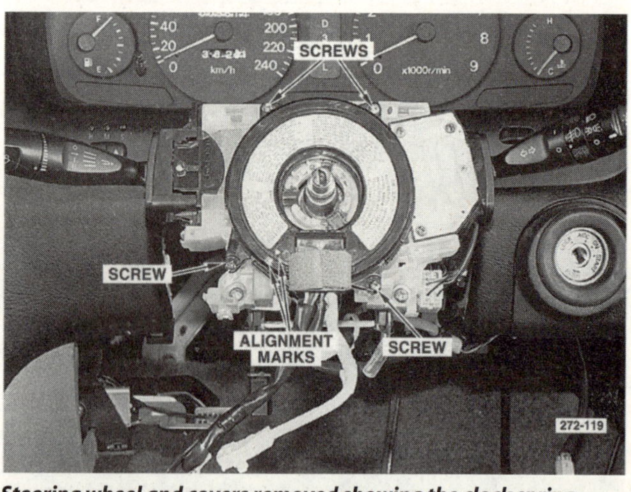

Steering wheel and covers removed showing the clock spring retaining screws and the two alignment marks.

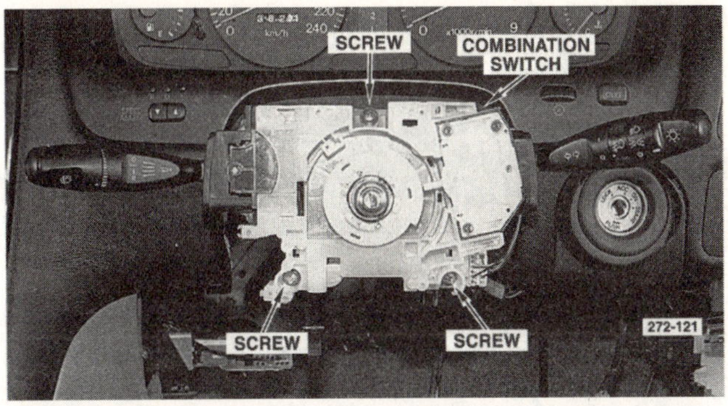

Steering column covers removed showing the combination switch retaining screws.

7 Mark the relationship between the steering column and steering wheel to ensure that the wheel is returned to the same position. Grasp the steering wheel at the twelve o'clock and six o'clock positions and rock the steering wheel while moving your hands around the steering wheel until it separates from the steering column.

If this fails to work, a steering wheel puller may be needed.

Avoid turning the steering column or clock spring to prevent damage to the clock spring when it is reinstalled.

8 Pass the wiring through the steering wheel and remove the wheel from the column.

9 Remove the screws retaining the steering column covers and remove the covers from the column.

10 Disconnect the horn wiring connector from beneath the steering column. Unclip the clock spring wiring from along the steering column. Remove the retaining screws and remove the clock spring. Do not rotate the clock spring.

11 Disconnect the wiring connectors from the combination switch assembly. Remove the retaining screws and remove the combination switch assembly from the vehicle.

12 If necessary, the wiper switch and the headlamp switch assemblies can be renewed separately after removing the retaining screws and removing the switches from the base plate. On the wiper switch, separate the cover from the switch assembly after removing the retaining screws.

The headlamp switch also includes the turn signal switch.

Installation is a reversal of the removal procedure with attention to the following points:

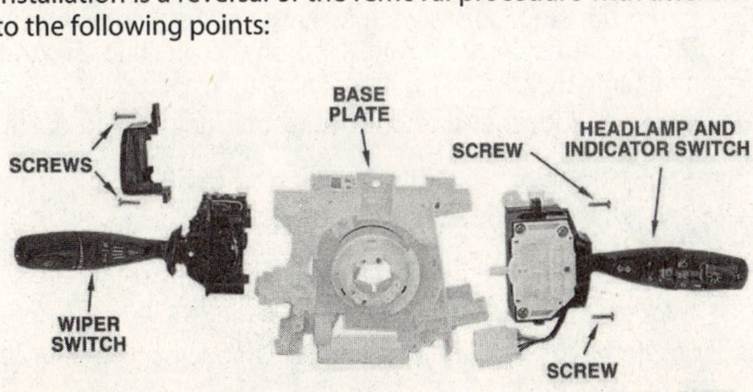

Combination switch components.

1 Where removed, install the wiper switch and headlamp switch to the base plate. Tighten the retaining screws.

2 Install the combination switch assembly to the steering column and connect the wiring connectors.

3 Install the combination switch retaining screws and tighten securely.

The clock spring will be damaged if rotated past its normal limits. Prior to installing the clock spring, hold the inner hub and rotate the outer hub in a clockwise direction until resistance is felt. Rotate the outer hub anti clockwise approximately 3 and 4/5 turns and then align the two alignment marks. Install the clock spring in this position.

4 Install the clock spring and tighten the retaining screws. Secure the wiring to the base of the steering column and connect the horn wiring connector. Do not connect the airbag wiring connector beneath the column at this stage.

5 Install the steering column covers, ensuring that the dowels on the top cover locates into the top of the combination switch base plate. Install the lower cover and tighten the retaining screws.

6 Install the steering wheel over the steering column. Pass the wiring through the hole in the base of the steering wheel and install the wheel onto the column, aligning the marks made on removal.

7 Install the steering wheel retaining nut and tighten to 30 Nm.

8 Connect the wiring to the horn switch and where fitted, the cruise control and remote sound system controls.

9 Where fitted, connect the wiring to the rear of the airbag. Install the airbag/horn pad and tighten the Torx bolts.

10 Where fitted, enable the drivers airbag. Refer to Body (chapter 17).

INSTRUMENT CLUSTER SWITCHES

To Remove and Install

1 Remove the instrument cluster surround panel, as described under the Instrument Cluster heading in this chapter.

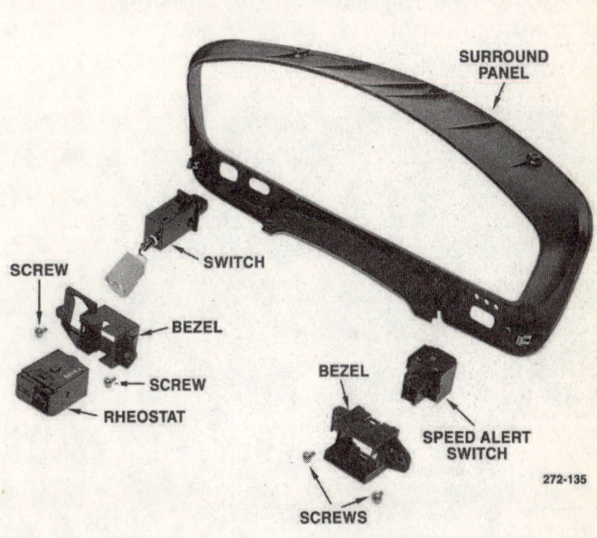

Instrument cluster surround panel switches and rheostat.

2 Remove the screws retaining the switches to the surround panel. Remove the switch and bezel from the rear of the instrument cluster surround panel.

3 Release the retaining tangs and remove the switches through the front of the bezel for all switches except the dashboard illumination rheostat, which is removed from the rear of the bezel.

Installation is a reversal of the removal procedure.

REVERSE LAMP SWITCH

On models equipped with an automatic transaxle, a combined inhibitor switch and reverse lamp switch is installed. The switch can be checked and adjusted as described in Automatic Transaxle (chapter 11).

On models equipped with a manual transaxle, use the following procedures.

To Test

1 Switch the ignition On and select reverse gear. The reverse lamps should operate.

2 If the reverse lamps do not operate, disconnect the wiring from the reverse lamp switch on the top of the transaxle.

3 Connect a jumper lead between the terminals of the reverse lamp switch wiring connector.

4 Switch the ignition On and check the reverse lamps. If the reverse lamps now operate, the reverse lamp switch can be considered faulty and should be renewed.

If the reverse lamps still fail to operate with the switch connector terminals bridged, check the condition of the lamps and fuses.

To Remove and Install

1 Working in the engine compartment, disconnect the wiring from the reverse lamp switch on the top of the transaxle.

2 Remove the reverse lamp switch from the transaxle housing and discard the washer.

Installation is a reversal of the removal procedure with attention to the following points:

1 Using a new washer, install the reverse lamp switch.

2 Tighten the switch to a torque of 32 Nm.

Reverse lamp switch location. Manual transaxle models.

STOP LAMP SWITCH

To Test

1 Depress the brake pedal and have an assistant check if the stop lamps operate.

2 If the stop lamps do not operate, remove the screws retaining the bonnet lever to the lower dash panel and position the lever clear of the work area.

3 Remove the screw cover from the top of the driver side lower dash panel. Remove the remaining screws from the lower dash panel, disconnect the panel from the top left side of the dashboard and partially remove the panel. Where fitted, rotate the footwell lamp bulb holder anti clockwise and remove the panel from the vehicle.

4 Disconnect the stop lamp switch wiring connector and connect a jumper lead between the wiring connector terminals.

If the stop lamps operate, check the stop lamp switch as described below. If the adjustment is correct, the switch requires renewal.

If the stop lamps do not operate, check the fuse, lamps and wiring.

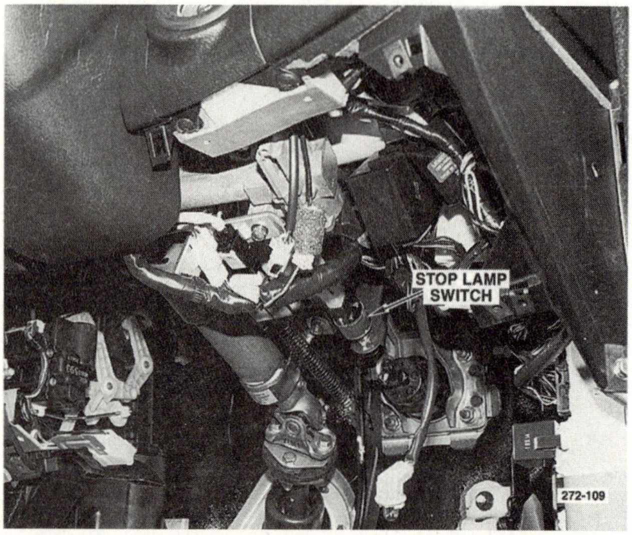

Lower dash panel removed showing the stop lamp switch location.

To Adjust

1 Remove the lower dash panel as described under To Test.

2 Loosen the stop lamp switch locknut and rotate the switch until the stop lamps extinguish when the pedal is released and illuminate as soon as the pedal is depressed.

3 Tighten the locknut and check that the adjustment is still correct.

To Renew

1 Remove the lower dash panel as described under To Test.

2 Disconnect the wiring connector from the stop lamp switch.

3 Loosen the locknut and unscrew the stop lamp switch from the brake pedal bracket.

Installation is a reversal of the removal procedure, ensuring that the stop lamp switch is adjusted as previously described.

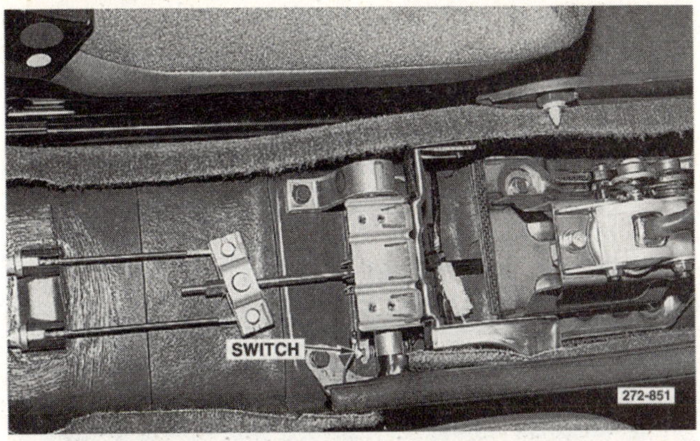

Console removed showing the handbrake warning lamp switch.

HANDBRAKE WARNING LAMP SWITCH

To Check

1. Remove the centre console from the vehicle. Refer to Body (chapter 17).
2. Disconnect the wire from the switch and bridge the wire to earth.
3. Renew the switch if the handbrake warning lamp illuminates with the ignition On.

   If the handbrake lamp does not illuminate, check the fuse, the wiring and the bulb. Refer to the wiring diagrams.

To Renew

1. Remove the centre console from the vehicle. Refer to Body (chapter 17).
2. Disconnect the wiring connector from the handbrake warning lamp switch.
3. Remove the screw retaining the switch to the handbrake lever and remove the switch.

If a screwdriver small enough to access the screw is not available, it may be necessary to remove the handbrake lever and then remove the screw. Refer to Brakes (chapter 15) if necessary.

Installation is a reversal of the removal procedure.

Courtesy lamp switch removed from the door pillar.

COURTESY LAMP DOOR SWITCH

To Remove and Install

1. With the door fully open, remove the retaining screw from the switch mounting flange.
2. Partially withdraw the switch from the door pillar.
3. Disconnect the wiring from the rear of the switch and remove the switch from the vehicle.

Installation is a reversal of the removal procedure.

HAZARD LAMP SWITCH

To Remove and Install

1. Remove the radio surround panel. Refer to the Sound System heading.
2. Separate the retaining tangs and remove the switch from the rear of the surround panel.

Installation is a reversal of the removal procedure.

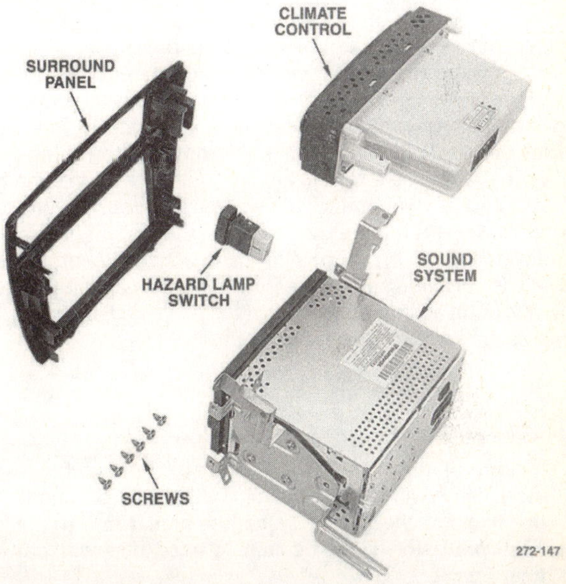

Sound system, climate control and hazard lamp switch, removed from the surround panel. Verada model shown.

BOOT RELEASE/FUEL DOOR SWITCH — VERADA Xi MODELS

To Remove and Install

1. Using a small screwdriver wrapped in cloth, remove the switch assembly from the door trim by levering between the edge of the switch and door trim.
2. Partially remove the switch and disconnect the wiring connector.

Installation is a reversal of the removal procedure.

Power Windows

The power window switch, when moved to the Down position, supplies battery power to the power window motor which winds the window down. Moving the power window switch to the

Removing the boot/fuel door release switch. Verada models.

Up position, reverses the polarity of the power to the window motor. This causes the window motor to rotate in the opposite direction, winding the window up. The drivers window has a slightly different switch. Depressing this switch further down makes the window wind all the way down without having to hold the switch. A short time after the window has reached the bottom of it's travel, the switch will automatically return to its rest position. This is done by measuring the current draw on the window motor. When the window has reached the bottom of it's travel the current draw on the power window motor increases, this acts upon an internal circuit within the switch assembly and releases the switch to the rest position.

TO CHECK

1 Attempt to operate the power windows.

If none of the windows will operate, remove the switch assembly from the drivers door as described later under this heading and check for power at the blue wire with the ignition On. Use the wiring diagram to trace the fault if power is unavailable.

If at least one window operates, proceed to step 2.

If power is available, check for continuity to earth at the power window switch black wire and repair as necessary.

2 Using the wiring diagram, check for the appropriate power and earth supplies to each motor or the motor that is not working.

If the appropriate power and earth supplies are not leaving the switch assembly, renew the power window switch assembly.

3 If the switch assembly is supplying power and earth as required, remove the door trim panel from the door with the faulty window. Refer to Body (chapter 17).

4 Disconnect the wiring connector from the power window motor and connect a test lamp between the terminals of the wiring connector. The test lamp should illuminate whenever the power window switch is pressed in either direction.

i *It will be necessary to connect the power window switch wiring connectors on the rear doors for this test to work. The test lamp should illuminate when the switch on the door is operated.*

If using an LED type test lamp, the lamp may only illuminate with the current flowing in one direction. If, for example, the test lamp only illuminates when the power window switch is moved to the Up position, alternate the test lamp leads and check if the test lamp now illuminates in the Down position. If so, this does not indicate a fault.

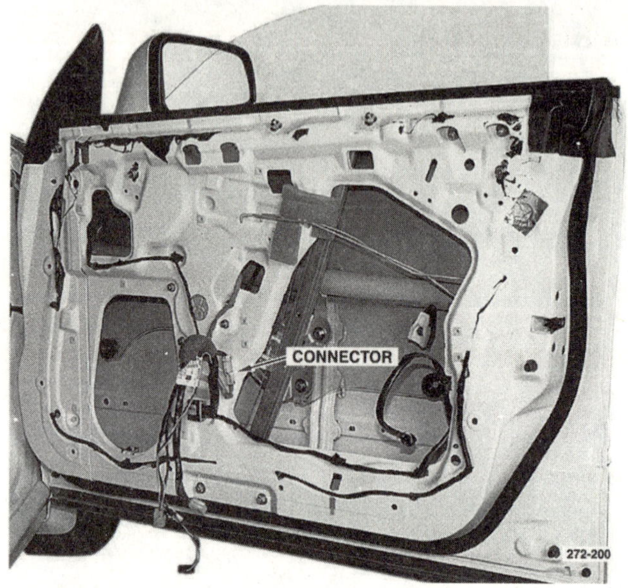

Power window motor wiring connector.

If the test lamp does not illuminate, check the wiring between the door and the power window switch assembly for an open circuit.

5 If the test lamp is illuminating as required, check for continuity between the motor wiring connector terminals, using an ohmmeter. If an open circuit is indicated, renew the power window regulator assembly as described in Body (chapter 17).

TO REMOVE AND INSTALL SWITCHES

1 Insert a small screwdriver with the blade wrapped in cloth between the door trim and the power window switch assembly. Gently lever the switch assembly upward.

2 Disconnect the wiring connector from the switch assembly and remove it from the vehicle.

Installation is a reversal of the removal procedure.

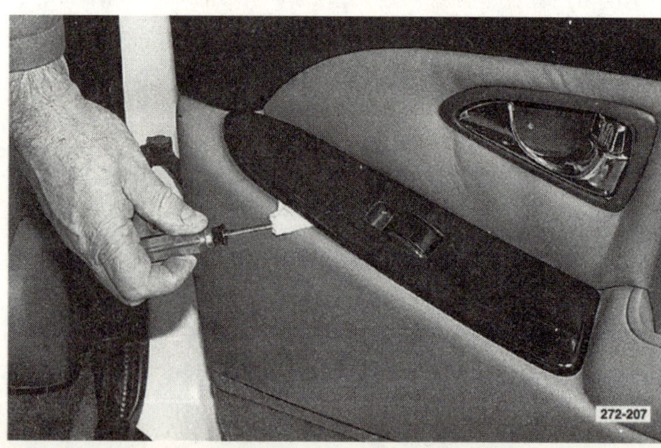

Removing the power window switch from the door.

TO REMOVE AND INSTALL MOTOR

The power window motor is an integral part of the window regulator and if faulty, must be renewed as an assembly. A procedure to remove the window regulator is in Body (chapter 17).

Power Mirrors

The power mirror is operated by two small motors. One motor controls horizontal movement and the other motor controls the vertical movement of the mirror.

Changing the polarity alters the direction of rotation of each motor, allowing the mirror to move in each direction.

Rotating the lever on the power mirror switch allows movement of the left and right hand mirrors using the one switch assembly.

TO CHECK

1 Remove the power mirror switch, as described under this heading and check for battery power with the ignition On at the blue wire. If battery voltage is not available, use the wiring diagrams to trace the fault.

2 Ensure that there is continuity to earth at the black wire. Repair as necessary.

3 Remove the mirror that is not working as described in Body (chapter 17).

4 Disconnect the mirror wiring connector and place a test lamp between the blue/yellow and the blue/black wires on the wiring connector.

If the fault is in the passenger side mirror place the test lamp between the blue/white wire and the blue/black wire.

Ensure that the lever on the power mirror switch is pointing toward the mirror being tested. For example, if working on the driver side mirror, the lever should be pointing toward the driver mirror.

5 The test lamp should illuminate when the power mirror switch is moved to either the In or Out position.

If using an LED type test lamp, the lamp may only illuminate with the current flowing in one direction. If, for example, the test lamp only illuminates when the power mirror switch is moved to the In position, alternate the test lamp leads and check if the test lamp now illuminates in the Out position. If so, this does not indicate a fault.

6 Repeat steps 3 and 4 between the white/blue wire and blue/black wire, ensuring that the test lamp illuminates when the power mirror switch is moved to either the Up or Down positions.

If the fault is in the passenger side mirror place the test lamp between the blue/orange wire and the blue/black wire.

7 If the test lamp does not illuminate at one or more of these checks, and the wiring and connections are not open circuited, renew the power mirror switch.

8 If no fault can be found, check the continuity between the blue/yellow wire (right hand mirror) or blue/white wire (left hand mirror) and the blue/black wires. Check the continuity

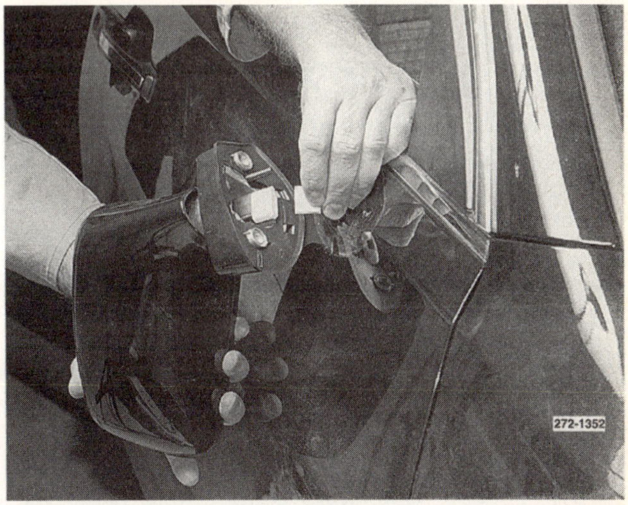

Removing the power mirror from the door.

between the blue/orange wire (right hand mirror) or white/blue wire (left hand mirror) and the blue/black wire at the faulty mirror. If an open circuit is indicated at either or both wires, renew the mirror.

TO REMOVE AND INSTALL SWITCH

1 Using a small screwdriver wrapped in cloth, remove the power mirror switch and bezel from the dashboard.

2 Disconnect the wiring from the rear of the switch and remove it from the vehicle.

Installation is a reversal of the removal procedure.

Disconnecting the wiring from the power mirror switch.

TO REMOVE AND INSTALL MIRRORS

The power mirror removal and installation procedure is covered in Body (chapter 17).

Lamps

BULBS

Different types of bulbs are used throughout the range of vehicles covered in this manual and it is important that if renewing bulbs, the correct voltage, wattage and type is used.

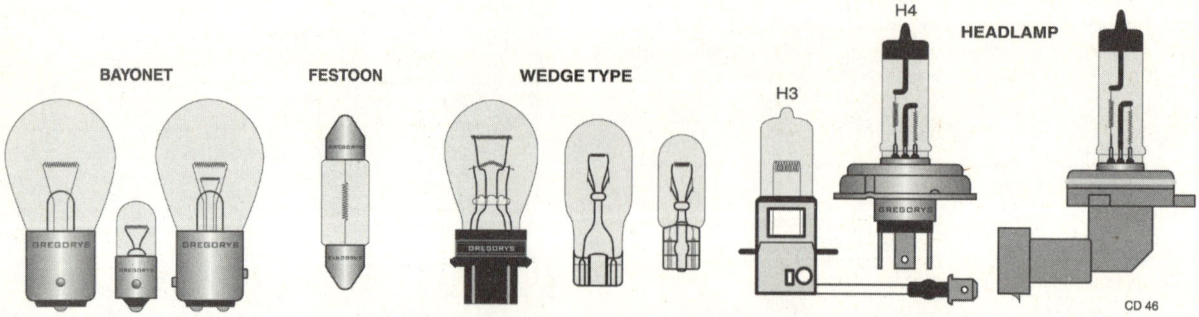

Common types of bulbs fitted to motor vehicles.

The voltage and wattage is usually stamped on the bulb base or glass. A bulb stamped 12V/21W means that the bulb is a 12 volt bulb with 21 watts of power.

Dual filament bulbs are used in the stop/tail lamps. These are basically two bulbs in one base. For example, if the bulb is stamped 12V 21/5W, then this means that the bulb is a 12 volt bulb with 21 watts of power at one filament and 5 watts of power at the second filament.

To Test Bulbs

1 Remove the bulb from the bulb holder.

2 Connect a jumper lead between the positive battery terminal and the contact at the base of the bulb.

3 Connect another jumper lead between the negative battery terminal and the base of the bulb. The bulb should illuminate.

If the bulb has two filaments, connect the positive jumper lead to the second contact on the base of the bulb. The second filament should illuminate.

If the bulb fails to illuminate, it should be renewed with a bulb of the correct voltage and wattage.

HEADLAMP

To Remove and Install

1 Using a small screwdriver, raise the tangs on the radiator grille retaining clips. When all of the clips are raised, remove the grille from the vehicle.

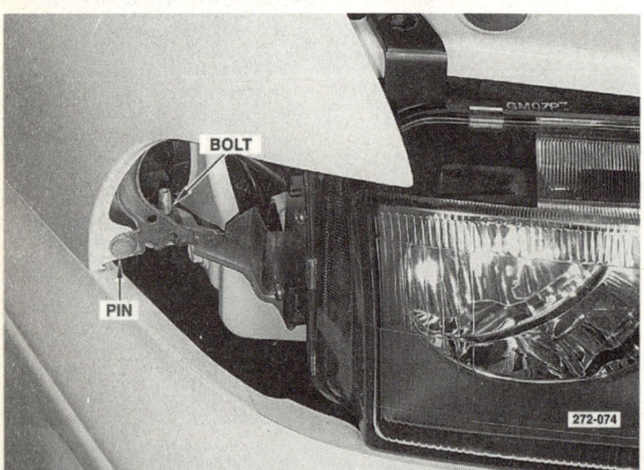

Turn signal removed showing the headlamp bracket bolt.

2 Working inside the engine compartment, disconnect the spring clip retaining the turn signal lamp assembly. Pull the turn signal lamp partially from the vehicle. Rotate the bulb holder anticlockwise and remove the lamp from the vehicle.

3 Raise the front of the vehicle and support on chassis stands. Refer to General Information (chapter 1) for the positions of the jacking points.

4 Working beneath the vehicle, remove the bolt retaining the headlamp bracket to the inner guard.

5 Disconnect the wiring connectors from the rear of the headlamp assembly and the bolts retaining the headlamp to the radiator support panel.

6 On TL/TW KL/KW models, release the top bumper bar retainers to allow the bar to be pulled down to facilitate removal of the headlamp.

7 Manoeuvre the headlamp from the vehicle.

Installation is a reversal of the removal procedure with attention the following points:

1 Hold the headlamp in position and install all bolts finger tight before tightening securely.

2 Install the turn signal lamp and front grille.

3 Check and if necessary, adjust the headlamp aim as described later in this chapter.

To Renew Headlamp Bulb

1 Disconnect the wiring connector from the rear of the bulb.

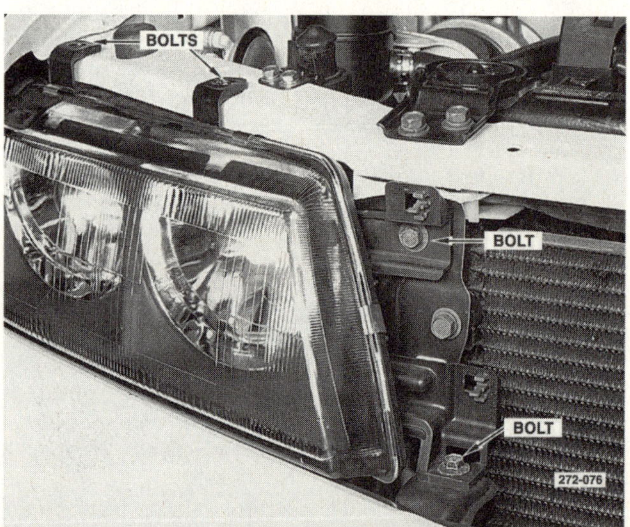

Radiator grille removed showing the headlamp bolts.

Renewing the high beam lamp bulb.

2 Unscrew the dust cover from the rear of the headlamp assembly. Be careful not to misplace the rubber dust boot.

3 Unclip the spring retainer from the bulb and withdraw the bulb and bulb holder.

4 Pull the headlamp bulb from the bulb holder.

Care must be taken not to handle the glass portion of the bulb as this will shorten the life of the bulb. Accidental fingerprints can be removed by applying methylated spirits to the bulb and drying with a clean tissue or cloth.

5 Carefully install the bulb into the bulb holder and install the bulb holder into the headlamp, ensuring correct alignment of the bulb tangs

6 Install the spring retainer to the headlamp body.

7 Install the rubber dust boot and dust cover to the headlamp body and connect the wiring connector to the bulb.

8 Check the operation of the headlamp and adjust the aim if necessary as described in this chapter.

To Adjust Headlamp Aim

1 Ensure that the tyres are inflated to the correct pressure and that the vehicle is positioned on level ground.

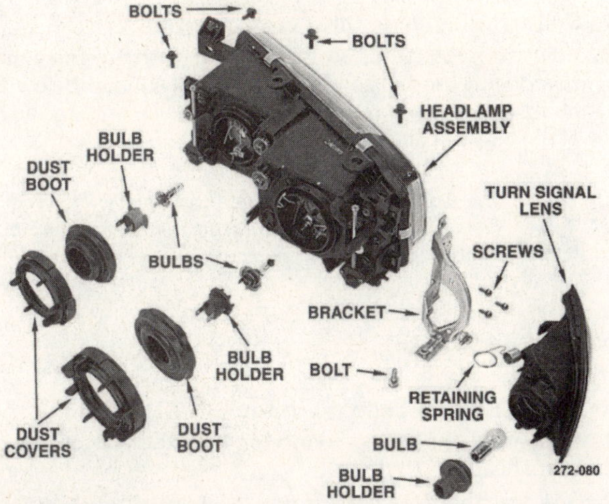

Headlamp components. Verada model shown.

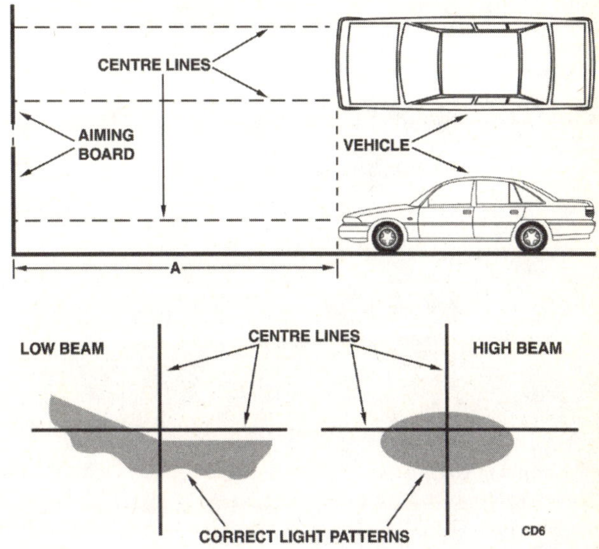

Headlamp adjustment using an aiming board. Dimension A to be 3 metres.

2 Ensure that the vehicle is unladen, except for a weight of 55 kg placed on the driver seat. On Verada models, disconnect the wiring connectors from the high beam lamp bulbs.

3 When using headlamp aiming equipment, refer to the manufacturers instructions for the equipment being used.

4 When using an aiming board, position the vehicle 3 metres from, and square to, the board.

5 Mark a reference line on the board to correspond with the horizontal centre line of the headlamps.

6 Mark reference lines on the board to correspond with the vertical centre lines of the headlamps.

7 Switch on the headlamps and select low beam.

8 Adjust the headlamps to comply with local regulations. As a guide, adjust the lamps so that the high beam pattern strikes the aiming board with the horizontal cut off point 22 mm below the horizontal centre line and the point where the beam raises from horizontal is on or to the left of the vertical centre line.

Adjusting the headlamp aim.

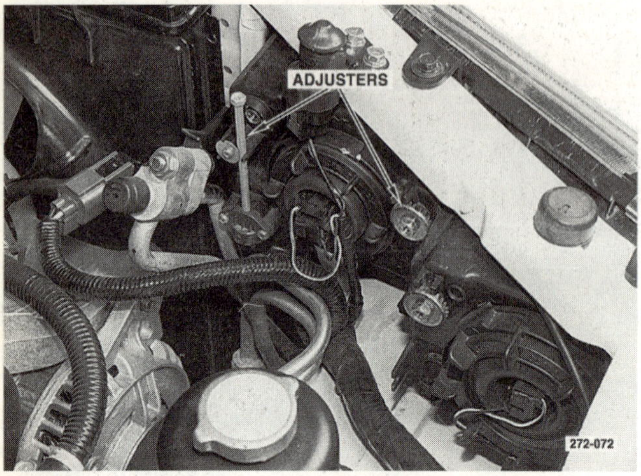

High beam lamp adjusters. Verada models.

9 On Verada models, proceed as follows:

a Disconnect the low beam lamp wiring connector and connect the high beam lamp connectors.

b Switch on the headlamps and select high beam.

c Adjust each high beam lamp so that the centre of the light pattern is the same as the low beam lamps on the aiming board.

> *By connecting the high beam lamp and then disconnecting it, an accurate high beam adjustment can be made.*

d Connect the low beam wiring connectors when satisfied with the high beam adjustment.

FOG LAMP

To Renew Bulb

1 Raise the front of the vehicle and support on chassis stands. Refer to General Information (chapter 1) for the positions of the jacking points.

2 Working beneath the vehicle, unscrew the cover from the rear of the fog lamp and raise the spring clip from the rear of the bulb using a small screwdriver.

3 Remove the bulb from the lamp and disconnect the wiring connector.

Installation is a reversal of the removal procedure.

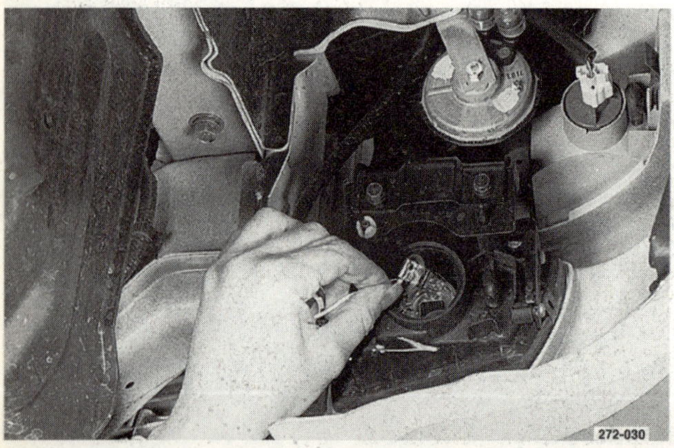

Renewing the fog lamp bulb.

To Remove and Install

1 Raise the front of the vehicle and support on chassis stands. Refer to General Information (chapter 1) for the positions of the jacking points.

2 Working beneath the vehicle, unscrew the cover from the rear of the fog lamp and raise the spring clip from the rear of the bulb using a small screwdriver.

3 Remove the screws and bolts retaining the fog lamp to the bumper and remove from the rear of the bumper.

Installation is a reversal of the removal procedure.

To Adjust Fog lamp Aim

1 Ensure that the tyres are inflated to the correct pressure and that the vehicle is positioned on level ground.

2 Ensure that the vehicle is unladen, except for a weight of 55 kg placed on the driver seat.

3 When using headlamp aiming equipment, refer to the manufacturers instructions for the equipment being used.

4 When using an aiming board, position the vehicle 10 metres from, and square to, the board.

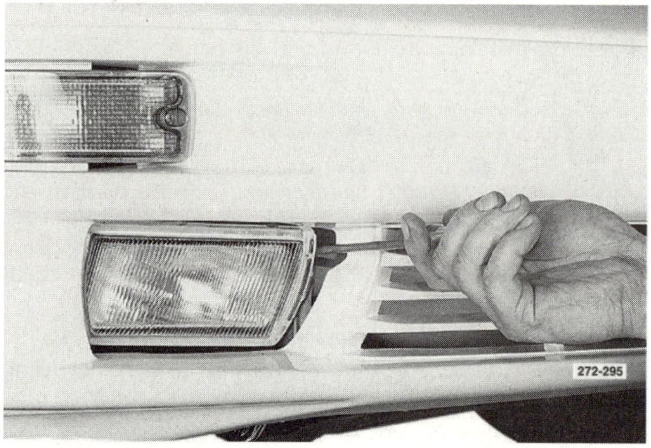

Adjusting the fog lamp aim.

5 Mark a reference line on the board to correspond with the horizontal centre line of the fog lamps.

6 Mark a reference line on the board to correspond with the vertical centre line of the fog lamps.

7 Switch the fog lamps On.

8 Adjust the fog lamps so that the beam strikes the aiming board with the horizontal cut off point 47 mm below the horizontal centre line and check that the beam is to the left of the vertical centre line.

> *The beam can only be adjusted horizontally. If the vertical beam is not to the left of the vertical line, check the fog lamp mounting screws and nuts.*

FRONT PARK LAMP

To Renew Bulb — Except Verada

1 Disconnect the wiring connector from the rear of the low beam headlamp bulb.

2 Unscrew the dust cover from the rear of the headlamp assembly. Be careful not to misplace the rubber dust boot.

3 Separate the retaining spring tangs and pull the park lamp bulb holder from the headlamp assembly.
4 Pull the bulb from the bulb holder.

Installation is a reversal of the removal procedure.

To Renew Bulb — Verada

1 Remove the screws retaining the park lamp to the front bumper and partially remove the lamp from the bumper.
2 Rotate the bulb holder anticlockwise and remove it from the park lamp. Depress and rotate the bulb in an anticlockwise direction and remove it from the bulb holder.
3 Install the new bulb to the bulb holder and turn the bulb holder clockwise to install it to the park lamp.

Installation is a reversal of the removal procedure.

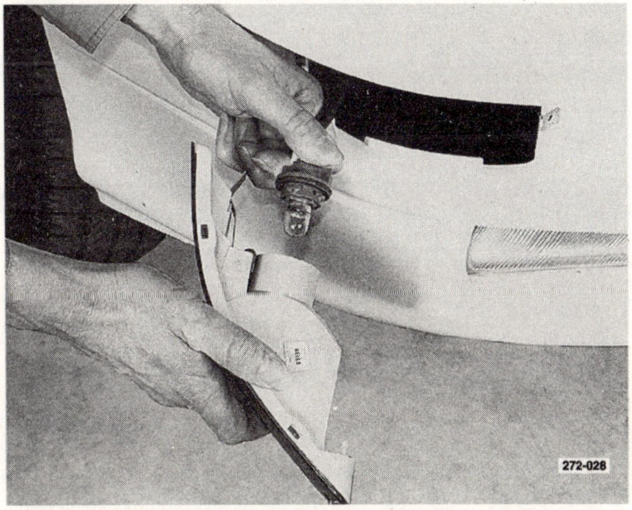

Removing the park lamp bulb holder from the lens. Verada models.

FRONT TURN SIGNAL LAMP

To Renew Bulb

1 Working in the engine compartment, disconnect the spring clip retaining the turn signal lamp assembly. Pull the turn signal lamp partially from the vehicle.

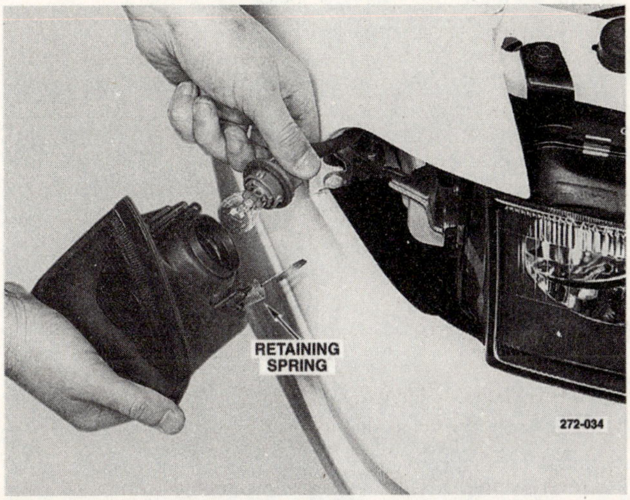

Removing the turn signal lamp bulb holder from the lens.

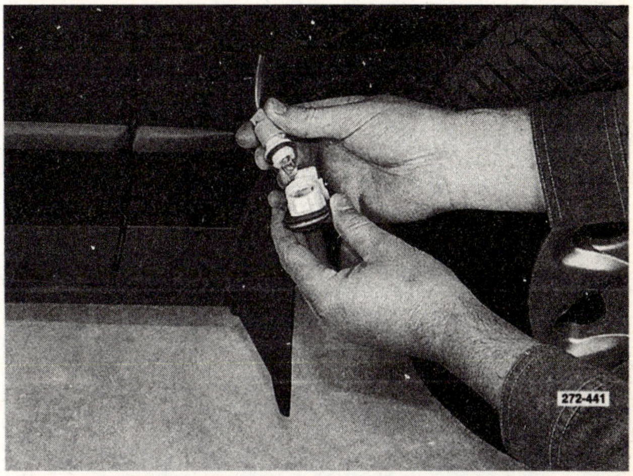

Removing the bulb holder from the side turn signal lamp.

2 Rotate the turn signal lamp bulb holder anticlockwise and remove the bulb holder from the lamp assembly.
3 Press the new bulb in and turn clockwise to install it to the bulb holder.

Installation is a reversal of the removal procedure.

SIDE TURN SIGNAL LAMP

To Renew Bulb

1 Using finger pressure, carefully lever the lamp outwards. Withdraw the lamp from the vehicle.

If the lamp is tight a small flat bladed screwdriver, wrapped in cloth can be used to lever the lamp from the guard.

On TL/TW and KL/KW models, carefully prise the rear edge of the side turn signal lamp housing out, where the housing meets the inner guard.

2 Rotate the bulb holder anticlockwise and withdraw it from the lamp.
3 Remove the bulb from the bulb holder.
4 Ensure that the contacts are clean and install a new bulb.
5 Install the bulb holder to the lamp and install the lamp to the vehicle.

Renewing a rear combination lamp bulb. Sedan models.

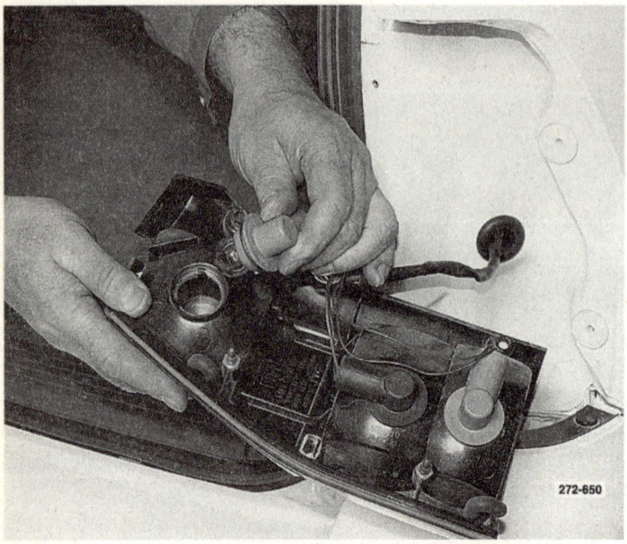

Renewing a rear combination lamp bulb. Station wagon models.

REAR COMBINATION LAMPS

To Renew Bulb — Sedan

1 Open the luggage compartment lid.

2 Working inside the boot, remove the plastic screw clips retaining the trim panel over the rear of the combination lamp assembly. Manoeuvre the trim panel clear of the work area.

3 Remove the nuts retaining the lamp assembly to the rear of the vehicle.

4 Depress and rotate the bulb in an anticlockwise direction and remove it from the bulb holder.

Installation is a reversal of the removal procedure ensuring that the bulb holder is free from corrosion.

To Renew Bulb — Station Wagon

1 With the tailgate open, remove the combination lamp retaining screws.

2 Remove the combination lamp assembly from the vehicle, disengaging the two pins from the outside of the lamp assembly.

3 Rotate the appropriate bulb holder anticlockwise and remove it from the lamp assembly.

4 Depress and rotate the bulb in an anticlockwise direction and remove it from the bulb holder.

Installation is a reversal of the removal procedure ensuring that the bulb holder is free from corrosion.

INBOARD COMBINATION LAMPS — VERADA SEDAN

To Renew Bulb

1 Remove the plastic buttons retaining the trim panel to the boot lid.

2 Separate the glued corners of the trim from the boot lid and remove it from the vehicle.

3 Remove the screw retaining the bulb holders to the rear of the lamp and partially remove it.

4 Depress and rotate the bulb in an anticlockwise direction and remove it from the bulb holder.

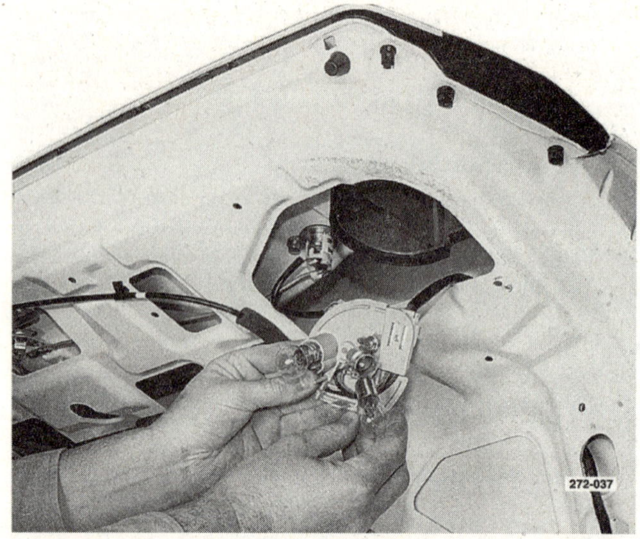

Renewing an inboard combination lamp bulb. Verada models. Trim panel removed.

Installation is a reversal of the removal procedure ensuring that the bulb holder is free from corrosion and where necessary, glue the corners of the trim to the boot lid.

To Remove and Install

1 Remove the bulb holder as previously described.

2 Remove the nuts retaining the inboard combination lamp to the boot lid.

3 Remove the lamp assembly from the boot lid, gathering the seal.

Installation is a reversal of the removal procedure, ensuring that the seal is renewed if damaged.

HIGH MOUNTED STOP LAMP

To Renew Bulb

1 On sedan models proceed as follows:

a Working in the luggage compartment, reach upwards to the underside of the lamp and rotate the bulb holder anticlockwise. Remove the holder from the lamp.

b Pull the bulb out of the bulb holder.

2 On station wagon models proceed as follows:

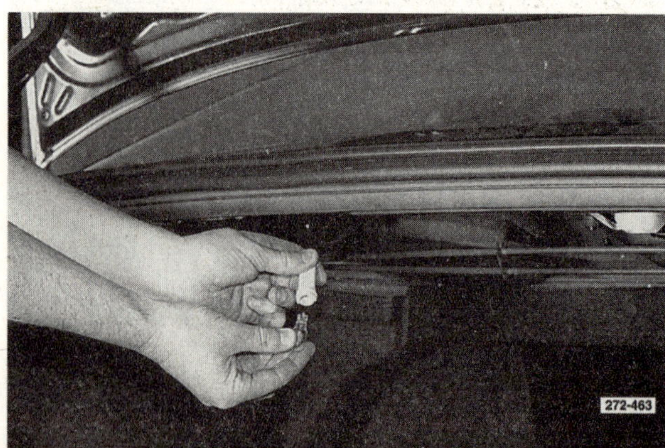

Renewing a high mount stop lamp bulb. Sedan models.

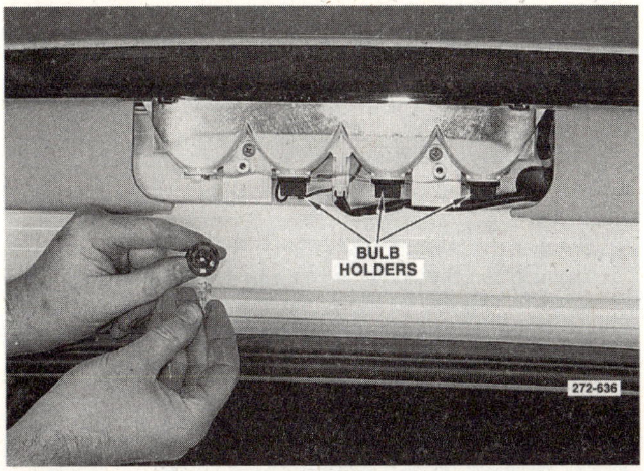

Renewing one of the high mount stop lamp bulbs. Station wagon models.

- **a** Remove the plugs from the high mount stop lamp trim cover and then remove the screws retaining the trim cover to the stop lamp.
- **b** Slide the cover toward the bottom of the tailgate and then remove it from the vehicle.
- **c** Rotate the relevant bulb holder anticlockwise and remove it from the lens.
- **d** Pull the bulb from the bulb holder.

Installation is a reversal of the removal procedure ensuring that the bulb is of the correct voltage and wattage.

NUMBER PLATE LAMP

To Renew Bulb

1. On sedan models, remove the screws retaining the number plate lamp to the boot lid and pull the lens down.
2. On station wagon models, depress the tang on the side of the number plate lamp assembly and manoeuvre it from the tailgate.
3. Twist the bulb holder assembly anticlockwise and remove it from the lens housing.
4. Pull the bulb from the bulb holder.

Installation is a reversal of the removal procedure ensuring that the bulb holder is free from corrosion and that the bulb is of the correct voltage and wattage.

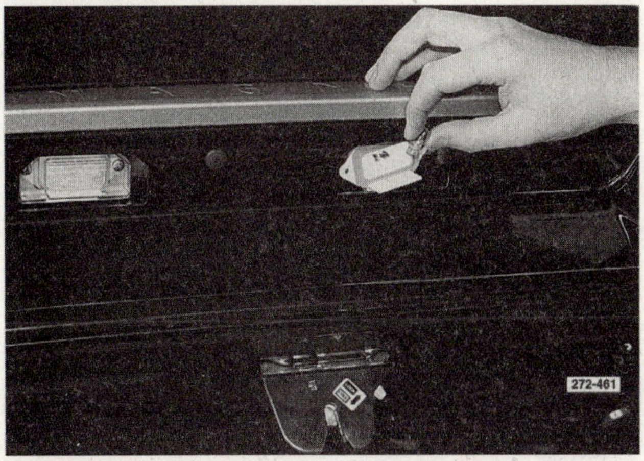

Renewing one of the number plate lamp bulbs. Sedan models.

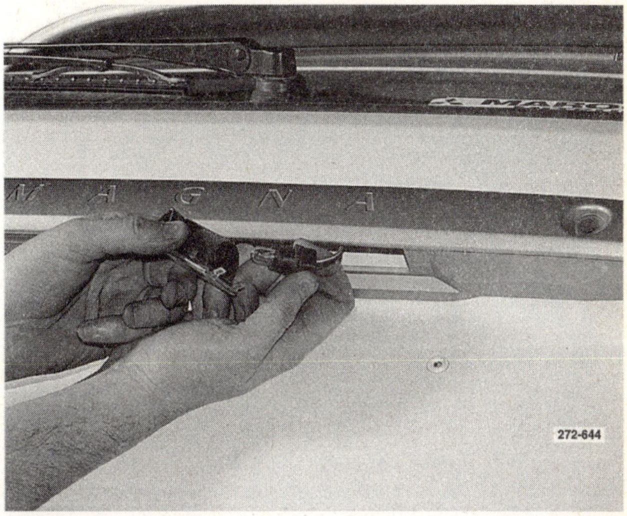

Renewing one of the number plate lamp bulbs. Station wagon models.

COURTESY LAMPS

To Renew Bulb(s)

1. To renew the courtesy lamp bulb, proceed as follows:
   - **a** Carefully prise the lens from the lamp assembly using a small screwdriver.
   - **b** Carefully remove the festoon bulb from the lamp.
   - **c** Install the bulb to the lamp ensuring that the terminals provide a good electrical contact.
2. To renew the reading lamp bulb on models with a combined courtesy lamp and reading lamp assembly, proceed as follows:
   - **a** Remove the courtesy lamp lens as previously described.
   - **b** Remove the retaining screws and partially remove the lamp from the roof of the vehicle.
   - **c** Disconnect the wiring from the rear of the lamp and remove from the vehicle.
   - **d** Remove the screw retaining the bulb holder to the rear of the lamp assembly and separate it from the lamp.
   - **e** Pull the bulb from the bulb holder.
   - **f** Install the bulb to the lamp ensuring that the terminals provide a good electrical contact.

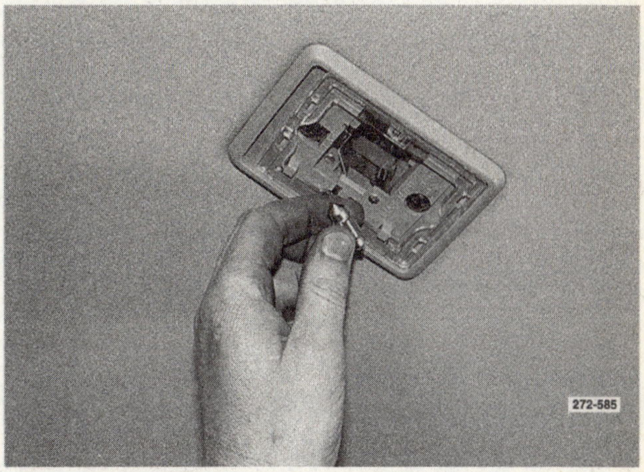

Renewing the courtesy lamp bulb.

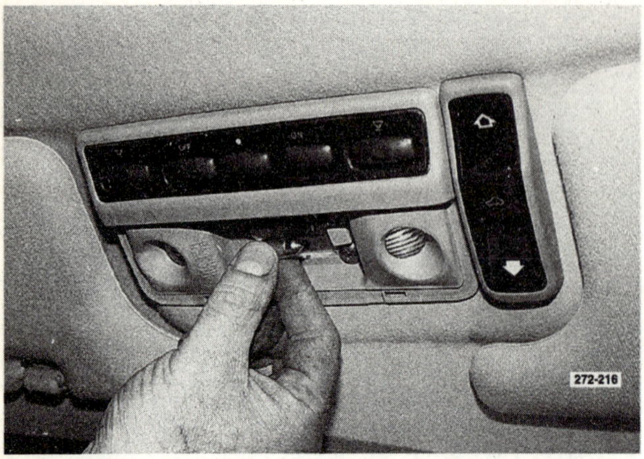

Renewing the courtesy lamp bulb on models with combined reading lamps. Note the screw retaining the reading lamp lens.

3 To renew the reading lamp bulb on models with a separate reading lamp assembly, proceed as follows:

a Prise the reading lamp assembly from the roof using a small screwdriver wrapped in cloth.

b Pull the bulb from the bulb holder.

4 To renew the upper door lamp bulb, where fitted, prise the lens from the door trim panel, rotate the bulb holder anticlockwise and pull the bulb from the bulb holder.

5 To renew the lower door lamp bulb, where fitted, remove the screws retaining the lamp assembly to the door trim panel. Rotate the bulb holder anticlockwise and remove it from the lamp assembly. Pull the bulb from the bulb holder.

6 On models with rear reading lamps, proceed as follows:

a Carefully prise the reading lamp assembly from the C pillar trim.

b Rotate the bulb holder anticlockwise and remove it from the reading lamp assembly. Pull the bulb from the bulb holder.

Installation is a reversal of the removal procedure ensuring that the bulb is of the correct voltage and wattage.

FOOTWELL LAMPS

To Renew Bulb

1 To renew the driver side footwell lamp bulb, proceed as follows:

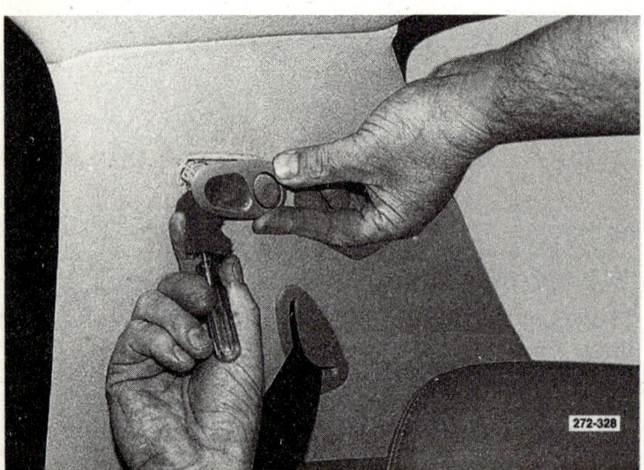

Removing the rear reading lamp.

a Remove the screws retaining the bonnet lever to the lower dash panel and position the lever clear of the work area.

b Remove the screw cover from the top of the driver side lower dash panel. Remove the remaining screws from the lower dash panel, disconnect the panel from the top left side of the dashboard and partially remove the panel.

2 To renew the passenger side footwell lamp bulb, proceed as follows:

a Remove the plastic screw clips retaining the lower dash panel. When the screw is removed, pull the clip from the panel. Repeat this on the remaining screw clips.

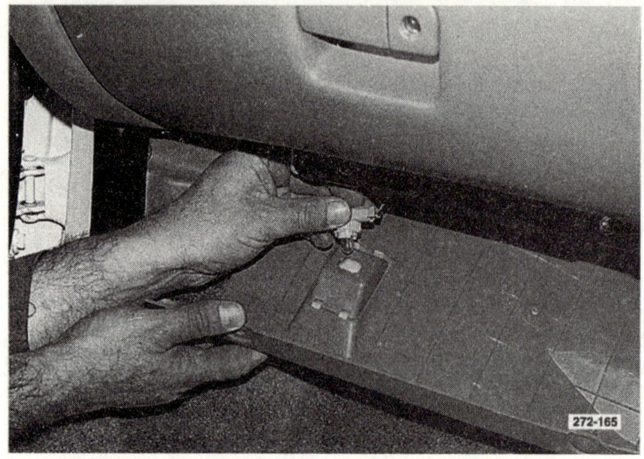

Renewing the passenger side footwell lamp bulb.

b Partially remove the panel and rotate the footwell lamp bulb holder anticlockwise and disconnect from the panel.

c Pull the bulb from the bulb holder.

GLOVE COMPARTMENT LAMP

To Renew Bulb

1 Unscrew the plastic screw clips retaining the lower dash panel on the passenger side of the vehicle. When the screw is removed, pull the clip from the panel. Repeat this on the remaining screw clips.

2 Partially remove the panel and where fitted, rotate the footwell lamp bulb holder anticlockwise and disconnect from the panel. Remove the panel from the vehicle.

3 Remove the screws retaining the glove compartment. Open the glove compartment and remove it from the vehicle.

4 Remove the screws retaining the glove compartment latch and remove the latch. Remove the remaining screws and bolts that retain the glove compartment surround panel.

5 Partially remove the panel, rotate the glove compartment lamp bulb holder anticlockwise and disconnect from the panel.

6 Pull the bulb from the bulb holder.

LUGGAGE COMPARTMENT LAMP

To Renew Bulb — Sedan

1 Working in the luggage compartment, carefully prise the lamp from the body and remove the festoon bulb from the lamp.

2 Install the festoon bulb ensuring that the terminals have sufficient tension to provide good electrical contact.

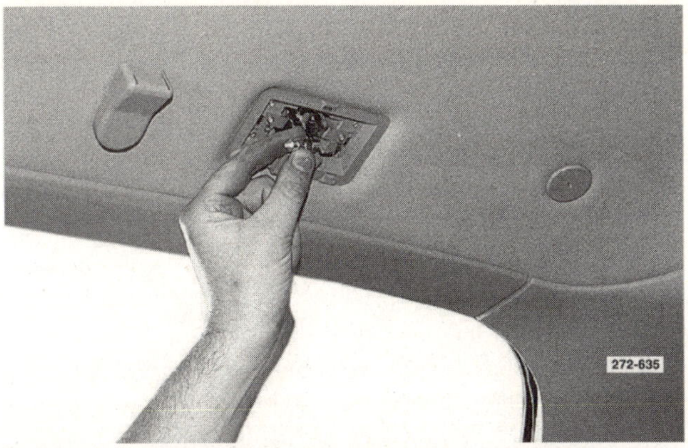

Renewing the courtesy lamp bulb. Station wagon models.

To Renew Bulb – Station Wagon

1. Carefully prise the lens from the lamp assembly using a small screwdriver.
2. Carefully remove the festoon bulb from the lamp.
3. Install the bulb to the lamp ensuring that the terminals have sufficient tension to provide a good electrical contact.

Windscreen Wipers

WIPER BLADE

To Remove and Install

1. Raise the windscreen wiper arm and blade from the windscreen until it locks in the vertical position.
2. Depress the clip at the centre of the wiper blade and slide the blade from the hook on the end of the wiper arm.

Installation is a reversal of the removal procedure, ensuring that the clip locks the blade securely.

WIPER BLADE RUBBER

To Renew — Spine Type

1. Remove the windscreen wiper blade as previously described.

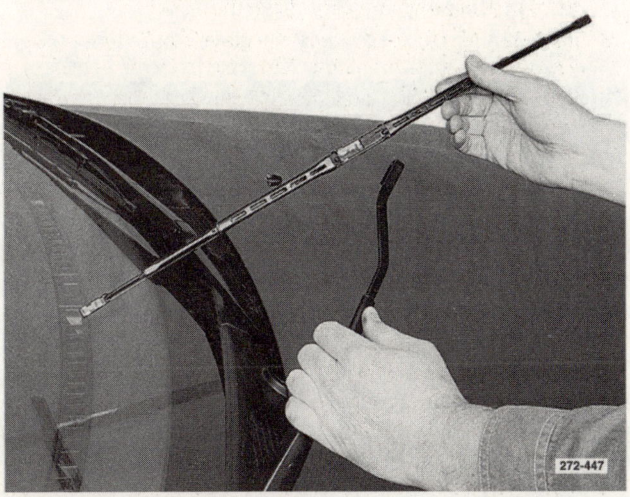

Check and renew the wiper blades at regular intervals.

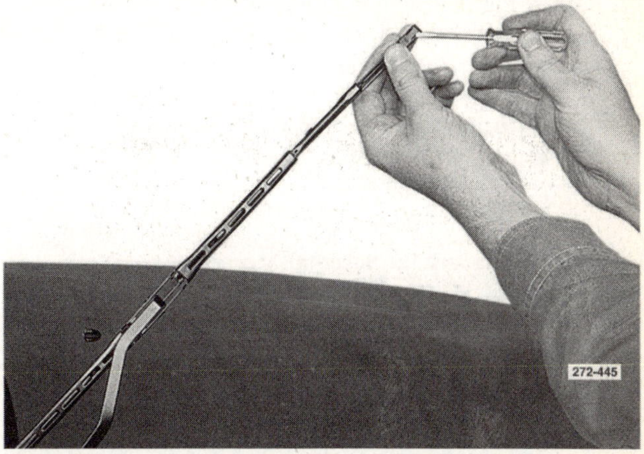

Removing the wiper blade rubber end cap using a small screwdriver.

2. Squeeze the sides of the retaining clip together and slide the wiper rubber and spine assembly from the wiper blade.

> *It is advisable to install a new rubber and spine as an assembly to ensure correct contact between the rubber and the windscreen.*

3. Slide the new rubber and spine into the wiper blade ensuring that the spine enters all the lugs on the blade and that the retaining clip locks into the wiper blade.
4. Install the wiper blade to the wiper arm.

To Renew — Rail Type

1. Remove the windscreen wiper blade as previously described.
2. Using a small screwdriver or similar, raise the tang and slide the end cap from the wiper arm.
3. Slide the wiper blade from the wiper arm.

Installation is a reversal of the removal procedure with attention to the following points:

1. Slide the wiper blade onto the wiper arm ensuring that the blade is located in all of the runners.
2. Ensure that the end cap tang is located correctly.

WIPER ARM

To Remove and Install

1. Ensure that the wipers are in the Park position and raise the bonnet.

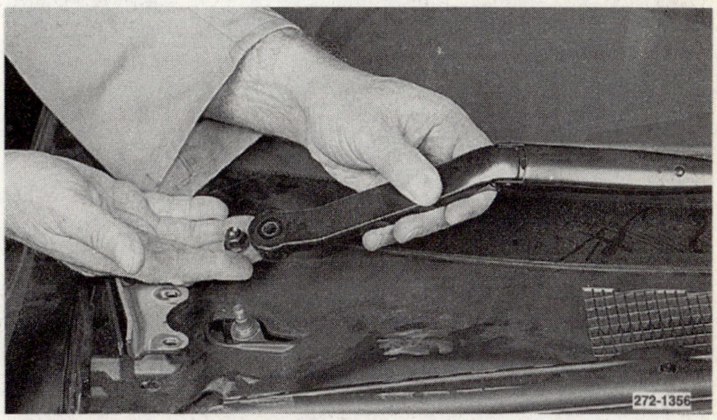

Removing the windscreen wiper arm.

2 Remove the wiper arm retaining nut.
3 Carefully prise the wiper arm from the wiper pivot shaft taking care not to damage the vehicle.

> TIP *It may be necessary to lift the arm from the windscreen and rock the arm back and forth with the nut loose to separate the components. Do not lever from the windscreen.*

Installation is a reversal of the removal procedure with attention to the following points:

1 With the wiper motor in the Park position, ensure that the wiper blades are parallel with the bottom edge of the windscreen.
2 Operate the windscreen wipers and ensure that the blades do not contact the windscreen surround.

WIPER MOTOR AND LINKAGE

To Remove and Install

1 Ensure that the wipers are in the Park position and raise the bonnet.
2 Remove the nuts retaining the wiper arms to the shafts and remove the wiper arms.
3 Remove the plastic screw clips retaining the driver side plenum chamber cover. Lift the end of the passenger side plenum cover to allow removal of the driver side cover from beneath it. Remove the driver side plenum chamber cover from the vehicle.
4 Disconnect the wiring connector from the wiper motor and separate the connector from the bracket on the firewall. Push the grommet through and feed the wiring through the firewall.
5 Remove the bolts retaining the wiper motor and linkage assembly to the plenum chamber. Manoeuvre the assembly from the plenum chamber.

> i *The wiper motor, linkages and the passenger side wiper pivot shaft housing are removed as an assembly. Two retaining bolts are on the passenger side wiper pivot shaft housing.*

6 Remove the nut from the wiper motor shaft and separate the wiper linkage from the motor.

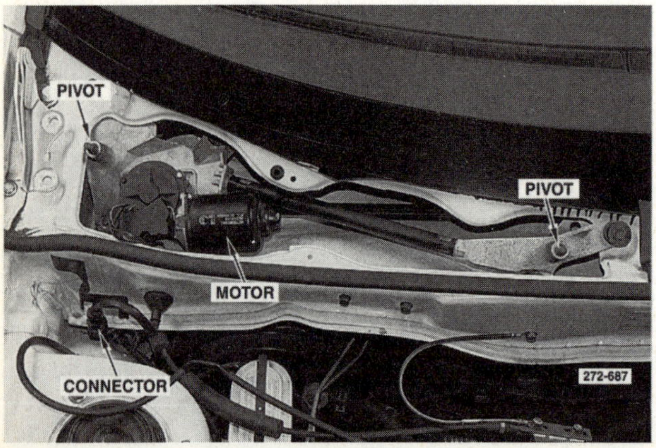

Wiper motor components, plenum chamber cover removed.

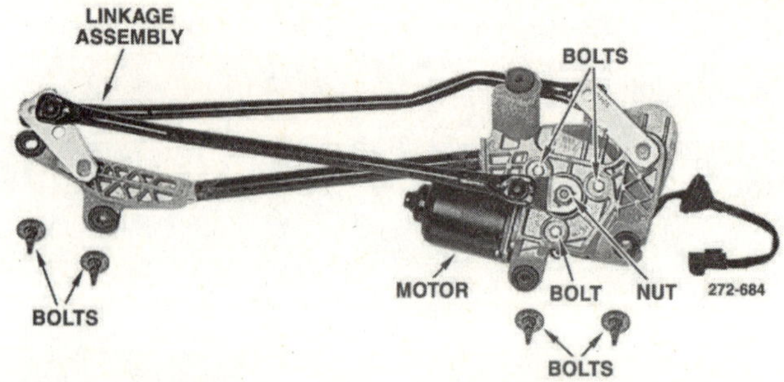

Wiper linkage assembly removed, showing the motor retaining bolts and shaft nut.

7 Remove the wiper motor retaining bolts and separate the wiper motor from the wiper linkage assembly.

Installation is a reversal of the removal procedure with attention to the following points:

1 Ensure that the wiper motor mounting bolts and the wiper linkage bolts are securely tightened.

> i *Ensure that the wiper linkage bolts with the oval shaped bushes are returned to the passenger side wiper pivot shaft housing.*

2 Install the driver side plenum chamber cover, ensuring that the two covers slot together.
3 Ensure that the wiper arms are installed to the correct position as previously described.

Tailgate Wiper

TO REMOVE AND INSTALL

Wiper Blade

1 Lift the wiper arm from the rear window until it locks in the vertical position.
2 Depress the retaining clip at the end of the wiper arm and slide the wiper blade from the wiper arm.
3 To remove the wiper rubber from the blade, depress the tangs on the clip at the end of the rubber, and slide the spine and rubber from the blade.

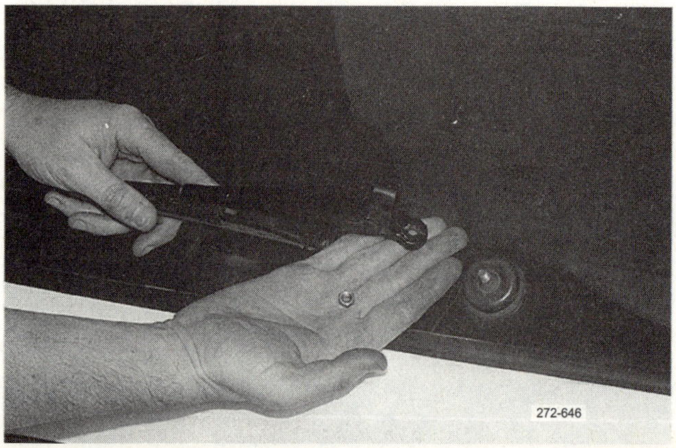

Removing the tailgate wiper arm.

It is advisable to install a new rubber and spine to ensure correct contact between the rubber and the rear window glass.

Installation is a reversal of the removal procedure with attention to the following point:

The wiper rubber retaining clip should be at the bottom when the wiper blade assembly is vertical.

Slide the wiper rubber and spine into the wiper blade ensuring that the spine enters all of the lugs on the wiper blade and that the clips securely lock into the lugs on the wiper blade.

Wiper Arm

1 Raise the cover on the wiper arm to gain access to the retaining nut.
2 Remove the nut retaining the arm to the pivot shaft.
3 Carefully prise the wiper arm from the pivot shaft taking care not to crack the rear window.

Installation is a reversal of the removal procedure ensuring that the wiper blade is parallel to the glass weatherstrip in the Park position.

Wiper Motor

1 Remove the wiper arm as previously described.
2 Remove the cap from the wiper pivot shaft.
3 Raise the tailgate, prise the covers open on the tailgate pull handle and remove the bolts retaining the handle to the tailgate. Remove the handle.
4 Remove the clips retaining the trim panels and remove the trim panels.
5 Disconnect the wiring connector from the wiper motor.
6 Remove the nuts retaining the wiper motor to the tailgate and remove the wiper motor from the vehicle.

Installation is a reversal of the removal procedure with attention to the following points:

1 Install the wiper motor to the tailgate.
2 Ensure that the wiper motor nuts are securely tightened.
3 Test the operation of the wiper motor before installing the tailgate trim panel.

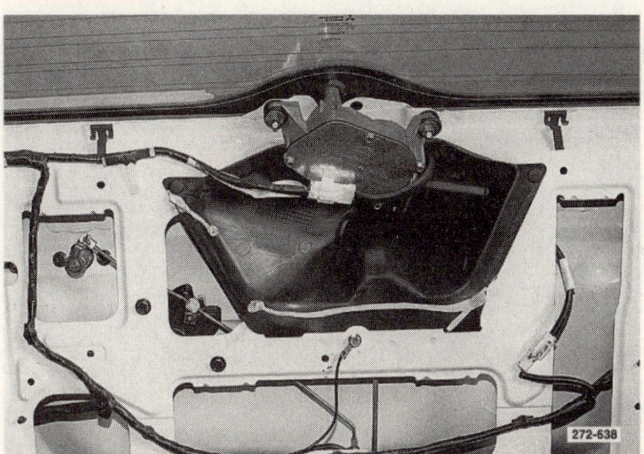

Tailgate trim panels removed showing the wiper motor assembly.

Sound System

To Remove and Install

1 Ensure that the ignition is Off and the sound system security code is available.

TIP *The security code is in the vehicle owners manual.*

2 On manual transaxle models, unscrew the gear lever knob.
3 On Verada models, proceed as follows:
a Open the coin tray lid and firmly pull the coin tray from the console. Disconnect the wiring connectors from the base of the accessory socket.
b Pull the cup holder from the front of the console.
4 Grasp the rear edges of the console fascia panel and pull upward. Working toward the front of the console, raise the fascia panel, releasing the retaining clips. When all of the clips are released, manoeuvre the fascia panel from the console.
5 On Magna models, remove the ashtray from the vehicle. Pull the ashtray surround panel partially from the dashboard, disengaging the clips retaining the surround to the dashboard. Disconnect the cigarette lighter wiring from the base of the socket and remove the surround.
6 On Verada models, pull the ashtray assembly, disengaging the clips retaining the surround to the dashboard. Disconnect the cigarette lighter wiring from the base of the socket and remove the ashtray.
7 Remove the screws retaining the bonnet lever to the driver side lower dash panel and position the lever clear of the work area. Remove the screw cover from the top of the panel. Remove the lower dash panel screws and disconnect the panel from the top left side of the dashboard. Partially remove the panel and if applicable, rotate the footwell lamp bulb holder anti clockwise and separate it from the panel. Remove the driver side lower dash panel from the vehicle.
8 Disengage the vent control inner cable from the linkage on the heater unit. Disconnect the clip retaining the outer cable.
9 Grasp the centre vent assembly and pull from the dashboard, disengaging the retaining clips. Ensure that the vent control cable does not get caught when removing the vent assembly from the dashboard.

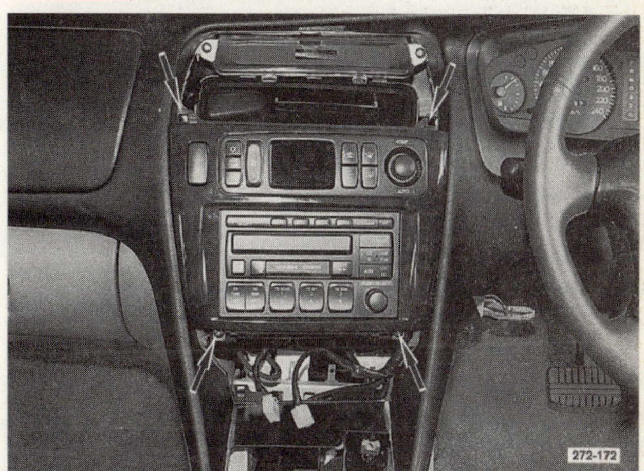

Radio surround panel screw locations. Verada model illustrated.

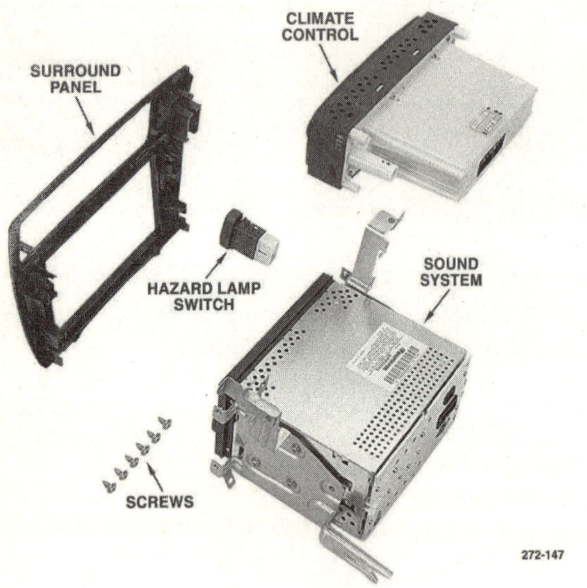

Sound system, climate control and hazard lamp switch, removed from the surround panel. Verada model shown.

10 Remove the screws securing the radio surround panel to the dashboard.

11 Partially remove the surround panel and sound system and disconnect the wiring and aerial lead from the rear of the sound system.

> *On models with climate control, the climate control will also come from the dashboard with the sound system and surround panel. Disconnect the wiring and sniffer motor hose from the rear of the climate control prior to removing the sound system and surround panel assembly from the dashboard.*

12 If necessary, remove the retaining screws and separate the sound system from the surround panel.

> *Take note of the number of washers placed between the surround panel and the mounting brackets and ensure that the same number of washers are returned to each side to ensure a flush fit when the surround panel is installed to the dashboard.*

Installation is a reversal of the removal procedure with attention to the following points:

1. Ensure that all wiring connectors including the aerial lead are securely connected.
2. Slide the radio and surround panel assembly into position in the dashboard, ensuring that the wires are not pinched between the radio and dashboard.
3. Ensure the vent cable is connected to the heater unit.
4. Enter the security code as described later in this chapter.

SPEAKERS

To Remove and Install

Removal of the front door speakers involves removing the front door trim as described in Body (chapter 17). The speakers are retained by screws.

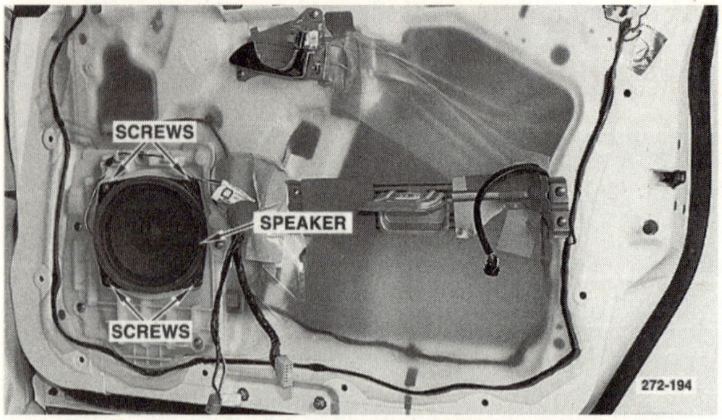

Front door trim removed showing the speaker retaining screws.

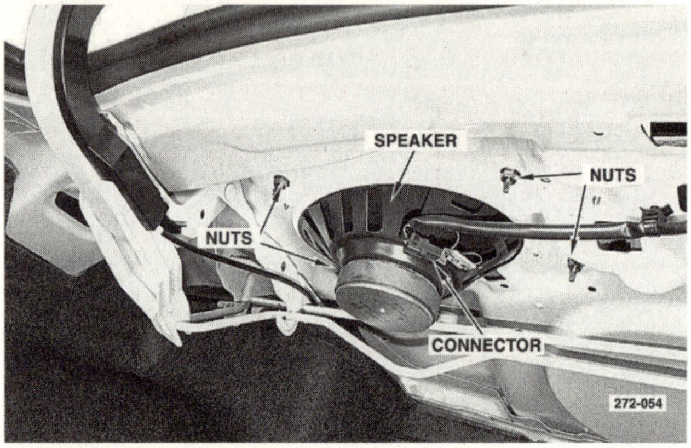

Rear parcel shelf speakers. Sedan models.

Removal of the rear parcel shelf speakers involves removing the rear parcel shelf. Removing the screws retaining the rear speakers, disconnecting the wiring connector and removing the speakers from the parcel shelf.

On station wagon models removal of the rear speakers involves removing the tailgate interior trim panel and removing the screws retaining the speakers to the tailgate.

Some models have two additional speakers (tweeters) placed either side of the front demister grille.

Where fitted, removal of the additional tweeters involves easing the speaker grille upwards and removing the speaker retaining screws.

When installing the speakers, ensure that all wiring connectors are secure and making good contact.

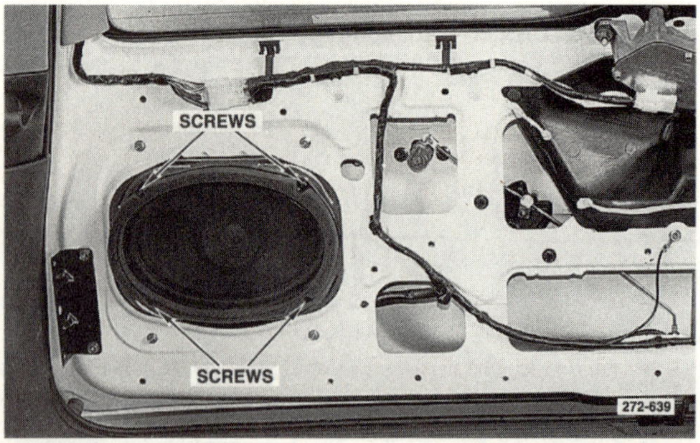

Tailgate trim panel removed showing the rear speaker retaining screws.

RADIO AERIAL

To Remove and Install — Sedan

1 Position the aerial mast in the fully retracted position.

2 Working in the luggage compartment, remove the plastic retainers from the left hand trim panel. Manoeuvre the panel from the vehicle.

3 On models with an electric boot and fuel door release, proceed as follows:

a Remove the bolts retaining the release actuator to the support bracket. Allow the actuator to hang, supported by the release cables.

b Remove the nuts and bolts retaining the support bracket. Partially remove the support bracket, disconnect the wiring connector from the rear of the power aerial and manoeuvre the support bracket clear of the work area. If necessary, disconnect the wiring from the boot and fuel door opening relay.

4 Remove the screw retaining the earth wire(s) to the inner guard and disconnect the aerial lead from the aerial mast.

5 Remove the nut and spacer from the top of the aerial mast. Remove the nut(s) from inside the luggage compartment and manoeuvre the aerial assembly from the vehicle.

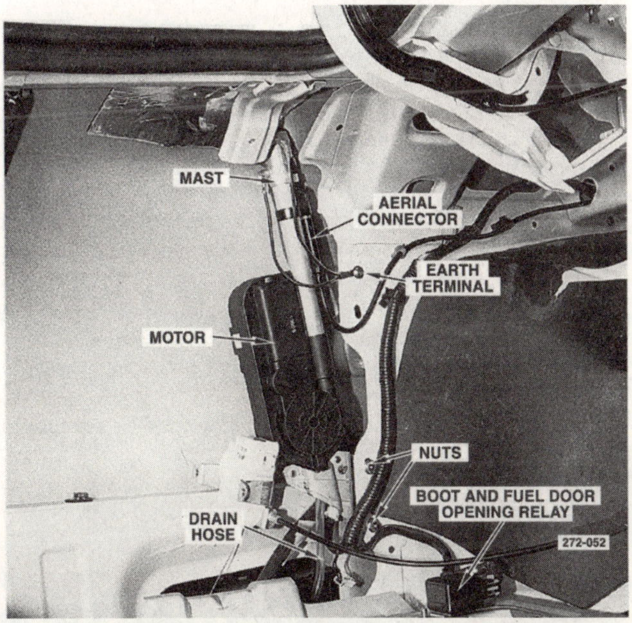

Trim panel and support bracket removed showing the power aerial components.

Installation is a reversal of the removal procedure with attention to the following points:

1 Install the aerial to the vehicle and loosely install the lower retaining nut(s).

2 Install the upper spacer and nut and tighten securely.

3 Tighten the lower retaining nut(s).

4 Install the wiring connectors and earth wire(s).

5 Where fitted, install the support bracket, ensuring the wiring is connected to the power aerial motor and the boot and fuel door opening relay. Install the boot and fuel door release actuator to the support bracket and tighten the retaining bolts securely.

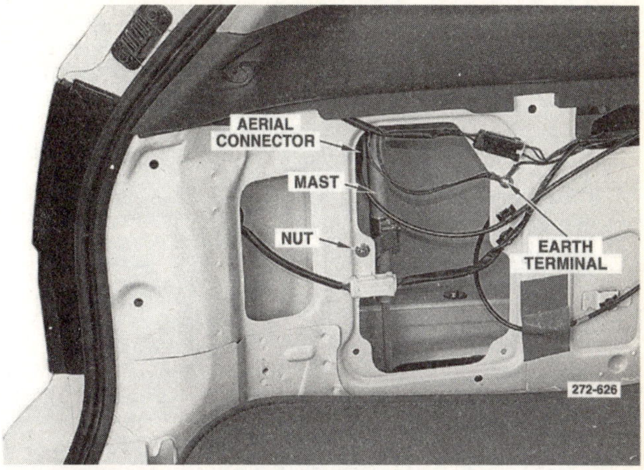

Trim panel removed showing the aerial components. Station wagon models.

To Remove and Install — Station Wagon

1 Position the aerial mast in the fully retracted position.

2 Open the tailgate and fold the carpet away from the left hand trim panel.

3 Remove the screws retaining the timber panel to the left of the spare wheel cover and remove the panel from the vehicle.

4 Pull the rear of the lower trim panel from the left side of the luggage compartment. It is only necessary to remove enough panel to allow access to the aerial.

5 Remove the screw retaining the earth wire(s) to the inner guard and disconnect the aerial lead from the aerial mast. On models with a power aerial, disconnect the wiring connector from the rear of the power aerial.

6 Working outside the vehicle, remove the nut and spacer from the top of the aerial mast.

7 Remove the nut(s) from inside the luggage compartment and manoeuvre the aerial assembly from the vehicle.

To Renew Power Aerial Mast

1 Remove the upper retaining nut from the aerial mast.

2 Access the power aerial as previously described.

3 On models with an electric boot and fuel door release, proceed as follows:

a Remove the bolts retaining the release motor to the support bracket. Allow the motor to hang, supported by the release cables.

b Remove the nuts and bolts retaining the support bracket. Partially remove the support bracket, disconnect the wiring connector from the rear of the power aerial and manoeuvre the support bracket clear of the work area. If necessary, disconnect the wiring from the electric boot and fuel door release module.

c Separate the power aerial wiring from the support bracket and reconnect it to the power aerial.

4 Place a bridging wire between the red/black wire and the white/green wire on the power aerial connector.

5 Switch the ignition to the Accessories position. This will operate the aerial motor, expelling the drive cable from the aerial assembly. When the cable is free of the power aerial motor, remove the bridging wire from the connector. Do not switch the ignition Off at this stage.

6 Uncoil the new aerial mast cable and insert the end of the cable into the aperture of the aerial mast hole with the serrations on the plastic end of the drive cable facing the motor. Feed the cable until resistance is felt.

7 Switch the ignition Off. This will load the drive serrations onto the aerial motor drive gear.

The cable should start to retract into the aerial housing.

If the plastic end of the drive cable does not engage with the gears on the motor, twist the cable slightly in both directions.

8 When the cable engages, feed the cable and aerial mast into the aerial assembly.

9 Install the aerial mast upper retaining nut.

SECURITY CODE

The sound system is protected by a security code, which must be entered whenever the power supply is interrupted, for example when the vehicle battery is disconnected.

The code is given to the original owner on delivery of the vehicle. If necessary, the code can be obtained from an authorised dealer by supplying the vehicle identification number (VIN) and proof of vehicle ownership.

Refer to General Information (chapter 1) for the location of the vehicle identification number (VIN).

To Enter Code

1 With the ignition in the accessories position, the number 1 should appear in the display.

2 Enter your security code by pressing the numbered preset station buttons.

3 If the correct security code has been entered, the sound system will function.

4 If an incorrect security code has been entered, the number 2 will appear on the display. If after four attempts, the correct security code has not been entered, "—" will appear on the display and the sound system will shut down for approximately one hour. When the display changes to 1, the security code can be re-entered.

5 If the code is entered incorrectly three times, a delay period is activated. The delay period increases every time an incorrect code is entered.

Ventilation Fan

The removal and installation procedure for the ventilation fan motor is described in Ventilation (chapter 18).

Instrument Cluster

TO REMOVE AND INSTALL

1 Set the front wheels to the straight ahead position.

2 Where fitted, disable and remove the drivers airbag. Refer to Body (chapter 17).

Instrument cluster surround panel screw locations.

3 On models without an airbag, remove the horn pad screws from each side of the steering wheel and partially remove the horn pad.

4 Disconnect the wiring from the horn switch and where fitted, the cruise control and sound system remote control assembly.

5 Unlock the steering column tilt mechanism and position the steering column to its lowest position. Lock the lever in that position.

6 Ensure that the steering wheel is in the straight ahead position and loosen the steering wheel retaining nut. Do not completely remove the nut.

7 Mark the relationship between the steering column and steering wheel to ensure that the wheel is returned to the same position. Grasp the steering wheel at the twelve o'clock and six o'clock positions and rock the steering wheel while moving your hands around the steering wheel until it separates from the steering column.

If this fails to work, a steering wheel puller may be needed.

8 Pass the wiring through the steering wheel and remove the wheel from the column.

Avoid turning the steering column or clock spring to prevent damage to the clock spring when it is reinstalled.

9 Remove the screws retaining the steering column covers and remove the covers from the column.

Surround panel removed showing the instrument cluster retaining screws.

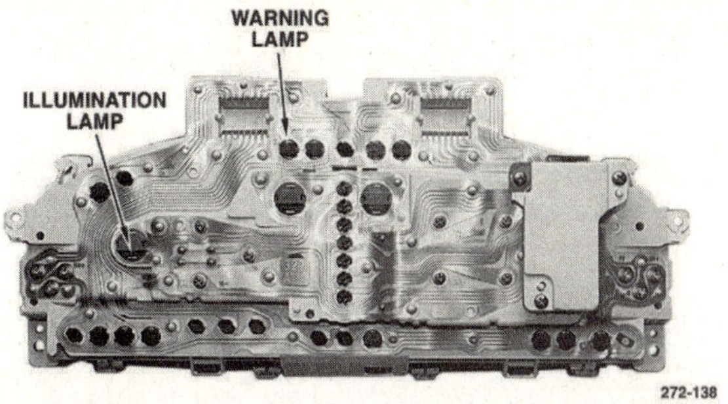

Rear of the instrument cluster showing various components. The different size of the warning lamp and illumination lamp bulb holders is shown.

10 Remove the two screws retaining the top of the instrument cluster surround panel. Pull the top of the panel partially out and then pull the bottom of the panel out, disengaging the clips from the bottom of the panel.

11 Disconnect the wiring connectors from the rear of the switches on the instrument cluster surround panel and remove the panel from the vehicle.

12 Remove the screws retaining the instrument cluster to the dashboard. Partially remove the instrument cluster and disconnect the wiring connector from the rear of the cluster.

13 Remove the instrument cluster from the vehicle.

14 If any of the warning lamp or illumination bulbs require renewal, rotate the bulb holder anticlockwise and remove it from the cluster assembly.

15 If renewing a gauge proceed as follows:

a Remove the screws retaining the top of the lens and mask, raise the tangs on the top and bottom of the lens and slide the lens from the front of the cluster.

b Raise the retaining clips and remove the mask from the front of the instrument cluster.

> *Do not touch the face of the mask or gauges with bare hands as the oil in the skin can leave a permanent stain.*

c Remove the screws retaining the gauge requiring removal and manoeuvre the gauge through the front of the cluster.

d Remove the screws retaining the printed circuit board to the rear of the cluster. Disconnect the wiring and remove the circuit board from the cluster.

Installation is a reversal of the removal procedure with attention to the following points:

1 Ensure that the wiring connector is securely connected to the rear of the cluster.

2 Squarely install the cluster to the dashboard, ensuring that the rear of the cluster engages the wiring connectors on the dashboard.

3 Connect the wiring connectors to the rear of the switches on the instrument cluster surround panel.

4 Install the surround panel and tighten the retaining screws securely.

5 Install the steering column covers and tighten the retaining screws and then install the steering wheel, aligning the marks made on removal. Tighten the retaining nut securely.

6 Install and enable the drivers airbag as described in Body (chapter 17).

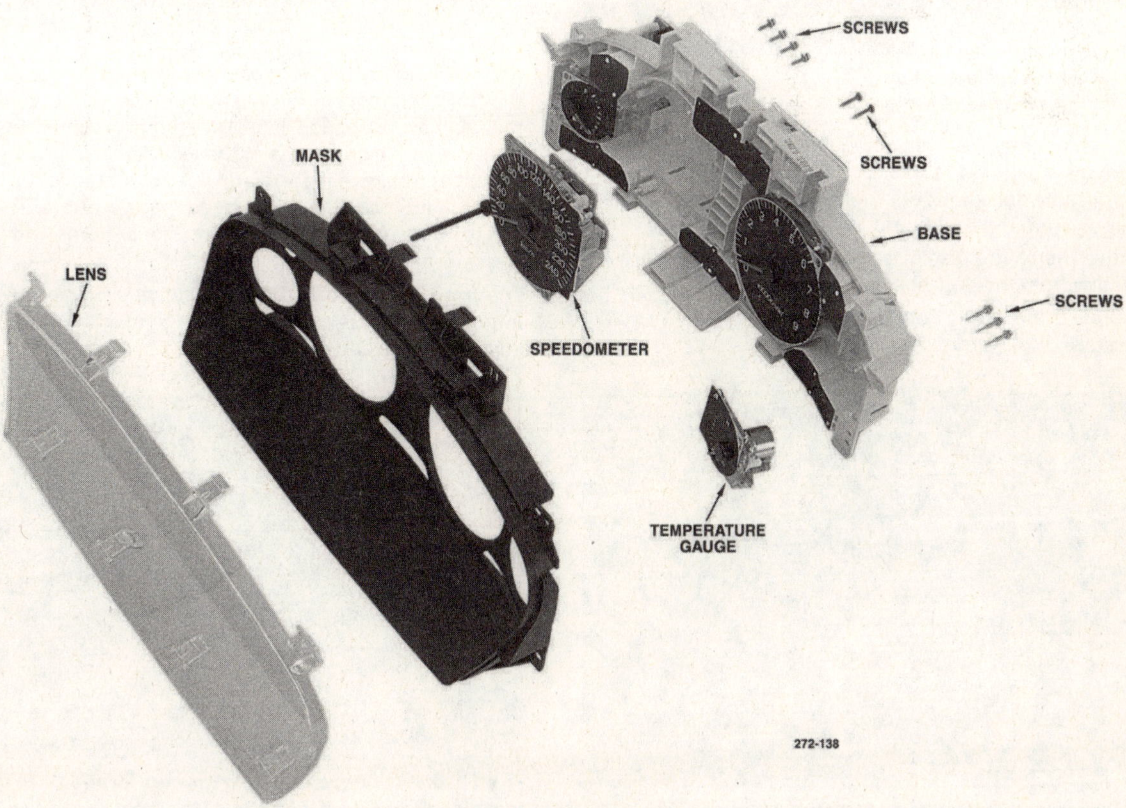

Instrument cluster dismantled.

Centre Display

The centre display on TE models, consists of the charge, low oil pressure and low fuel warning lamps. On Verada models, the clock is also part of the centre display. On TF models, a speed alert is also incorporated within the centre display assembly.

TO REMOVE AND INSTALL

1 Remove the screws retaining the bonnet lever to the driver side lower dash panel and position the lever clear of the work area. Remove the screw cover from the top of the panel. Remove the lower dash panel screws and disconnect the panel from the top left side of the dashboard. Partially remove the panel, rotate the footwell lamp bulb holder anti clockwise, where fitted, and separate it from the panel. Remove the driver side lower dash panel from the vehicle.

2 Disengage the vent control inner cable from the linkage on the heater unit. Disconnect the clip retaining the outer cable.

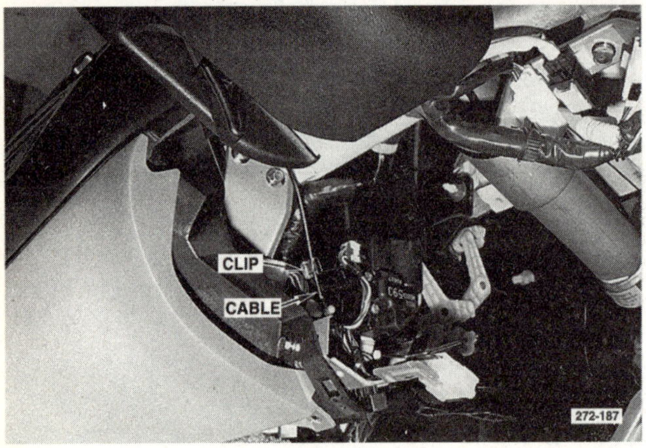

Centre vent control cable location.

3 Grasp the centre vent assembly and pull from the dash-board, disengaging the retaining clips. Ensure that the vent control cable does not get caught when removing the vent assembly from the dashboard.

4 Remove the centre display screws and partially remove the assembly from the dashboard.

5 Disconnect the wiring connector from the rear and remove the centre display from the vehicle.

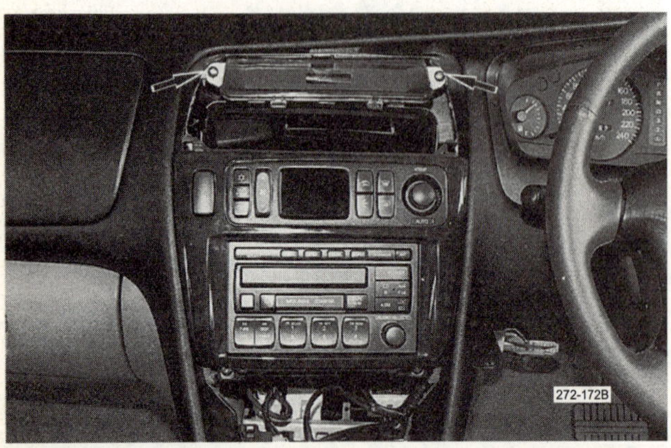

Centre vent assembly removed showing the centre display retaining screws.

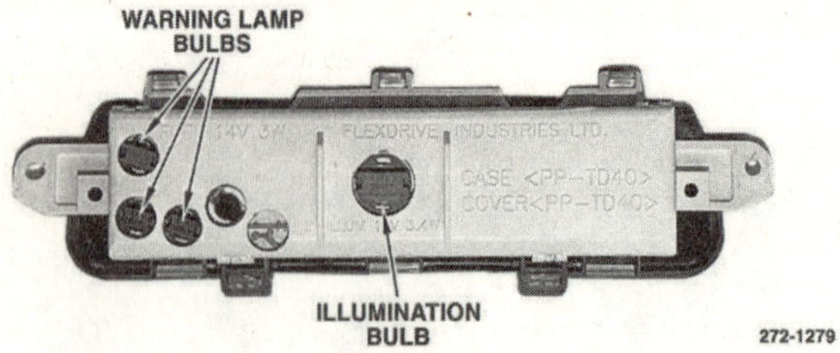

Rear of the centre display showing the bulbs.

Installation is a reversal of the removal procedure with attention to the following points:

1 Connect the wiring to the centre display and install it to the dashboard, ensuring that the wiring is not pinched. Install the retaining screws and tighten securely.

2 Install the surround panel, ensuring that the cable is not fouled when feeding down the dashboard.

3 Align the centre vents with the vent ducts and push the surround panel into the dashboard, engaging the retaining clips.

4 Connect the inner vent cable to the lever and the outer cable to the bracket, ensuring that the clip is secured to the side of the heater unit.

5 Install the lower dashboard panel.

Horn

To Remove and Install Horn Switches — T30 Torx socket

All models are fitted with one low tone horn mounted behind the front bumper on the passenger side of the vehicle. Some models have a second high tone horn mounted in the same position on the driver side of the vehicle. Models with a security system and alarm have an alarm horn mounted to the right hand inner guard, near the windscreen washer bottle. The removal and testing procedures are the same for all horns, with exception of the location.

The horn pad, mounted on the steering wheel energises the horn relay, which is located in the engine compartment relay box adjacent to the battery.

Front bumper removed showing the horn location. On some models a second horn is located in the same position on the driver side of the vehicle.

TO REMOVE AND INSTALL

1 Raise the front of the vehicle and support on chassis stands. Refer to General Information (chapter 1) for the positions of the jacking points.
2 Remove the screw clips retaining the splash guard in the front of the wheel arch. Pull the splash guard clear of the work area.
3 Remove the bolt retaining the horn to the front of the vehicle.
4 Disconnect the wiring connector and remove the horn from the vehicle.
5 To check the horn, use jumper leads to connect the positive terminal of a 12 volt battery to the horn terminal. Connect the body of the horn to the negative terminal, the horn should sound. Renew the horn if necessary.

Installation is a reversal of the removal procedure.

HORN SWITCHES

To Remove, Check and Renew

1 Remove the horn fuse from the driver side fuse compartment.
2 Where fitted, disable and remove the drivers airbag as described in Body (chapter 17).
3 On models not fitted with an airbag, remove the screws from either side of the horn pad. Partially remove the horn pad and disconnect the wiring connector.
4 To test the switch, connect an ohmmeter to the wiring terminal of the switch, and the other lead to the metal body of the horn pad.

   Continuity should exist when the switch is depressed. When released, the ohmmeter should indicate an open circuit. Renew the horn switch if necessary.

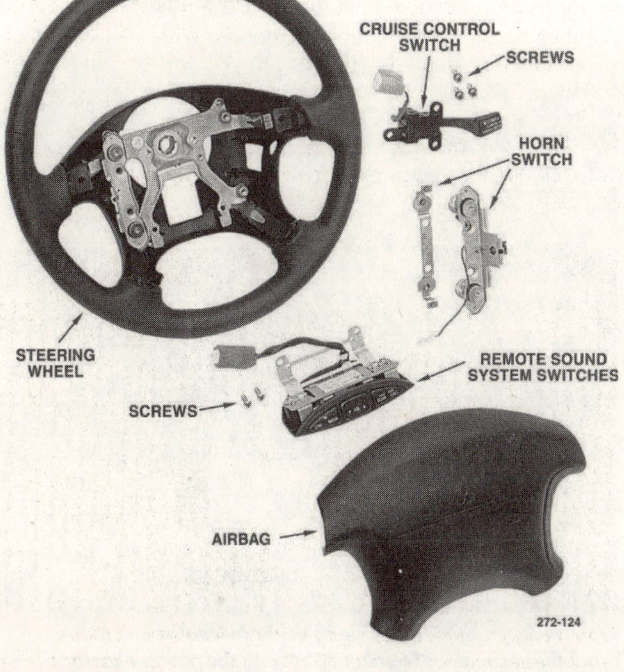

Steering wheel components.

5 If necessary, remove the Torx screws retaining the horn switches to the steering wheel.
6 Remove the horn switches noting the installed positions of the switch contacts and springs to aid assembly.

Installation is a reversal of the removal procedure, ensuring that the airbag is enabled as described in Body (Chapter 17).

HORN RELAY

The relay is located in the engine compartment fuse and relay box.

Refer to the Switches and Relays heading for a procedure to test the horn relay.

Power Door Locks and Keyless Entry

The keyless entry system, where fitted, operates independantly of the engine immobiliser system. Up to four remote control transmitters can be programmed to the keyless entry system. The signals from the transmitters are received by the keyless entry module which then signals the module controlling the power door locks to lock or unlock the doors. On some sedans, the luggage compartment can also be opened using the remote control transmitter.

If the vehicle can be locked and unlocked using the remote control transmitters or key but the vehicle will not start and it is suspected that the engine immobiliser is at fault, refer to the Engine Immobiliser System heading in Part 1 of Fuel and Engine Management (chapter 7) for procedures on diagnosing and testing this system.

On Magna models, the power door lock control module is contained within the keyless entry module located above the throttle pedal.

On Verada models, the keyless entry module is located on the rear parcel shelf and the power door lock control module is mounted above the throttle pedal. On models with an alarm system, the keyless entry system operates in conjunction with the ETACS control unit, which is the alarm system that works in conjunction with the power door locks.

Lower dashboard panel removed showing the power door lock control module. Magna models.

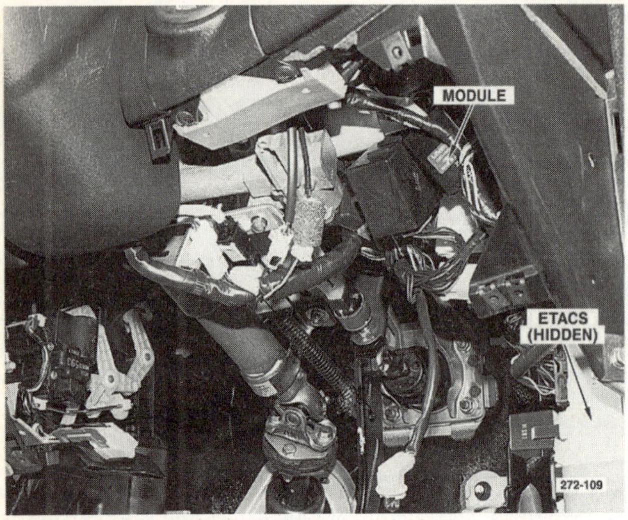

Lower dash panel removed showing the power door lock control module and ETACS location. Verada models.

Due to the variations and complexity of these systems, the following operations are aimed to assist the operator in isolating any faults to a particular component or system. If the following operations fail to find the fault, it is recommended that the vehicle be referred to an authorised dealer who will have the necessary equipment to diagnose the fault.

DOOR ACTUATOR

To Remove and Install

The door actuator is part of the door lock assembly on the front doors and can be removed and installed using the door lock procedure in Body (chapter 17). The rear door actuators can be removed and installed using the procedure described below.

1 Remove the trim panel from the rear door. Refer to Body (chapter 17).
2 Mark the control rod on either side of the clip of the power door lock actuator. Open the clip, using a small screwdriver and separate the control rod from the clip.
3 Remove the nut and bolts retaining the actuator to the door. Disconnect the wiring connector and manoeuvre the actuator from the door.

Installation is a reversal of the removal procedure, ensuring that the actuator clip aligns with the marks previously marked on the control rod.

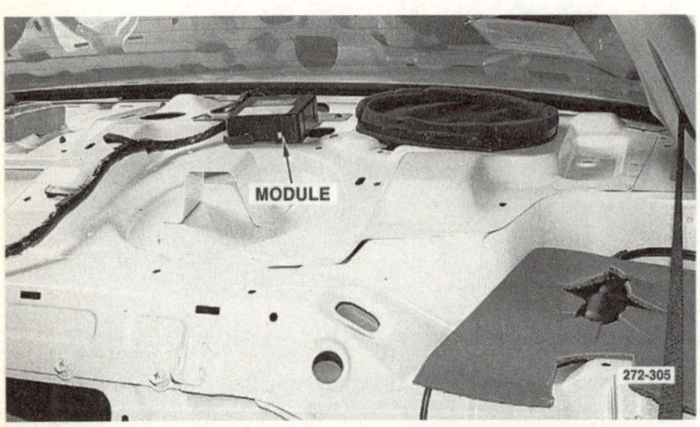

Rear parcel shelf removed showing the keyless entry module. Verada Models.

To Check

1 Ensure that all of the doors are locked. Place the snipper buttons in the lock position if necessary.
2 Disconnect the wiring connector from the power door lock control module.
3 Using jumper leads connected to battery power, connect the positive lead to the blue/black wire and the negative lead to the blue/red wire on the power door lock control module.

   All of the door actuators should cycle to the unlock position.
4 Swap the positive and negative wires and all of the door locks should cycle to the lock position.
5 If all of the door actuators cycle in both the lock and unlock positions, the actuators can be considered serviceable.

   If one of the doors will only lock or unlock, the actuator may require adjustment. Refer to the heading below.

> *The adjustment procedure can only be done on the rear doors (and the tailgate, on station wagon models). If one of the front doors will only lock or unlock, the most probable cause would be a seized or damaged door lock linkage. Remove the door trim panel and sealing sheet and clean and lubricate the lock assembly. Refer to Body (chapter 17).*

If any of the door actuators did not cycle, proceed as follows:

a Remove the door trim panel and the sealing sheet from the door with the faulty actuator as described in Body (chapter 17).
b Check that the door lock linkages are not seized or damaged, causing the lock actuator to fail.
c Disconnect the wiring connector from the door actuator. Using an ohmmeter, check the wiring for continuity between the door lock actuator connector and the power door lock control module. Repair the wiring or connectors as necessary.
d Check for continuity between the actuator wires at the door actuator connector, referring to the wiring diagrams.

   If an open circuit is indicated, renew the door lock assembly as described in Body (chapter 17).

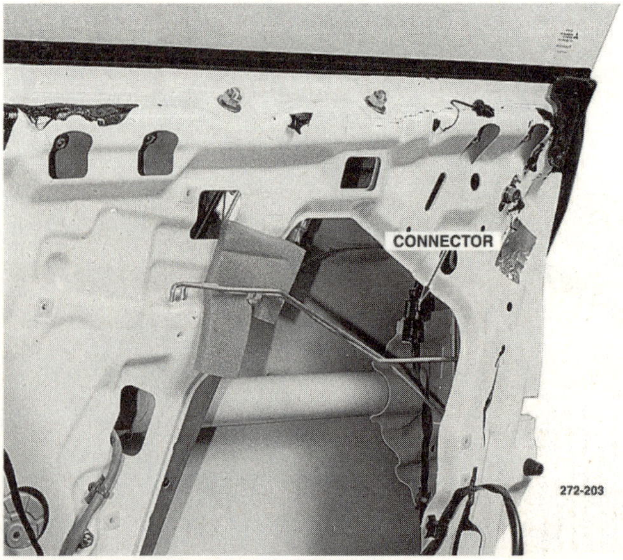

Front door trim panel removed showing the power door lock actuator connector.

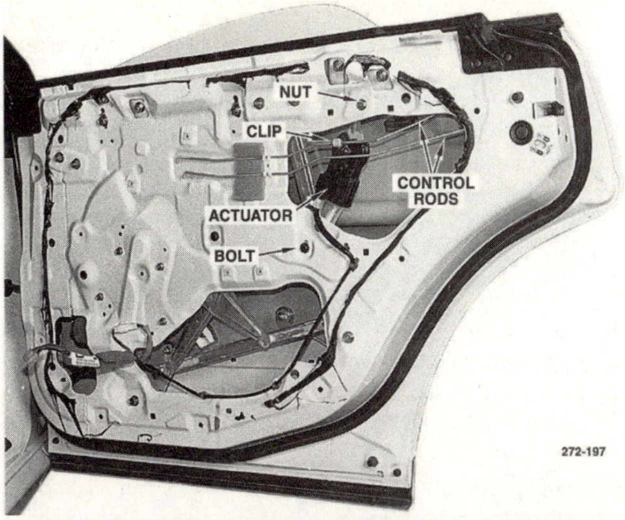

Rear door trim panel removed showing the power door lock actuator.

e Where fitted, check that there is continuity between the door actuator switch wire and the earth wire in the door lock actuator only when the door is unlocked. Refer to the wiring diagram.

Renew the door lock assembly as described in Body (chapter 17), if necessary.

To Adjust

1. Remove the door trim panel and sealing sheet from the rear door with the faulty actuator.
2. If working on the rear door, install the interior door handle, ensuring that the linkages are connected.
3. Mark the control rod on either side of the clip on the actuator. Open the clip, using a small screwdriver and separate the control rod from the clip.
4. Move the clip further along the control rod, clip into the actuator clip and clip the cover into place.

> *If the door will not lock, move the actuator clip slightly toward the interior door handle. If the door will not unlock, move the actuator clip toward the lock assembly. Only move the clip in increments of 2 – 3 mm at a time.*

5. Operate the power door locks and check the operation of the door locks. Readjust if necessary.

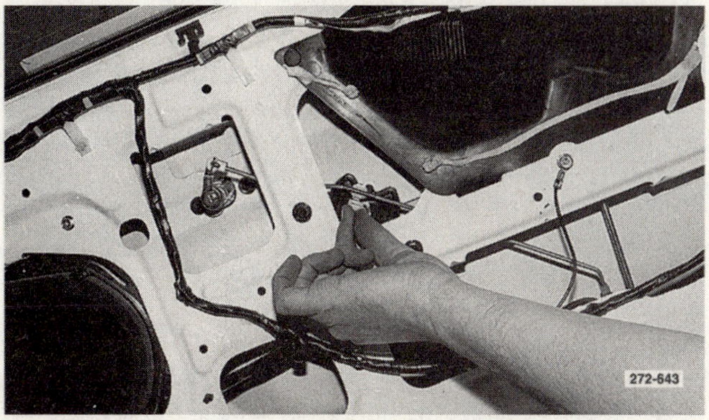

Disconnecting the control rod from the power door lock actuator. Station wagon models.

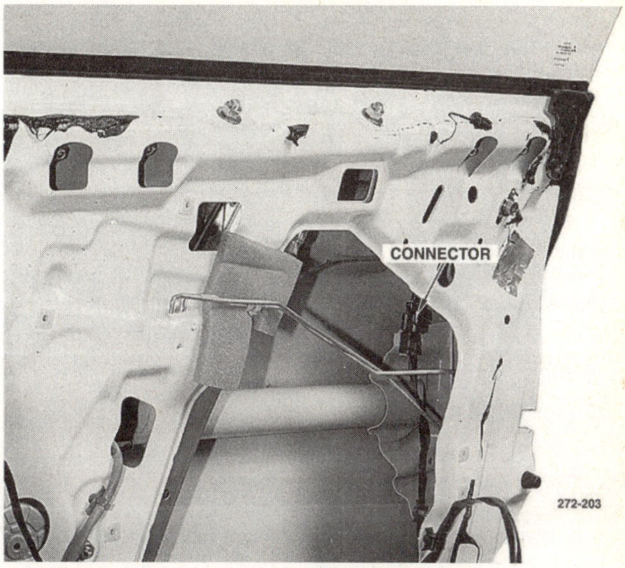

Door lock cylinder switch wiring connector location.

DOOR LOCK CYLINDER SWITCHES

The door lock cylinder switches are not fitted to all models.

To Remove and Install

Remove the door lock cylinder as described in Body (chapter 17).

To Check

1. Remove the door trim panel and the sealing sheet from the door with the faulty door lock cylinder switch.
2. Separate the lock cylinder connector and using the wiring diagrams, check that there is continuity between the earth wire and the unlock wire with the key unlocking the door.
3. Check that there is continuity between the earth wire and the lock wire with the key locking the door.

   Renew the lock cylinder switch if necessary.

DOOR LOCK CONTROL SWITCH — FRONT DOORS

The door lock control switch is an integral part of the door lock assembly. Refer to Body (chapter 17) for the door lock assembly removal and installation procedures.

To Check

1. Remove the door trim panel and the sealing sheet from the door with the faulty door lock control switch.
2. Separate the door lock control connector and using the wiring diagrams, check that there is continuity between the earth wire and the unlock wire with the switch in the unlock position.
3. Check that there is no continuity between the two wires with the switch in the lock position.

   Renew the door lock assembly, if necessary.

REMOTE CONTROL TRANSMITTER

The keyless entry system can operate up to 4 remote control transmitters. Each transmitter emits its own frequency to activate the power door locks, locking and unlocking the doors. The transmitters are designed to function within 5 metres of the vehicle. A watch type battery is located within the transmitter and can be renewed separately. On late model vehicles, a three button remote control transmitter, on sedan models, and a two button remote control transmitter on station wagon models locks and unlocks the doors. The third button on sedan models will open the luggage compartment after pressing the button twice.

Remote Control Transmitter Programming

Use the following procedure when renewing or adding an extra transmitter, to the keyless entry system. The procedure can also be used to exclude a transmitter that has been lost, preventing anyone unlocking the vehicle using the lost transmitter.

1 Ensure that all of the remote control transmitters being programmed to the keyless entry system are at hand.
2 Place an ignition key into the ignition and turn the ignition from Off to On three times within a five second period, leaving the ignition On.
3 Within 5 seconds, press the door button on the remote control transmitter. The turn signals will flash to indicate the system storing the code for that particular remote control transmitter.
4 Within 5 seconds, press the door button on the next remote control transmitter, ensuring that the turn signal lamps flash.
5 Repeat step 4 on any remaining remote control transmitters that are to be programmed to this vehicle and then switch the ignition Off.
6 Ensure that all of the transmitters operate the vehicle.

Renewing the Remote Control Transmitter Battery

The battery within the remote control transmitter has approximately a five year life span. The battery can be renewed as follows:

1 Remove the screw from the rear of the remote control transmitter and separate the two halves of the assembly.
2 Using a small screwdriver or tweezers, remove the battery from the rear of the transmitter assembly.
3 Insert the new battery with the positive side of the battery facing toward the rear of the transmitter assembly.

TIP *Do not touch the battery terminals with bare fingers.*

4 Ensure that the seal between the two halves of the assembly is installed correctly. Install the two halves together and tighten the retaining screw.

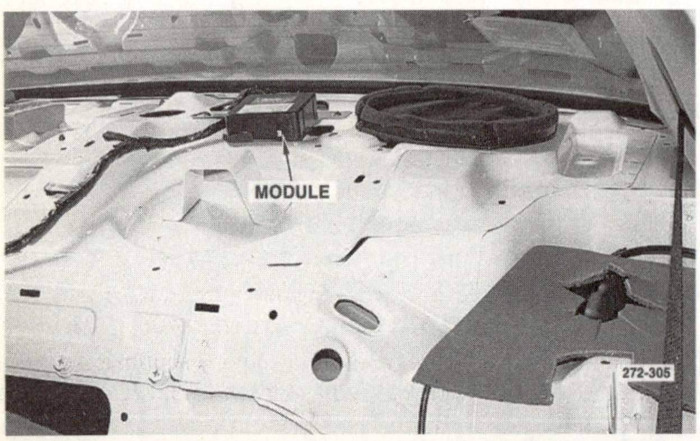

Rear parcel shelf removed showing the keyless entry module. Verada models.

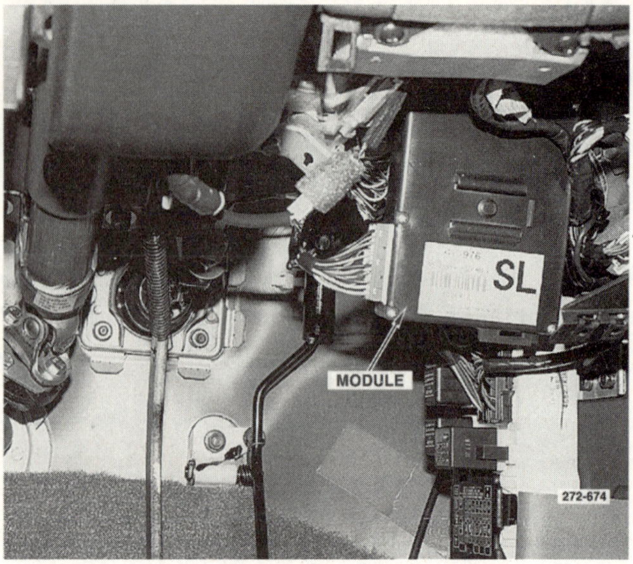

Lower dashboard panel removed showing the power door lock control module. Magna models.

KEYLESS ENTRY MODULE

The keyless entry module is combined within the power door lock module on Magna models. If renewing the keyless entry module (Verada models) or power door lock module (Magna models), the remote control transmitters will have to be reprogrammed to the module. Refer to the Remote Control Transmitter Programming heading for the reprogramming procedure.

To Remove and Install — Verada Models

1 Remove the rear seat and parcel shelf. Refer to Body (chapter 17).
2 Remove the screws retaining the module and disconnect the wiring connector.
3 Remove the keyless entry module from the vehicle.

Installation is a reversal of the removal procedure.

To Remove and Install — Magna Models

1 Remove the screws retaining the bonnet lever to the driver side lower dash panel and position the lever clear of the work area. Remove the screw cover from the top of the panel. Remove the lower dash panel screws and disconnect the panel from the top left side of the dashboard. Partially remove the panel, rotate the footwell lamp bulb holder anticlockwise and separate it from the panel. Remove the driver side lower dash panel from the vehicle.
2 Depress the dimple retaining the module to the bracket and pull the module from the bracket.
3 Disconnect the wiring connector from the module and remove the module from the vehicle.

Installation is a reversal of the removal procedure.

POWER DOOR LOCK CONTROL MODULE — VERADA Xi

To Remove and Install

The location and removal and installation procedures for the power door lock control module is the same as the Keyless Entry Module procedure for the Magna model.

ETACS CONTROL UNIT — VERADA Xi

The Electronic Time Alarm Control System (ETACS) is fitted to Verada Xi models. It is an additional module that works in conjunction with the power door lock and keyless entry systems to provide an audible alarm if the vehicle is entered while the system is armed. A horn is located in the engine compartment adjacent to the windscreen washer bottle.

Testing the ETACS control unit is limited without proper diagnostic equipment. Before referring the vehicle to an authorised dealer, use the wiring diagrams as a guide and check the condition of the wiring connectors and fuses and relays that affect the operation of the ETACS unit.

To Remove and Install

1 Disconnect the negative battery terminal.
2 Remove the screws retaining the bonnet lever to the driver side lower dash panel and position the lever clear of the work area. Remove the screw cover from the top of the panel. Remove the lower dash panel screws and disconnect the panel from either side of the steering column. Partially remove the panel, rotate the footwell lamp bulb holder anticlockwise and separate it from the panel. Remove the driver side lower dash panel from the vehicle.
3 Remove the screw from between the kick panel and the sill panel. Lift the sill panel to disengage the retaining clips and remove the panel from the vehicle.
4 Remove the plastic screw from the carpet adjacent to the throttle pedal, lift the front carpet retainer from the sill panel and remove the carpet and kick panel from beneath the fuse and relay box. Position the carpet and kick panel clear of the work area, beneath the throttle pedal.
5 Remove the nut retaining the fuse and relay box to the side of the vehicle. Remove the fuse and relay box from the stud and pull it down to disengage the clip retaining the top of the box.
6 Working at the rear of the fuse and relay box, disconnect the wiring connectors to gain access to the ETACS control unit. Remove the retaining screw and pull the ETACS control unit from the rear of the fuse and relay box.

Installation is a reversal of the removal procedure, ensuring that the wiring connectors are securely installed.

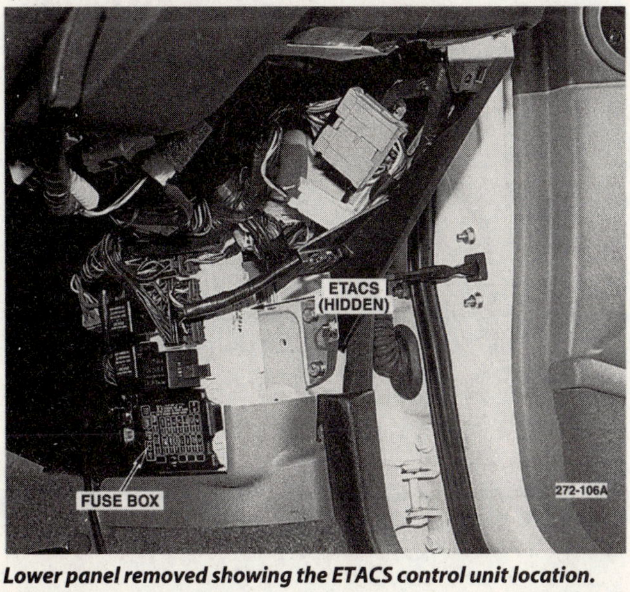

Lower panel removed showing the ETACS control unit location.

Cruise Control

To Check System — Hand vacuum pump, vacuum gauge
To Remove and Install Horn Switches — T30 Torx Socket

The electronic cruise control system consists of an electronic control module and a control switch assembly. The electronic control module receives a signal from the speedometer and activates an electric vacuum pump. The pump controls vacuum acting on a diaphragm type actuator, operating the throttle lever via a cable and maintaining the preset speed.

The cruise control system will not activate until the vehicle speed is above 40 km/h. The system is deactivated when the brake pedal, clutch pedal (manual transaxle) the Off switch on the instrument cluster or the cancel switch on the control lever is deactivated.

The cruise control system incorporates complex electronic circuitry, therefore it is recommended that the vehicle be taken to an authorised dealer for testing and repair should the operation of the system be in doubt and the following procedures fail to rectify the problem.

CRUISE CONTROL MODULE

To Remove and Install

1 On manual transaxle models, unscrew the gear lever knob.
2 On Verada models, open the coin tray lid and firmly pull the coin tray from the console. Disconnect the wiring connectors from the base of the accessory socket and then pull the cup holder from the front of the console.
3 Grasp the rear edges of the console fascia panel and pull upward. Working toward the front of the console, raise the fascia panel, releasing the retaining clips. When all of the clips are released, manoeuvre the fascia panel from the console.
4 On Magna models, remove the ashtray from the vehicle. Pull the ashtray surround panel partially from the dashboard, disengaging the clips retaining the top of the surround to the dashboard. Disconnect the cigarette lighter wiring from the base of the socket and remove the surround.
5 On Verada models, pull the ashtray assembly, disengaging the clips retaining the top of the ashtray assembly to the dashboard. Disconnect the cigarette lighter wiring from the base of the socket and remove the ashtray.
6 Remove the cruise control module retaining bolts and manoeuvre the module partially from the console.
7 Disconnect the wiring from the module and remove it from the vehicle.

Installation is a reversal of the removal procedure.

VACUUM ACTUATOR

To Check

1 Disconnect the vacuum hose from the vacuum actuator and connect a hand vacuum pump to the actuator.
2 Check that the throttle valve opens when vacuum is supplied to the actuator and that it stays open until the vacuum is released.

   Renew the vacuum actuator if it does not open or hold vacuum.

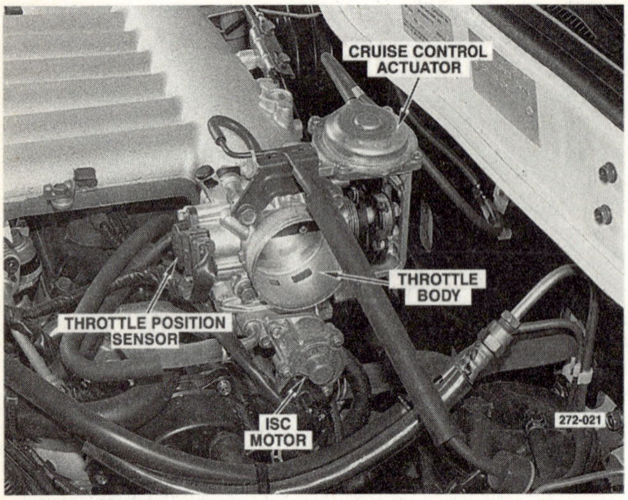

Cruise control vacuum actuator location.

To Remove and Install

1. Disconnect the vacuum hose from the vacuum actuator.
2. Remove the screws retaining the actuator bracket to the throttle body.
3. Hold the throttle lever open and twist the actuator linkage until it can be removed from the throttle lever.

Installation is a reversal of the removal procedure.

VACUUM PUMP

The vacuum pump is a sealed unit consisting of a pump motor, a control valve, which regulates the amount of vacuum applied to the actuator, and a release valve which exhausts the vacuum from the system.

To Check

1. Disconnect the vacuum hose from the vacuum actuator and connect a vacuum gauge to the hose.
2. Disconnect the wiring connector from the vacuum pump and separate the connector from the pump bracket.
3. Connect the positive lead of a 12 volt power supply to the red wire and the negative lead to the other three wires.

   The vacuum pump should operate and retain a vacuum of at least 53 kPa.

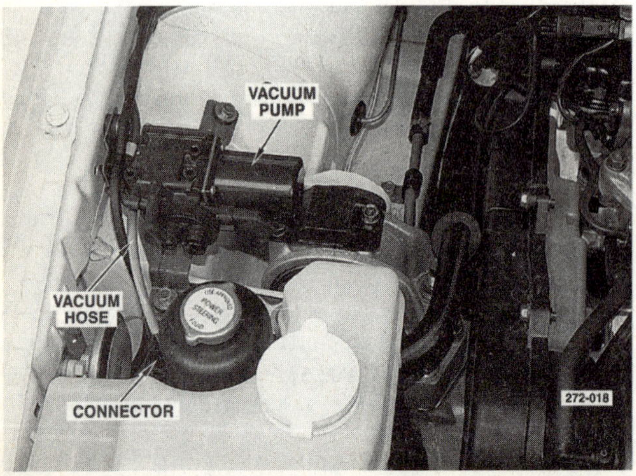

Cruise control vacuum pump location.

This is checking the vacuum pumps maximum output.

4. Remove the negative lead from the blue wire and check that the vacuum drops to less than 20 kPa.
5. Connect the negative lead back onto the blue wire and after the vacuum builds up, disconnect the negative lead from the green wire. The vacuum should drop to less than 20 kPa.

   Renew the vacuum pump if it doesn't operate as specified.

To Remove and Install

1. Remove the bolts retaining the pump to the vehicle.
2. Disconnect the wiring connector and hose from the pump and remove it from the vehicle.

Installation is a reversal of the removal procedure.

CRUISE CONTROL SWITCHES

To Check — Main Switch

1. Remove the cruise control main switch as described later in this chapter.
2. Using jumper leads connected to a 12 volt battery, connect the positive lead to terminal 2 of main switch wiring connector and the negative lead to terminal 5. Also connect an ohmmeter between terminals 1 and 2.
3. Depress the switch and check that there is continuity between terminals1 and 2. Also check that the warning lamp in the switch illuminates. Renew the switch if necessary.

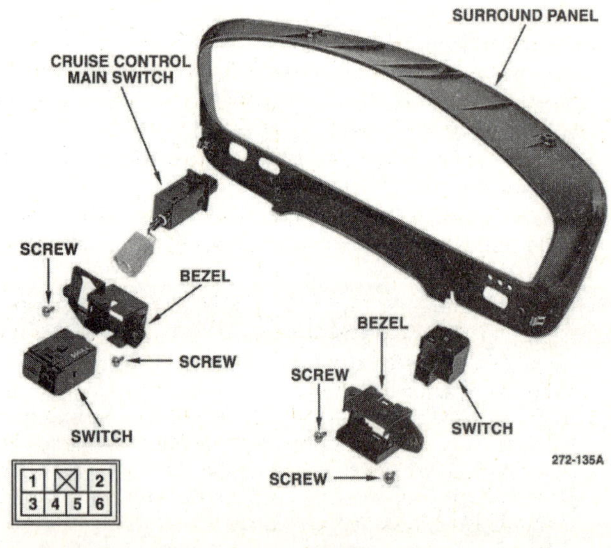

Instrument cluster surround panel and switches. Inset shows terminal identification for the cruise control main switch.

To Check — Control Switch

1. Where fitted, disable and remove the airbag. Refer to Body (chapter 17).
2. On models not fitted with an airbag, remove the horn pad. Refer to the Horn heading in this chapter.
3. Disconnect the cruise control switch connector and connect an ohmmeter between the switch terminals.

Renew the switch assembly if the resistances do not match the following table.

Switch position	Resistance
At rest	Open circuit
Cancel position	Approximately 0 ohms
Resume position	Approximately 820 ohms
Set position	Approximately 2 700 ohms

Cruise control switch and connector location.

To Remove and Install

Refer to the Instrument Cluster Switches heading in this chapter for a procedure to remove the cruise control main switch, located on the instrument cluster surround panel. To remove and install the control switch on the steering wheel, proceed as follows:

1 Disable and remove the drivers airbag. Refer to Body (chapter 17).

2 Disconnect the wiring from the horn switch, cruise control switch and where fitted, sound system remote control assembly.

3 Ensure that the steering wheel is in the straight ahead position and loosen the steering wheel retaining nut. Do not completely remove the nut.

4 Mark the relationship between the steering column and steering wheel to ensure that the wheel is returned to the same position. Grasp the steering wheel at the twelve o'clock and six o'clock positions and rock the steering wheel while moving your hands around the steering wheel until it separates from the steering column.

If this fails to work, a steering wheel puller may be needed.

5 Pass the wiring through the steering wheel and remove the wheel from the column.

6 Remove the Torx screws retaining the horn switch over the cruise control switch assembly. Remove the horn switch, springs, and lower plate from the steering wheel, noting the installed positions of the components.

7 Remove the retaining screws and manoeuvre the cruise control switch assembly from the steering wheel.

Installation is a reversal of the removal procedure with attention to the following points:

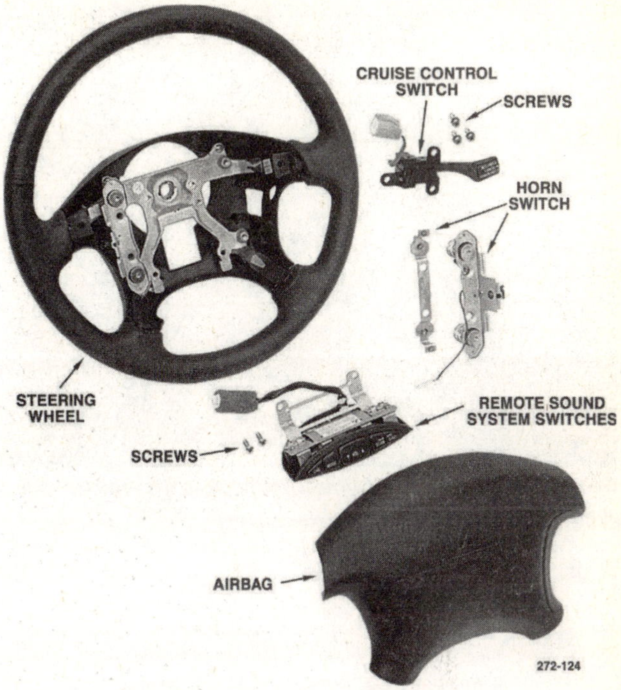

Steering wheel components.

1 Install the cruise control switch and tighten the retaining screws.

2 Install the horn switch, ensuring the components are returned to their original position.

3 Install the steering wheel and enable the airbag. Refer to Steering (chapter 12) and Body (chapter 17).

Trailer Wiring

There are several brands of trailer wiring connectors available with a variety of pin configurations. The most common is the seven pin type which provides for two auxiliary circuits, usually used for interior lighting and electric trailer brakes.

When installing a trailer socket, it may be necessary to replace the original vehicle flasher unit with a heavy duty unit to maintain the flasher rate in accordance with local regulations. Local regulations may also require the installation of a dashboard mounted repeater lamp.

Should the trailer socket and plug not be matched, adaptors can be made or obtained to facilitate connection.

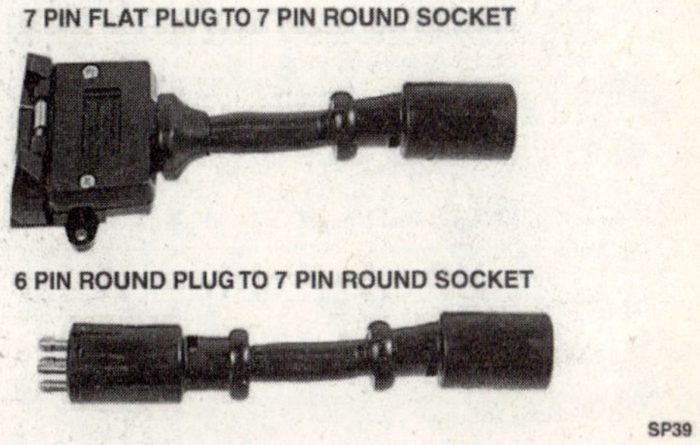

Various adapters are available if the trailer plug differs from the vehicle socket.

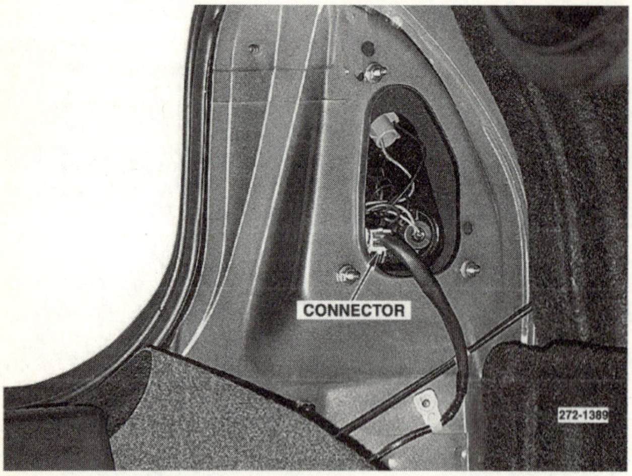

The genuine trailer wiring harness, connects to the vehicle at the point shown. Sedan models.

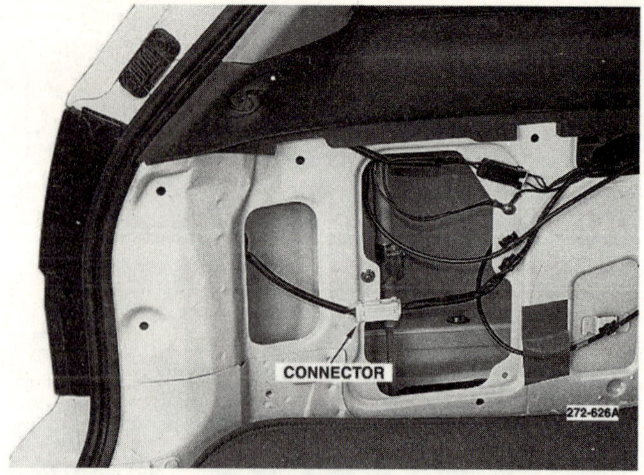

The genuine trailer wiring harness connects to the vehicle at the point shown. Station wagon models.

TO INSTALL TRAILER SOCKET

A special wiring harness is available from the manufacturer that plugs into the left hand tail lamp wiring. The wiring harness shortens the installation process.

The minimum materials required are a trailer socket, mounting bracket, seven core flex, insulated wiring connectors and a rubber grommet.

1. Disconnect the negative battery terminal.
2. Locate the wiring harness for the lamps at the rear of the vehicle.

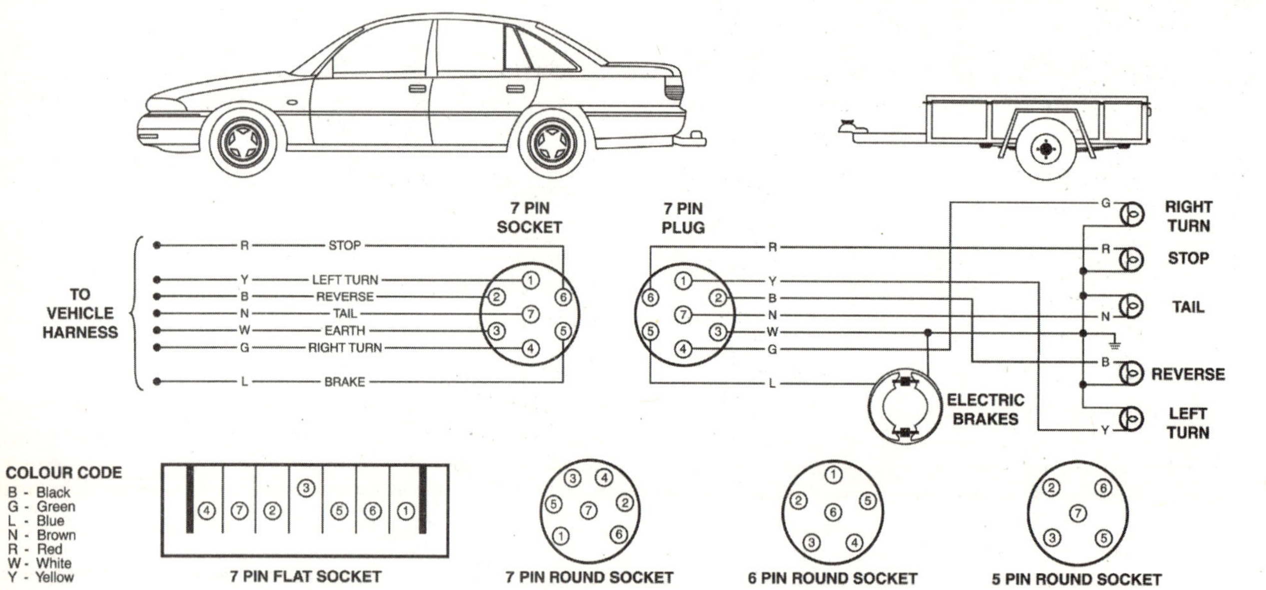

TERMINAL	SEVEN PIN SOCKET		TERMINAL	SIX PIN SOCKET		TERMINAL	FIVE PIN SOCKET	
1	Left turn signal lamp	Yellow	1	Tail lamp	Brown	2	Left turn signal lamp	Yellow
2	Reversing lamp	Black	2	Left turn signal lamp	Yellow	3	Earth	White
3	Earth	White	3	Right turn signal lamp	Green	5	Right turn signal lamp	Green
4	Right turn signal lamp	Green	4	Stop lamp	Red	6	Stop lamp	Red
5	Electric brakes	Blue	5	Auxiliary circuit	Blue	7	Tail lamp	Brown
6	Stop lamp	Red	6	Earth	White			
7	Tail lamp	Brown						

CD29A

Vehicle and trailer wiring diagram to conform to Australian Standard AS2513. Sockets and plugs shown are as viewed from the terminal side of component.

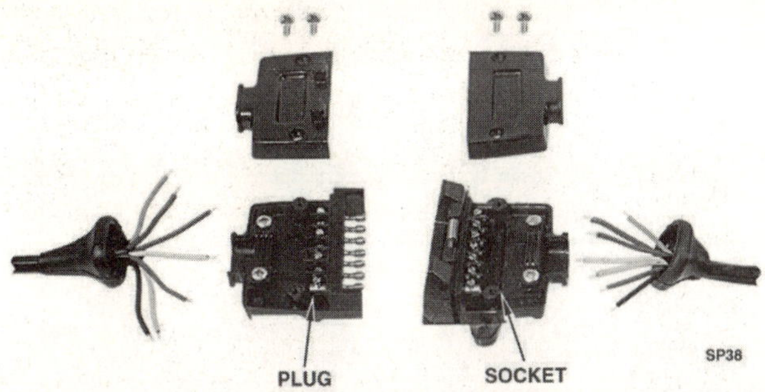

A Brylite seven pin trailer plug and socket ready to be wired up.

3 Identify the wires for the right turn signal lamp, left turn signal lamp, tail lamp, stop lamp, reverse lamp and earth. Use the vehicle wiring diagram in this chapter to determine the wire colour for each lamp circuit.

4 Connect the seven core flex to the various lamp wires using insulated connectors.

For seven pin sockets and plugs, use the following standard trailer wiring code as shown in the illustration.

5 If necessary, drill a suitably sized hole through the vehicle floor and install the rubber grommet to the hole.

Before drilling the hole, check whether there is an existing hole which may be used. Ensure that a rubber grommet is installed to prevent chafing of the wires.

6 Pass the seven core flex through the rubber grommet.

7 Cut the seven core flex to the required length and slide the dust cover onto the flex.

8 Remove approximately 15 mm of insulation from each wire in the flex.

9 Connect the wires of the seven core flex to the terminals of the socket using the standard wiring code and following the socket manufacturers connection procedure.

10 Assemble the socket and install the dust cover to the rear of the socket.

11 Install the mounting bracket as close as possible to the tow bar ball and high enough to maintain vehicle ground clearance.

12 Install the socket to the mounting bracket.

13 Connect a trailer plug to the socket and test for correct lamp operation.

TO INSTALL TRAILER PLUG

The minimum materials required are a trailer plug, seven core flex and insulated wiring connectors.

The installation procedure for the trailer plug is similar to that for the socket with attention to the following points:

1 Connect the trailer wiring to the plug according to the standard wiring colour code and the plug manufacturers instructions.

Use a suitable length of seven core flex and insulated connectors should the trailer wiring not be long enough.

2 If the trailer wiring does not conform to the standard colour code, identify each circuit as follows:

a Locate the trailer earth wire.

b Connect a jumper lead from the negative terminal of a 12 volt battery to the earth wire.

c Connect a jumper lead to the positive terminal of the 12 volt battery and connect each of the trailer lamp wires in turn to the positive battery terminal jumper lead, noting the colour of the wire and lamp circuit operated.

3 After installation, connect the trailer plug to the socket on the vehicle and check all lamps for correct lamp operation.

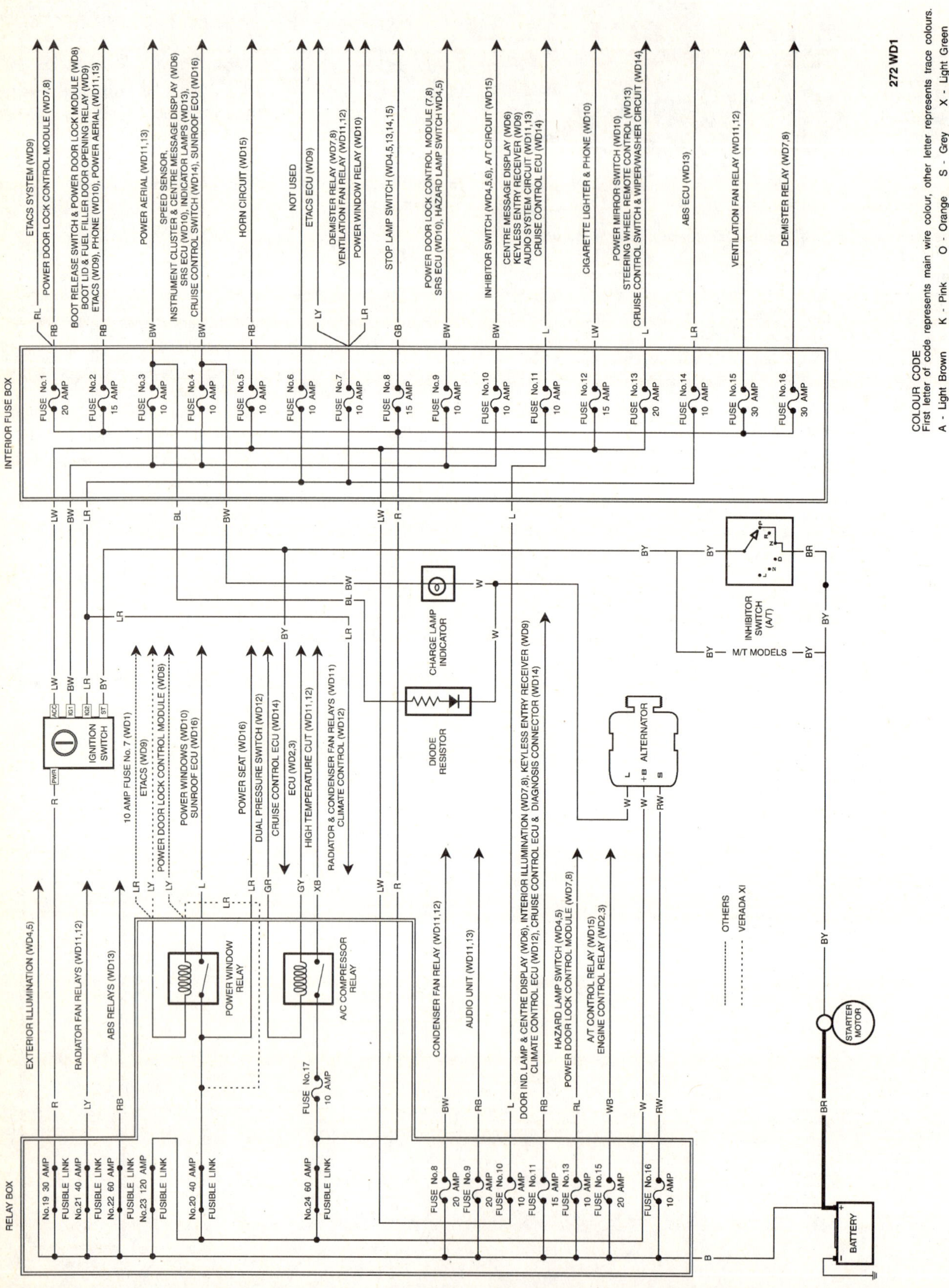

Power distribution, starting and charging systems. TE/TF/TH & KE/KF/KH models.

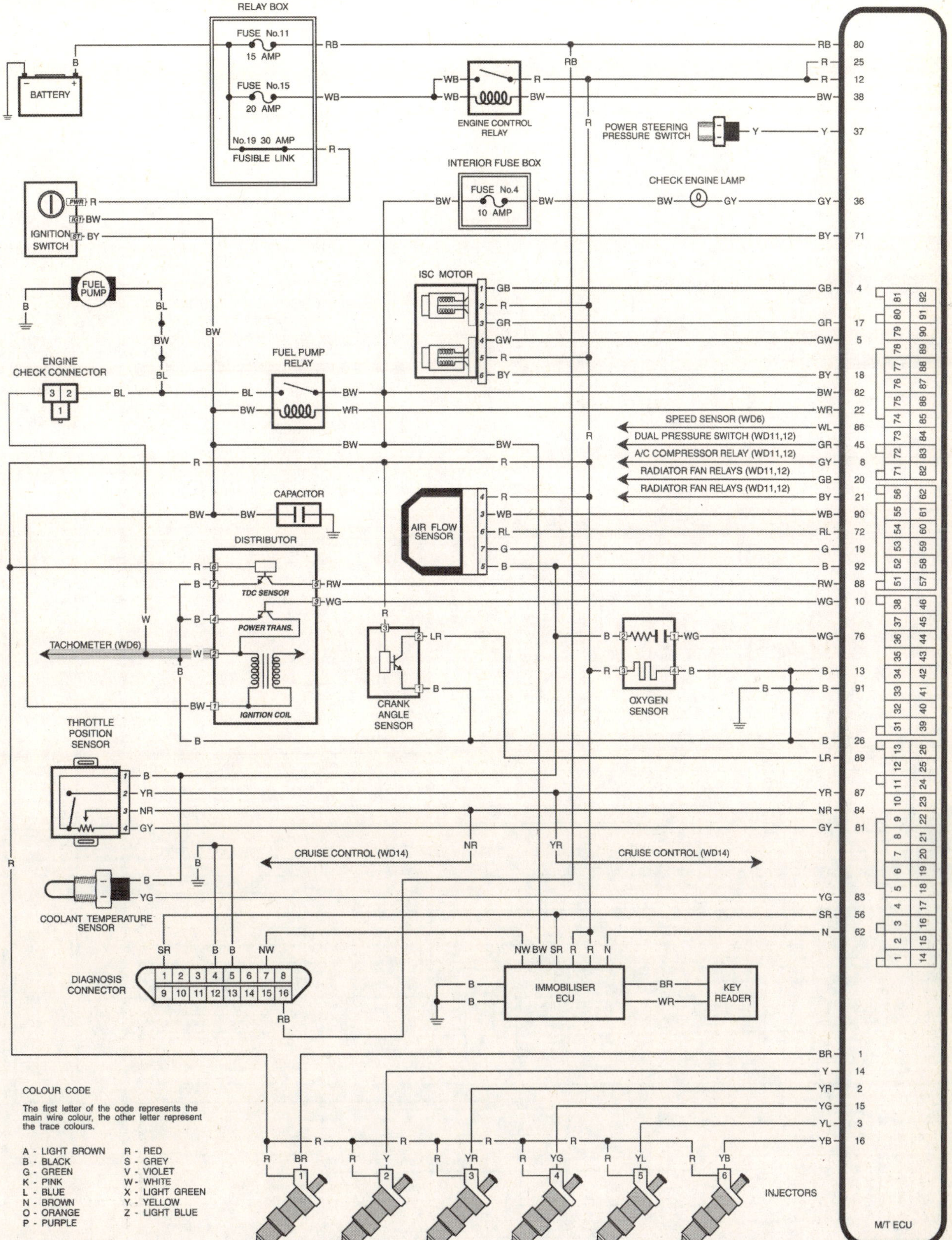

272 WD2

Fuel and engine management system – manual transaxle. TE/TF/TH & KE/KF/KH models.

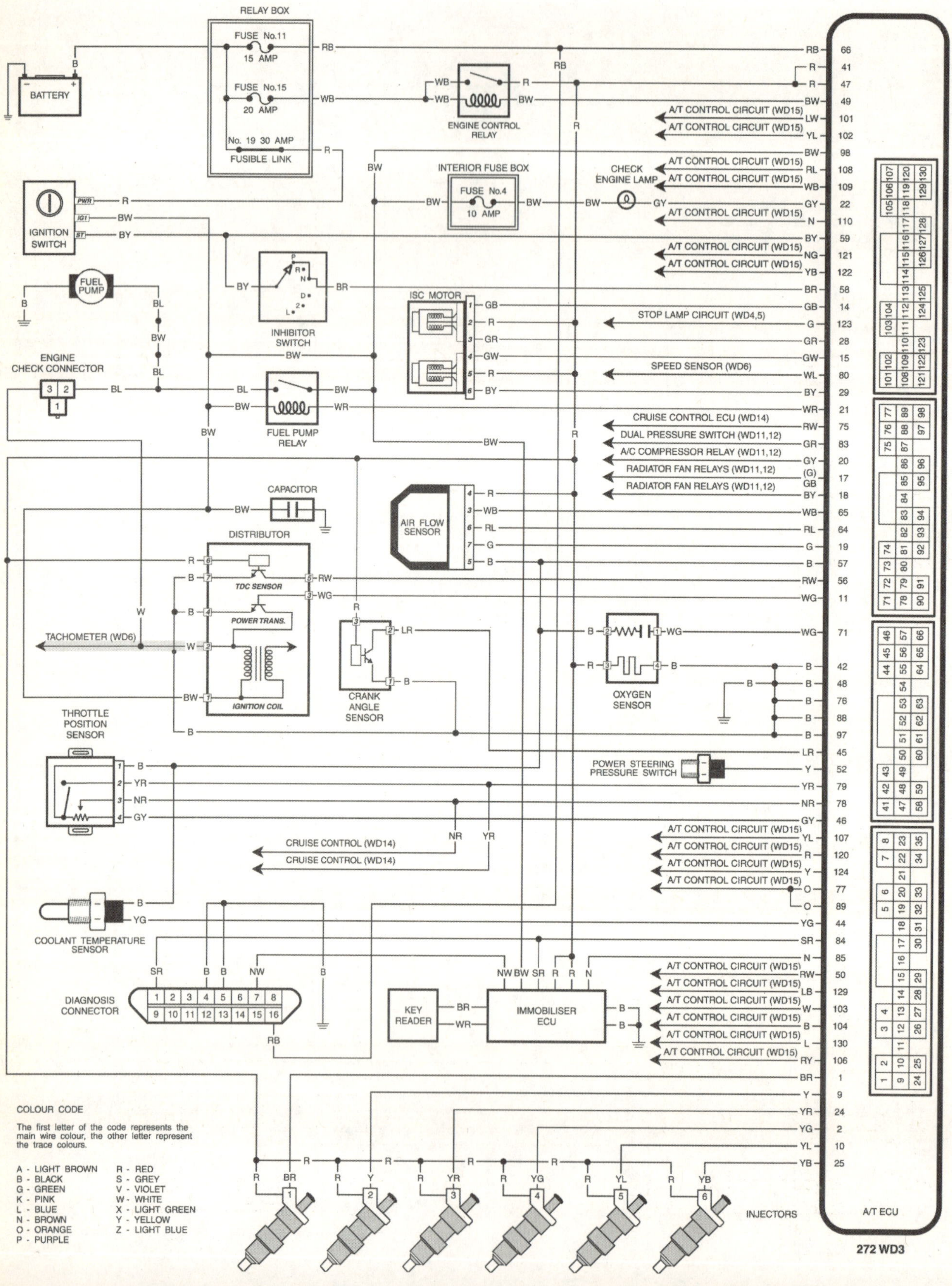

Fuel and engine management system – automatic transaxle. TE/TF/TH & KE/KF/KH models.

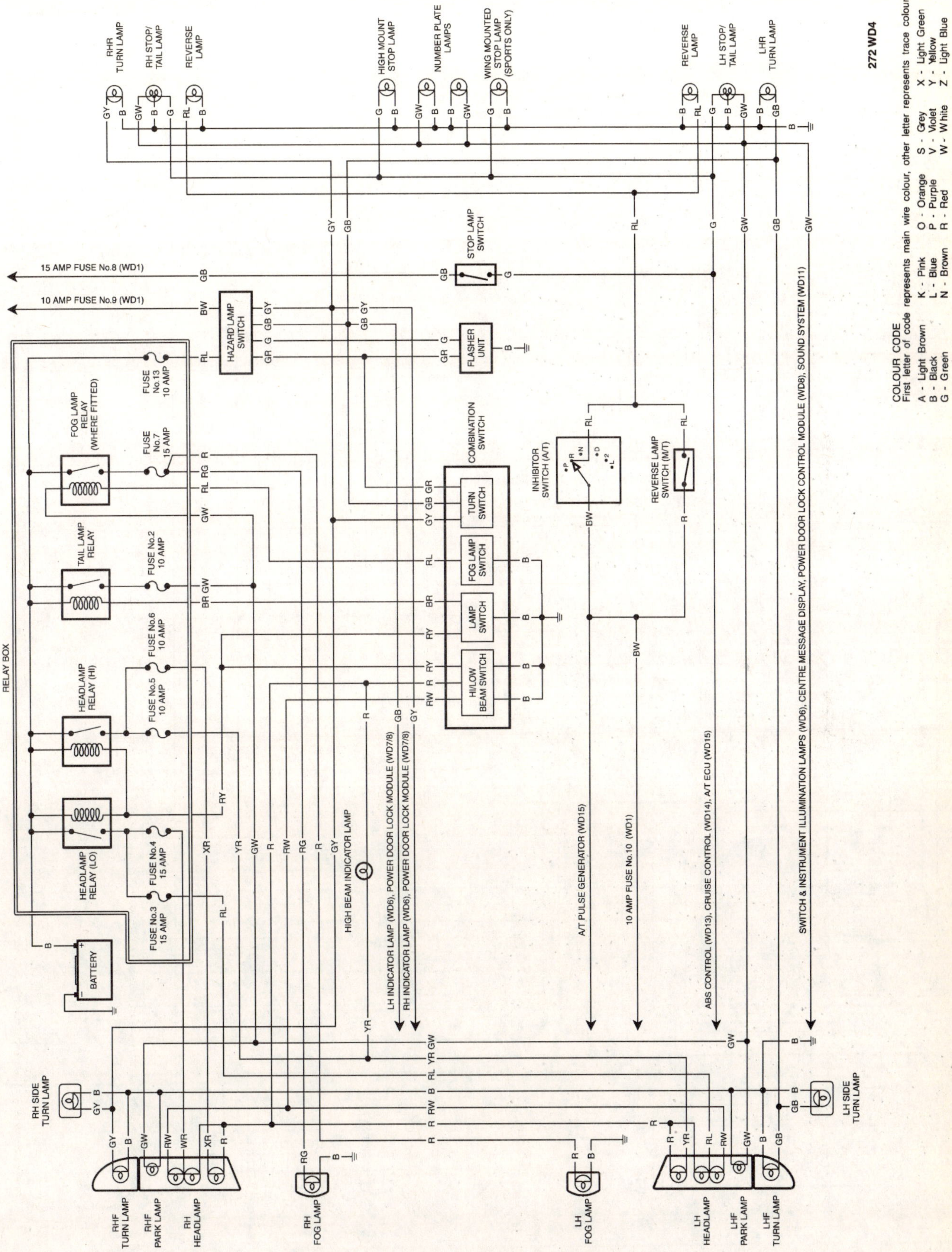

Exterior lighting system – Magna. TE/TF/TH models.

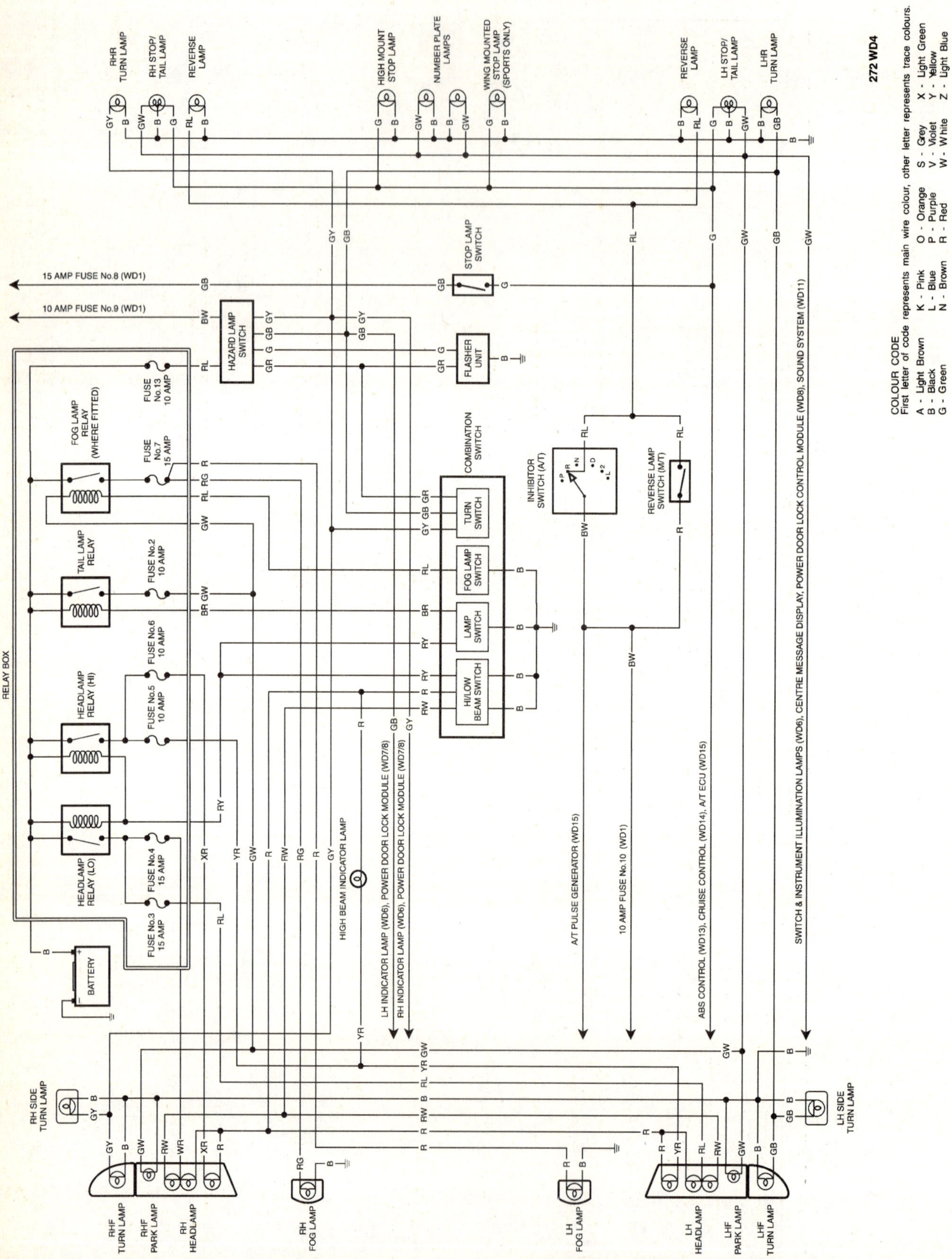

Exterior lighting system – Verada. KE/KF/KH models.

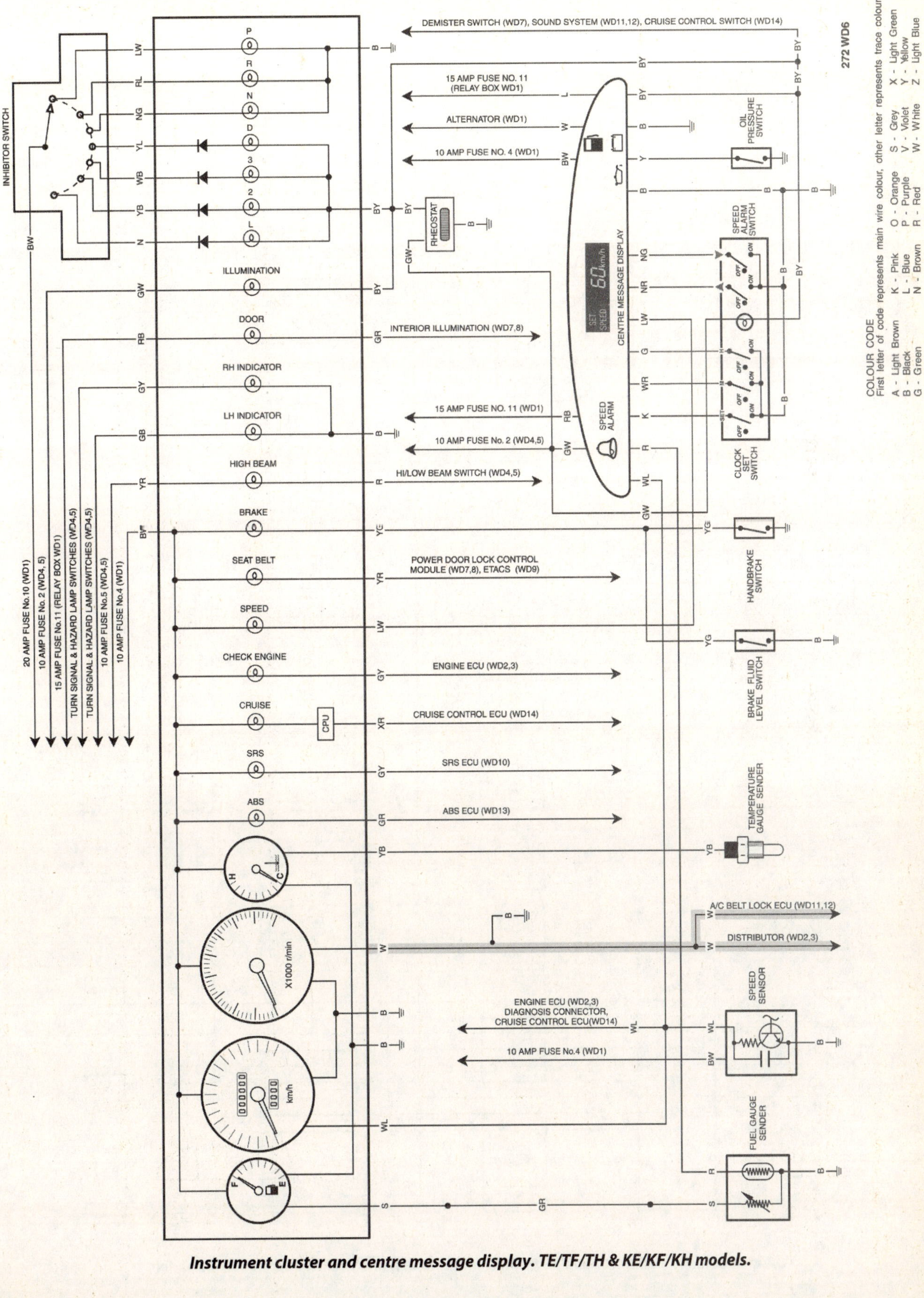

Instrument cluster and centre message display. TE/TF/TH & KE/KF/KH models.

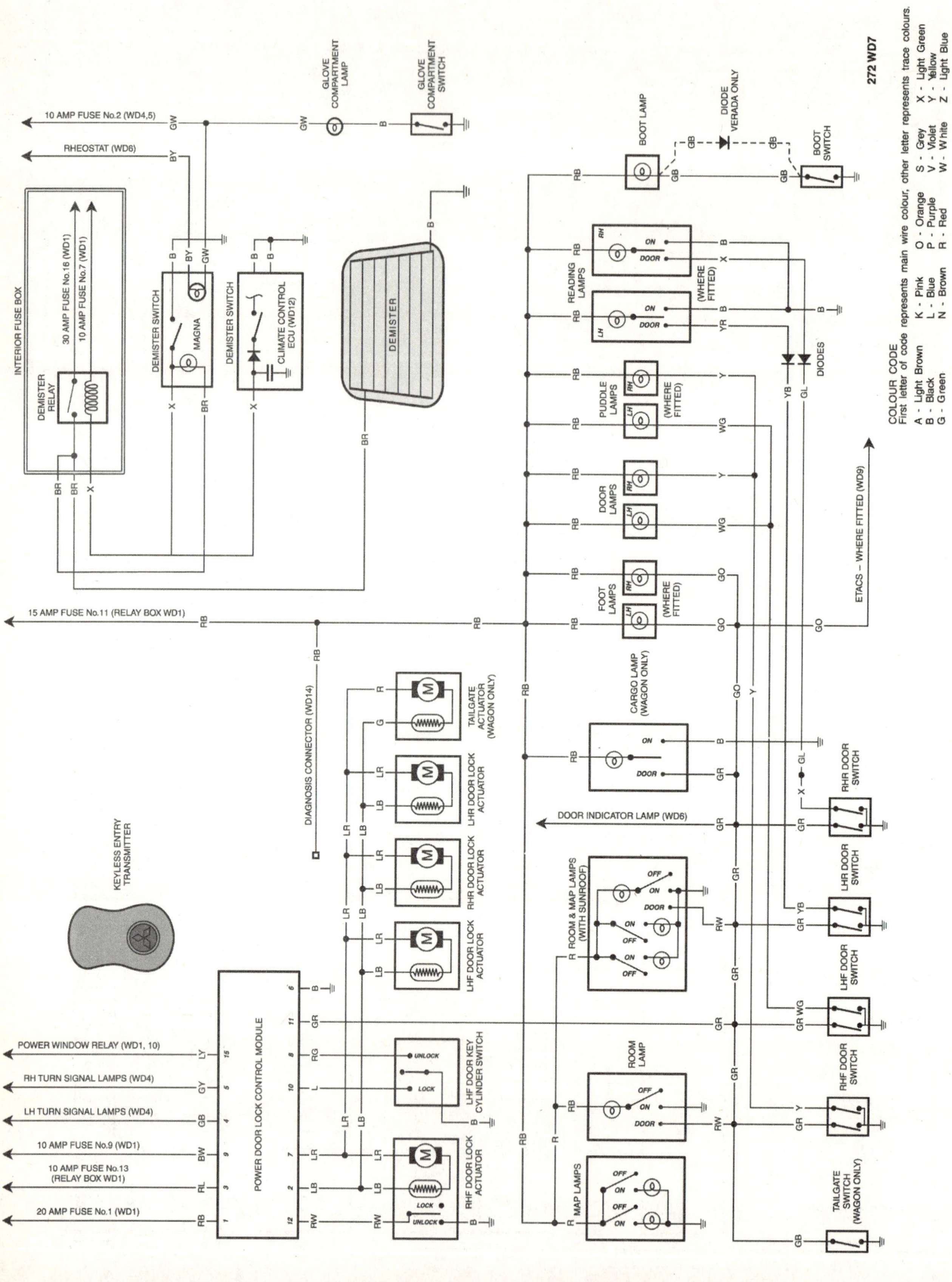

Power door locks except Verada Xi. Interior illumination and rear demister. TE/KE models.

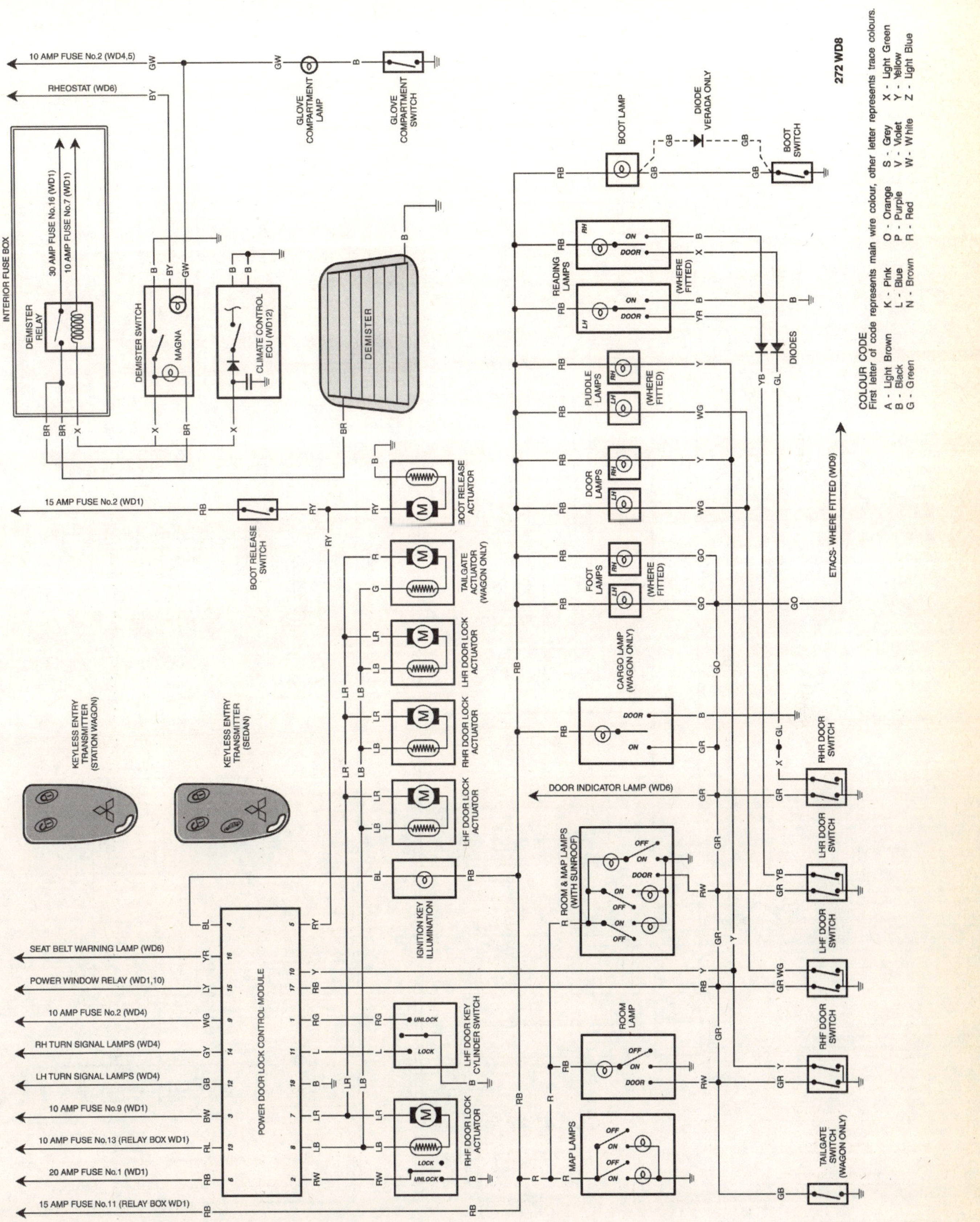

Power door locks except Verada Xi. Interior illumination and rear demister. TF/TH & KF/KH models.

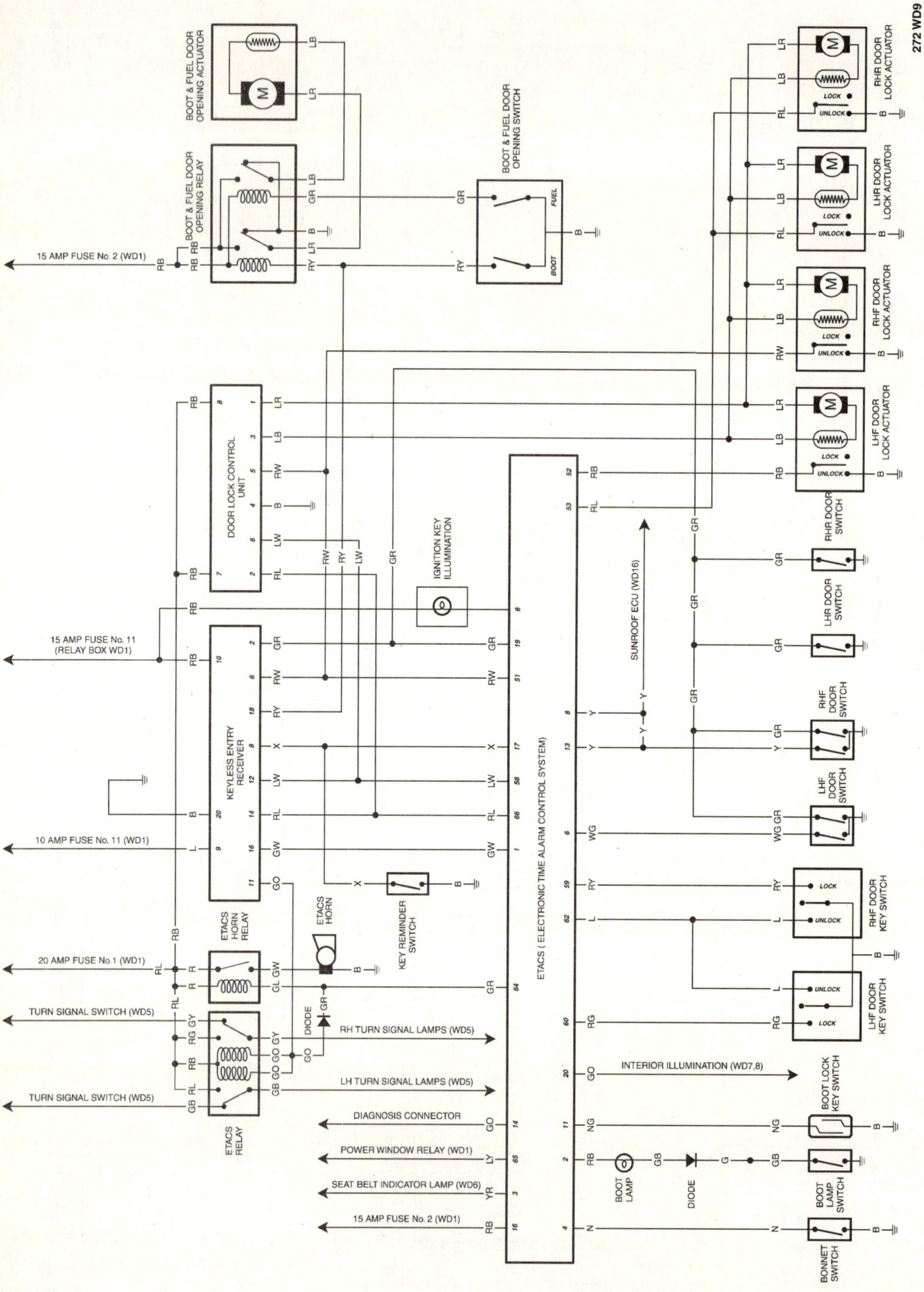

Keyless entry and alarm system – Verada Xi. KE/KF/KH models.

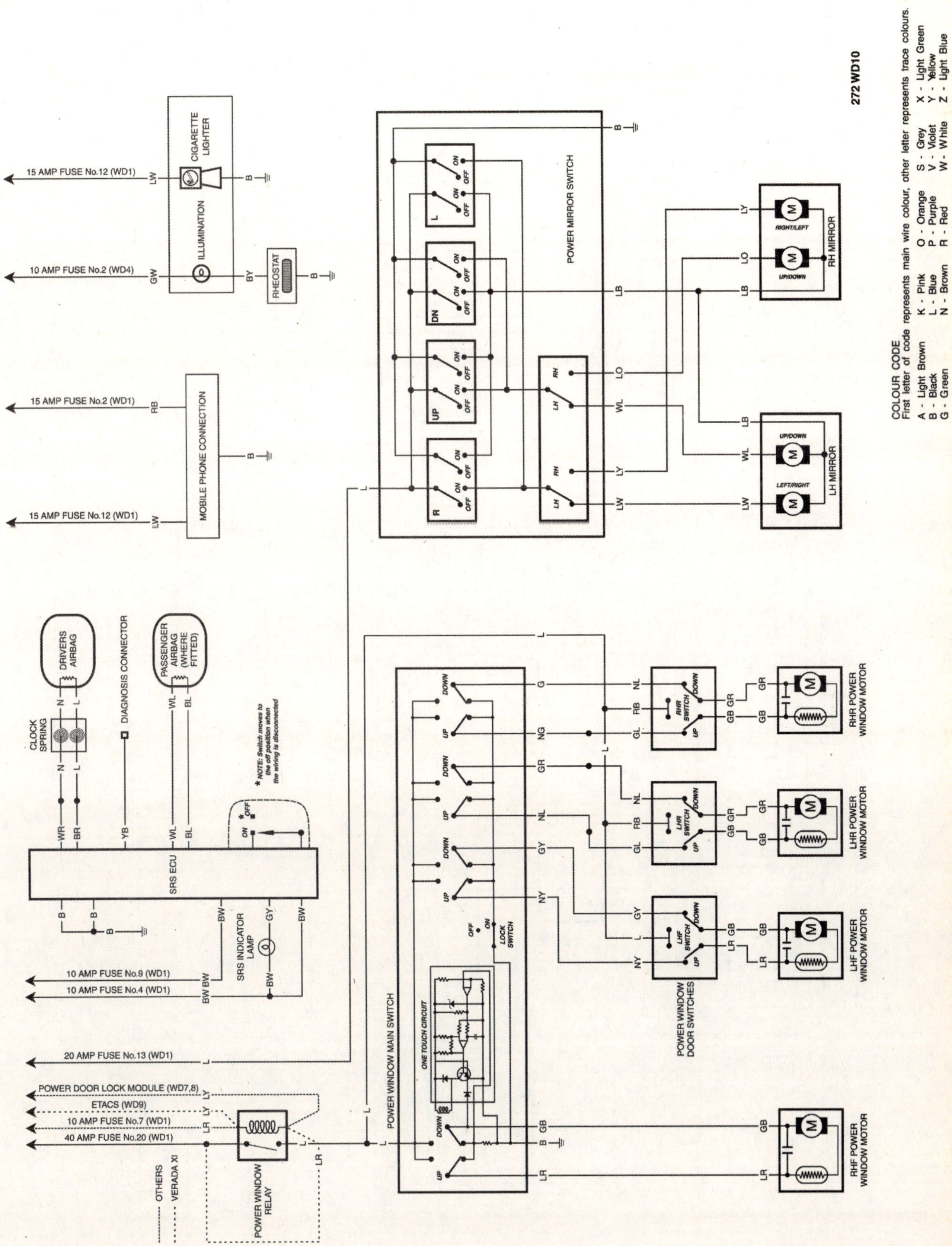

Power windows, power mirrors, SRS system, cigarette lighter and mobile phone connection. TE/TF/TH & KE/KF/KH models.

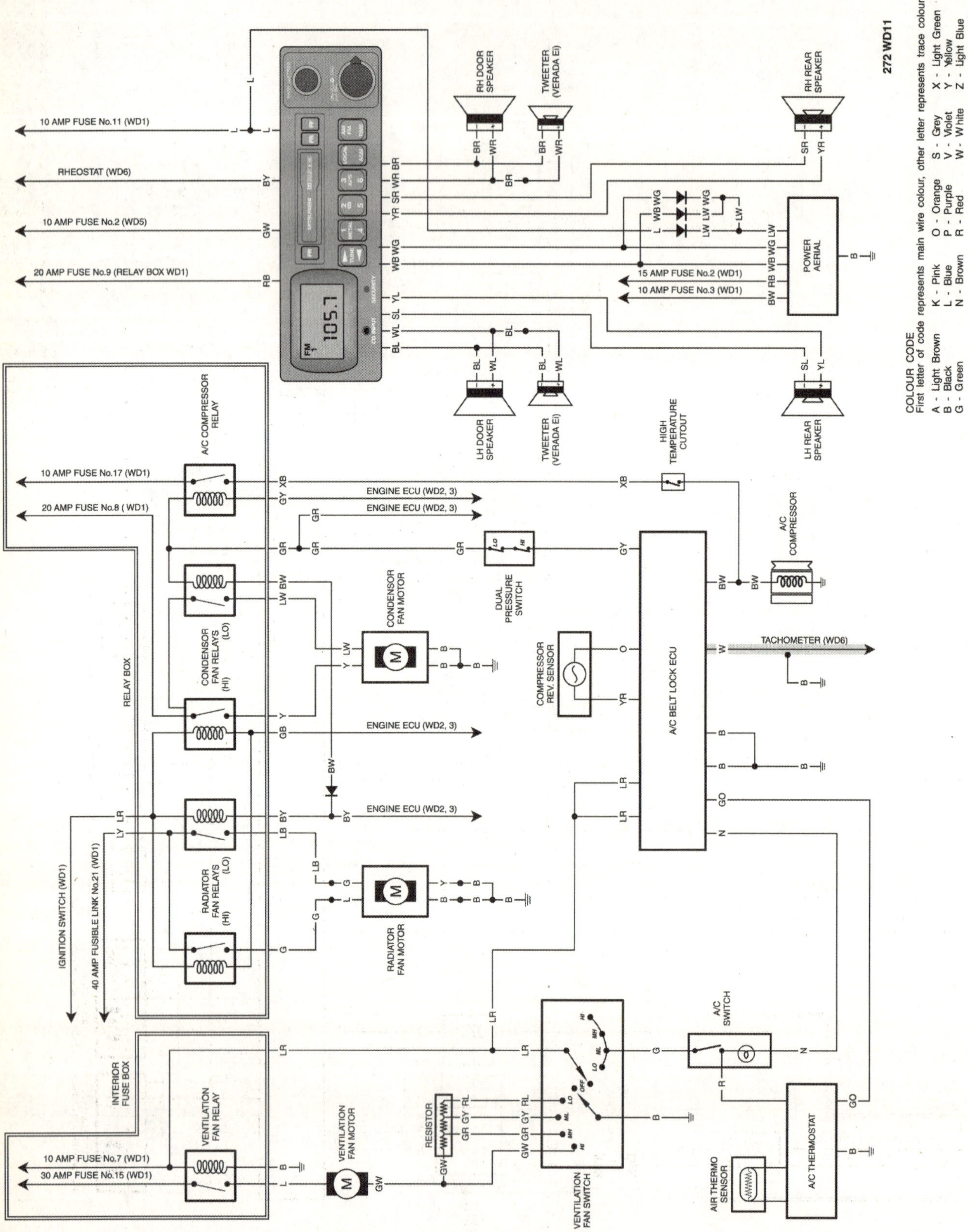

Cooling fans, heater/ventilation – Magna. Audio system – except Verada Xi. TE/TF/TH & KE/KF/KH models.

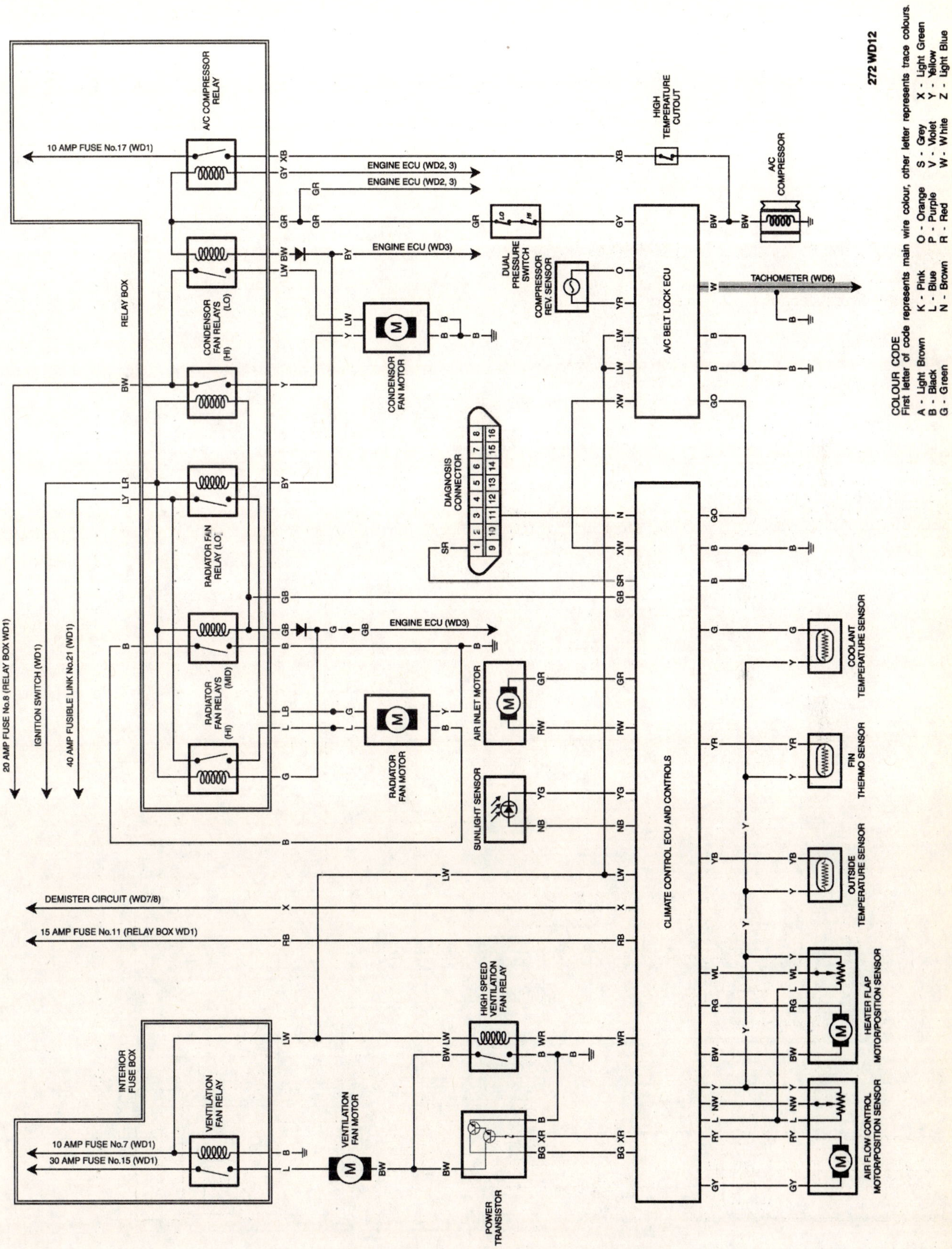

Cooling fans and climate control – Verada. KE/KF/KH models.

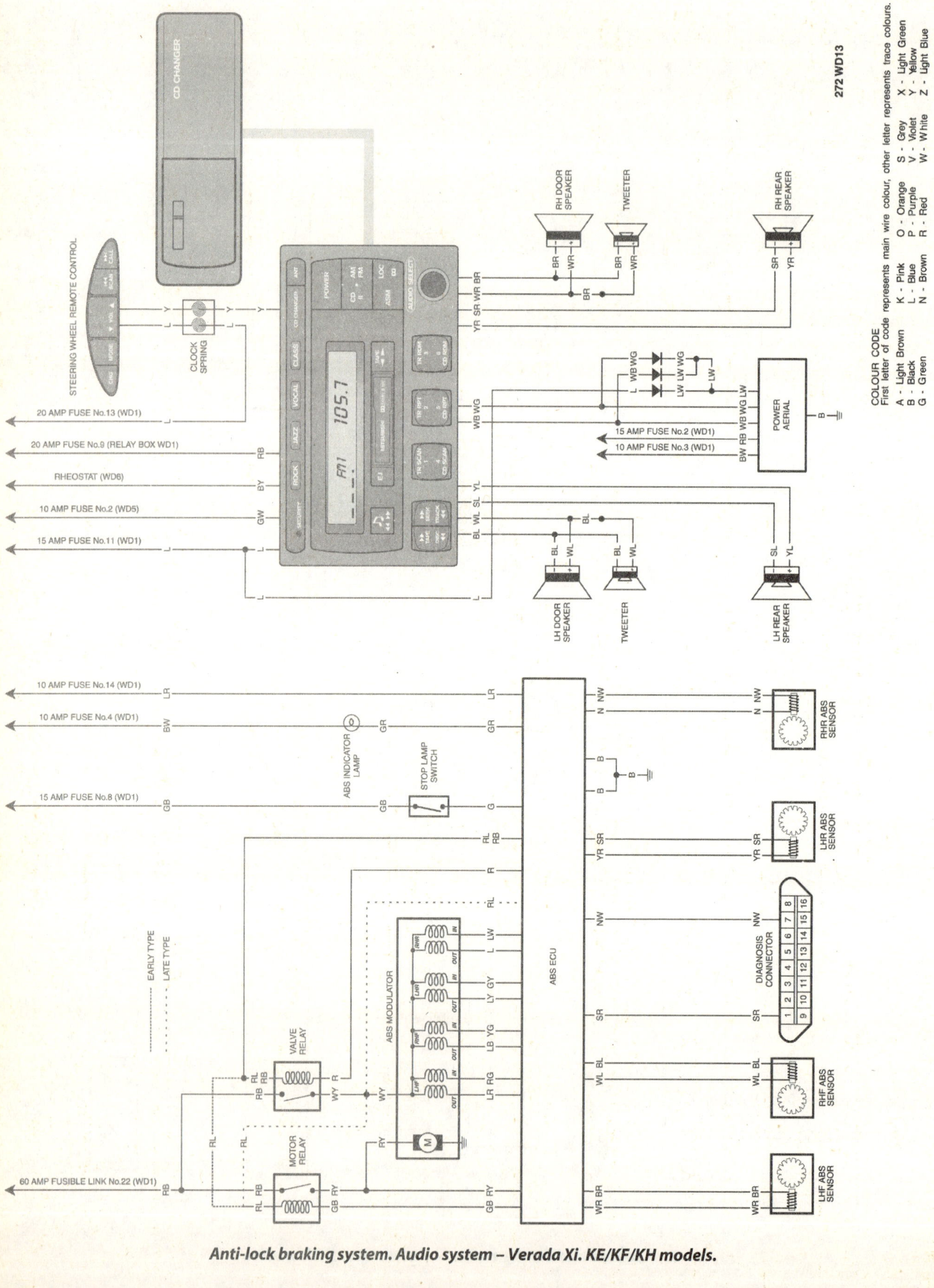

Anti-lock braking system. Audio system – Verada Xi. KE/KF/KH models.

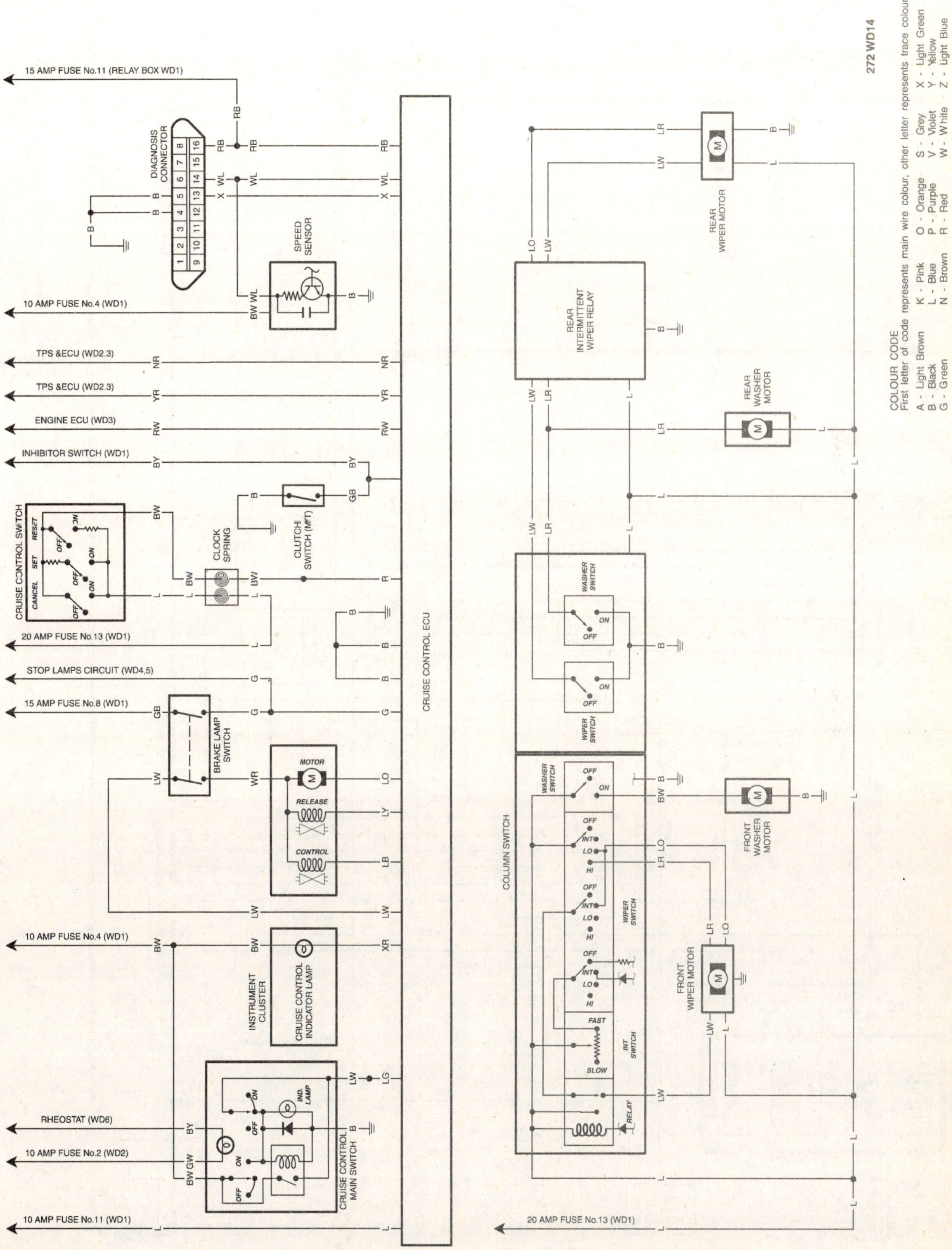

Cruise control and front/rear wiper/washers. TE/TF/TH & KE/KF/KH models.

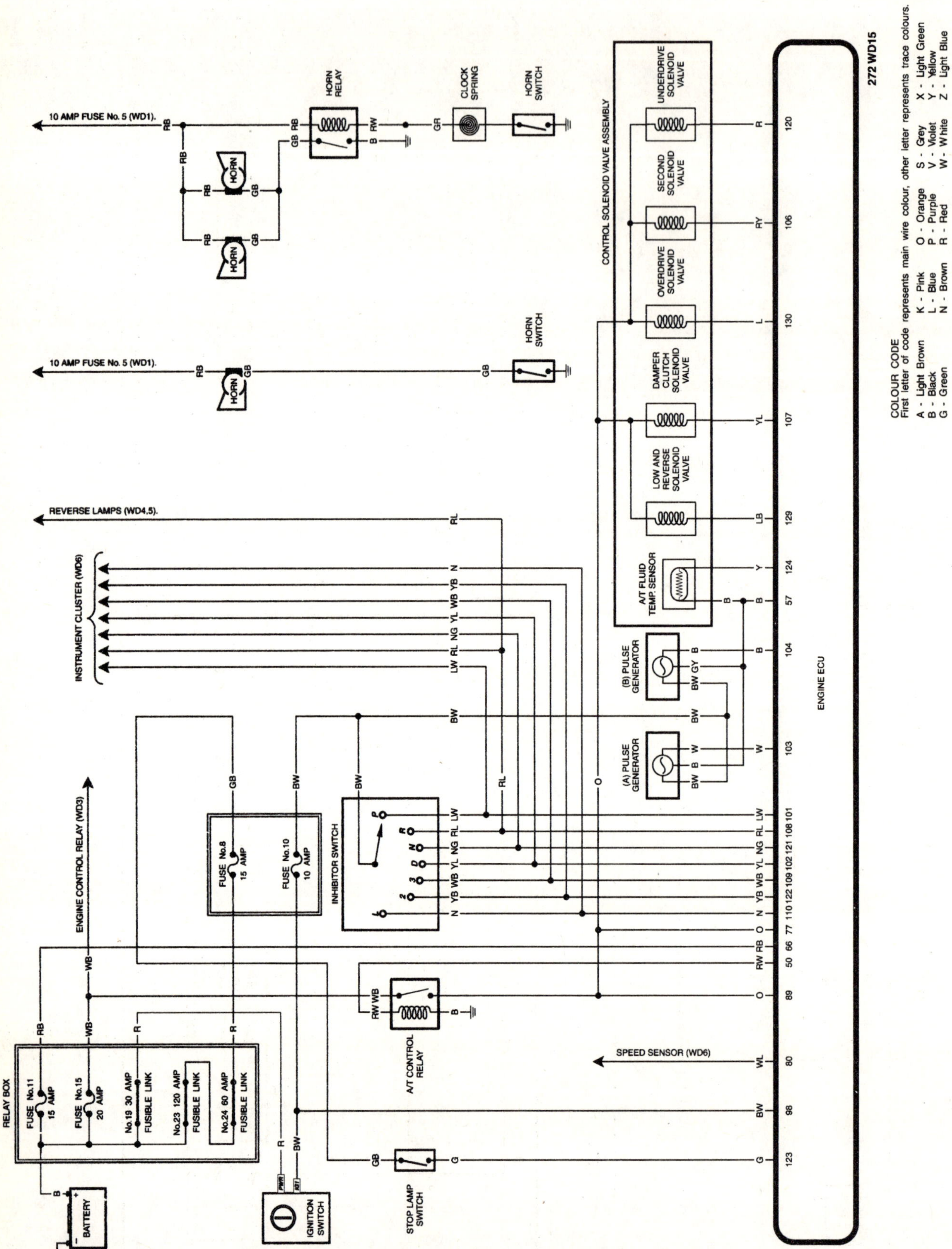

Automatic transaxle and horn circuit. TE/TF/TH & KE/KF/KH models.

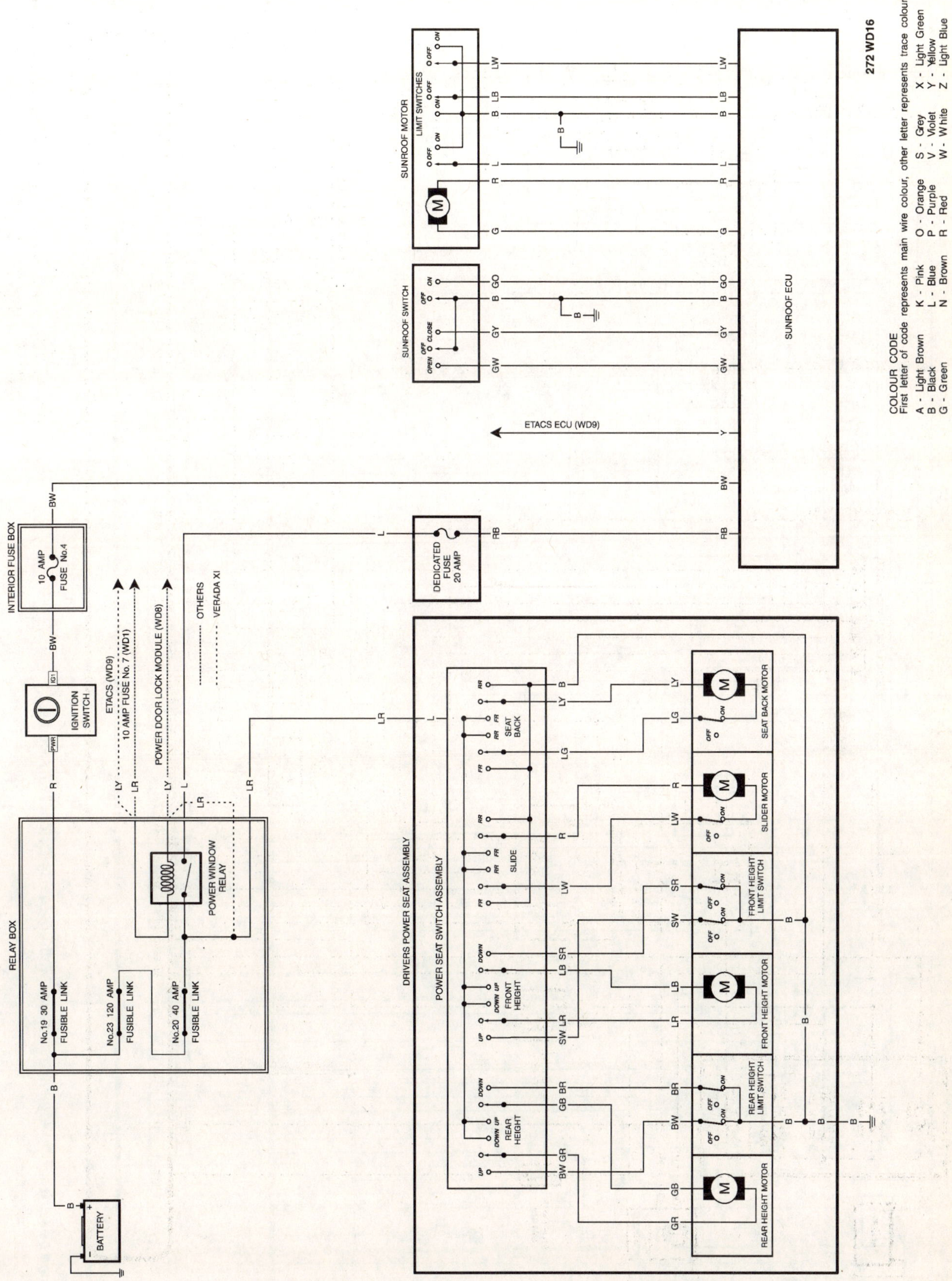

Drivers power seat and sunroof – Verada Xi. KE/KF/KH models.

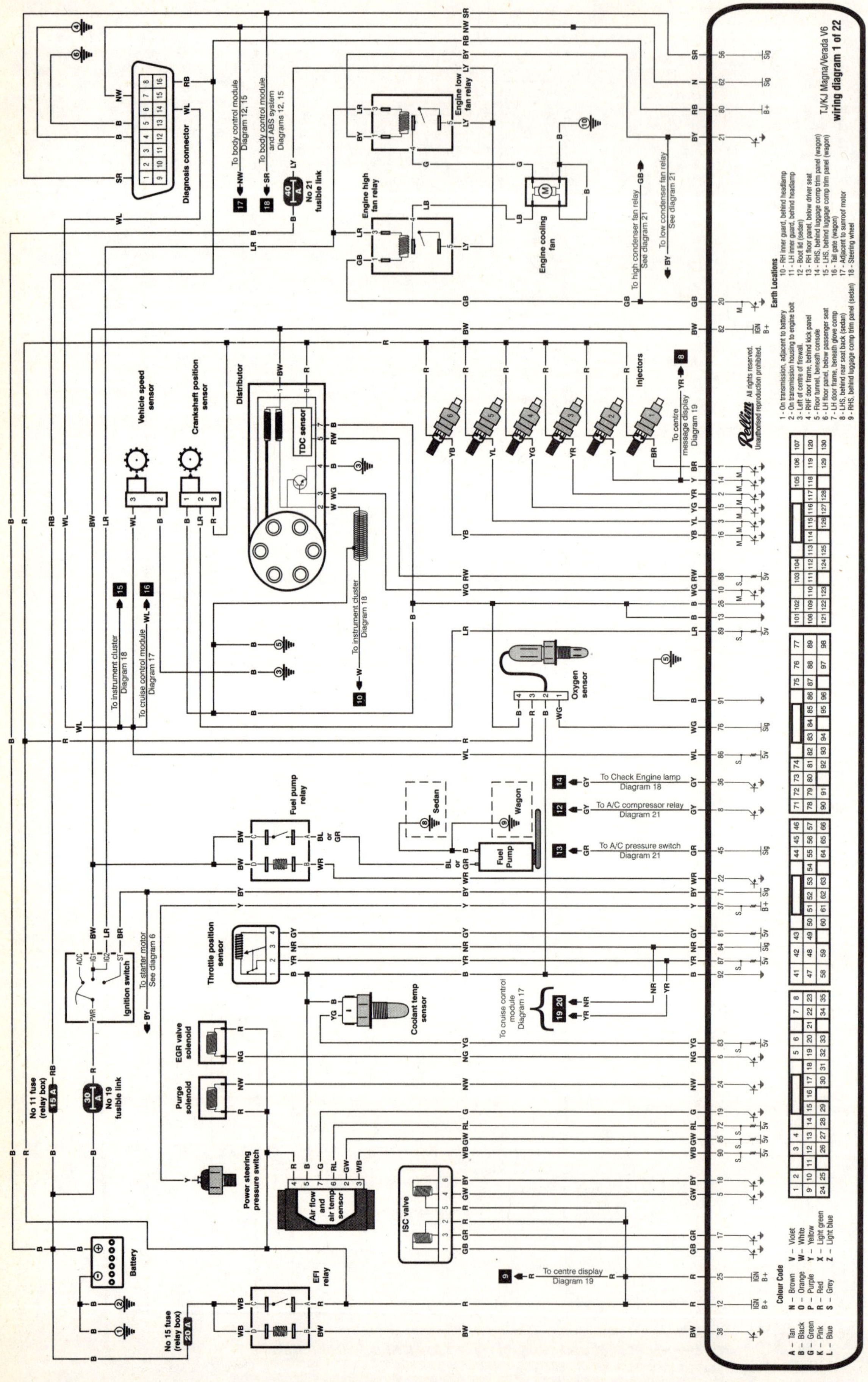

EFI and powertrain management systems. Manual transmission. TJ/KJ models.

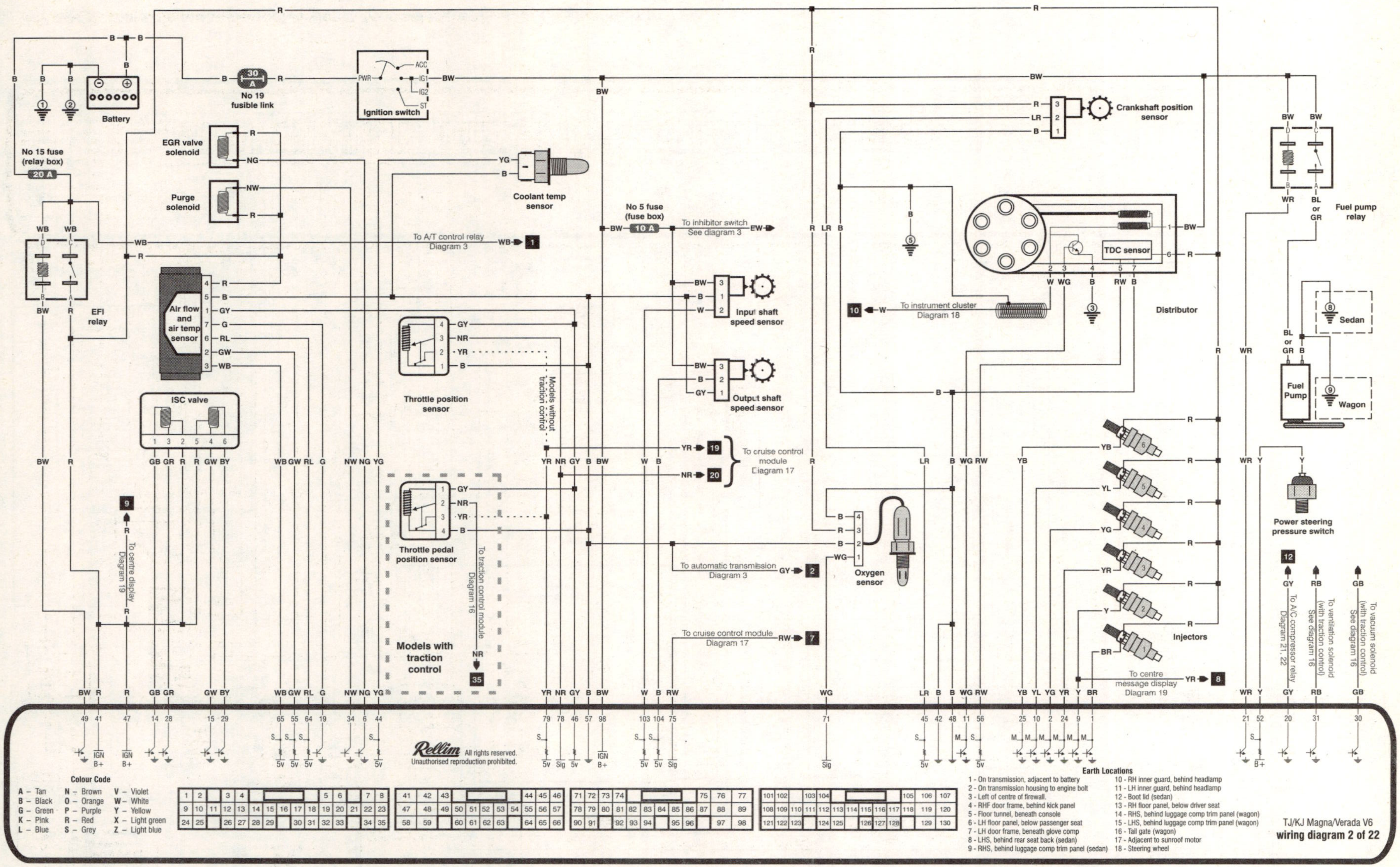

EFI and powertrain management systems. Standard automatic transmission. TJ/KJ models.

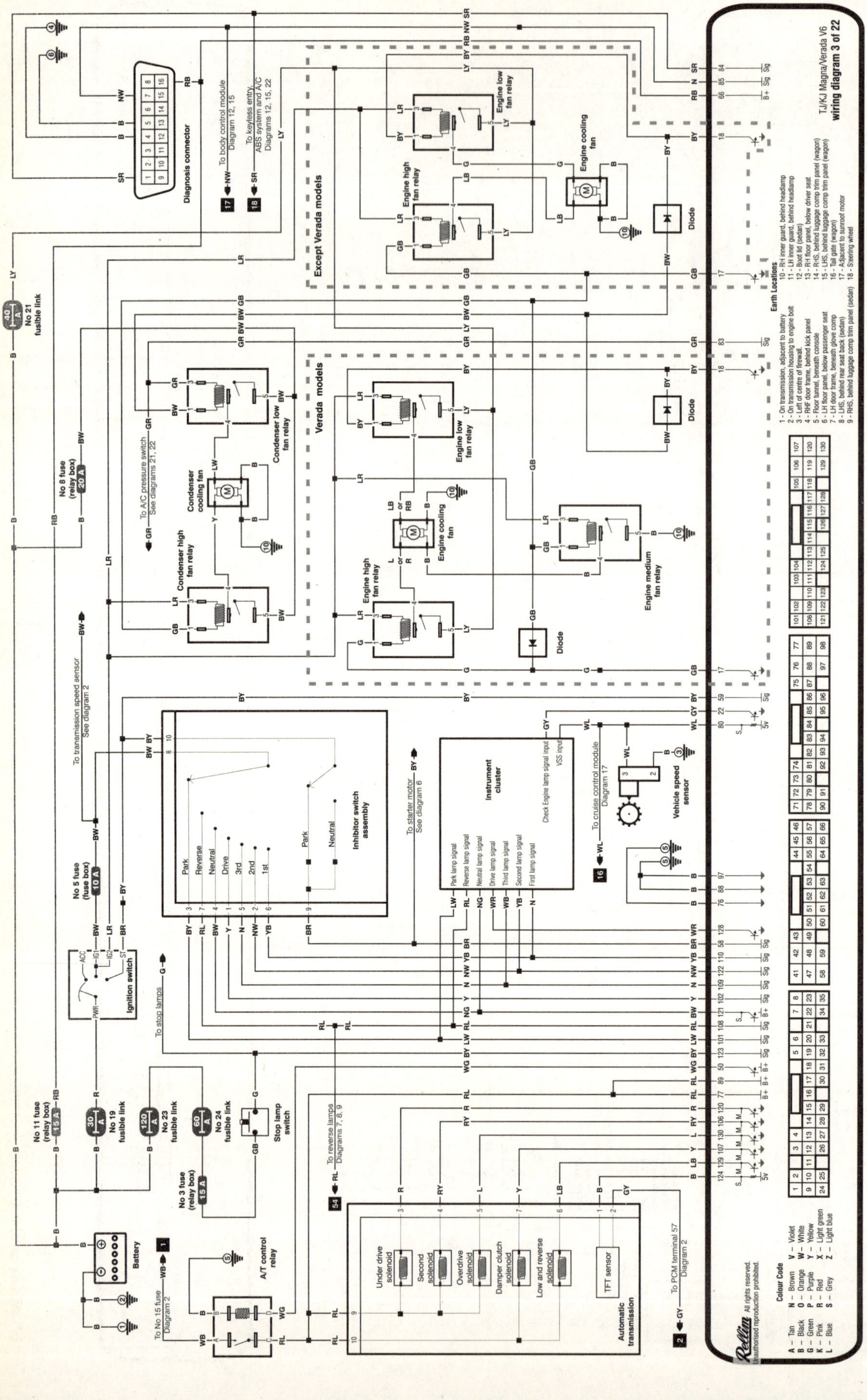

EFI and powertrain management systems. Standard automatic transmission. TJ/KJ models.

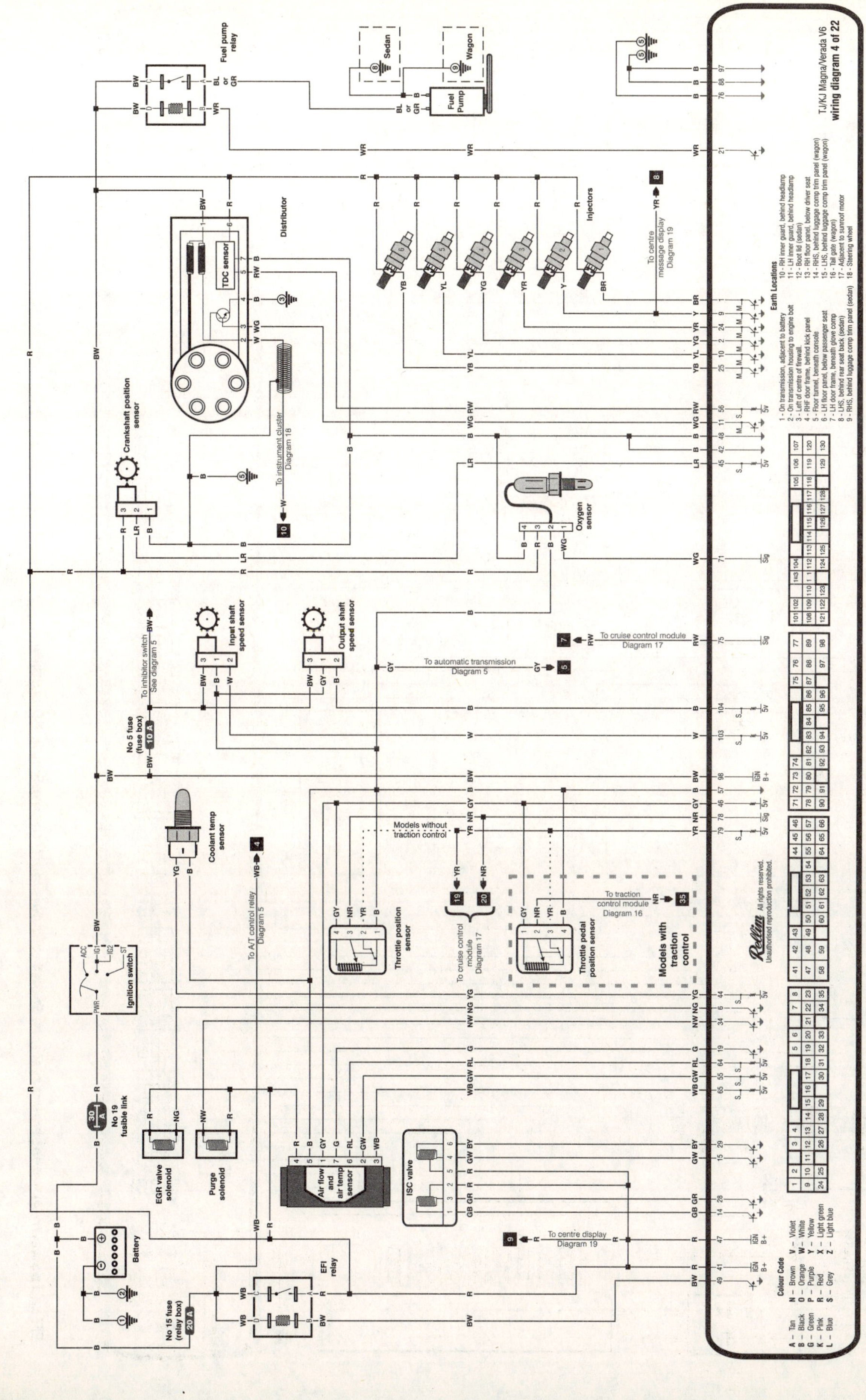

EFI and powertrain management systems. Sports shift automatic transmission. TJ/KJ models.

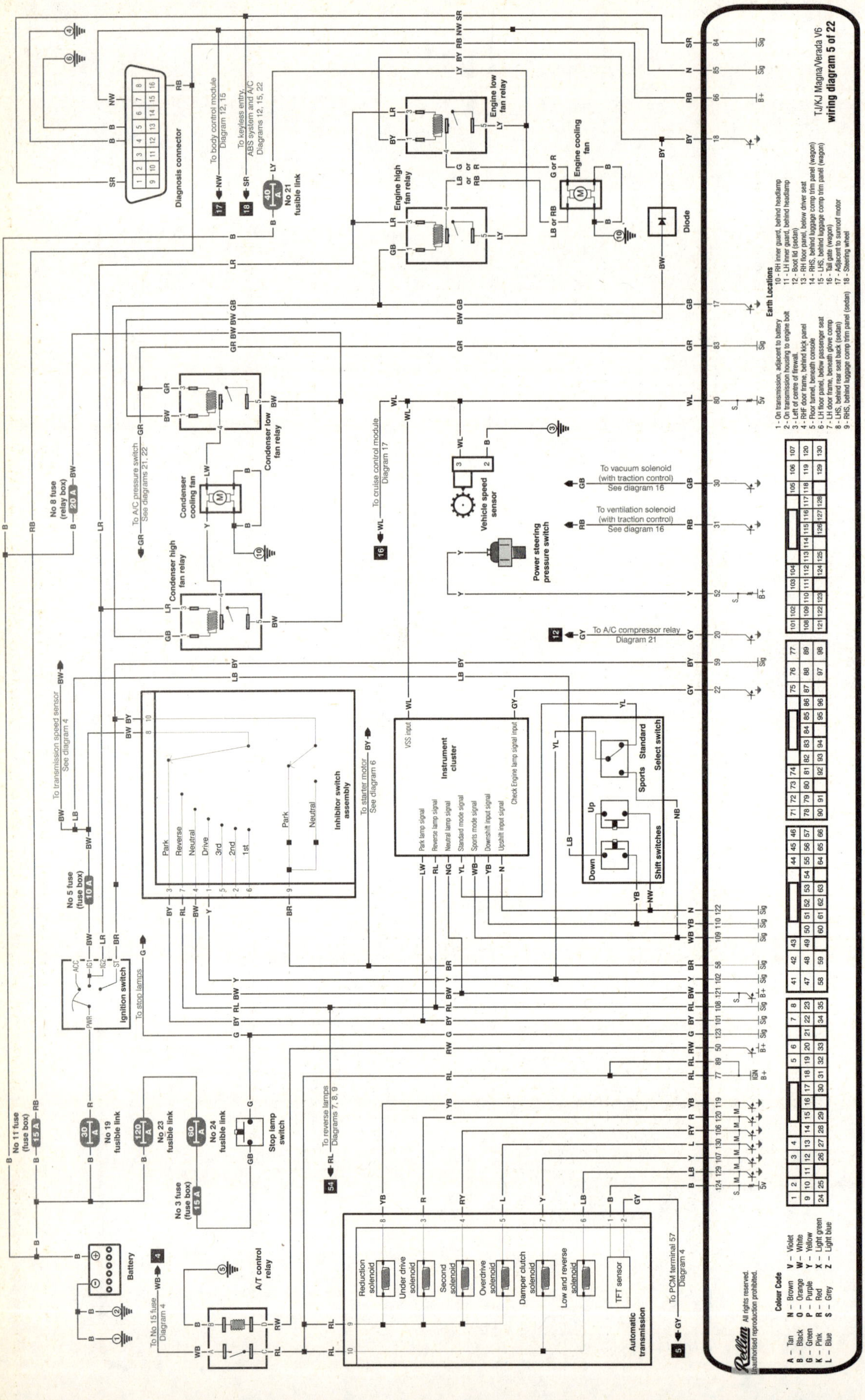

EFI and powertrain management systems. Sports shift automatic transmission. TJ/KJ models.

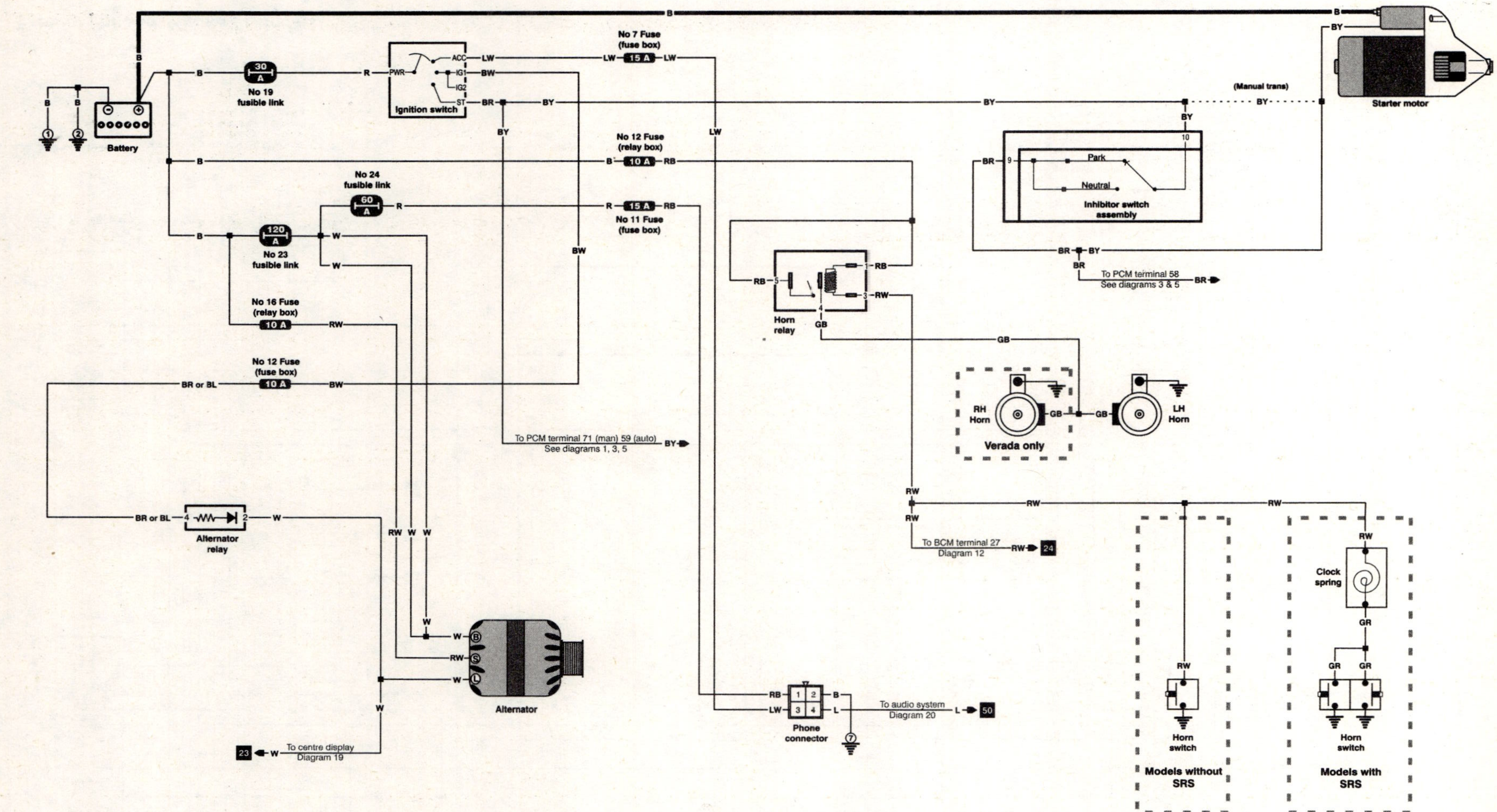

Colour Code

A – Tan	N – Brown	V – Violet
B – Black	O – Orange	W – White
G – Green	P – Purple	Y – Yellow
K – Pink	R – Red	X – Light green
L – Blue	S – Grey	Z – Light blue

Earth Locations

1 - On transmission, adjacent to battery
2 - On transmission housing to engine bolt
3 - Left of centre of firewall.
4 - RHF door frame, behind kick panel
5 - Floor tunnel, beneath console
6 - LH floor panel, below passenger seat
7 - LH door frame, beneath glove comp
8 - LHS, behind rear seat back (sedan)
9 - RHS, behind luggage comp trim panel (sedan)
10 - RH inner guard, behind headlamp
11 - LH inner guard, behind headlamp
12 - Boot lid (sedan)
13 - RH floor panel, below driver seat
14 - RHS, behind luggage comp trim panel (wagon)
15 - LHS, behind luggage comp trim panel (wagon)
16 - Tail gate (wagon)
17 - Adjacent to sunroof motor
18 - Steering wheel

TJ/KJ Magna/Verada V6
wiring diagram 6 of 22

Starting system, charging system and horn. TJ/KJ models.

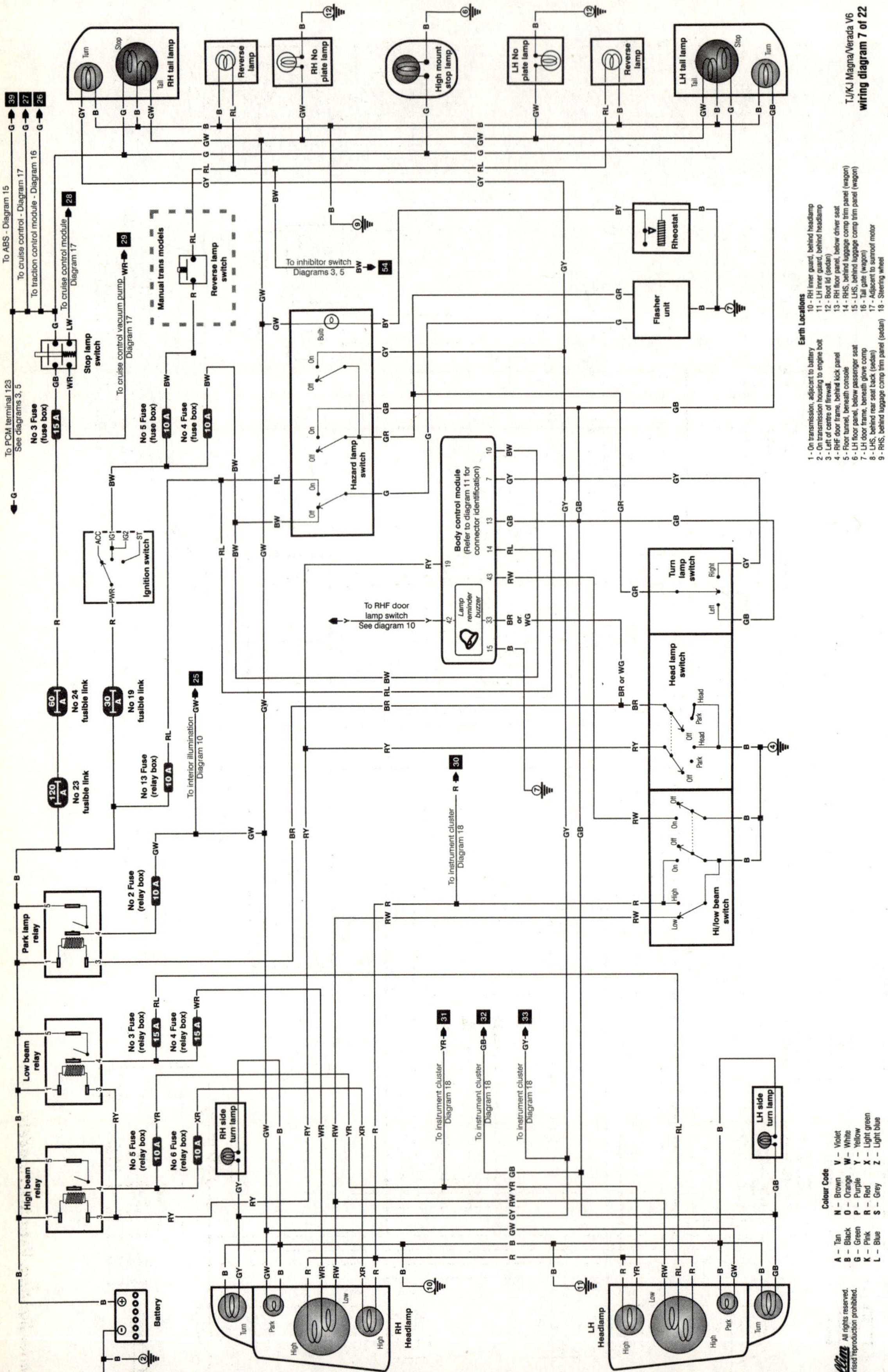

Exterior lighting. Magna TJ models.

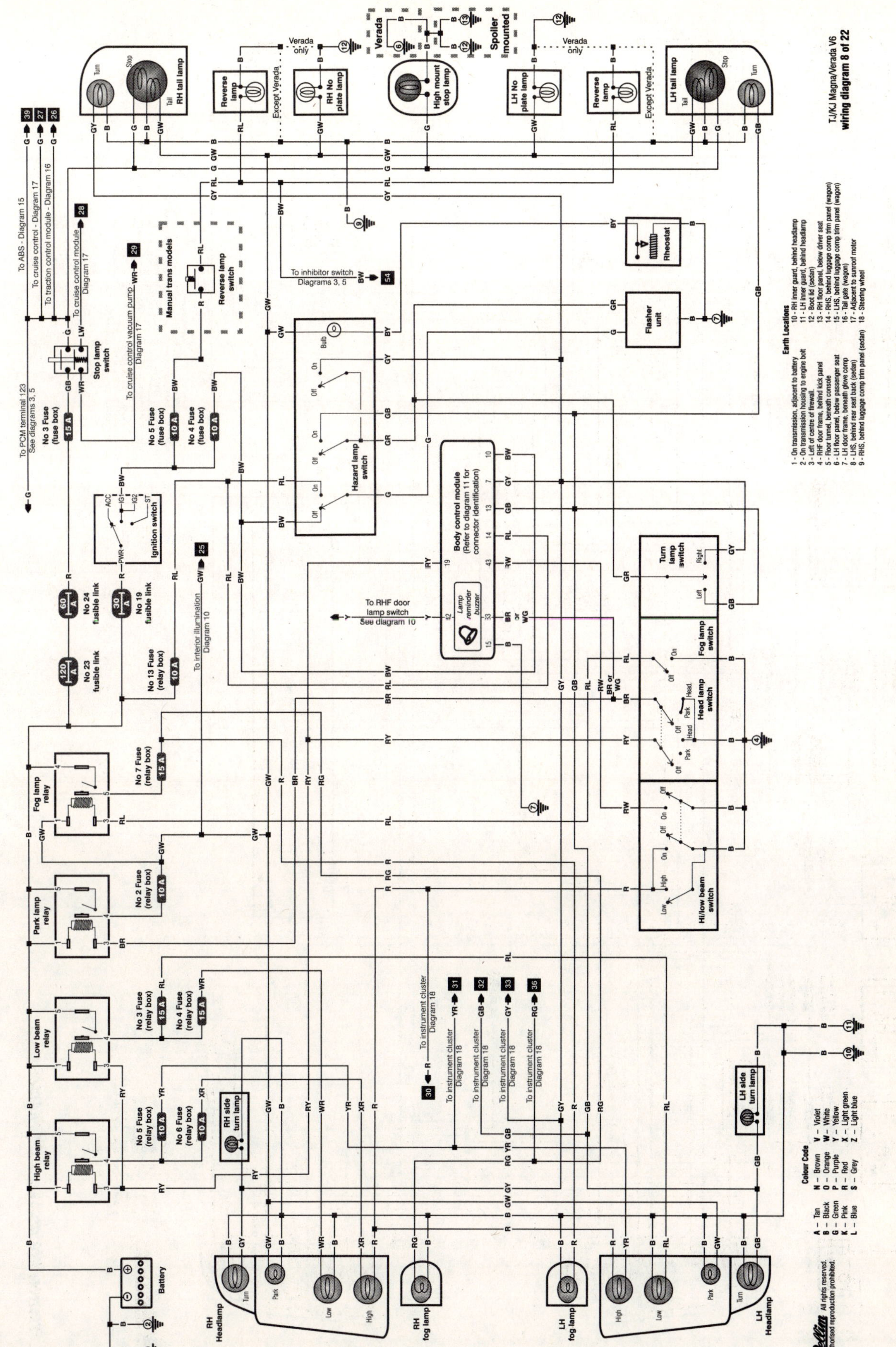

Exterior lighting. Verada KJ models.

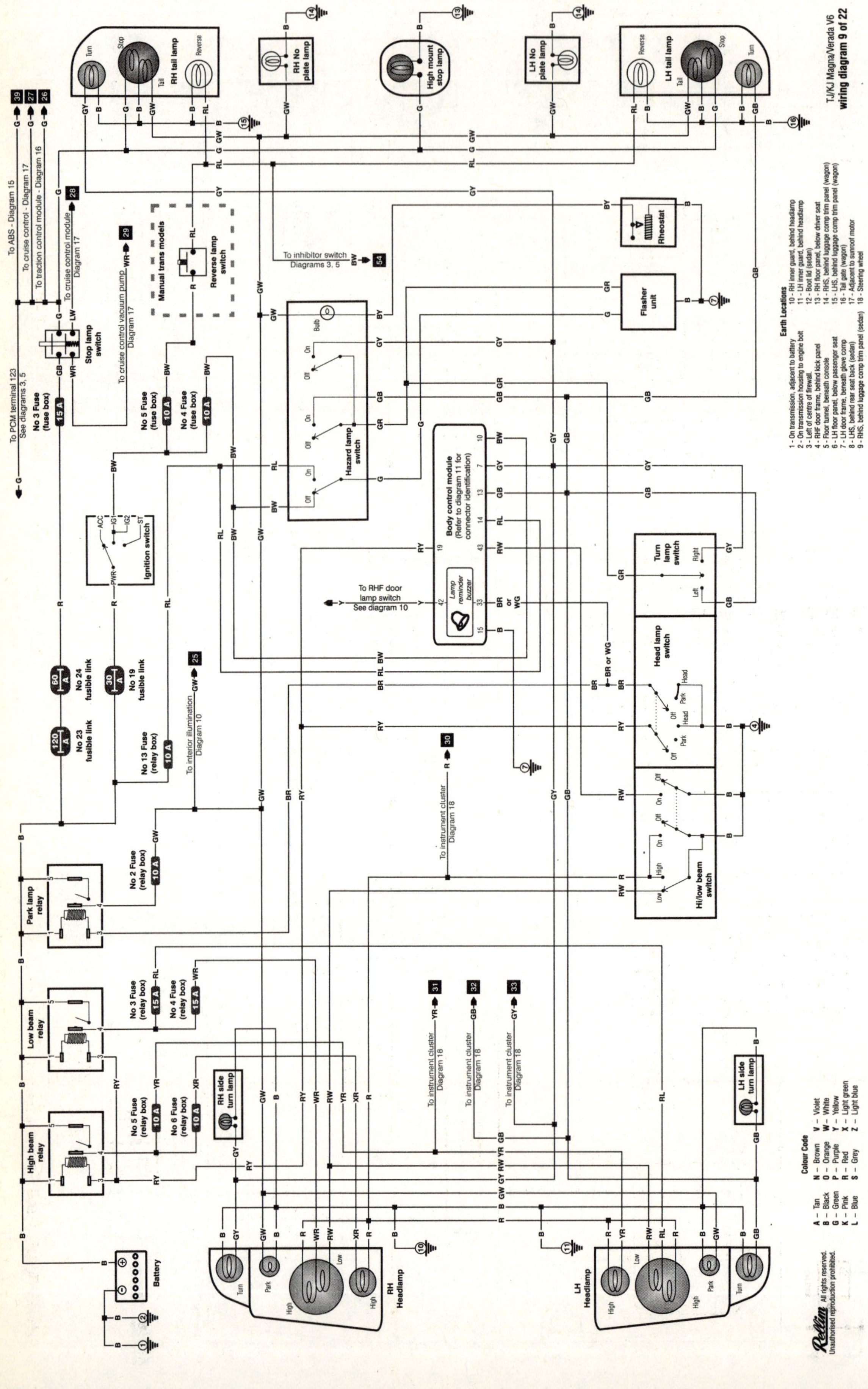

Exterior lighting. Wagon. TJ/KJ models.

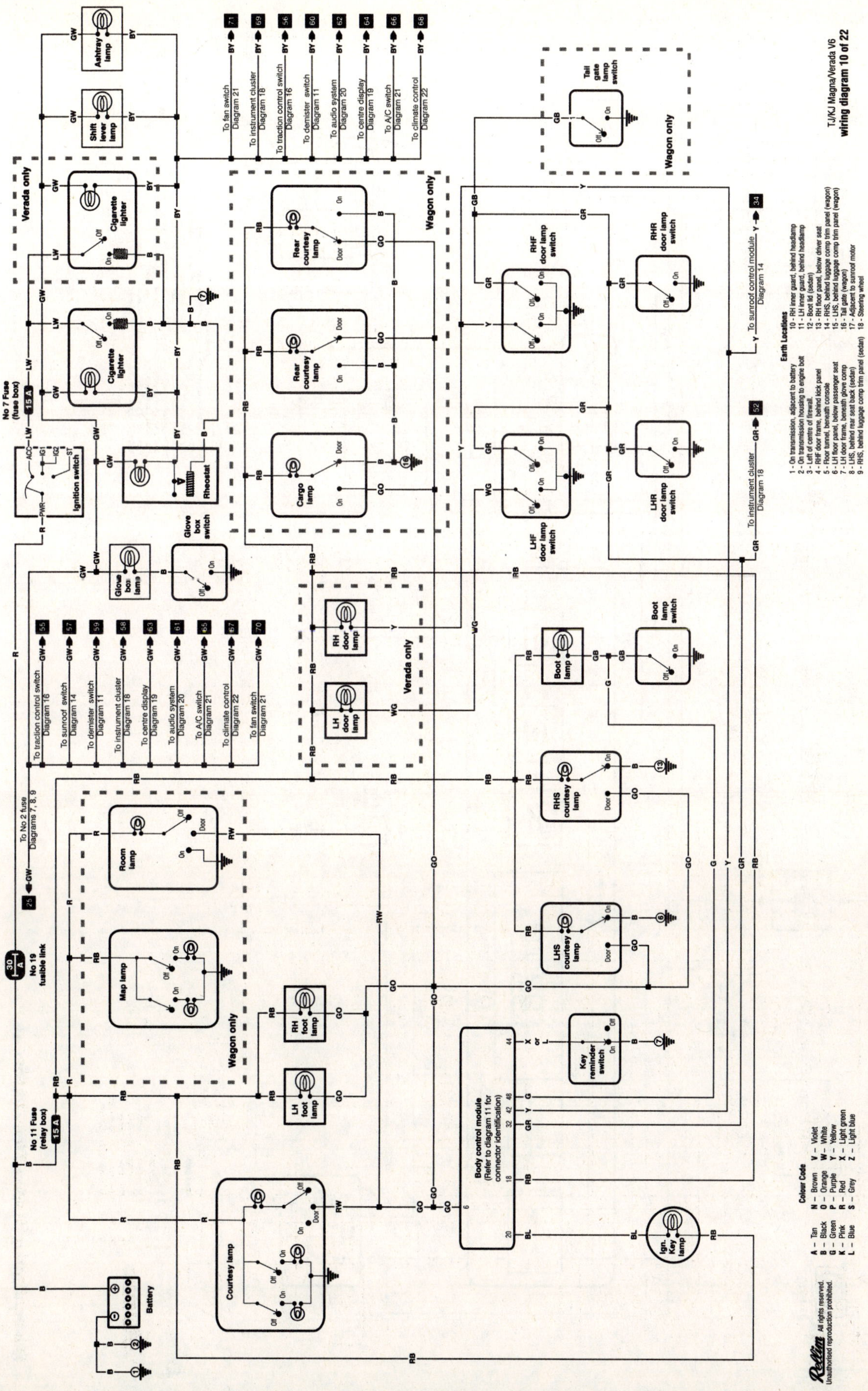

Interior lighting and cigarette lighter. TJ/KJ models.

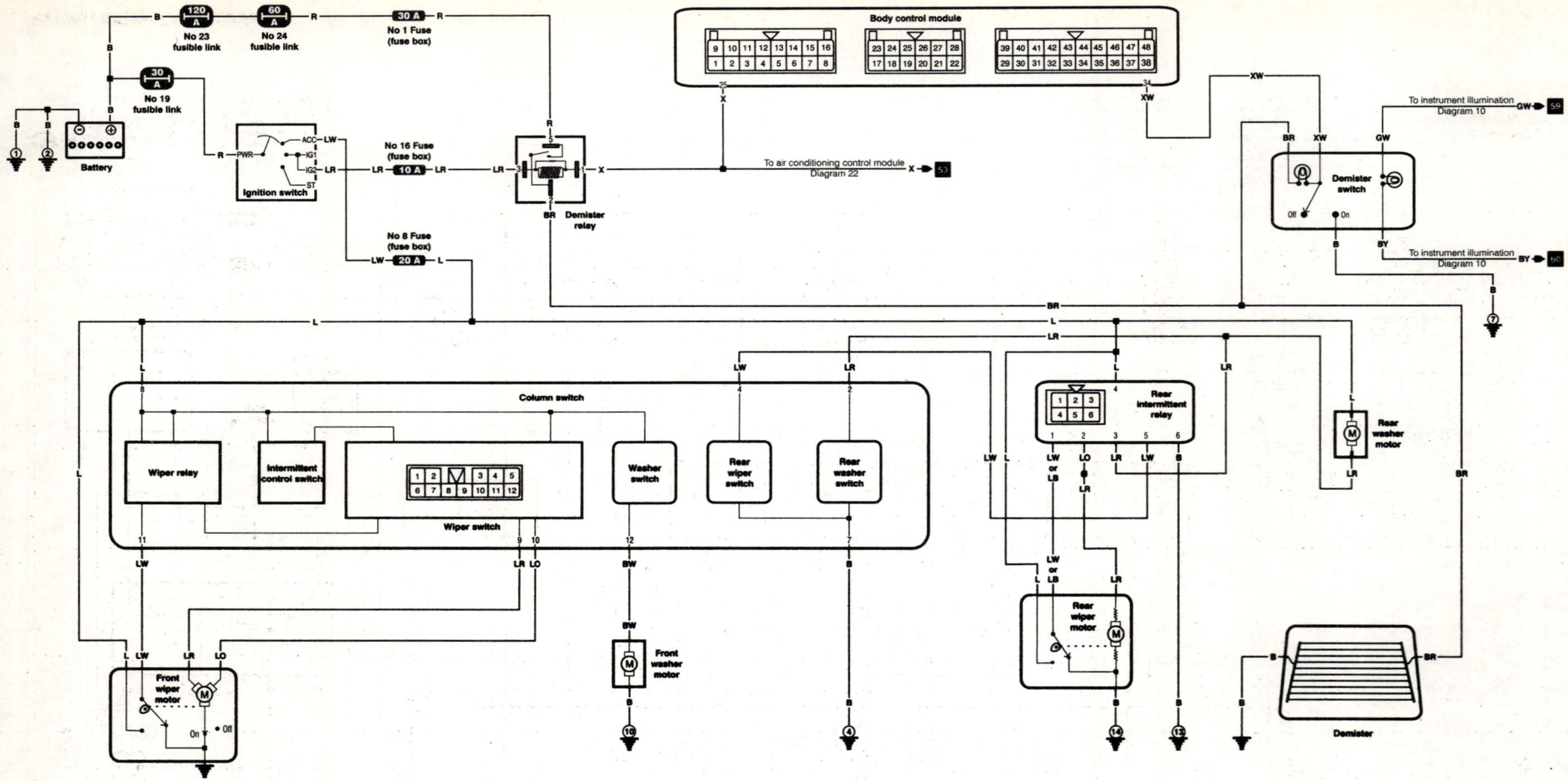

Colour Code

A – Tan N – Brown V – Violet
B – Black O – Orange W – White
G – Green P – Purple Y – Yellow
K – Pink R – Red X – Light green
L – Blue S – Grey Z – Light blue

Earth Locations

1 - On transmission, adjacent to battery
2 - On transmission housing to engine bolt
3 - Left of centre of firewall
4 - RHF door frame, behind kick panel
5 - Floor tunnel, beneath console
6 - LH floor panel, below passenger seat
7 - LH door frame, beneath glove comp
8 - LHS, behind rear seat back (sedan)
9 - RHS, behind luggage comp trim panel (sedan)
10 - RH inner guard, behind headlamp
11 - LH inner guard, behind headlamp
12 - Boot lid (sedan)
13 - RH floor panel, below driver seat
14 - RHS, behind luggage comp trim panel (wagon)
15 - LHS, behind luggage comp trim panel (wagon)
16 - Tail gate (wagon)
17 - Adjacent to sunroof motor
18 - Steering wheel

TJ/KJ Magna/Verada V6
wiring diagram 11 of 22

Wipers, washers and demister. TJ/KJ models.

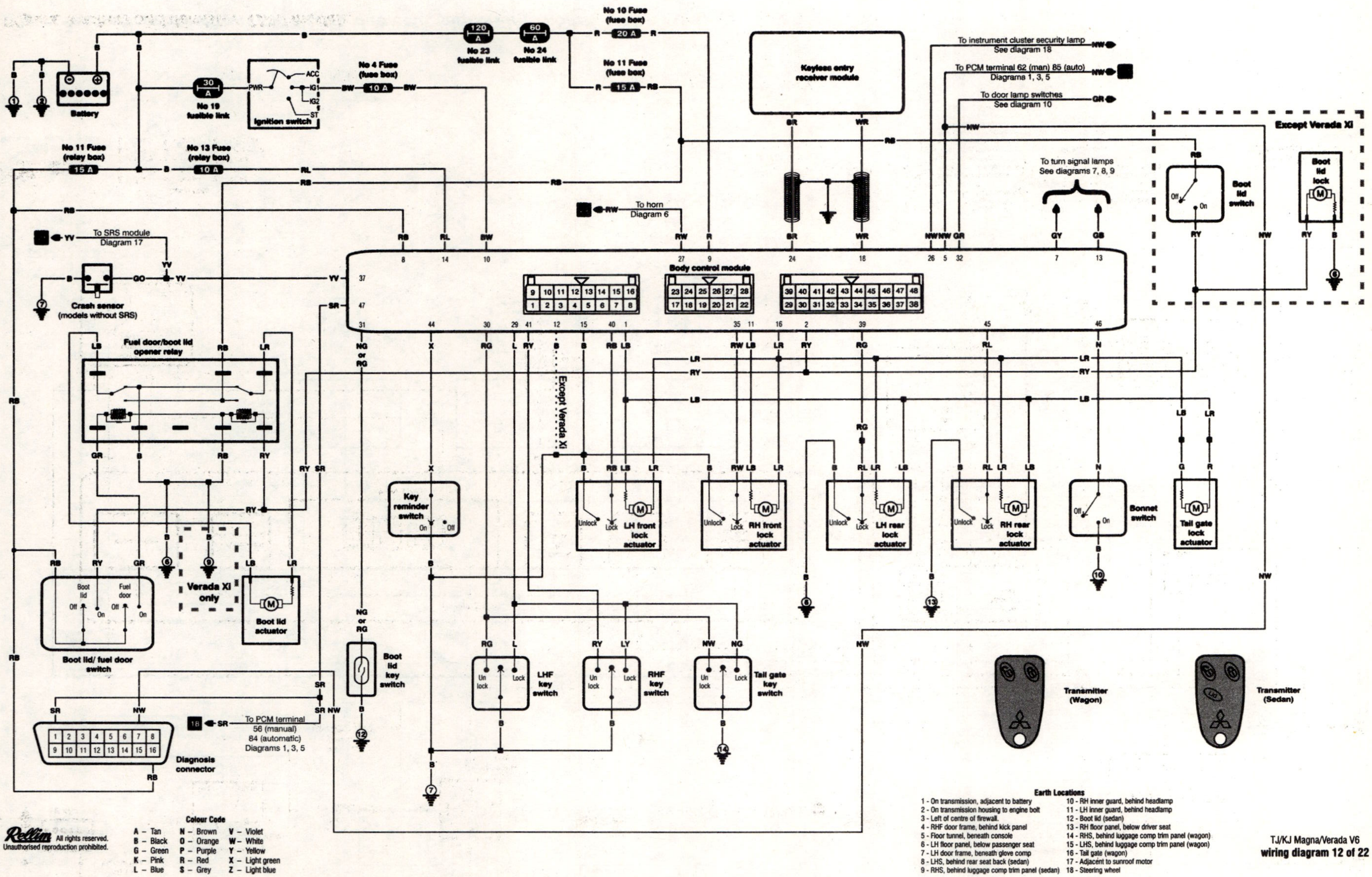

Keyless entry. TJ/KJ models.

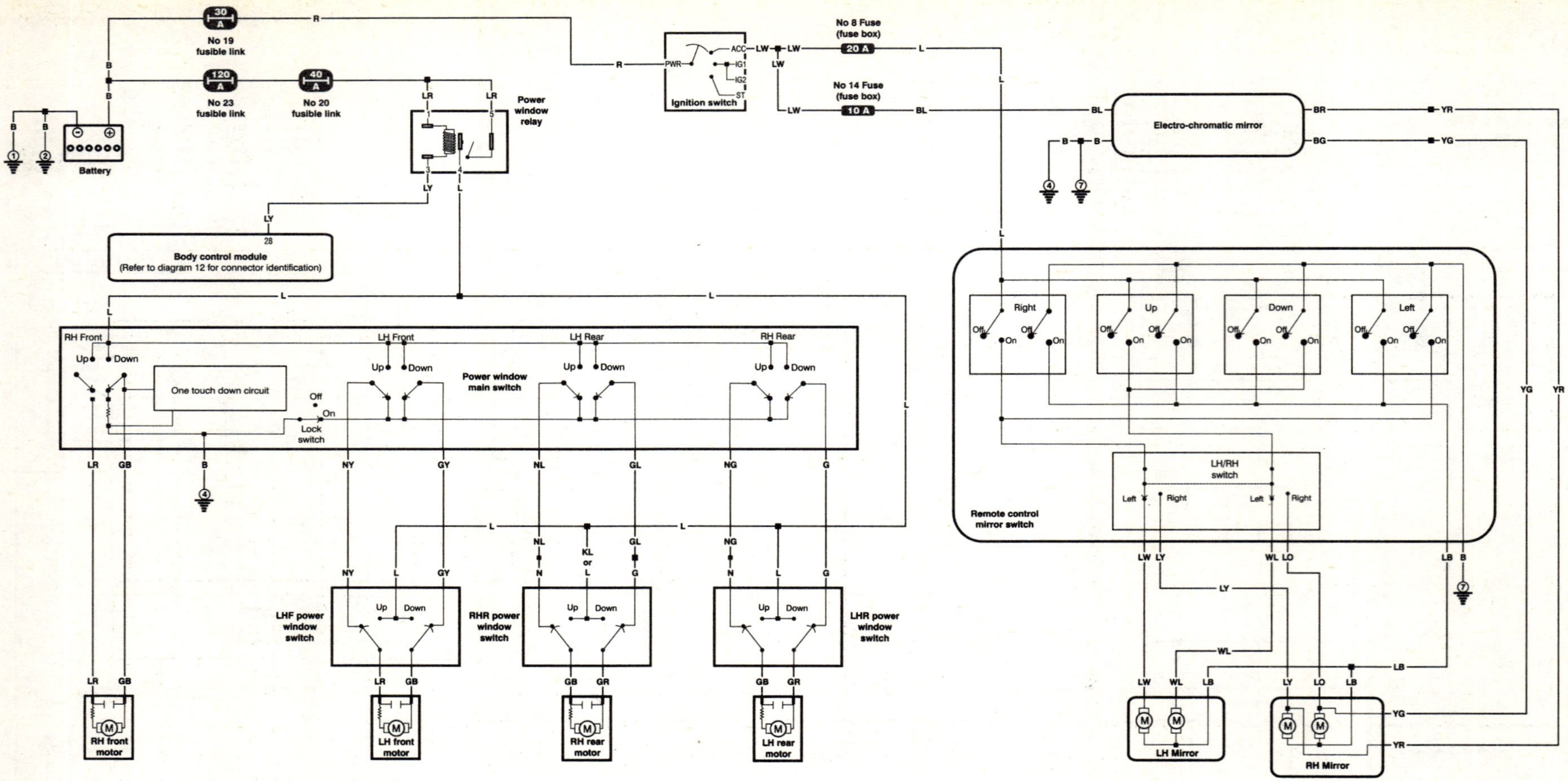

Colour Code

A – Tan	N – Brown	V – Violet
B – Black	O – Orange	W – White
G – Green	P – Purple	Y – Yellow
K – Pink	R – Red	X – Light green
L – Blue	S – Grey	Z – Light blue

Earth Locations

1 - On transmission, adjacent to battery
2 - On transmission housing to engine bolt
3 - Left of centre of firewall.
4 - RHF door frame, behind kick panel
5 - Floor tunnel, beneath console
6 - LH floor panel, below passenger seat
7 - LH door frame, beneath glove comp
8 - LHS, behind rear seat back (sedan)
9 - RHS, behind luggage comp trim panel (sedan)
10 - RH inner guard, behind headlamp
11 - LH inner guard, behind headlamp
12 - Boot lid (sedan)
13 - RH floor panel, below driver seat
14 - RHS, behind luggage comp trim panel (wagon)
15 - LHS, behind luggage comp trim panel (wagon)
16 - Tail gate (wagon)
17 - Adjacent to sunroof motor
18 - Steering wheel

TJ/KJ Magna/Verada V6
wiring diagram 13 of 22

Power windows, power mirrors and electro-chromatic mirror. TJ/KJ models.

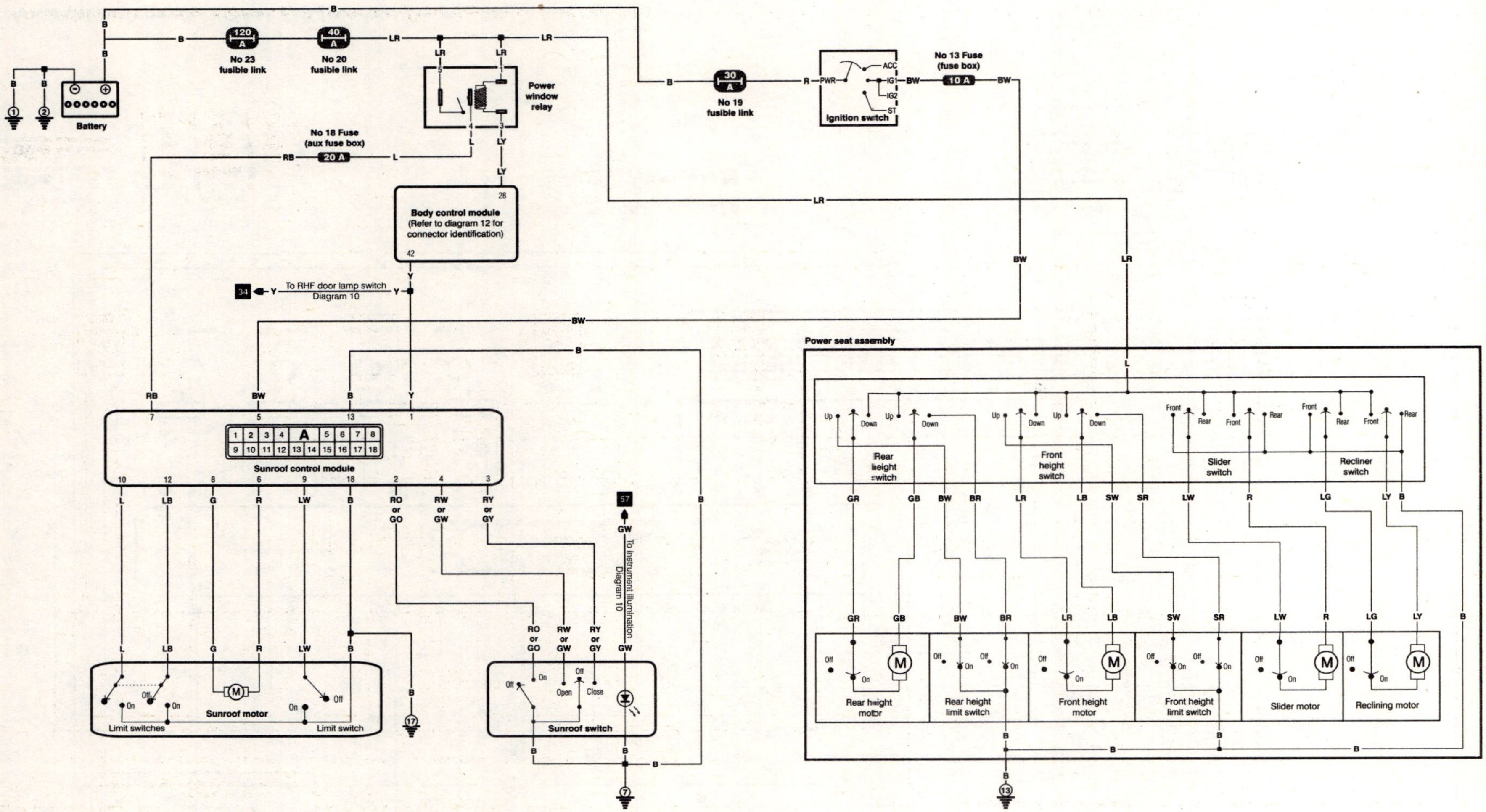

Colour Code

A – Tan
B – Black
G – Green
K – Pink
L – Blue
N – Brown
O – Orange
P – Purple
R – Red
S – Grey
V – Violet
W – White
Y – Yellow
X – Light green
Z – Light blue

Earth Locations

1 - On transmission, adjacent to battery
2 - On transmission housing to engine bolt
3 - Left of centre of firewall.
4 - RHF door frame, behind kick panel
5 - Floor tunnel, beneath console
6 - LH floor panel, below passenger seat
7 - LH door frame, beneath glove comp
8 - LHS, behind rear seat back (sedan)
9 - RHS, behind luggage comp trim panel (sedan)
10 - RH inner guard, behind headlamp
11 - LH inner guard, behind headlamp
12 - Boot lid (sedan)
13 - RH floor panel, below driver seat
14 - RHS, behind luggage comp trim panel (wagon)
15 - LHS, behind luggage comp trim panel (wagon)
16 - Tail gate (wagon)
17 - Adjacent to sunroof motor
18 - Steering wheel

TJ/KJ Magna/Verada V6
wiring diagram 14 of 22

Power seats and sunroof. TJ/KJ models.

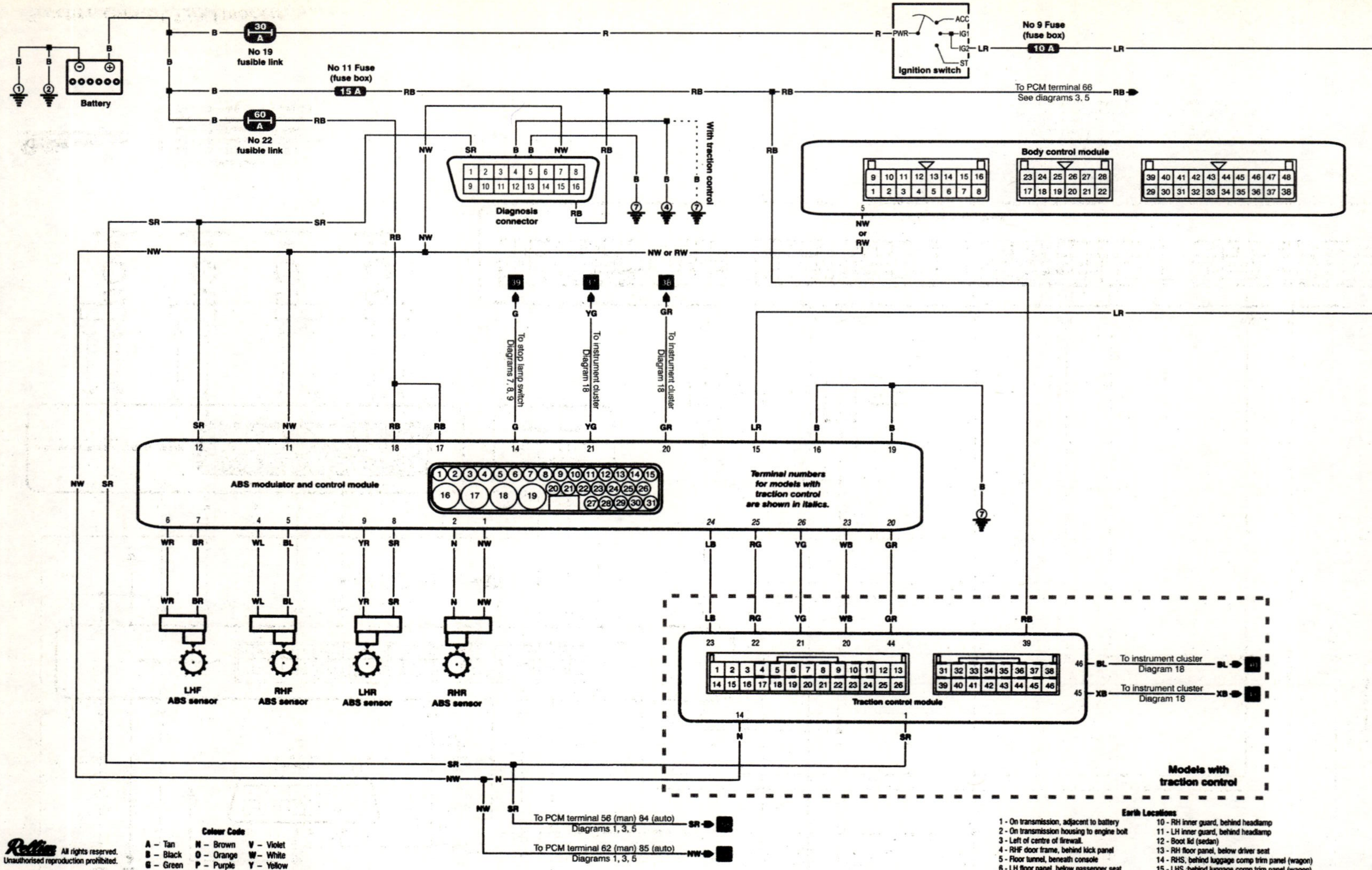

Colour Code

A – Tan
B – Black
G – Green
K – Pink
L – Blue
N – Brown
O – Orange
P – Purple
R – Red
S – Grey
V – Violet
W – White
Y – Yellow
X – Light green
Z – Light blue

Earth Locations

1 - On transmission, adjacent to battery
2 - On transmission housing to engine bolt
3 - Left of centre of firewall.
4 - RHF door frame, behind kick panel
5 - Floor tunnel, beneath console
6 - LH floor panel, below passenger seat
7 - LH door frame, beneath glove comp
8 - LHS, behind rear seat back (sedan)
9 - RHS, behind luggage comp trim panel (sedan)
10 - RH inner guard, behind headlamp
11 - LH inner guard, behind headlamp
12 - Boot lid (sedan)
13 - RH floor panel, below driver seat
14 - RHS, behind luggage comp trim panel (wagon)
15 - LHS, behind luggage comp trim panel (wagon)
16 - Tail gate (wagon)
17 - Adjacent to sunroof motor
18 - Steering wheel

TJ/KJ Magna/Verada V6
wiring diagram 15 of 22

Anitlock braking system (ABS). TJ/KJ models.

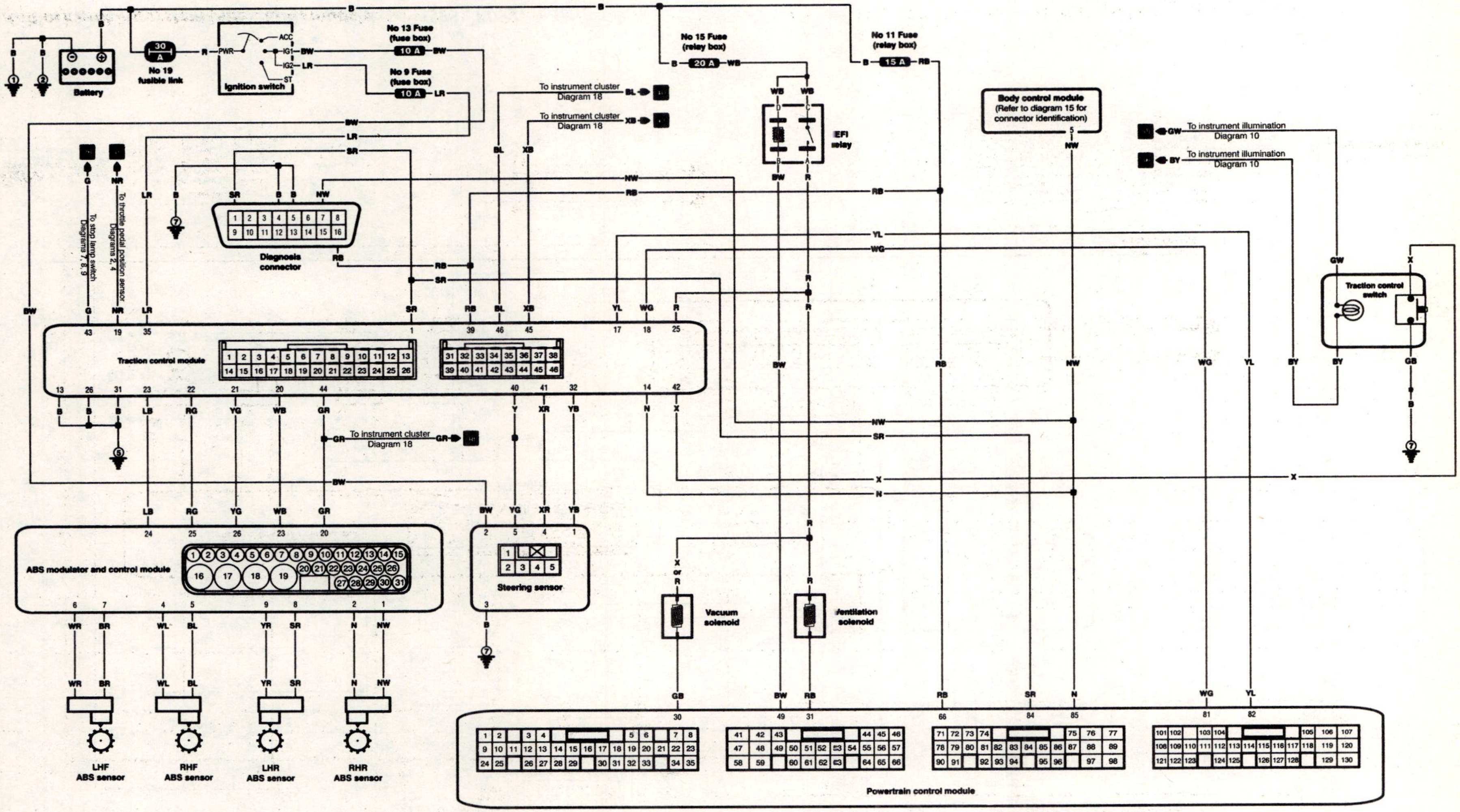

Colour Code

A – Tan	N – Brown	V – Violet
B – Black	O – Orange	W – White
G – Green	P – Purple	Y – Yellow
K – Pink	R – Red	X – Light green
L – Blue	S – Grey	Z – Light blue

Earth Locations

1 - On transmission, adjacent to battery
2 - On transmission housing to engine bolt
3 - Left of centre of firewall.
4 - RHF door frame, behind kick panel
5 - Floor tunnel, beneath console
6 - LH floor panel, below passenger seat
7 - LH door frame, beneath glove comp
8 - LHS, behind rear seat back (sedan)
9 - RHS, behind luggage comp trim panel (sedan)
10 - RH inner guard, behind headlamp
11 - LH inner guard, behind headlamp
12 - Boot lid (sedan)
13 - RH floor panel, below driver seat
14 - RHS, behind luggage comp trim panel (wagon)
15 - LHS, behind luggage comp trim panel (wagon)
16 - Tail gate (wagon)
17 - Adjacent to sunroof motor
18 - Steering wheel

TJ/KJ Magna/Verada V6
wiring diagram 16 of 22

Traction control. TJ/KJ models.

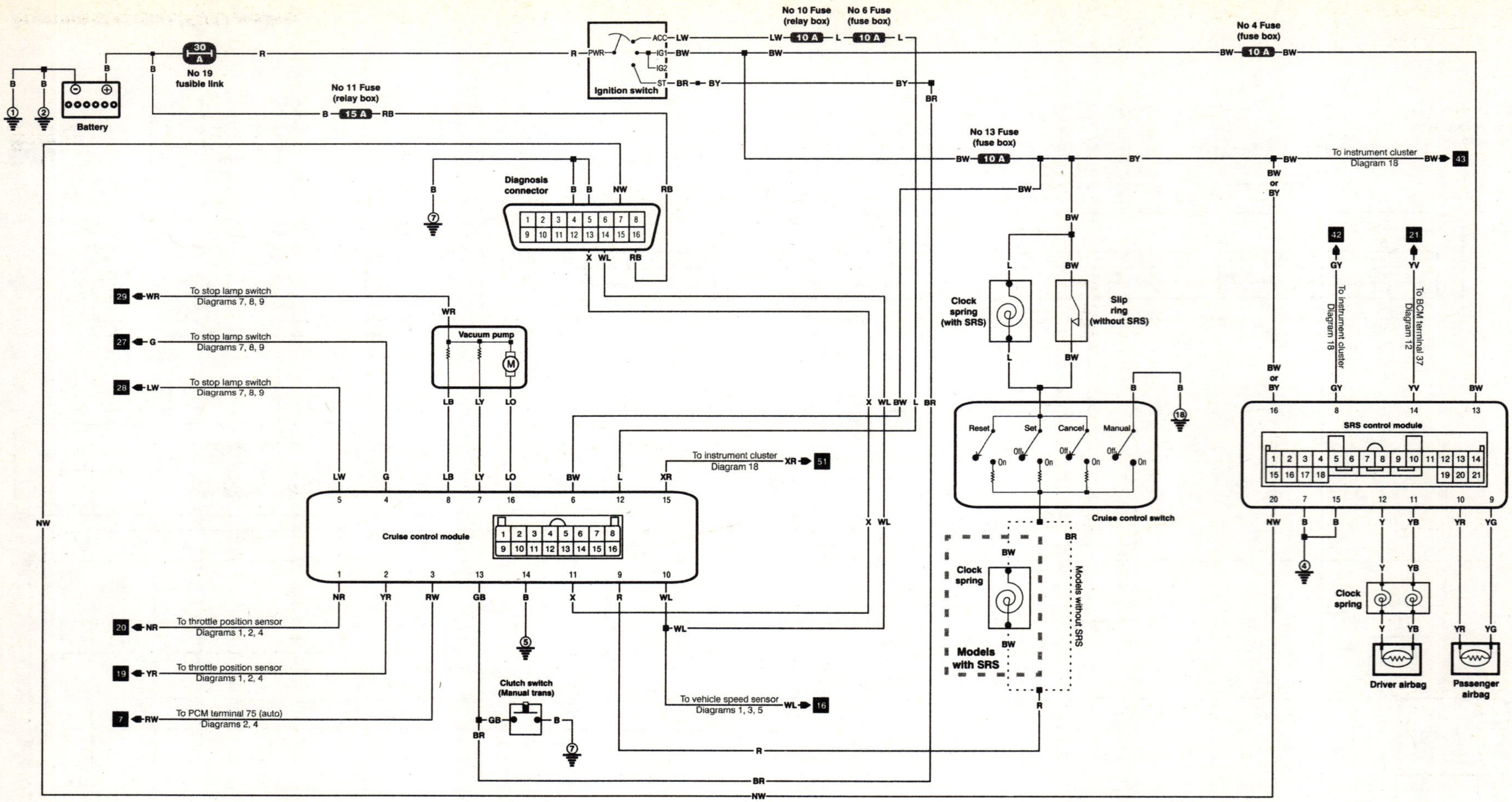

Colour Code

A – Tan
B – Black
G – Green
K – Pink
L – Blue
N – Brown
O – Orange
P – Purple
R – Red
S – Grey
V – Violet
W – White
Y – Yellow
X – Light green
Z – Light blue

Earth Locations

1 - On transmission, adjacent to battery
2 - On transmission housing to engine bolt
3 - Left of centre of firewall.
4 - RHF door frame, behind kick panel
5 - Floor tunnel, beneath console
6 - LH floor panel, below passenger seat
7 - LH door frame, beneath glove comp
8 - LHS, behind rear seat back (sedan)
9 - RHS, behind luggage comp trim panel (sedan)
10 - RH inner guard, behind headlamp
11 - LH inner guard, behind headlamp
12 - Boot lid (sedan)
13 - RH floor panel, below driver seat
14 - RHS, behind luggage comp trim panel (wagon)
15 - LHS, behind luggage comp trim panel (wagon)
16 - Tail gate (wagon)
17 - Adjacent to sunroof motor
18 - Steering wheel

TJ/KJ Magna/Verada V6
wiring diagram 17 of 22

Cruise control and supplemental restraint system (SRS). TJ/KJ models.

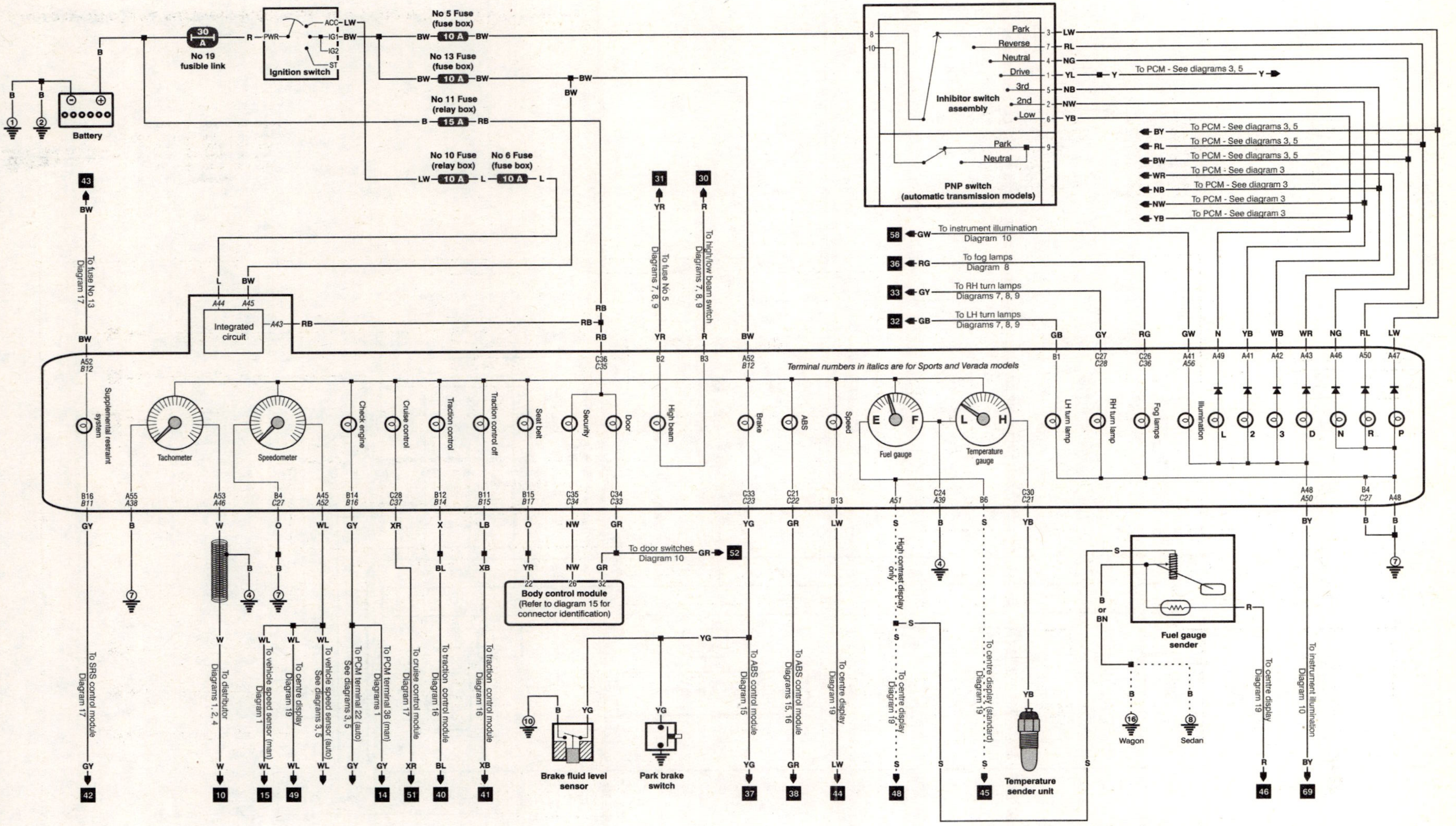

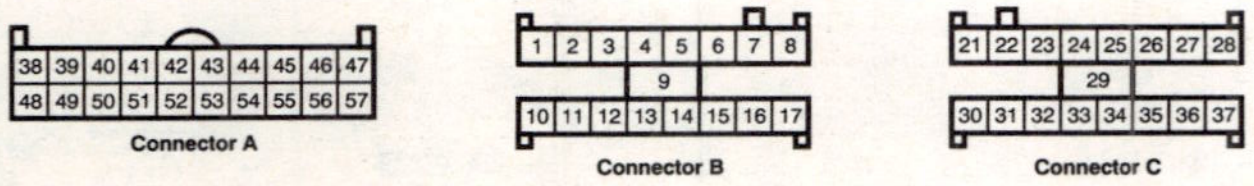

Colour Code

A – Tan	N – Brown	V – Violet
B – Black	O – Orange	W – White
G – Green	P – Purple	Y – Yellow
K – Pink	R – Red	X – Light green
L – Blue	S – Grey	Z – Light blue

Earth Locations

1 - On transmission, adjacent to battery	10 - RH inner guard, behind headlamp
2 - On transmission housing to engine bolt	11 - LH inner guard, behind headlamp
3 - Left of centre of firewall	12 - Boot lid (sedan)
4 - RHF door frame, behind kick panel	13 - RH floor panel, below driver seat
5 - Floor tunnel, beneath console	14 - RHS, behind luggage comp trim panel (wagon)
6 - LH floor panel, below passenger seat	15 - LHS, behind luggage comp trim panel (wagon)
7 - LH door frame, beneath glove comp	16 - Tail gate (wagon)
8 - LHS, behind rear seat back (sedan)	17 - Adjacent to sunroof motor
9 - RHS, behind luggage comp trim panel (sedan)	18 - Steering wheel

TJ/KJ Magna/Verada V6
wiring diagram 18 of 22

Instrument cluster. TJ/KJ models.

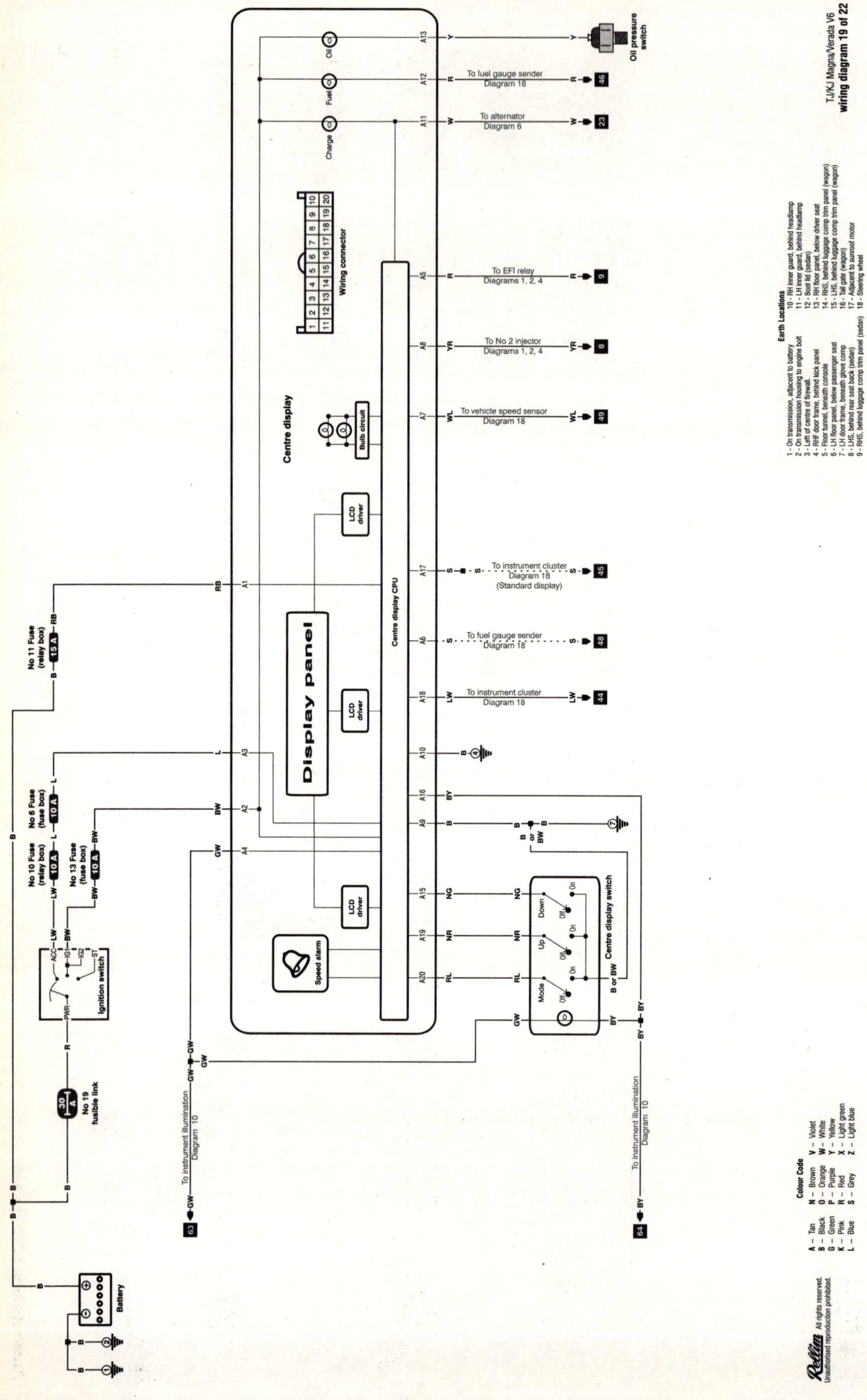

Centre display. TJ/KJ models.

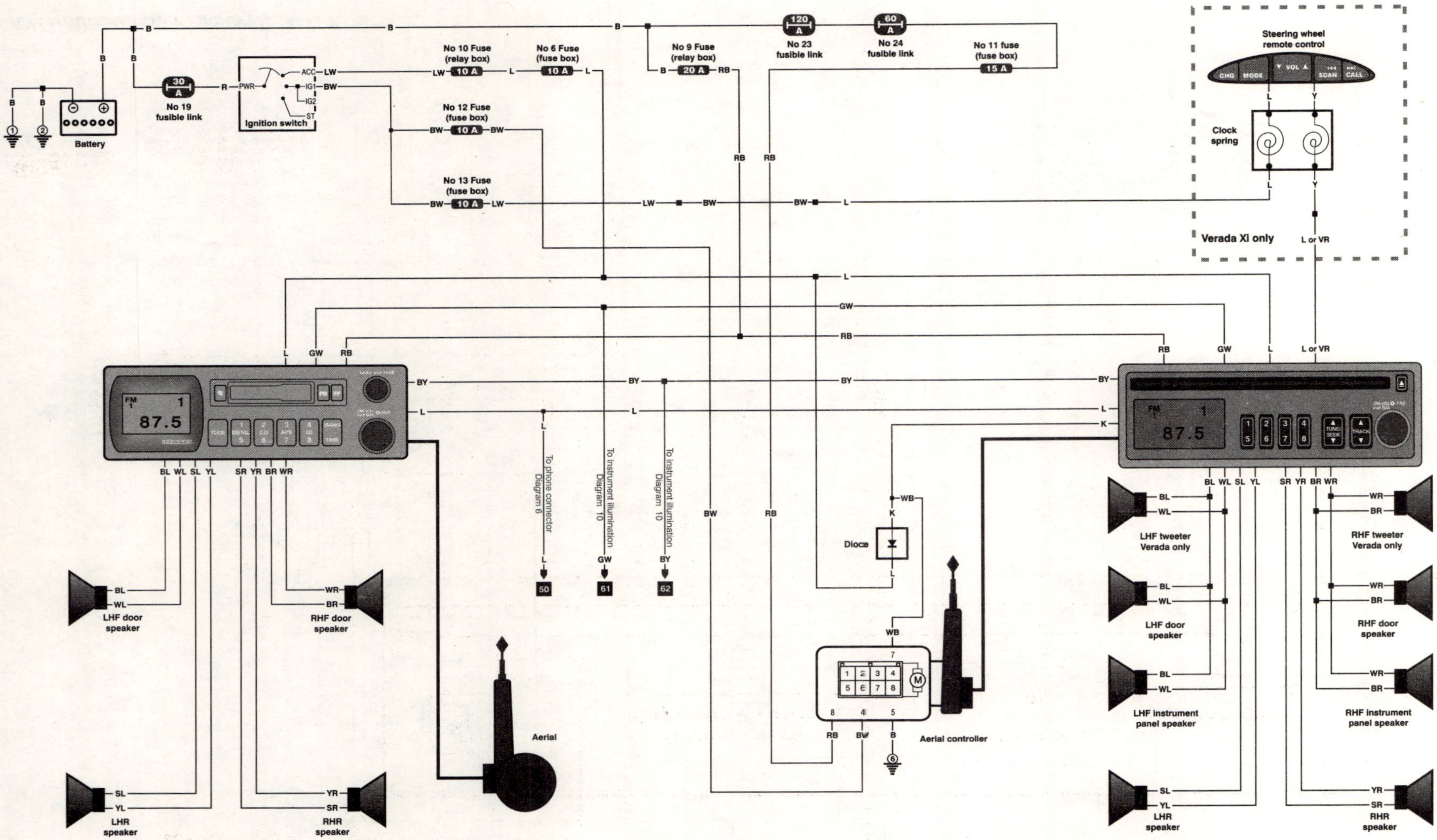

Colour Code

A – Tan	N – Brown	V – Violet
B – Black	O – Orange	W – White
G – Green	P – Purple	Y – Yellow
K – Pink	R – Red	X – Light green
L – Blue	S – Grey	Z – Light blue

Earth Locations

- 1 - On transmission, adjacent to battery
- 2 - On transmission housing to engine bolt
- 3 - Left of centre of firewall.
- 4 - RHF door frame, behind kick panel
- 5 - Floor tunnel, beneath console
- 6 - LH floor panel, below passenger seat
- 7 - LH door frame, beneath glove comp
- 8 - LHS, behind rear seat back (sedan)
- 9 - RHS, behind luggage comp trim panel (sedan)
- 10 - RH inner guard, behind headlamp
- 11 - LH inner guard, behind headlamp
- 12 - Boot lid (sedan)
- 13 - RH floor panel, below driver seat
- 14 - RHS, behind luggage comp trim panel (wagon)
- 15 - LHS, behind luggage comp trim panel (wagon)
- 16 - Tail gate (wagon)
- 17 - Adjacent to sunroof motor
- 18 - Steering wheel

TJ/KJ Magna/Verada V6
wiring diagram 20 of 22

Audio system. TJ/KJ models.

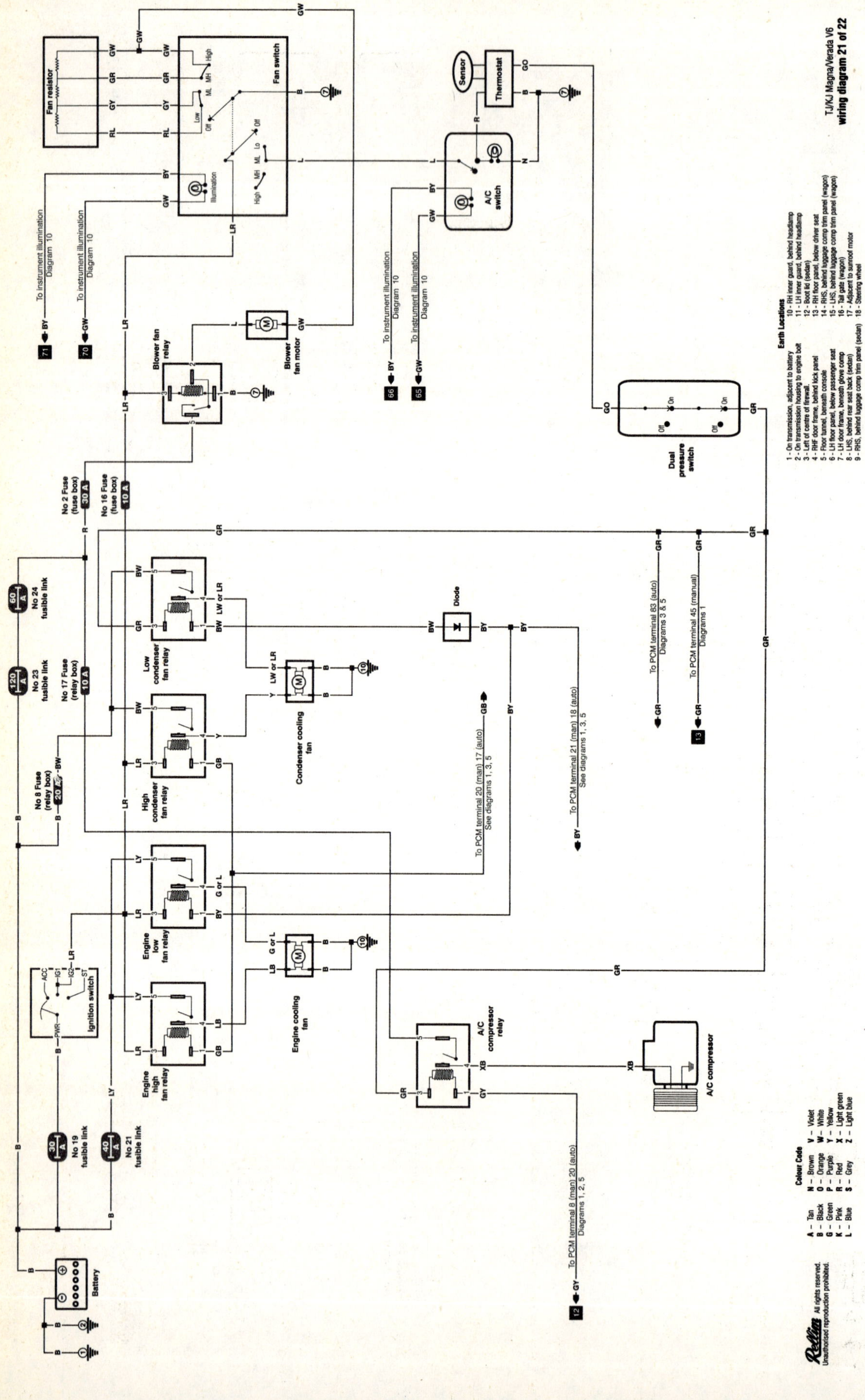

Air conditioning and cooling fans. TJ/KJ models.

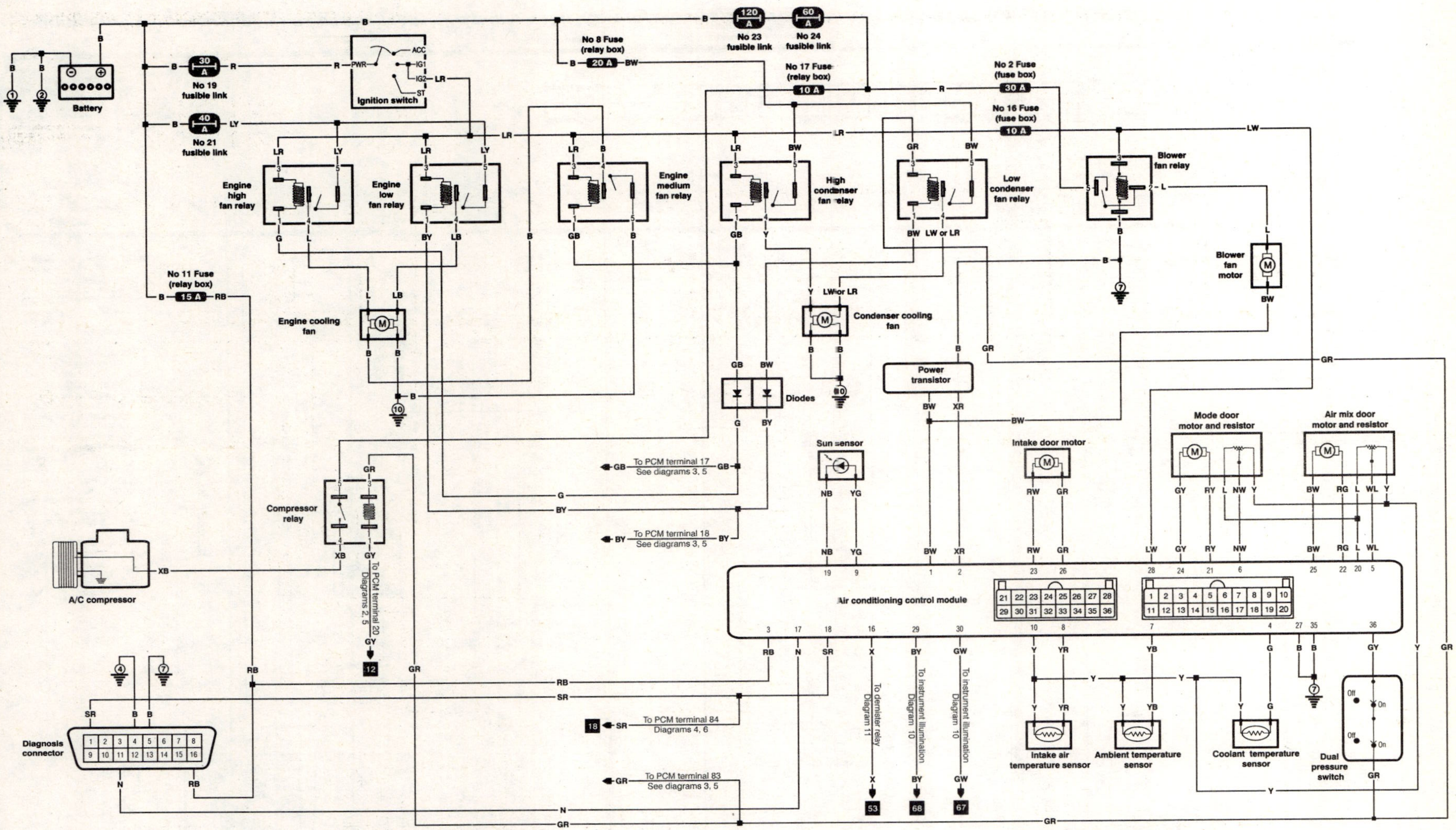

Colour Code

A – Tan	N – Brown	V – Violet
B – Black	O – Orange	W – White
G – Green	P – Purple	Y – Yellow
K – Pink	R – Red	X – Light green
L – Blue	S – Grey	Z – Light blue

Earth Locations

1 - On transmission, adjacent to battery
2 - On transmission housing to engine bolt
3 - Left of centre of firewall.
4 - RHF door frame, behind kick panel
5 - Floor tunnel, beneath console
6 - LH floor panel, below passenger seat
7 - LH door frame, beneath glove comp
8 - LHS, behind rear seat back (sedan)
9 - RHS, behind luggage comp trim panel (sedan)
10 - RH inner guard, behind headlamp
11 - LH inner guard, behind headlamp
12 - Boot lid (sedan)
13 - RH floor panel, below driver seat
14 - RHS, behind luggage comp trim panel (wagon)
15 - LHS, behind luggage comp trim panel (wagon)
16 - Tail gate (wagon)
17 - Adjacent to sunroof motor
18 - Steering wheel

TJ/KJ Magna/Verada V6
wiring diagram 22 of 22

Climate control and cooling fans. TJ/KJ models.

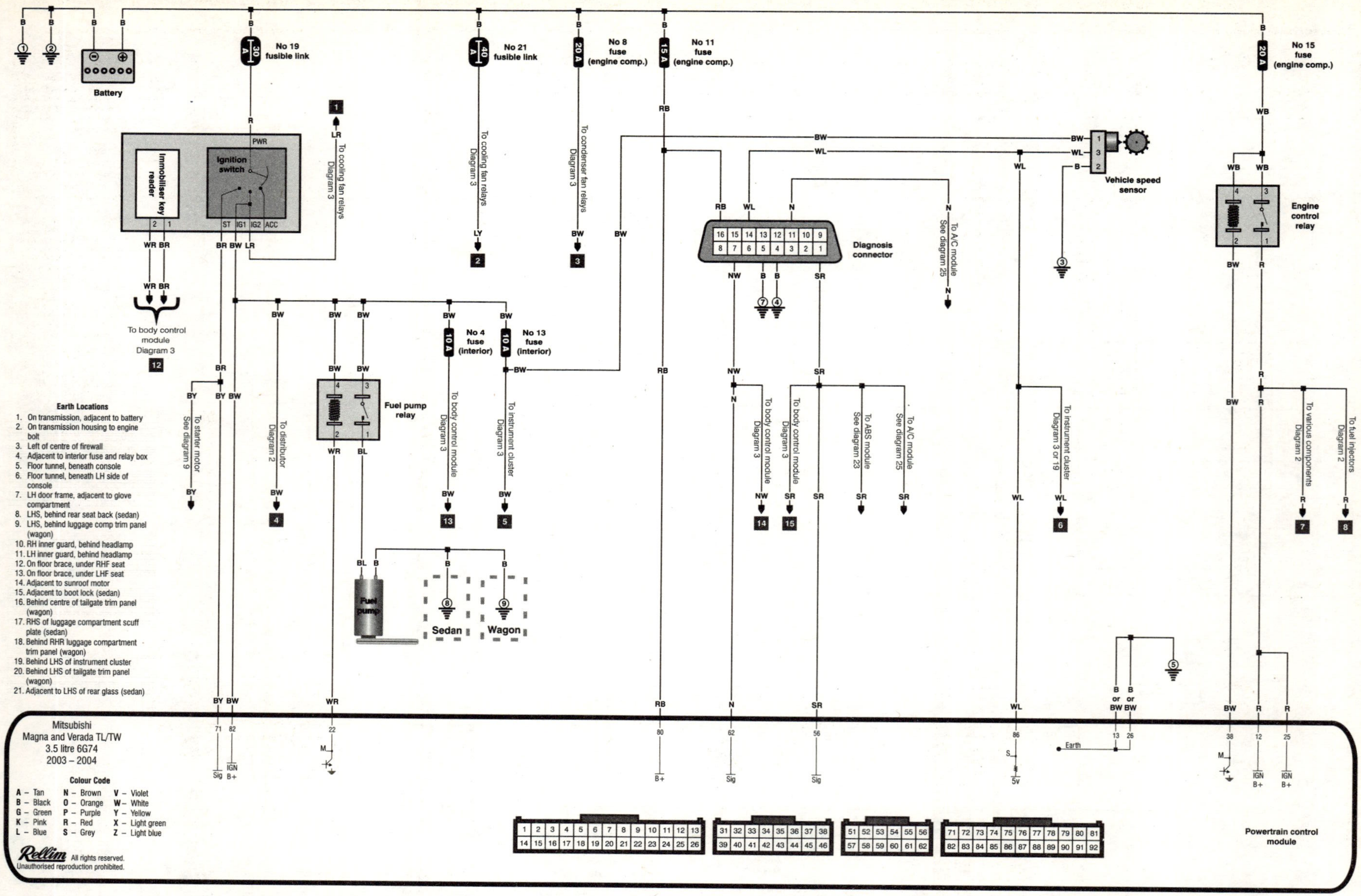

Diagram 1 — PCM power supply and earth circuits, fuel pump, vehicle speed sensor, diagnosis connector. Manual transmission. TL/KL & TW/KW models.

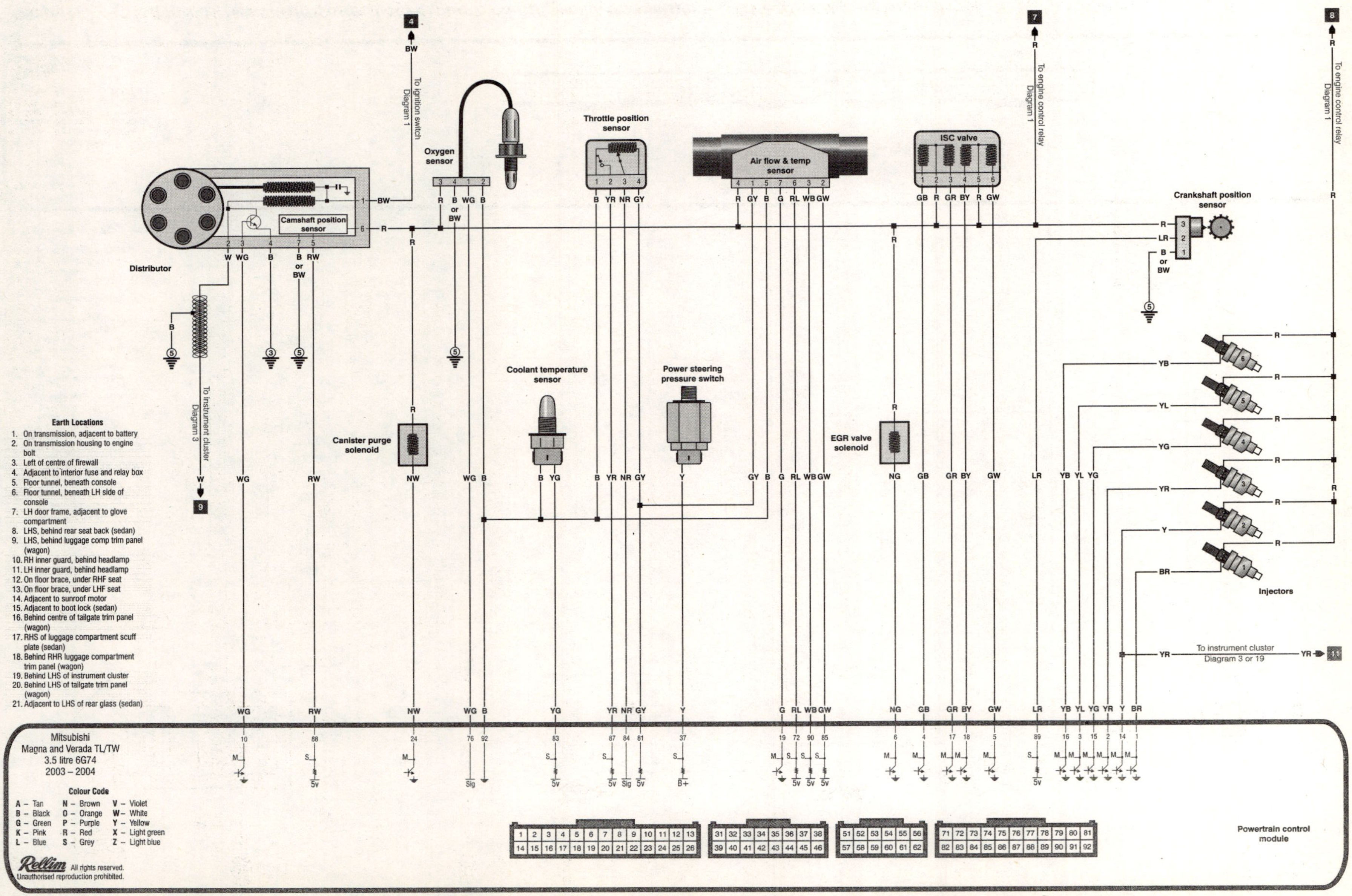

Earth Locations

1. On transmission, adjacent to battery
2. On transmission housing to engine bolt
3. Left of centre of firewall
4. Adjacent to interior fuse and relay box
5. Floor tunnel, beneath console
6. Floor tunnel, beneath LH side of console
7. LH door frame, adjacent to glove compartment
8. LHS, behind rear seat back (sedan)
9. LHS, behind luggage comp trim panel (wagon)
10. RH inner guard, behind headlamp
11. LH inner guard, behind headlamp
12. On floor brace, under RHF seat
13. On floor brace, under LHF seat
14. Adjacent to sunroof motor
15. Adjacent to boot lock (sedan)
16. Behind centre of tailgate trim panel (wagon)
17. RHS of luggage compartment scuff plate (sedan)
18. Behind RHR luggage compartment trim panel (wagon)
19. Behind LHS of instrument cluster
20. Behind LHS of tailgate trim panel (wagon)
21. Adjacent to LHS of rear glass (sedan)

Mitsubishi
Magna and Verada TL/TW
3.5 litre 6G74
2003 – 2004

Colour Code

A – Tan	N – Brown	V – Violet
B – Black	O – Orange	W – White
G – Green	P – Purple	Y – Yellow
K – Pink	R – Red	X – Light green
L – Blue	S – Grey	Z – Light blue

Diagram 2 — Various input sensors, distributor, canister purge and EGR valve solenoids, fuel injectors, ISC valve. Manual transmission. TL/KL & TW/KW models.

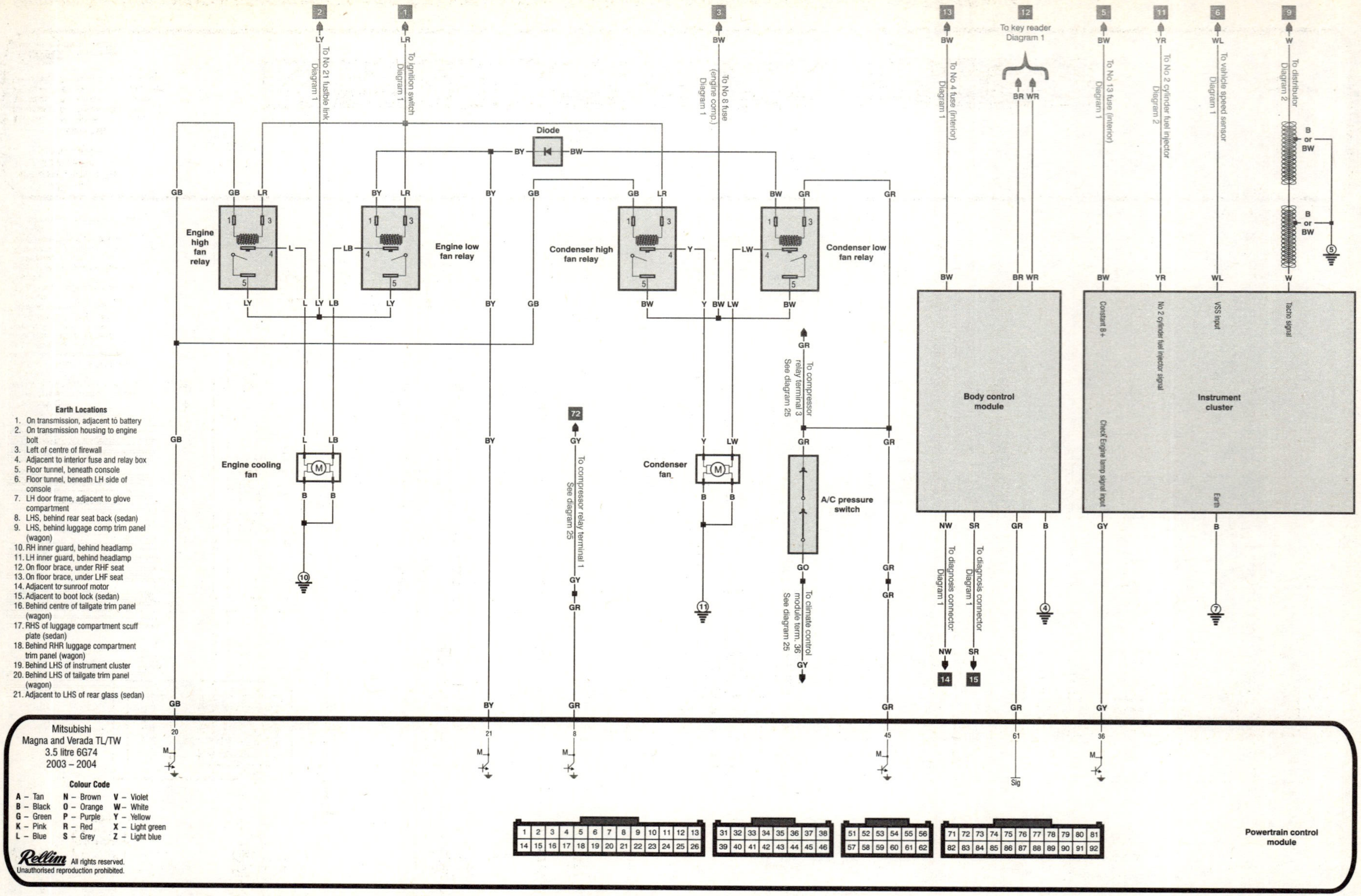

Diagram 3 — Cooling fans, instrument cluster, body control module. Manual transmission. TL/KL & TW/KW models.

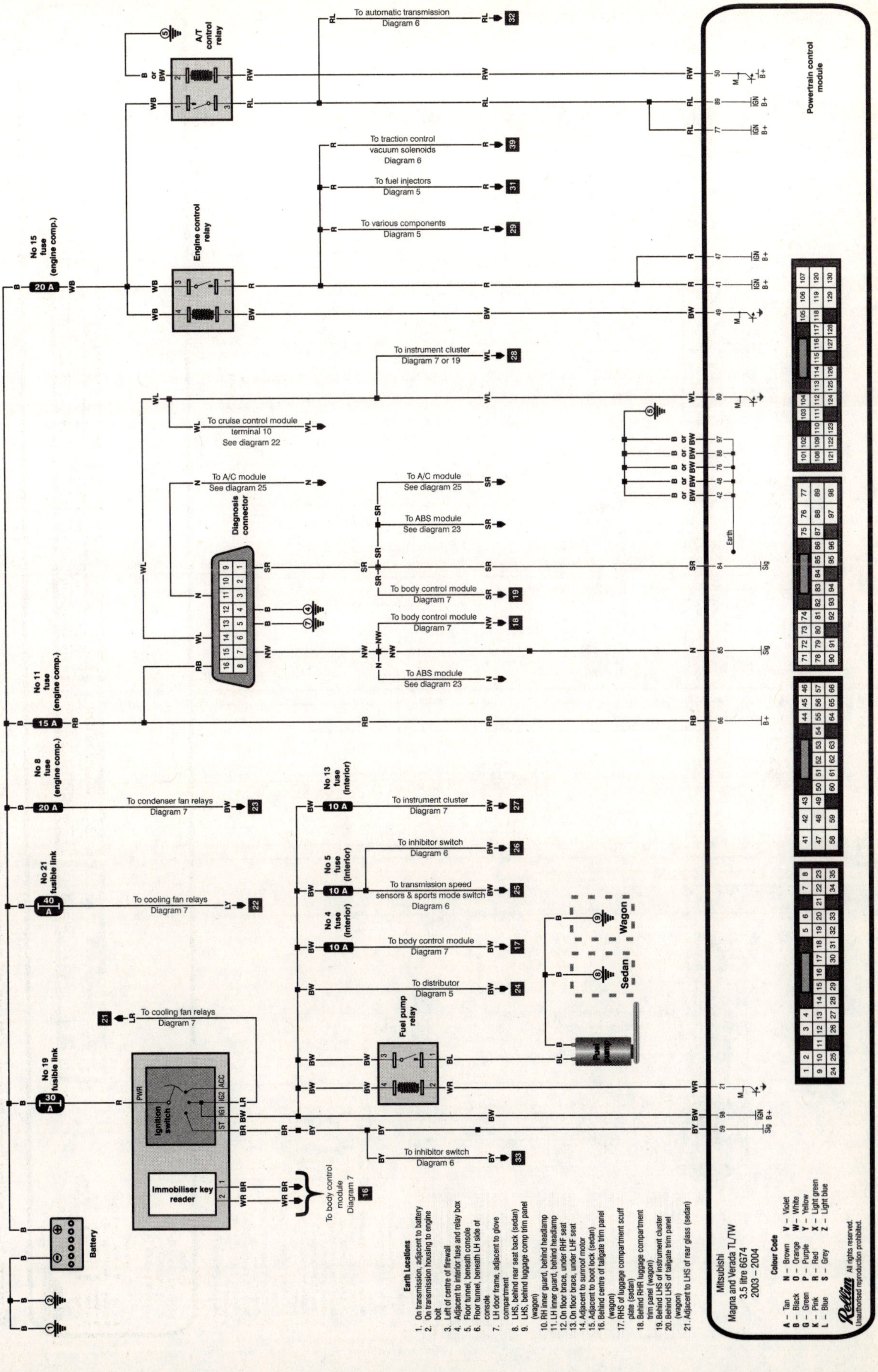

Diagram 4 — PCM power supply and earth circuits, fuel pump, diagnosis connector. Automatic transmission. TL/KL & TW/KW models.

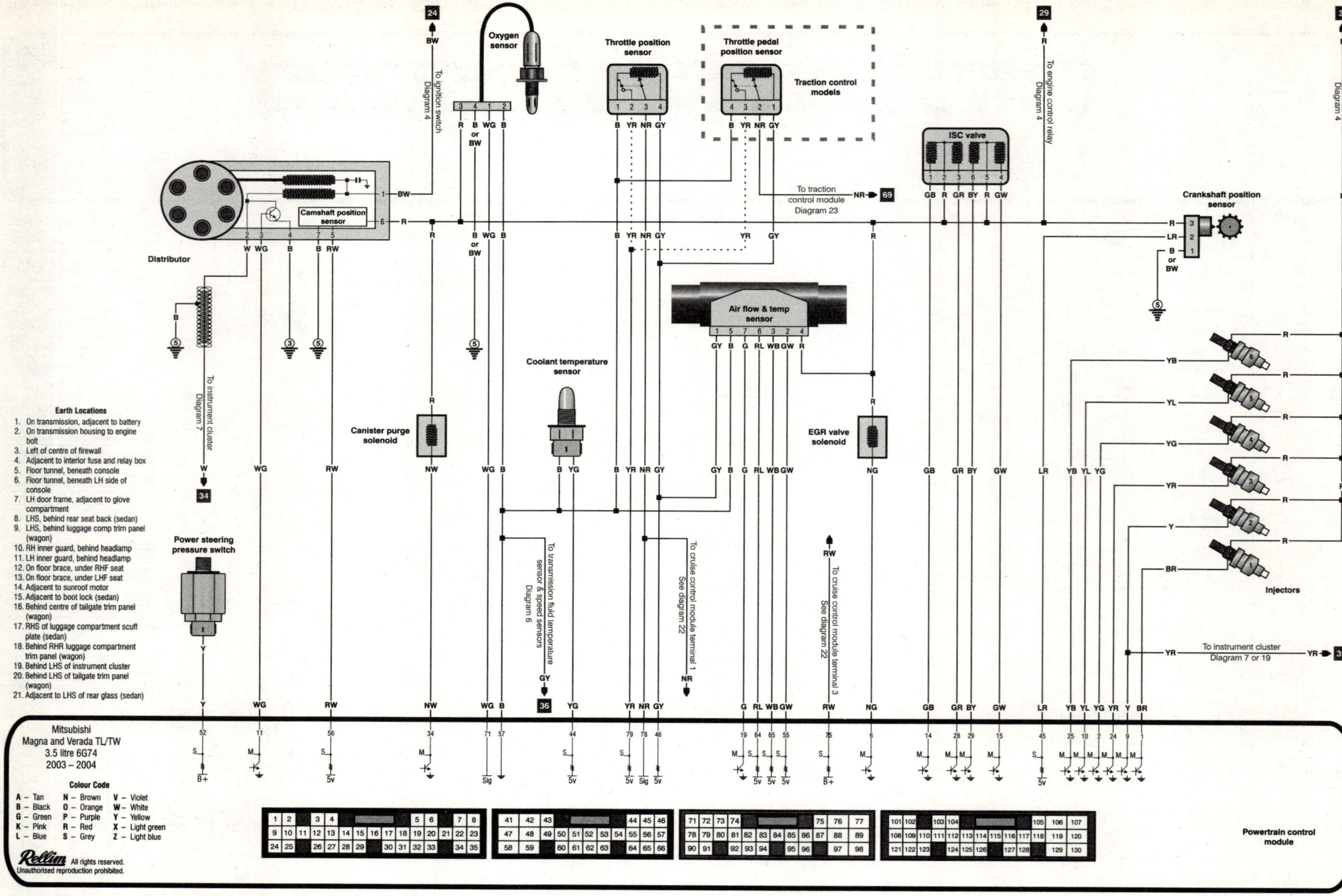

Diagram 5 — Various input sensors, distributor, canister purge and EGR valve solenoids, fuel injectors, ISC valve. Automatic transmission. TL/KL & TW/KW models.

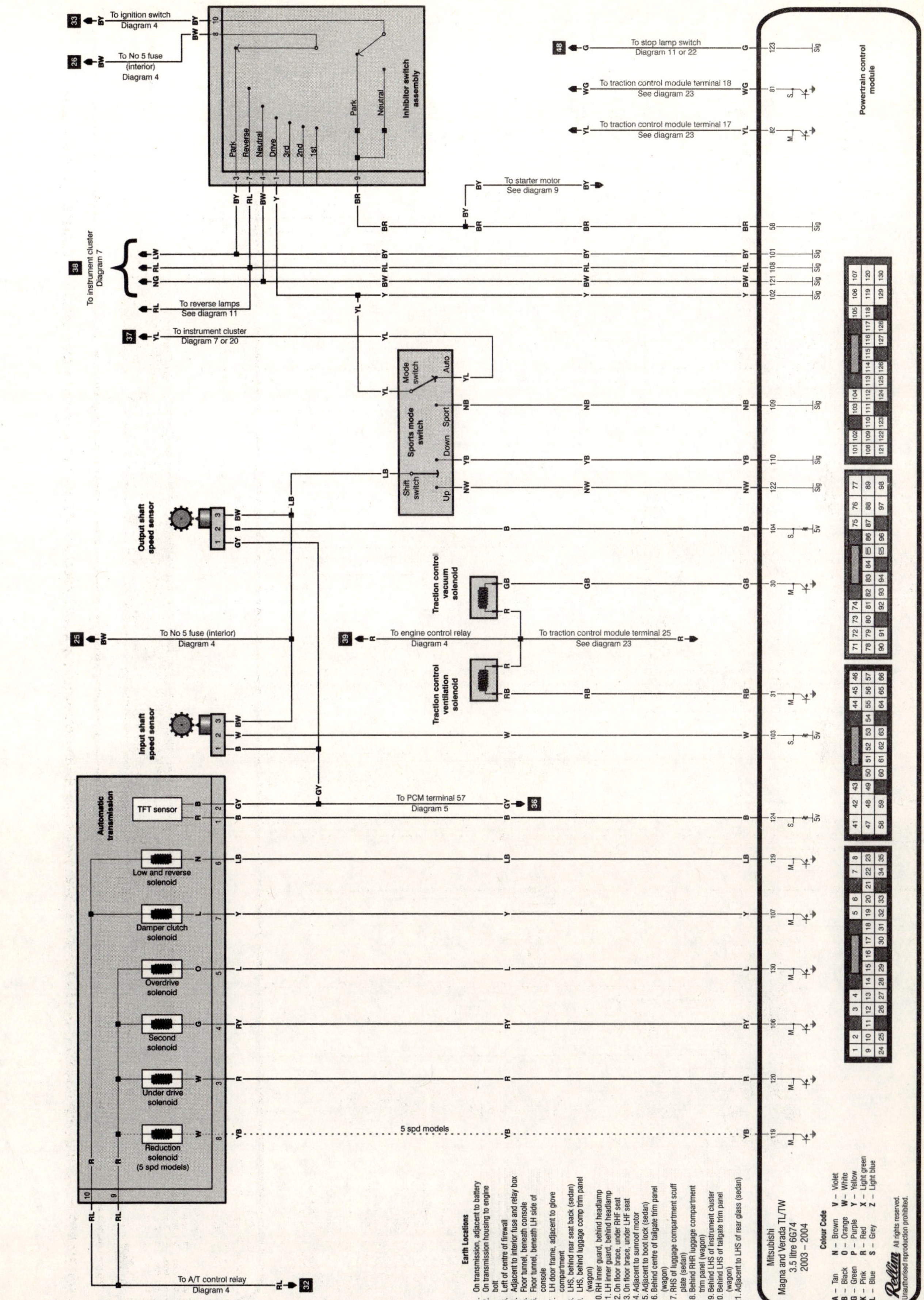

Diagram 6 — Auto trans, trans speed sensors, sports switch, inhibitor switch, traction control ventilation & vacuum solenoids. Automatic transmission. TL/KL & TW/KW models.

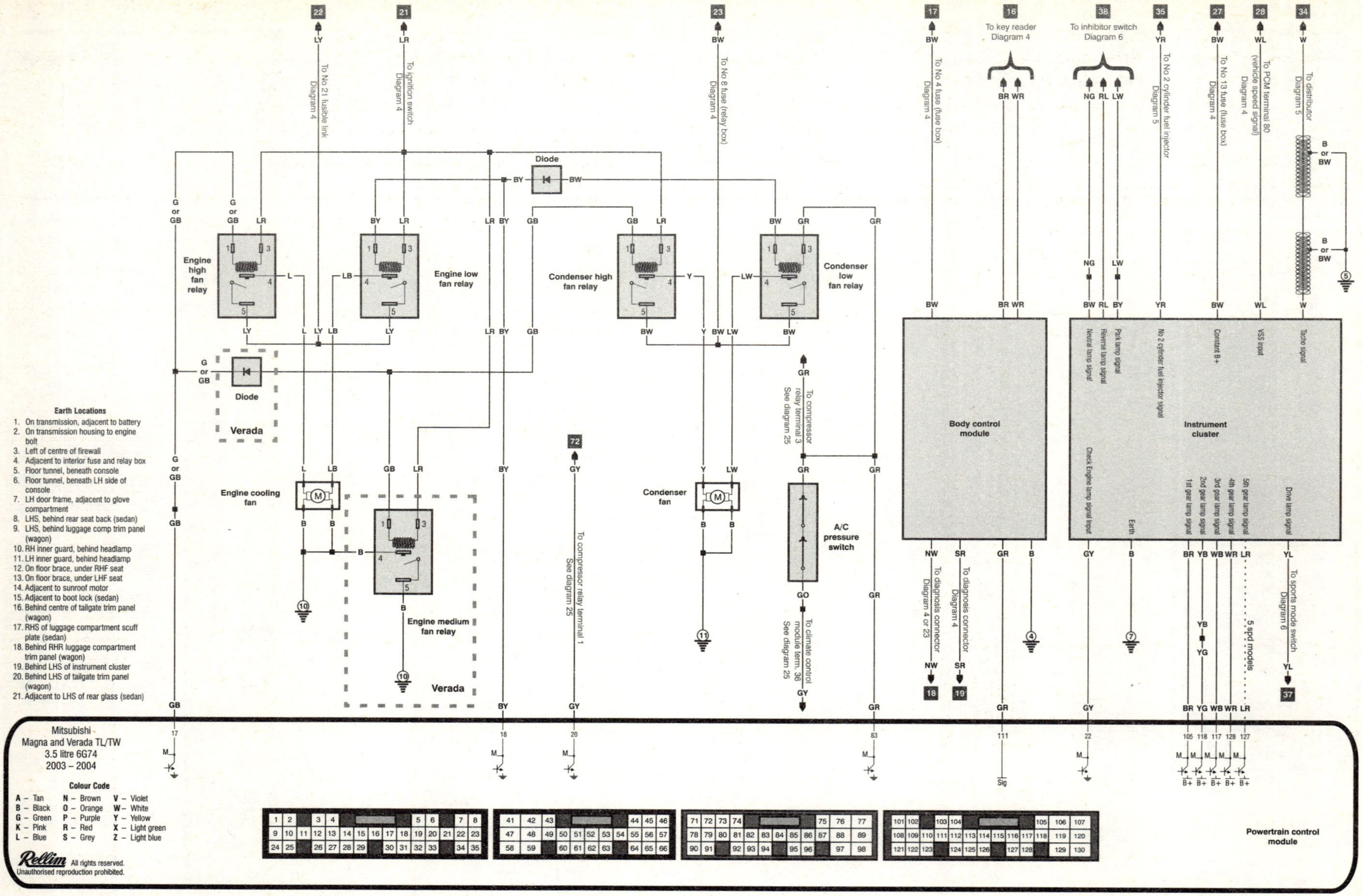

Diagram 7 — Cooling fans, instrument cluster. Automatic transmission. TL/KL & TW/KW models.

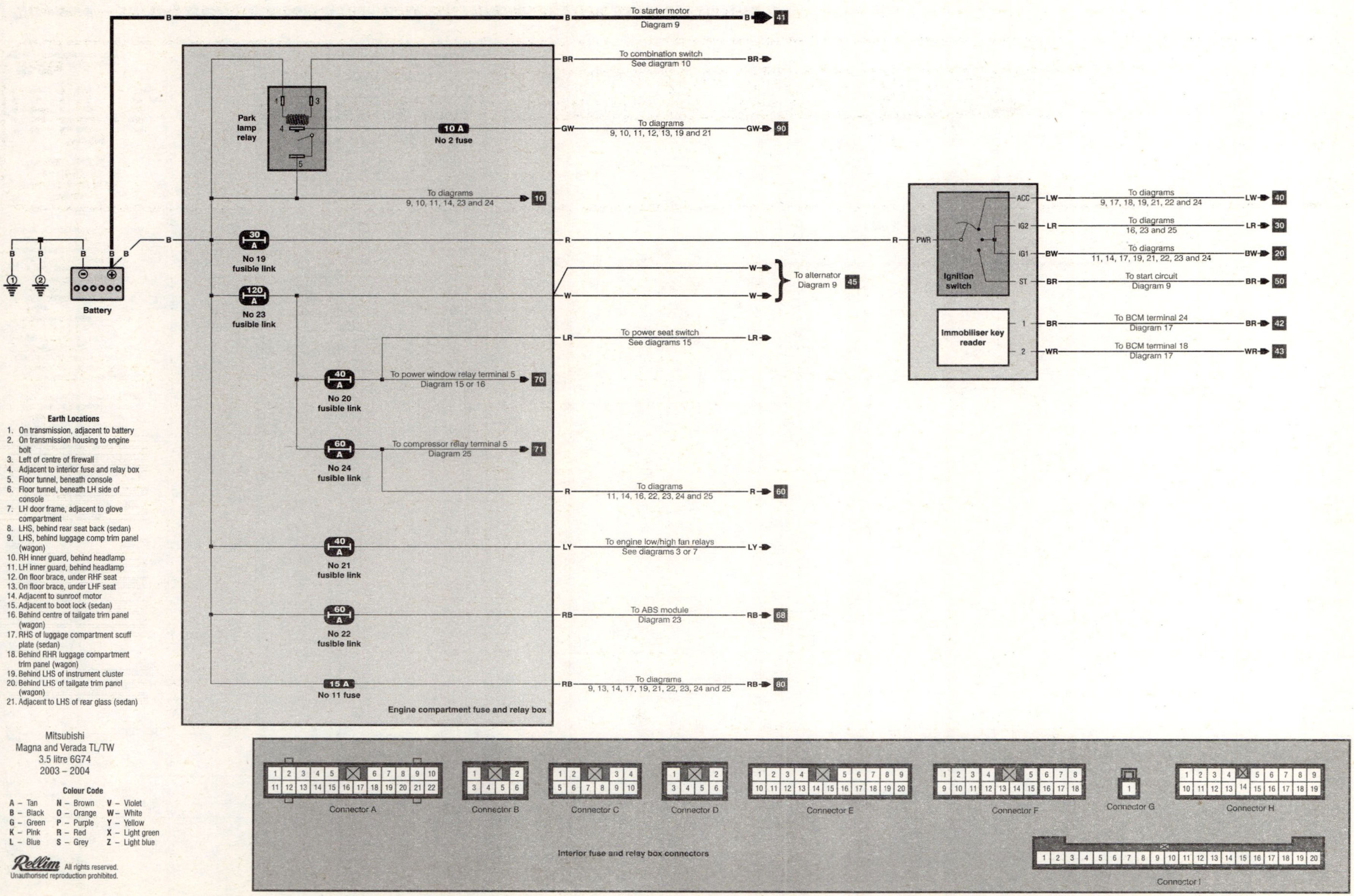

Diagram 8 — Power distribution, ignition switch. TL/KL & TW/KW models.

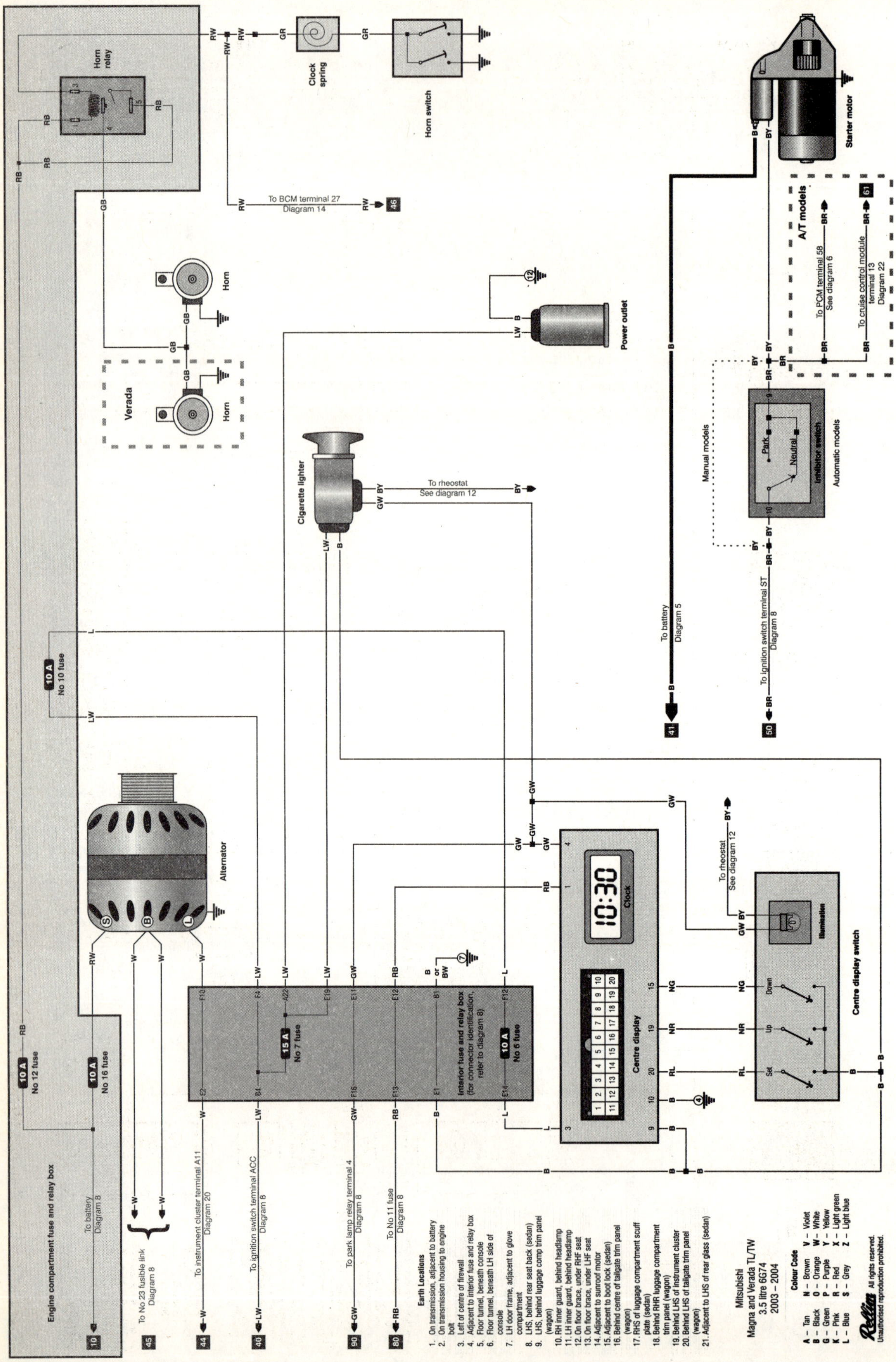

Diagram 9 — Starting system, charging system, horn, cigarette lighter, clock, power outlet. TL/KL & TW/KW models.

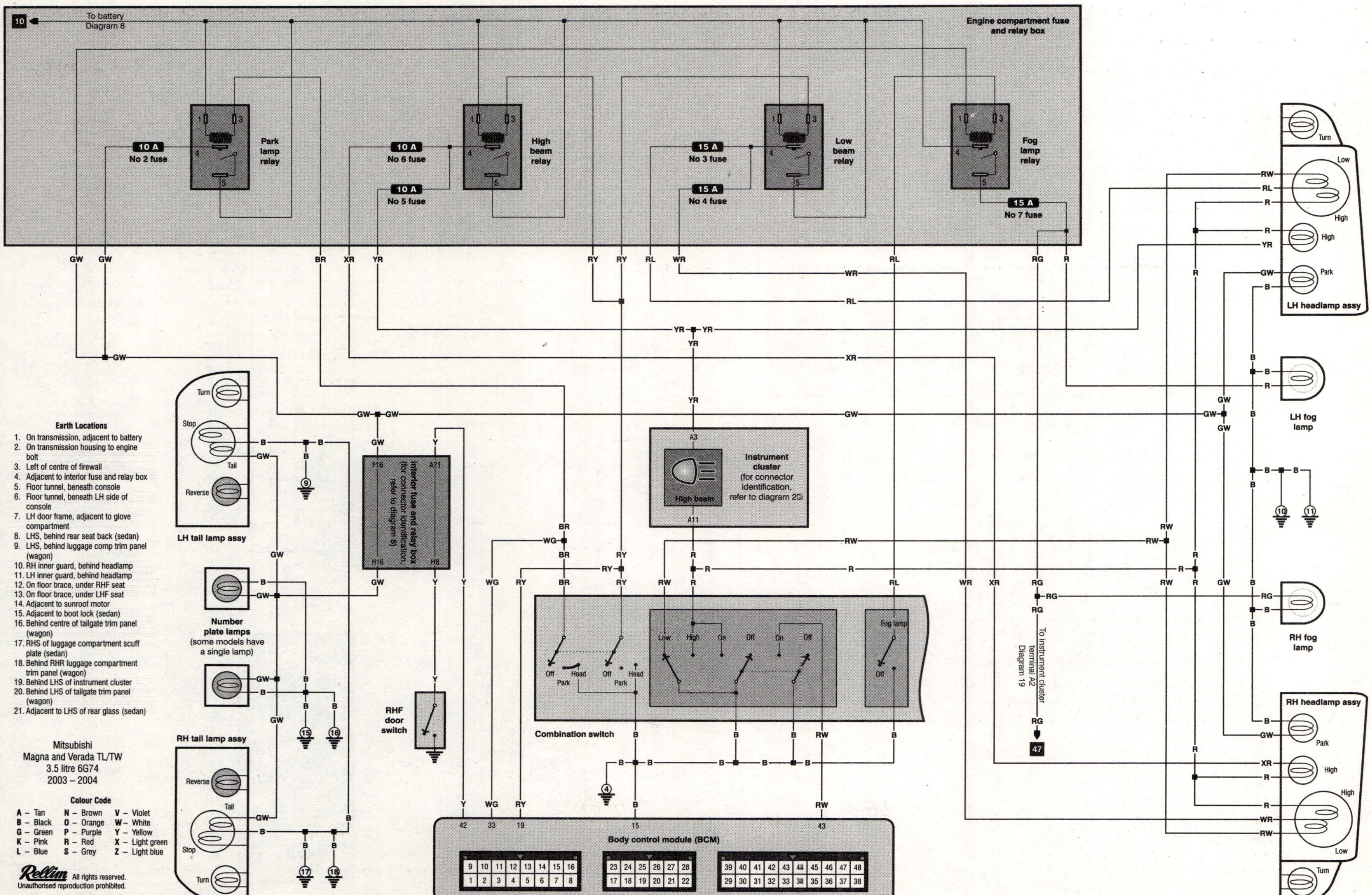

Diagram 10 — Headlamps, tail lamps, fog lamps, lamp reminder system. TL/KL & TW/KW models.

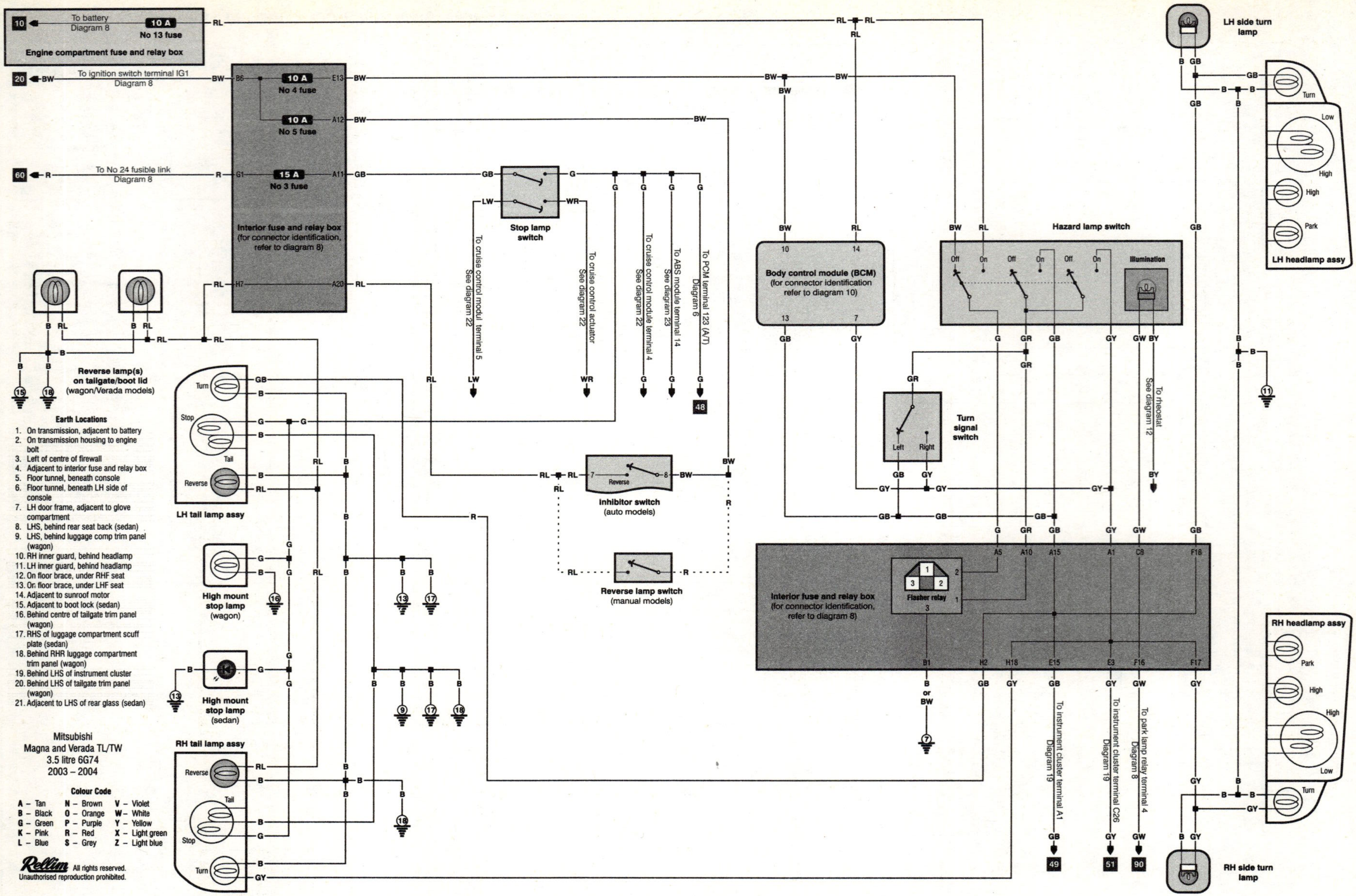

Diagram 11 — Turn signal/hazard lamps, stop lamps, reverse lamps. TL/KL & TW/KW models.

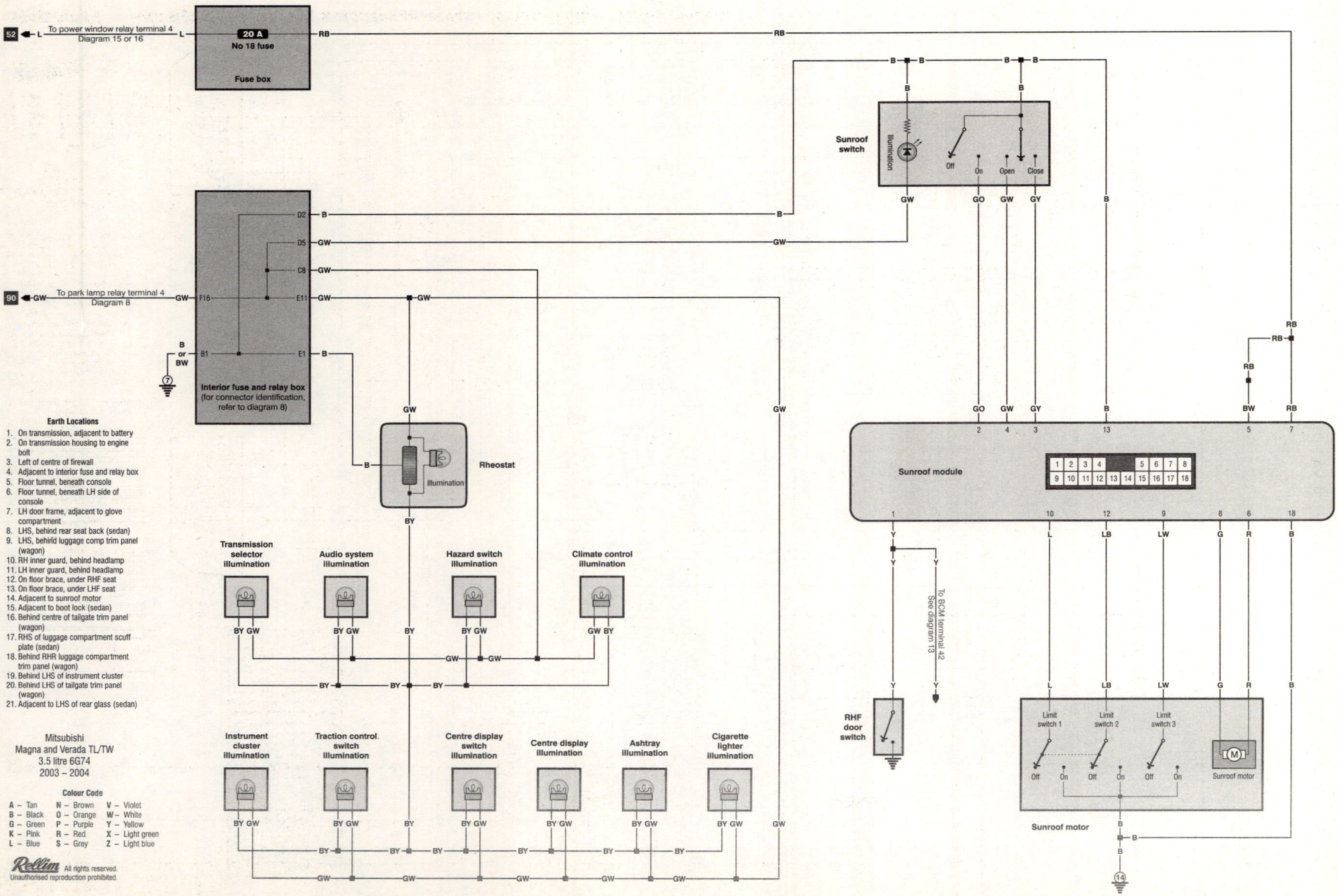

Earth Locations

1. On transmission, adjacent to battery
2. On transmission housing to engine bolt
3. Left of centre of firewall
4. Adjacent to interior fuse and relay box
5. Floor tunnel, beneath console
6. Floor tunnel, beneath LH side of console
7. LH door frame, adjacent to glove compartment
8. LHS, behind rear seat back (sedan)
9. LHS, behind luggage comp trim panel (wagon)
10. RH inner guard, behind headlamp
11. LH inner guard, behind headlamp
12. On floor brace, under RHF seat
13. On floor brace, under LHF seat
14. Adjacent to sunroof motor
15. Adjacent to boot lock (sedan)
16. Behind centre of tailgate trim panel (wagon)
17. RHS of luggage compartment scuff plate (sedan)
18. Behind RHR luggage compartment trim panel (wagon)
19. Behind LHS of instrument cluster
20. Behind LHS of tailgate trim panel (wagon)
21. Adjacent to LHS of rear glass (sedan)

Mitsubishi
Magna and Verada TL/TW
3.5 litre 6G74
2003 – 2004

Colour Code

A – Tan	N – Brown	V – Violet
B – Black	O – Orange	W – White
G – Green	P – Purple	Y – Yellow
K – Pink	R – Red	X – Light green
L – Blue	S – Grey	Z – Light blue

Diagram 12 — Instrument illumination, sunroof. TL/KL & TW/KW models.

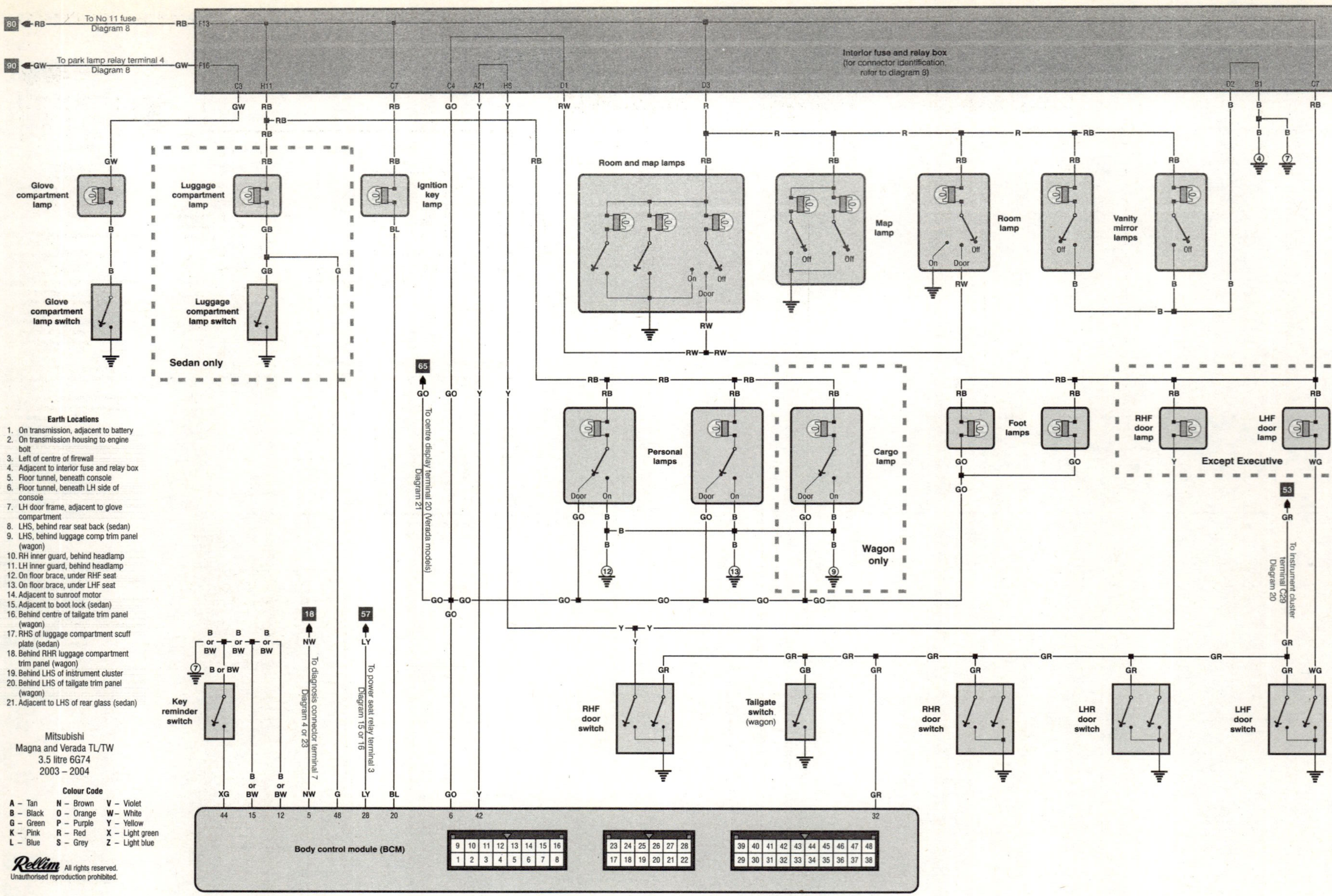

Earth Locations

1. On transmission, adjacent to battery
2. On transmission housing to engine bolt
3. Left of centre of firewall
4. Adjacent to interior fuse and relay box
5. Floor tunnel, beneath console
6. Floor tunnel, beneath LH side of console
7. LH door frame, adjacent to glove compartment
8. LHS, behind rear seat back (sedan)
9. LHS, behind luggage comp trim panel (wagon)
10. RH inner guard, behind headlamp
11. LH inner guard, behind headlamp
12. On floor brace, under RHF seat
13. On floor brace, under LHF seat
14. Adjacent to sunroof motor
15. Adjacent to boot lock (sedan)
16. Behind centre of tailgate trim panel (wagon)
17. RHS of luggage compartment scuff plate (sedan)
18. Behind RHR luggage compartment trim panel (wagon)
19. Behind LHS of instrument cluster
20. Behind LHS of tailgate trim panel (wagon)
21. Adjacent to LHS of rear glass (sedan)

Mitsubishi
Magna and Verada TL/TW
3.5 litre 6G74
2003 – 2004

Colour Code

A – Tan	N – Brown	V – Violet
B – Black	O – Orange	W – White
G – Green	P – Purple	Y – Yellow
K – Pink	R – Red	X – Light green
L – Blue	S – Grey	Z – Light blue

Diagram 13 — Interior lighting. TL/KL & TW/KW models.

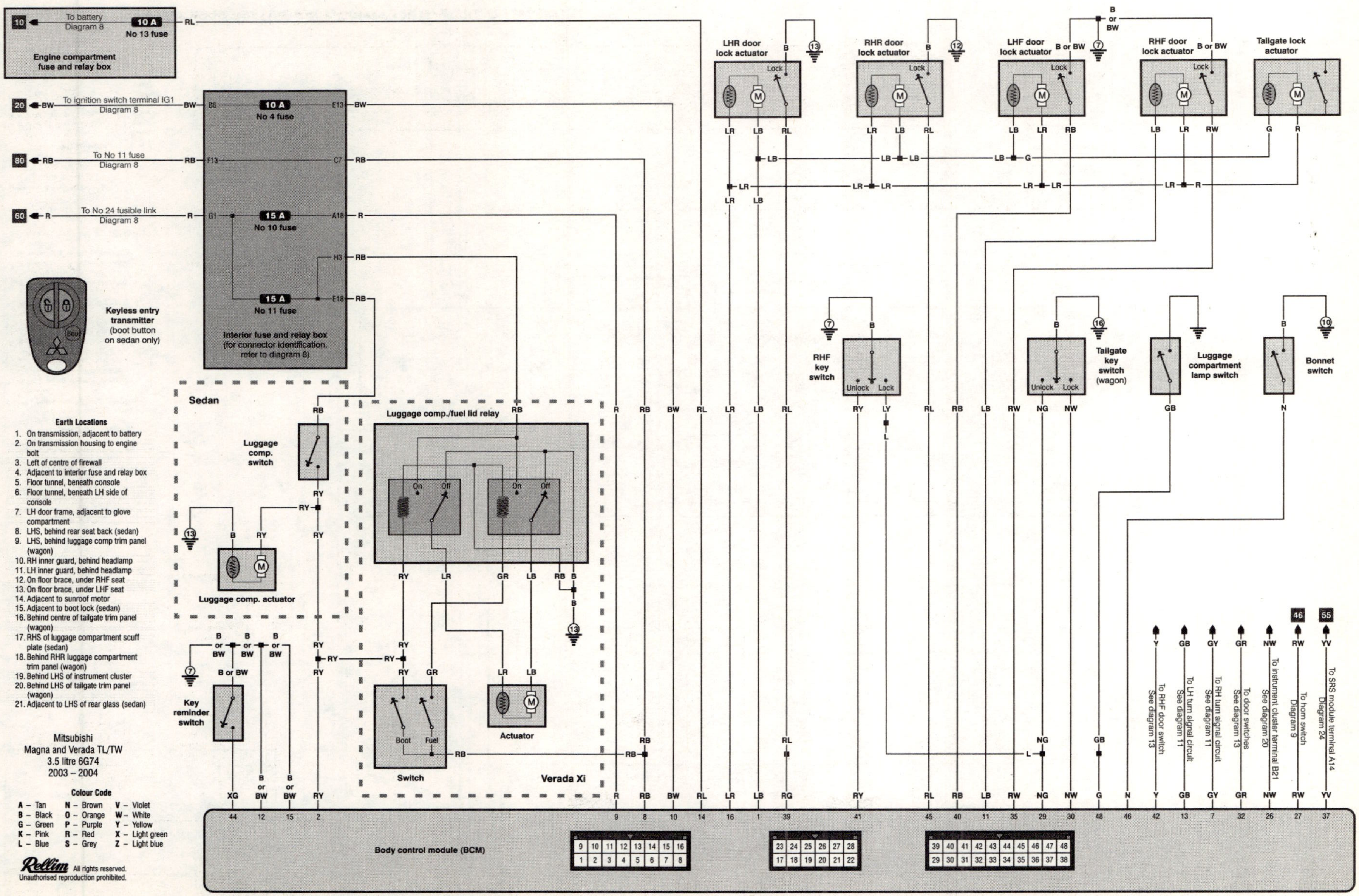

Diagram 14 — Power door locks, keyless entry, theft alarm. TL/KL & TW/KW models.

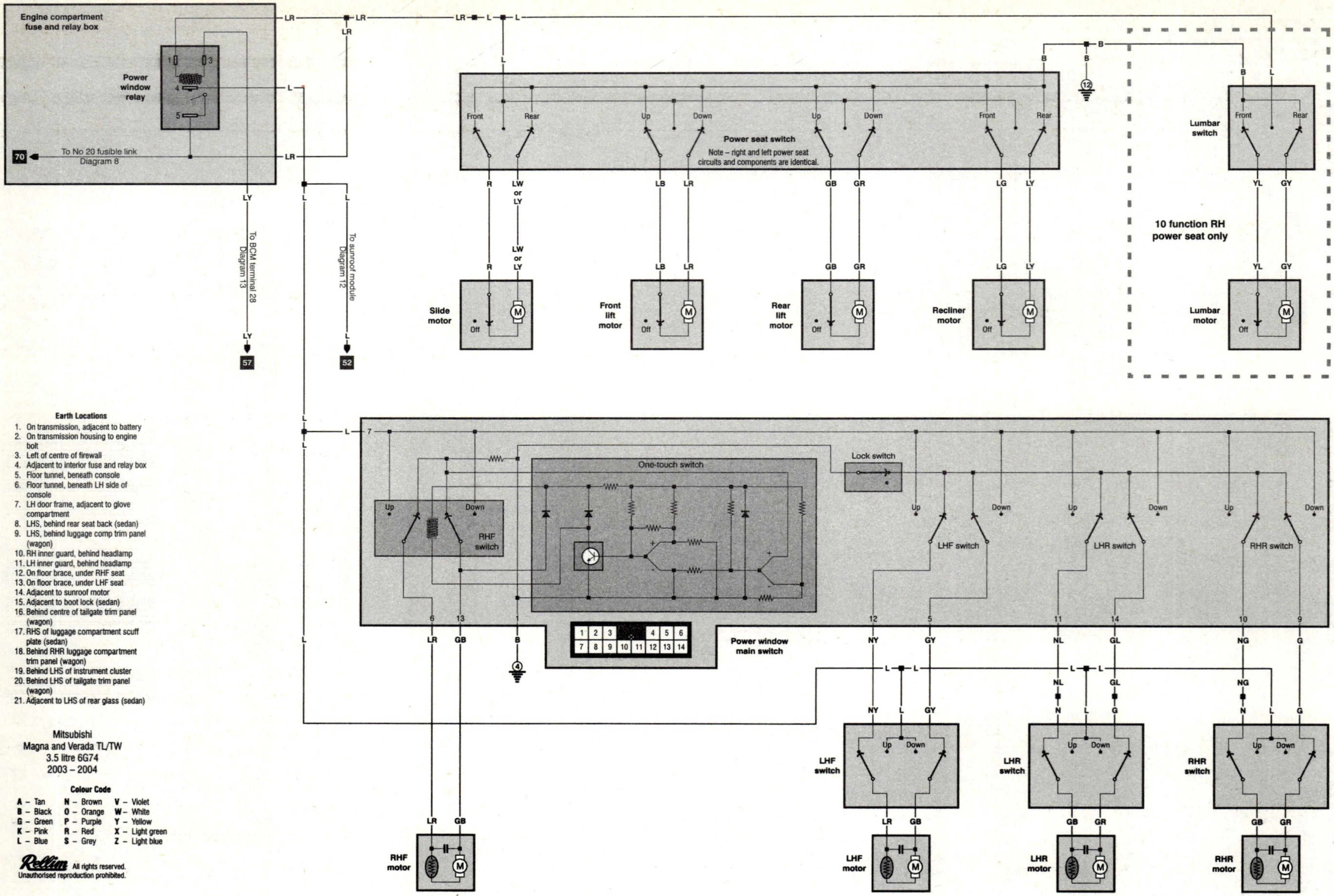

Earth Locations

1. On transmission, adjacent to battery
2. On transmission housing to engine bolt
3. Left of centre of firewall
4. Adjacent to interior fuse and relay box
5. Floor tunnel, beneath console
6. Floor tunnel, beneath LH side of console
7. LH door frame, adjacent to glove compartment
8. LHS, behind rear seat back (sedan)
9. LHS, behind luggage comp trim panel (wagon)
10. RH inner guard, behind headlamp
11. LH inner guard, behind headlamp
12. On floor brace, under RHF seat
13. On floor brace, under LHF seat
14. Adjacent to sunroof motor
15. Adjacent to boot lock (sedan)
16. Behind centre of tailgate trim panel (wagon)
17. RHS of luggage compartment scuff plate (sedan)
18. Behind RHR luggage compartment trim panel (wagon)
19. Behind LHS of instrument cluster
20. Behind LHS of tailgate trim panel (wagon)
21. Adjacent to LHS of rear glass (sedan)

Mitsubishi
Magna and Verada TL/TW
3.5 litre 6G74
2003 – 2004

Colour Code

A – Tan	N – Brown	V – Violet
B – Black	O – Orange	W – White
G – Green	P – Purple	Y – Yellow
K – Pink	R – Red	X – Light green
L – Blue	S – Grey	Z – Light blue

Diagram 15 — Power windows (Verada), power seats. TL/KL & TW/KW models.

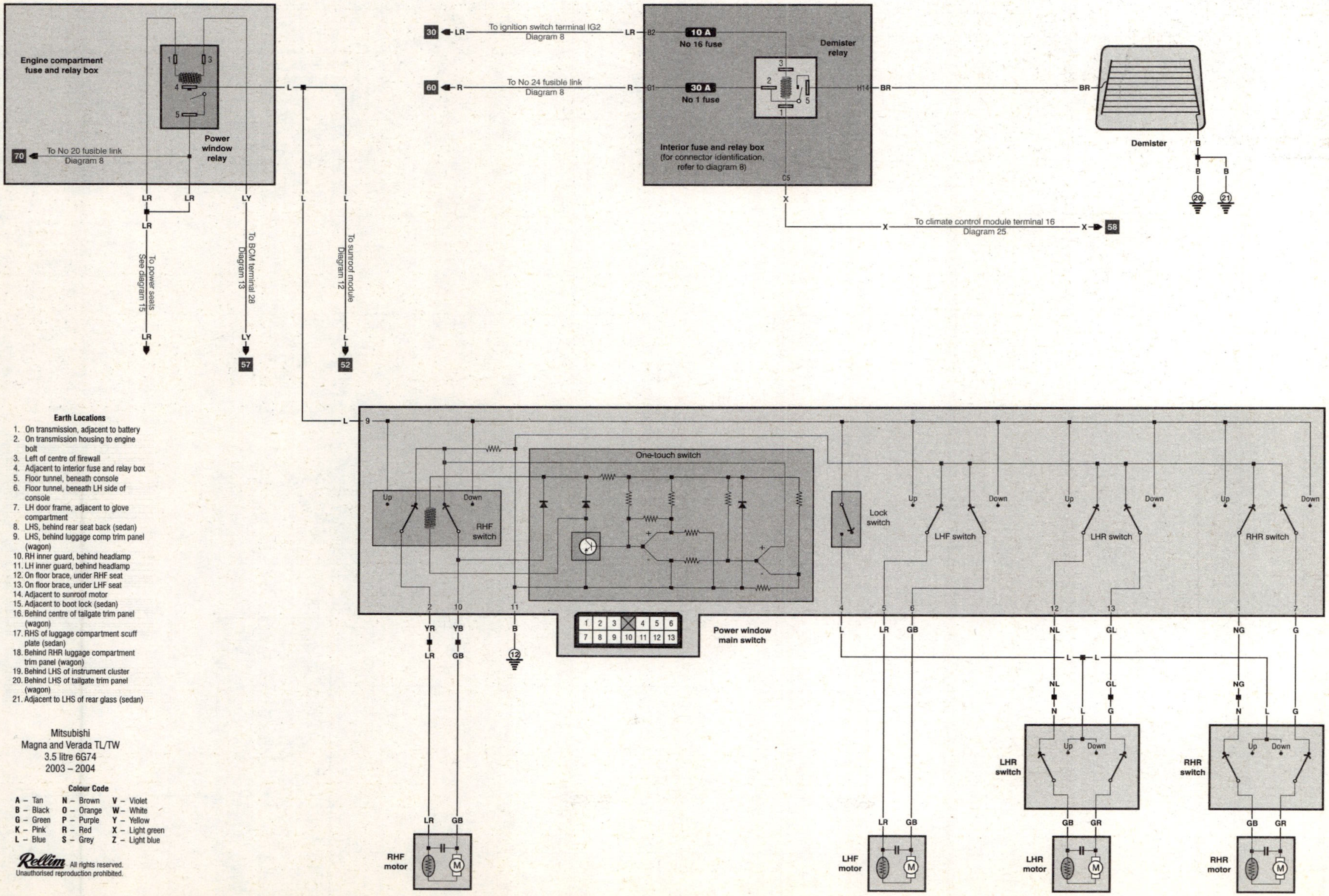

Earth Locations

1. On transmission, adjacent to battery
2. On transmission housing to engine bolt
3. Left of centre of firewall
4. Adjacent to interior fuse and relay box
5. Floor tunnel, beneath console
6. Floor tunnel, beneath LH side of console
7. LH door frame, adjacent to glove compartment
8. LHS, behind rear seat back (sedan)
9. LHS, behind luggage comp trim panel (wagon)
10. RH inner guard, behind headlamp
11. LH inner guard, behind headlamp
12. On floor brace, under RHF seat
13. On floor brace, under LHF seat
14. Adjacent to sunroof motor
15. Adjacent to boot lock (sedan)
16. Behind centre of tailgate trim panel (wagon)
17. RHS of luggage compartment scuff plate (sedan)
18. Behind RHR luggage compartment trim panel (wagon)
19. Behind LHS of instrument cluster
20. Behind LHS of tailgate trim panel (wagon)
21. Adjacent to LHS of rear glass (sedan)

Mitsubishi
Magna and Verada TL/TW
3.5 litre 6G74
2003 – 2004

Colour Code

A – Tan	N – Brown	V – Violet
B – Black	O – Orange	W – White
G – Green	P – Purple	Y – Yellow
K – Pink	R – Red	X – Light green
L – Blue	S – Grey	Z – Light blue

Diagram 16 — Power windows (except Verada), demister. TL/KL & TW/KW models.

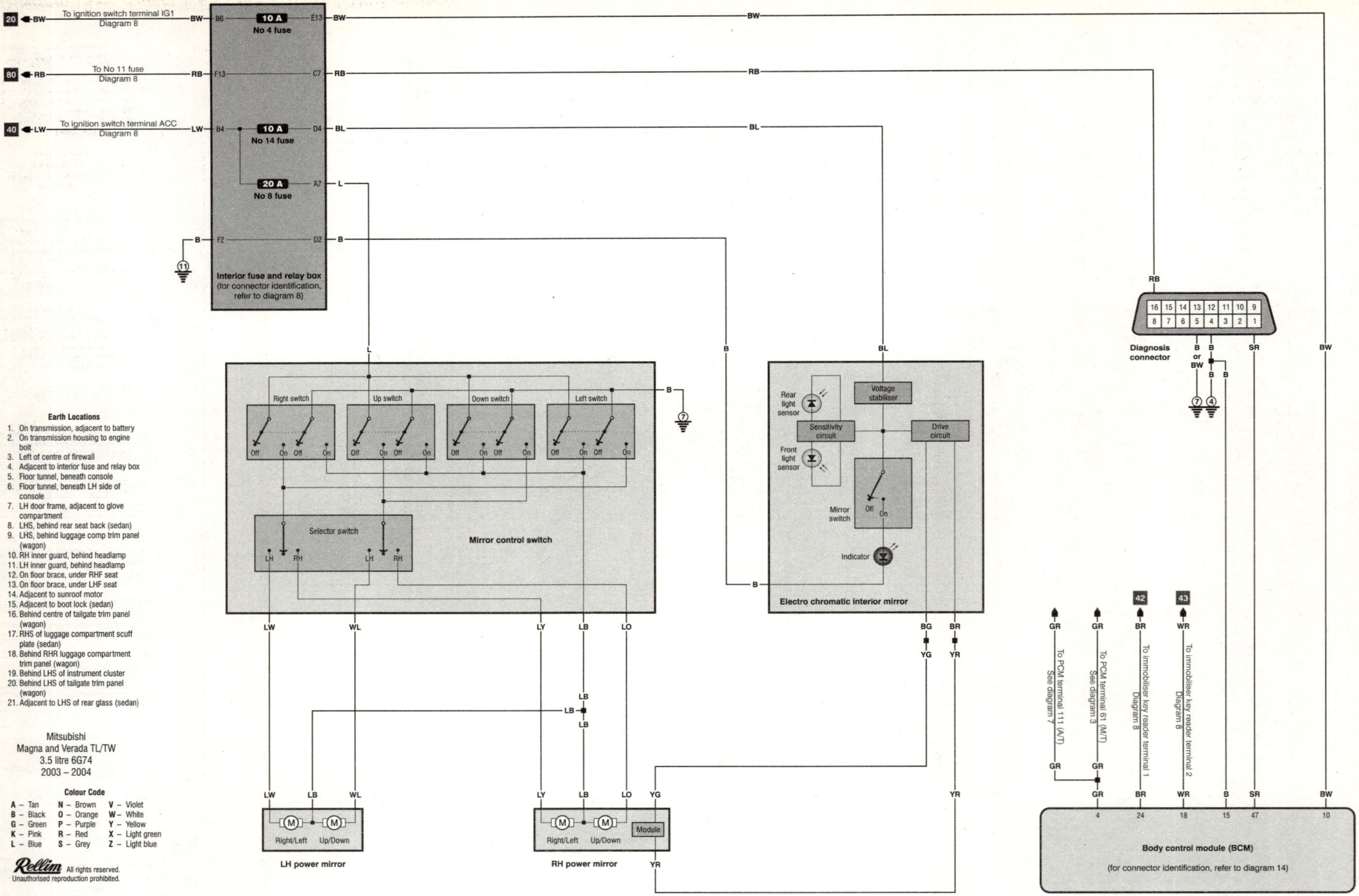

Diagram 17 — Power mirrors, electro chromatic interior mirror, immobiliser. TL/KL & TW/KW models.

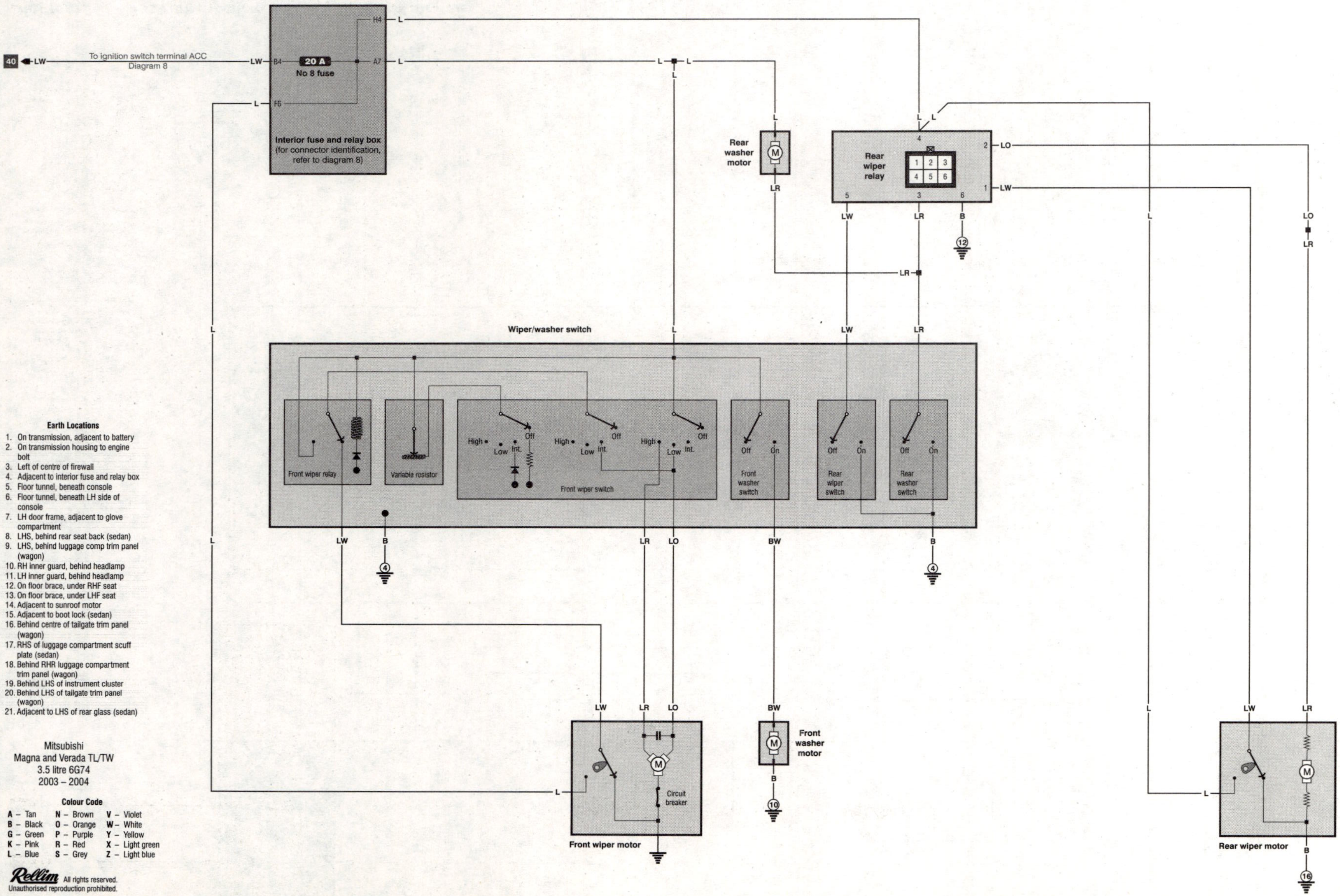

Diagram 18 — Wipers and washers. TL/KL & TW/KW models.

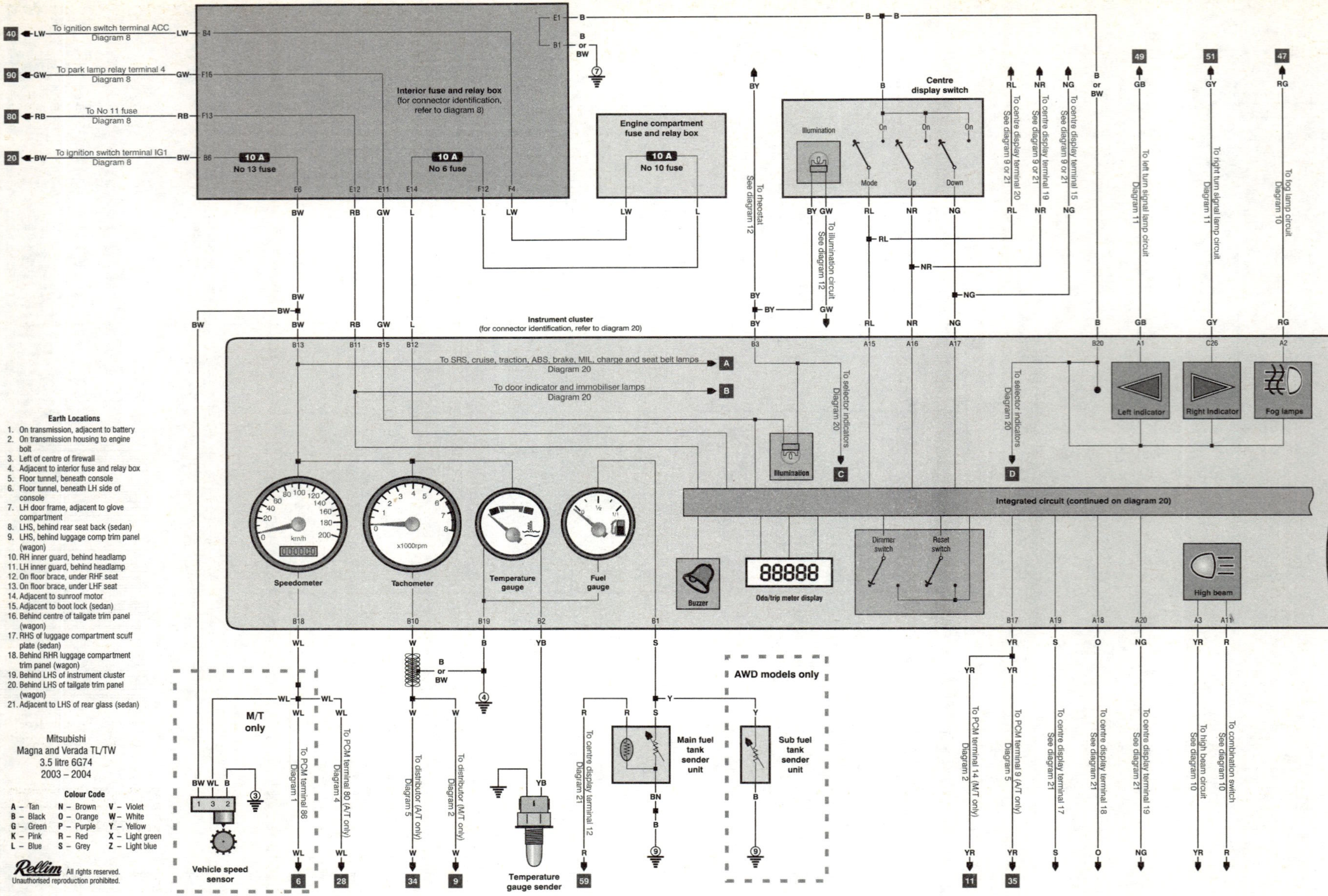

Diagram 19 — Instrument cluster. TL/KL & TW/KW models.

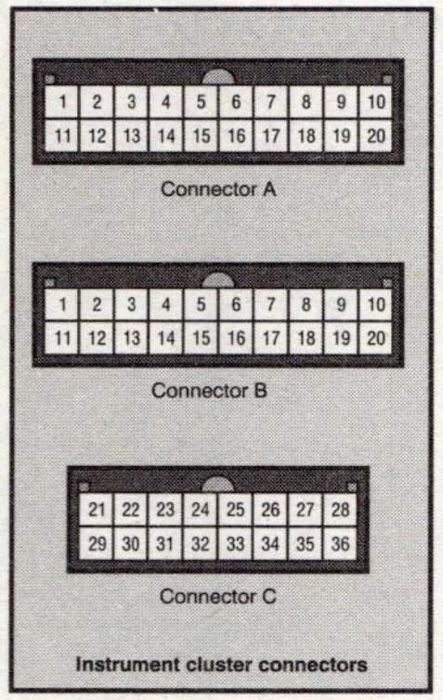

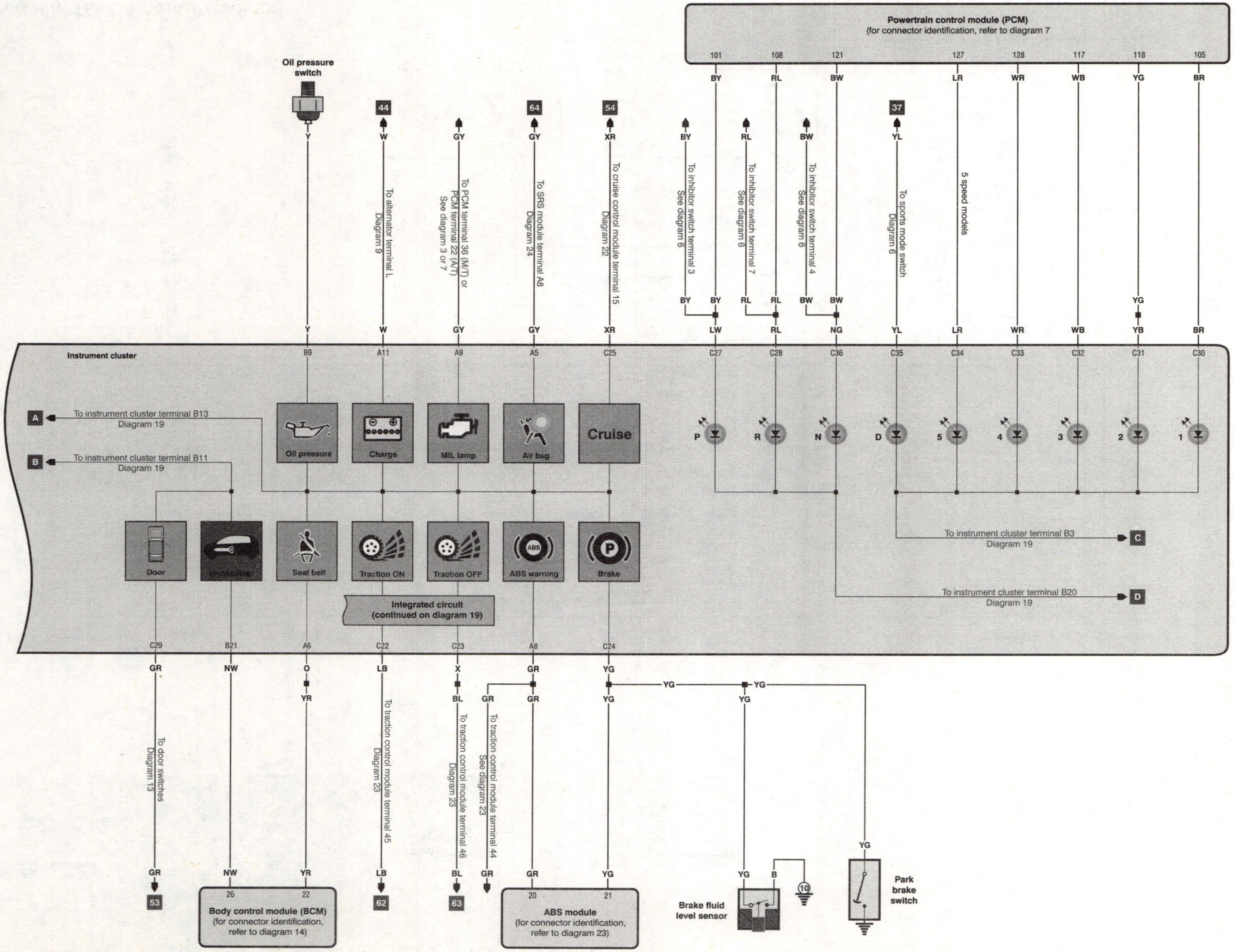

Earth Locations

1. On transmission, adjacent to battery
2. On transmission housing to engine bolt
3. Left of centre of firewall
4. Adjacent to interior fuse and relay box
5. Floor tunnel, beneath console
6. Floor tunnel, beneath LH side of console
7. LH door frame, adjacent to glove compartment
8. LHS, behind rear seat back (sedan)
9. LHS, behind luggage comp trim panel (wagon)
10. RH inner guard, behind headlamp
11. LH inner guard, behind headlamp
12. On floor brace, under RHF seat
13. On floor brace, under LHF seat
14. Adjacent to sunroof motor
15. Adjacent to boot lock (sedan)
16. Behind centre of tailgate trim panel (wagon)
17. RHS of luggage compartment scuff plate (sedan)
18. Behind RHR luggage compartment trim panel (wagon)
19. Behind LHS of instrument cluster
20. Behind LHS of tailgate trim panel (wagon)
21. Adjacent to LHS of rear glass (sedan)

Mitsubishi
Magna and Verada TL/TW
3.5 litre 6G74
2003 – 2004

Colour Code

A – Tan	N – Brown	V – Violet
B – Black	O – Orange	W – White
G – Green	P – Purple	Y – Yellow
K – Pink	R – Red	X – Light green
L – Blue	S – Grey	Z – Light blue

Diagram 20 — Instrument cluster. TL/KL & TW/KW models.

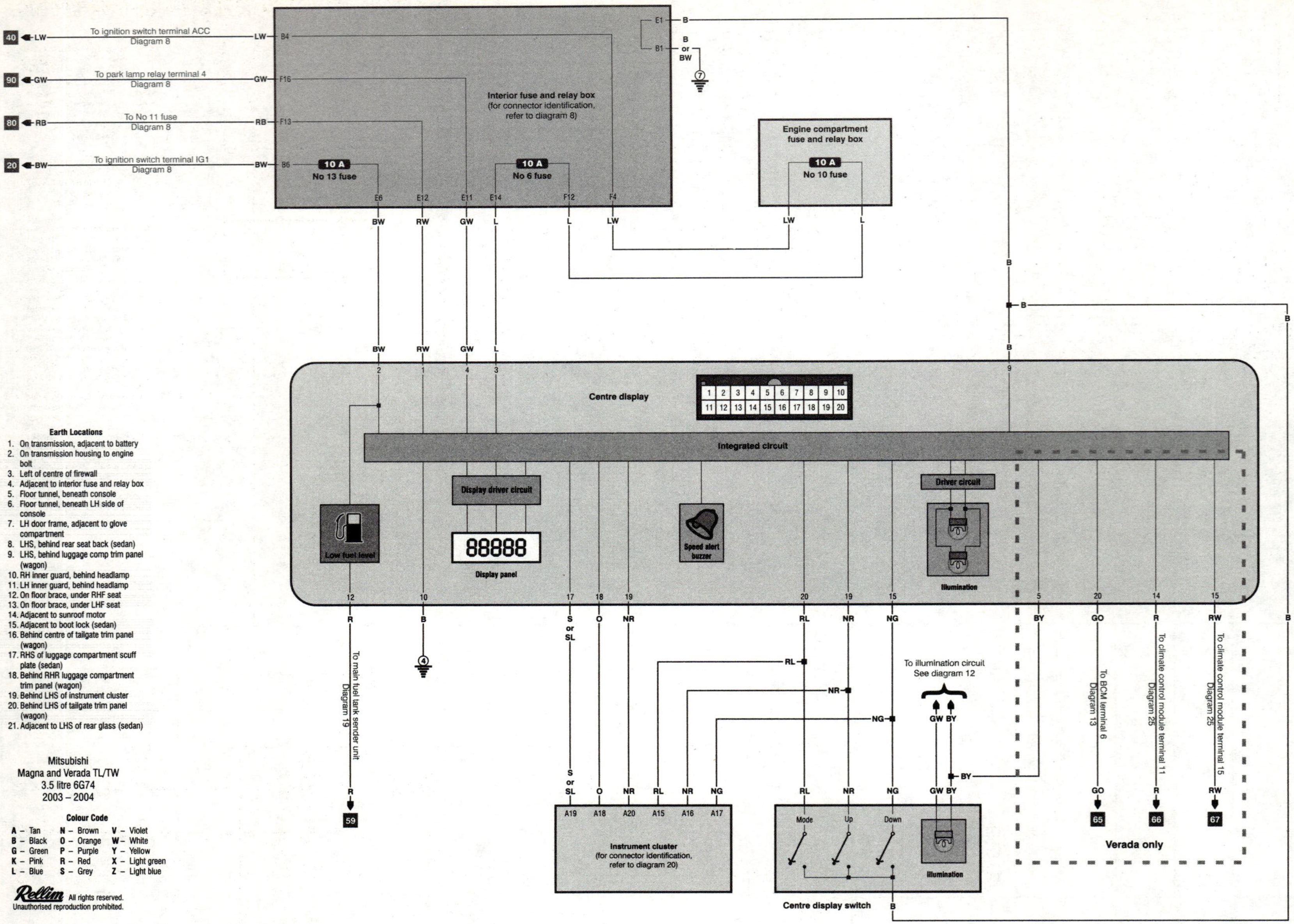

Earth Locations

1. On transmission, adjacent to battery
2. On transmission housing to engine bolt
3. Left of centre of firewall
4. Adjacent to interior fuse and relay box
5. Floor tunnel, beneath console
6. Floor tunnel, beneath LH side of console
7. LH door frame, adjacent to glove compartment
8. LHS, behind rear seat back (sedan)
9. LHS, behind luggage comp trim panel (wagon)
10. RH inner guard, behind headlamp
11. LH inner guard, behind headlamp
12. On floor brace, under RHF seat
13. On floor brace, under LHF seat
14. Adjacent to sunroof motor
15. Adjacent to boot lock (sedan)
16. Behind centre of tailgate trim panel (wagon)
17. RHS of luggage compartment scuff plate (sedan)
18. Behind RHR luggage compartment trim panel (wagon)
19. Behind LHS of instrument cluster
20. Behind LHS of tailgate trim panel (wagon)
21. Adjacent to LHS of rear glass (sedan)

Mitsubishi
Magna and Verada TL/TW
3.5 litre 6G74
2003 – 2004

Colour Code

A – Tan	N – Brown	V – Violet
B – Black	O – Orange	W – White
G – Green	P – Purple	Y – Yellow
K – Pink	R – Red	X – Light green
L – Blue	S – Grey	Z – Light blue

Diagram 21 — Centre display. TL/KL & TW/KW models.

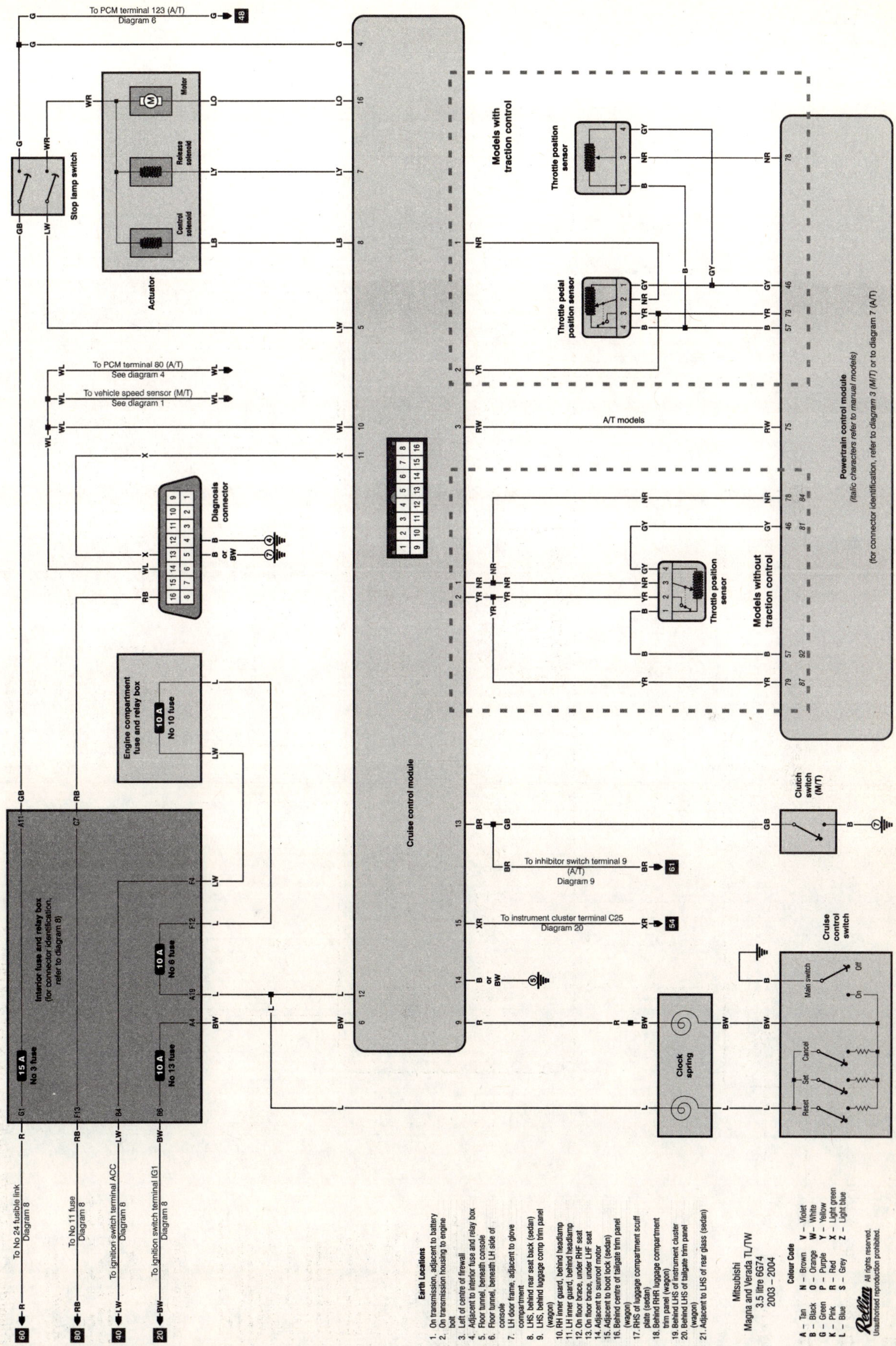

Diagram 22 — Cruise control. TL/KL & TW/KW models.

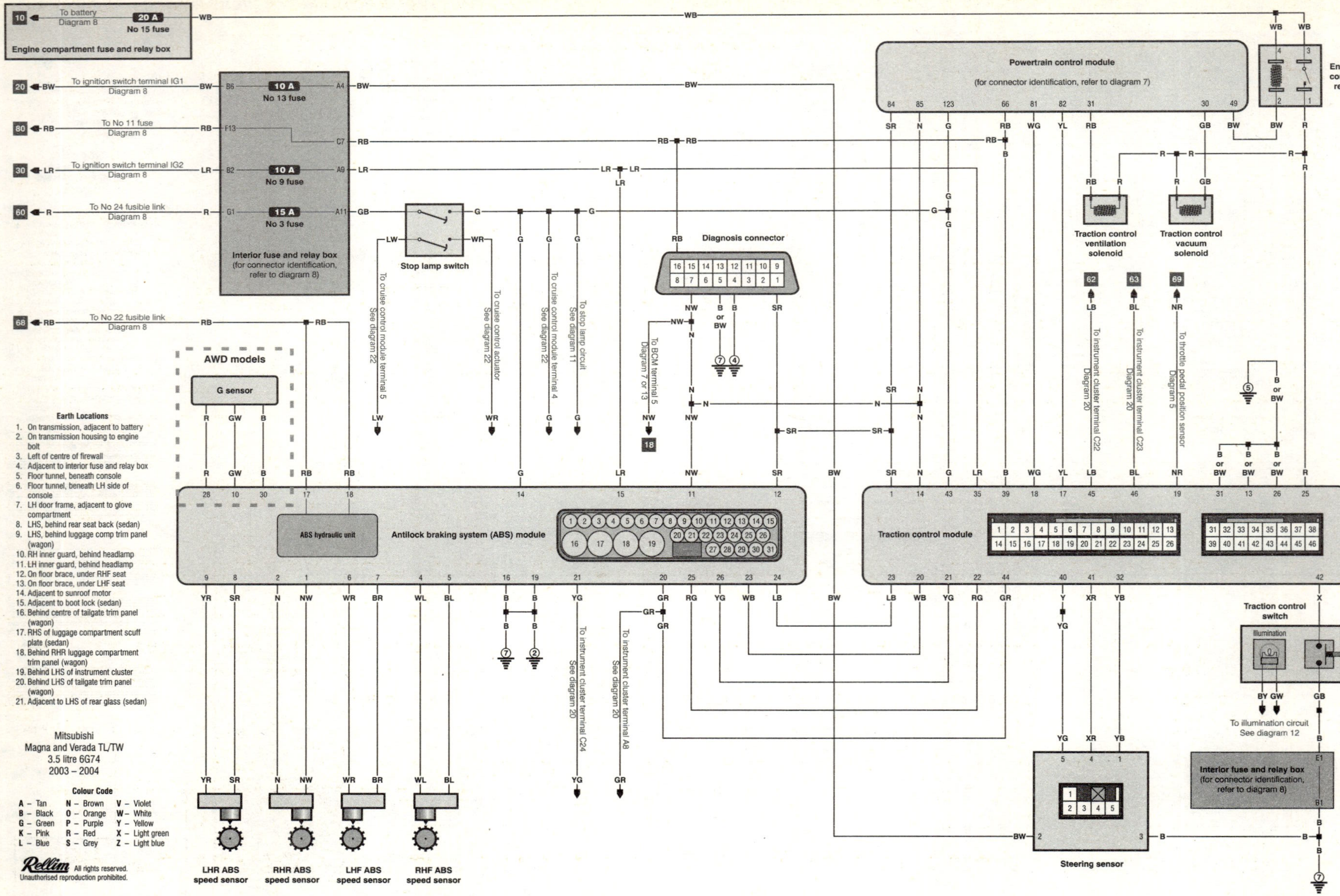

Diagram 23 — Antilock braking system (ABS), traction control. TL/KL & TW/KW models.

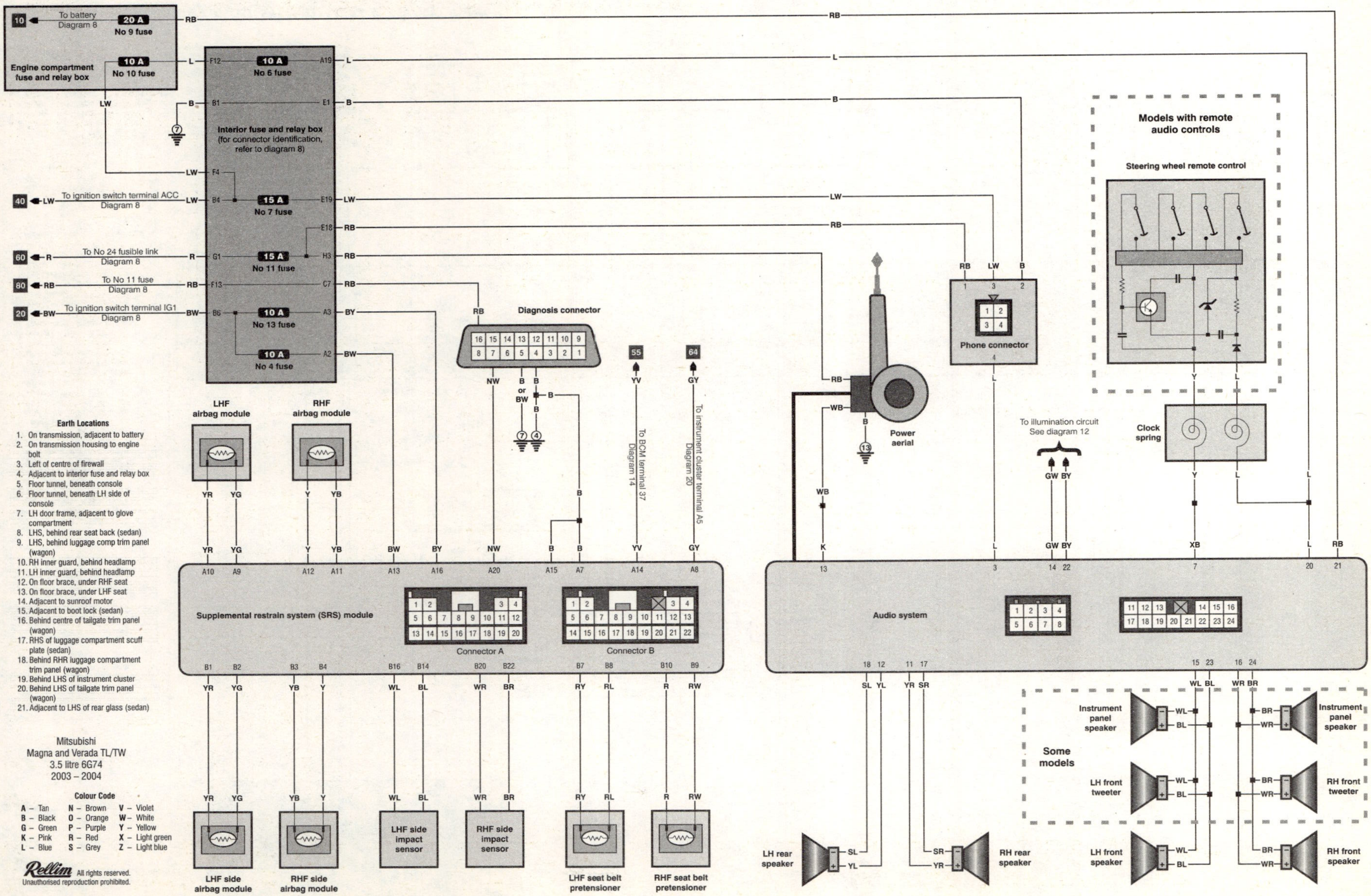

Earth Locations

1. On transmission, adjacent to battery
2. On transmission housing to engine bolt
3. Left of centre of firewall
4. Adjacent to interior fuse and relay box
5. Floor tunnel, beneath console
6. Floor tunnel, beneath LH side of console
7. LH door frame, adjacent to glove compartment
8. LHS, behind rear seat back (sedan)
9. LHS, behind luggage comp trim panel (wagon)
10. RH inner guard, behind headlamp
11. LH inner guard, behind headlamp
12. On floor brace, under RHF seat
13. On floor brace, under LHF seat
14. Adjacent to sunroof motor
15. Adjacent to boot lock (sedan)
16. Behind centre of tailgate trim panel (wagon)
17. RHS of luggage compartment scuff plate (sedan)
18. Behind RHR luggage compartment trim panel (wagon)
19. Behind LHS of instrument cluster
20. Behind LHS of tailgate trim panel (wagon)
21. Adjacent to LHS of rear glass (sedan)

Mitsubishi
Magna and Verada TL/TW
3.5 litre 6G74
2003 – 2004

Colour Code

A – Tan	N – Brown	V – Violet
B – Black	O – Orange	W – White
G – Green	P – Purple	Y – Yellow
K – Pink	R – Red	X – Light green
L – Blue	S – Grey	Z – Light blue

Diagram 24 — Supplemental restraint system (SRS), audio system. TL/KL & TW/KW models.

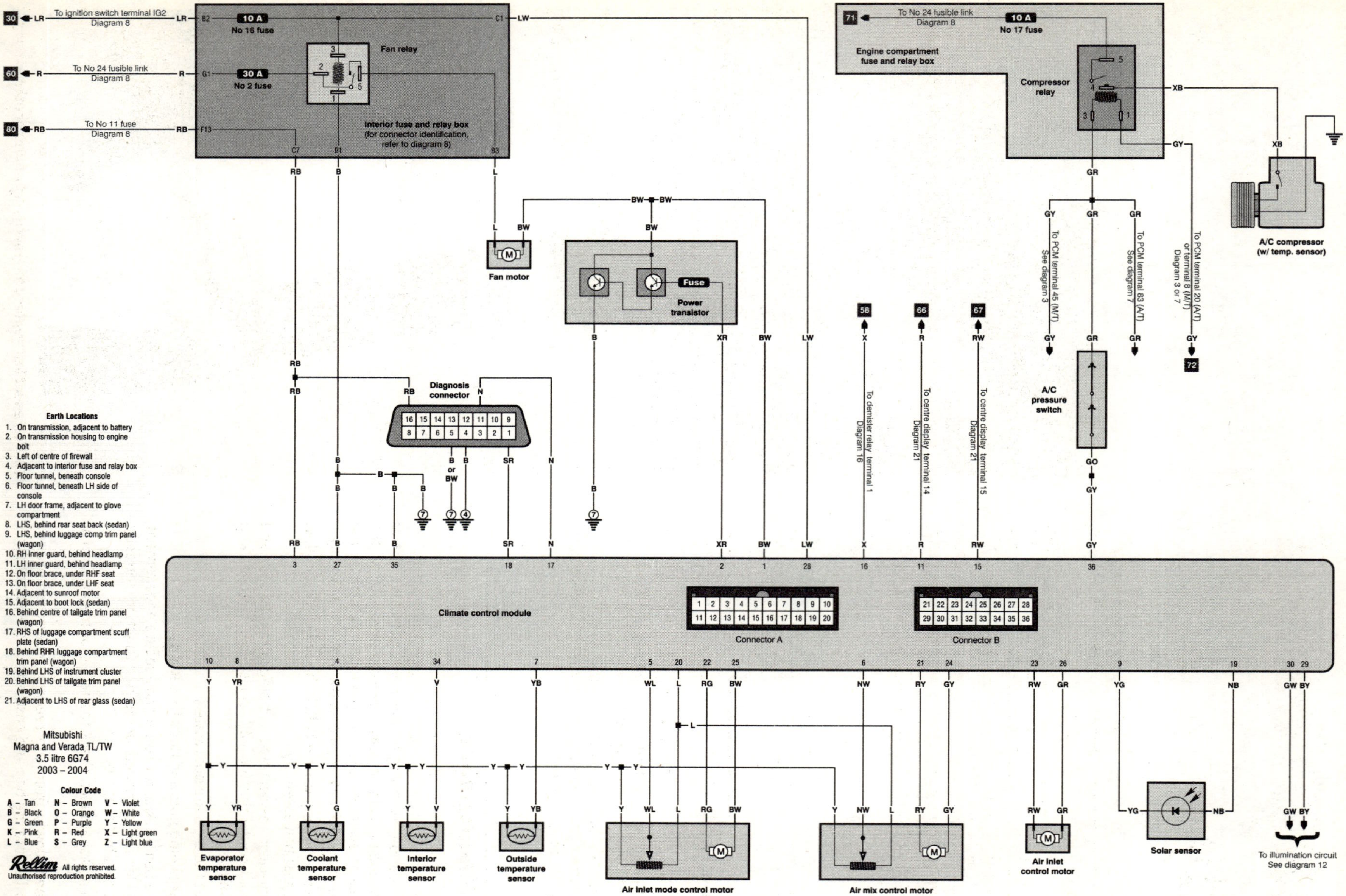

Earth Locations

1. On transmission, adjacent to battery
2. On transmission housing to engine bolt
3. Left of centre of firewall
4. Adjacent to interior fuse and relay box
5. Floor tunnel, beneath console
6. Floor tunnel, beneath LH side of console
7. LH door frame, adjacent to glove compartment
8. LHS, behind rear seat back (sedan)
9. LHS, behind luggage comp trim panel (wagon)
10. RH inner guard, behind headlamp
11. LH inner guard, behind headlamp
12. On floor brace, under RHF seat
13. On floor brace, under LHF seat
14. Adjacent to sunroof motor
15. Adjacent to boot lock (sedan)
16. Behind centre of tailgate trim panel (wagon)
17. RHS of luggage compartment scuff plate (sedan)
18. Behind RHR luggage compartment trim panel (wagon)
19. Behind LHS of instrument cluster
20. Behind LHS of tailgate trim panel (wagon)
21. Adjacent to LHS of rear glass (sedan)

Mitsubishi
Magna and Verada TL/TW
3.5 litre 6G74
2003 – 2004

Colour Code

A – Tan	N – Brown	V – Violet
B – Black	O – Orange	W – White
G – Green	P – Purple	Y – Yellow
K – Pink	R – Red	X – Light green
L – Blue	S – Grey	Z – Light blue

Diagram 25 — Climate control system. TL/KL & TW/KW models.

Chapter 17
Body

Contents

Airbag Warning

Use extreme caution when working around the airbag system. When carrying a live airbag assembly, always carry it with the trim cover facing away from you.

When storing the assembly on a bench, always store it with the trim cover facing up and avoid placing anything near or on the assembly.

Routine Maintenance

SEAT BELTS

1 Check the seat belt webbing for fraying, cuts, burns or other damage. On inertia reel seat belts, slowly pull the belt from the retractor reel to allow the entire length to be examined. If the webbing is damaged or worn, renew the seat belt assembly.

2 Check the rip stitching on the belt. If the red label is visible, renew the seat belt assembly.

3 If necessary, clean the seat belt webbing with mild soap and water. Avoid contamination with polish, oil, bleaches and other chemicals.

4 Ensure that the belts are not twisted.

5 Check all retaining nuts and bolts for security.

6 Inspect the belt buckles for correct and secure engagement. If the buckles are damaged or do not function correctly, renew the seat belt assembly.

7 On inertia reel seat belts, check the retractor mechanism as follows:

a Pull the belt slowly from the reel and check for smooth operation. Release the belt and ensure that it retracts fully.

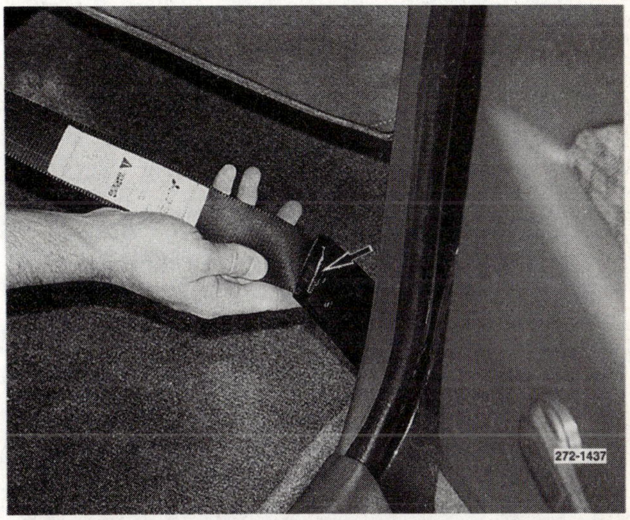

Seat belt rip stitching location, normally concealed by the plastic sleeve. When the red 'replace belt' label becomes visible at the point indicated by the arrow, the belt must be renewed.

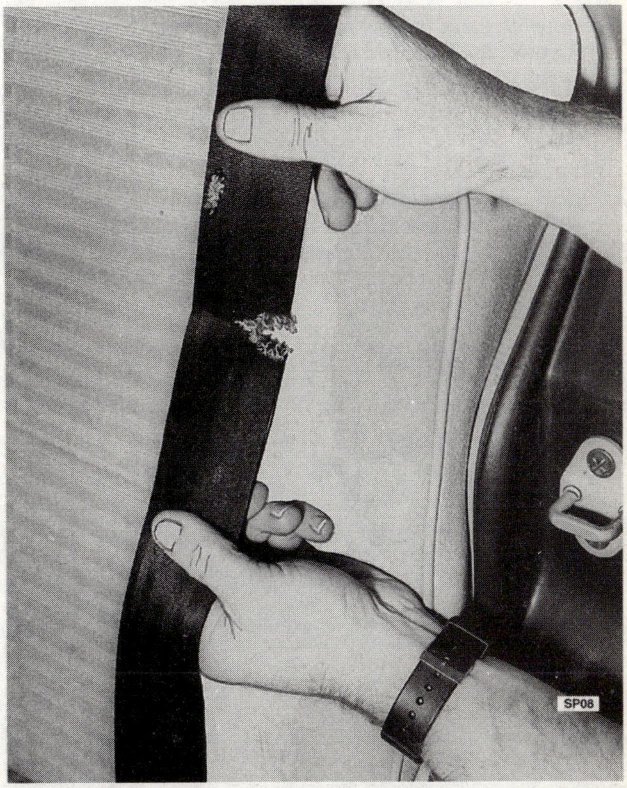

Check the seat belt webbing for fraying, cuts, burns or other damage.

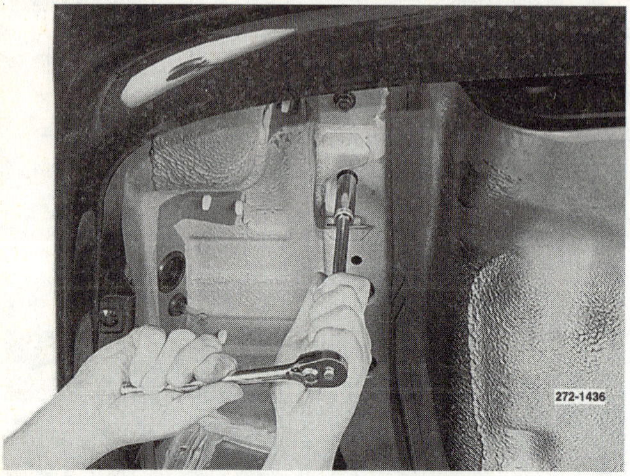

Checking the tightness of the rear bumper retaining bolts.

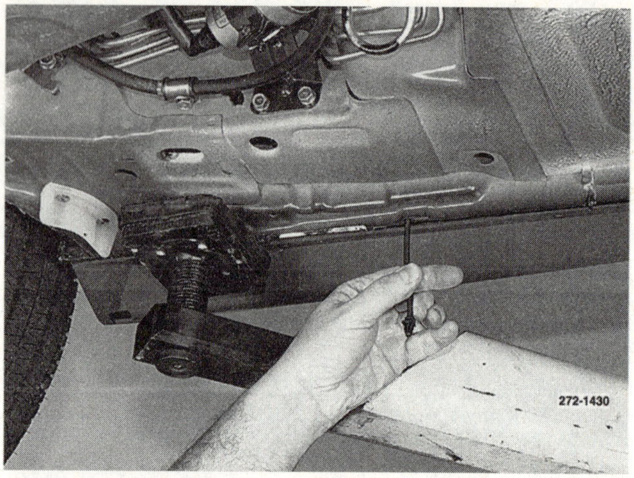

Cleaning the body panel drain holes using a nylon cable tie.

b Pull the belt rapidly from the reel. The reel should lock.

c Drive the vehicle at walking pace and attempt to withdraw the belt slowly while braking sharply. The reel should lock.

If the retractor mechanism does not function correctly, renew the seat belt assembly.

BODY

1. Periodically check the tightness of nuts and bolts retaining various body components.
2. Lubricate the door strikers and the bonnet lock with Dri Lube or similar. Avoid using grease on these components, as it can stain your clothes.
3. Lubricate the door hinges sparingly with engine oil or spray grease.
4. Apply graphite powder to the lock cylinders. Do not use oil or other liquid lubricants.
5. To prevent rust, periodically clean the wheel arches and the underbody of accumulated mud and dirt, using a strong jet of water.

After washing the underbody, the brakes may operate less efficiently due to wet brake pads. Apply the brakes for a short period of time while driving slowly. Be prepared for uneven brake operation and/or increased stopping distances.

6. Clean the exterior of the vehicle, using good quality car care products. Avoid using dishwashing detergent. Add a few drops of kerosene to the rinsing water to prevent rust developing.
7. Clean the drain holes, located in the bottom of the doors and also in the sill and in the quarter panels. Use a blunt instrument made of wood or plastic to prevent damage to the paint.
8. Ensure that the rubber plugs sealing the underbody access holes are in serviceable condition and correctly installed.
9. Inspect the paintwork for small scratches and chips. Apply the appropriate colour touch-up paint to the effected area, following the paint manufacturers instructions.

Airbag (SRS)

To Remove and Install Driver Airbag – T30 Torx socket

DESCRIPTION

The airbag is a supplemental restraint system (SRS), designed to be used in conjunction with the seat belt. The system uses two crash sensors, in the SRS control module that is located forward of the centre console.

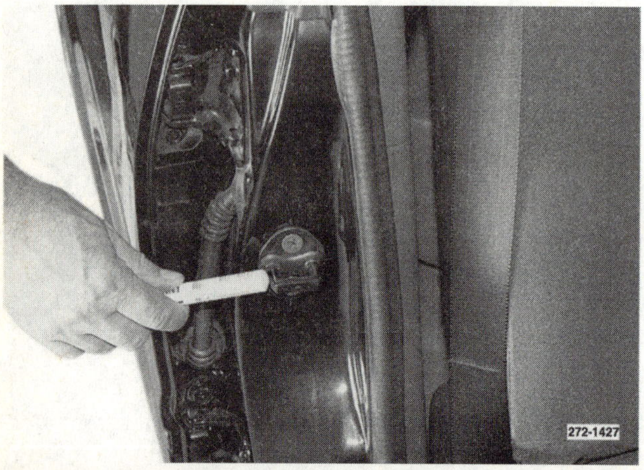

Lubricating the door lock striker with Dri Lube.

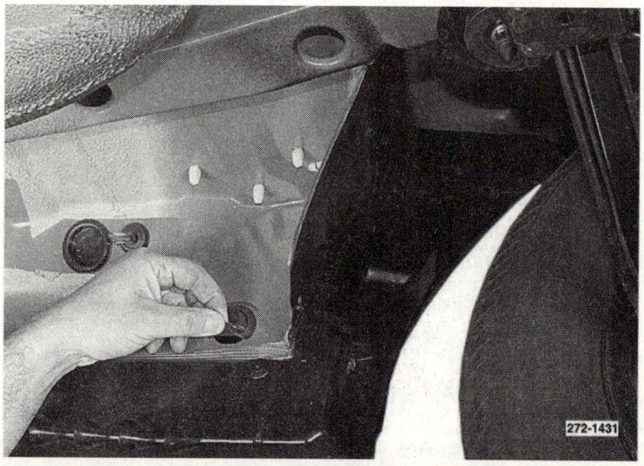

Checking for trapped water within the body panels.

In the event of a frontal collision in which a pre determined speed has been exceeded and the two sensors signal an impact, the airbag is activated. On models with passenger side airbag both the driver and passenger side bags are deployed simultaneously.

The airbag on the drivers side is located in the centre of the steering wheel. On models with passenger side airbag, the airbag is located behind the dashboard, in a special compartment above the glove compartment.

The deployment of the airbag is very rapid and the inflated airbag slows the forward motion of the occupant. After the deployment the airbag quickly deflates.

The airbag is electrically activated, but in the event of damage to the battery or its wiring, the module has its own backup power supply which it uses to deploy the system.

PRECAUTIONS

When working on the airbag or components of the vehicle which may effect the airbag, the following precautions must be observed.

Disable and remove the airbag when working on the steering wheel, the steering column, the dashboard or the wiring in the vicinity of SRS components (module or airbag).

When removed, always carry the airbag with the trim cover away from the person handling the device.

When storing the airbag, ensure that the trim cover is facing upwards.

When towing a vehicle with major front damage, disconnect the battery as a precaution.

Disconnecting the battery alone will not deactivate the airbag. The airbag must be disconnected from the vehicle wiring harness. On models with a passenger airbag, both airbags must be disconnected to disable the system.

TO DISABLE

1 Disconnect the negative battery terminal and wait at least one minute before commencing to remove the airbag assembly from the vehicle.

Location of the driver side airbag connector. Lower dashboard panel removed.

Because the battery and battery cables are confined by the battery box and cover, the manufacturer recommends that the negative battery cable be wrapped in cloth or insulation tape to prevent it from contacting with the negative battery terminal.

2 Remove the screws retaining the bonnet lever to the lower dash panel and position the lever clear of the work area.

3 Remove the screw cover from the top of the driver side lower dash panel. Remove the remaining screws from the lower dash panel, disconnect the panel from each side of the steering column and partially remove the panel.

4 Where fitted, rotate the footwell lamp bulb holder anti clockwise and remove the panel from the vehicle.

5 Lift the red tab on the airbag connector and separate the wiring connectors.

6 Where a passenger airbag is fitted, proceed as follows:

a Open the glove compartment and pull down the cover from the top of the glove compartment surround panel.

b Lift the red tab on the airbag connector and separate the wiring connectors.

Passenger airbag connector location. The connector can be accessed without removing the glove box surround.

TO ENABLE

1 Connect the driver airbag wiring connector, ensuring that the red tab on the connector is securely engaged.

2 Install the lower dash panel and bonnet lever, tightening the retaining screws securely.

3 Where a passenger airbag is installed, connect the airbag wiring connector to the harness connector, ensuring that the red tab on the connector is securely engaged. Securely install the cover to the top of the glove compartment surround panel.

4 Remove the insulation tape and connect the negative battery terminal.

TO REMOVE AND INSTALL

Driver Airbag

1 Disable the driver airbag as previously described.

2 Ensure that the steering wheel is in the straight ahead position and the key is removed from the ignition.

3 Remove the two T30 Torx bolts retaining the airbag.

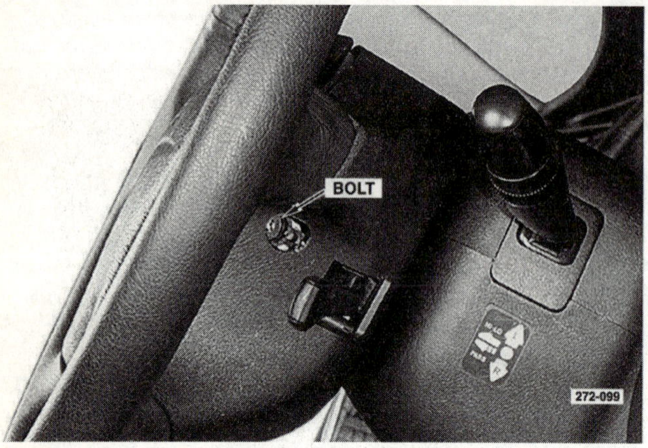

Airbag retaining bolt location. A second bolt is located on the opposite side of the steering wheel.

4 Partially remove the airbag, spread the tangs on each side of the airbag connector and pull the connector from the airbag.

5 Remove the drivers airbag from the vehicle and place it on a clean surface with the pad facing upward.

Installation is a reversal of the removal procedure with attention to the following points:

> *If the steering column or steering shaft has been disconnected for any reason make absolutely sure that the clock spring is centered to avoid damage to the fine ribbon cable inside. Refer to Steering (chapter 12) for more information on the clock spring.*

1 Connect the airbag wiring connector to the airbag ensuring that the paint marks on the airbag and connector align.

2 Position the airbag onto the steering wheel and tighten the Torx bolts securely.

3 Enable the airbag as previously described and connect the negative battery terminal.

Passenger Airbag

1 Disable the passenger airbag as previously described.

2 Unscrew the plastic scrivets retaining the lower dash panel on the passenger side of the vehicle. When the screw is removed, pull the clip from the panel. Repeat this on the remaining scrivets.

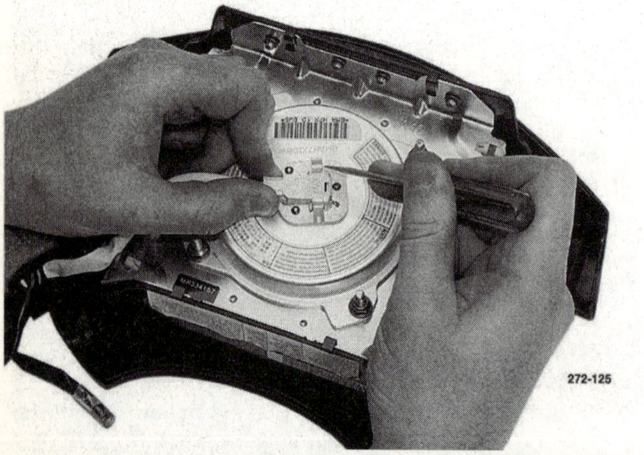

Spread the tangs to disconnect the wiring from the driver airbag.

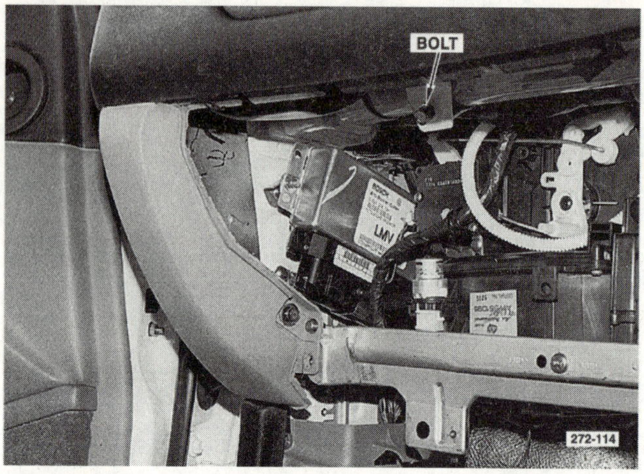

Glove compartment surround panel removed showing one of the passenger airbag retaining bolts.

3 Partially remove the panel and where fitted, rotate the footwell lamp bulb holder anticlockwise and disconnect from the panel. Remove the panel from the vehicle.

4 Remove the screws retaining the glove compartment. Open the glove compartment and remove it from the vehicle.

5 Remove the screws retaining the glove compartment latch and remove the latch. Remove the remaining screws and bolts that retain the glove compartment surround panel.

6 Partially remove the panel, rotate the glove compartment lamp bulb holder anticlockwise and disconnect from the panel. Pass the airbag wiring and cover through the opening in the surround panel and remove the surround panel from the vehicle.

7 Remove the bolts retaining the passenger airbag to the dashboard. Manoeuvre the airbag from the dashboard.

Installation is a reversal of the removal procedure with attention to the following points:

1 Manoeuvre the passenger airbag into position and install the retaining bolts securely.

2 Partially install the glove compartment surround panel and install the bulb holder and pass the wiring for the airbag through the opening.

3 Install the surround panel retaining screws and bolts and install the glove compartment latch.

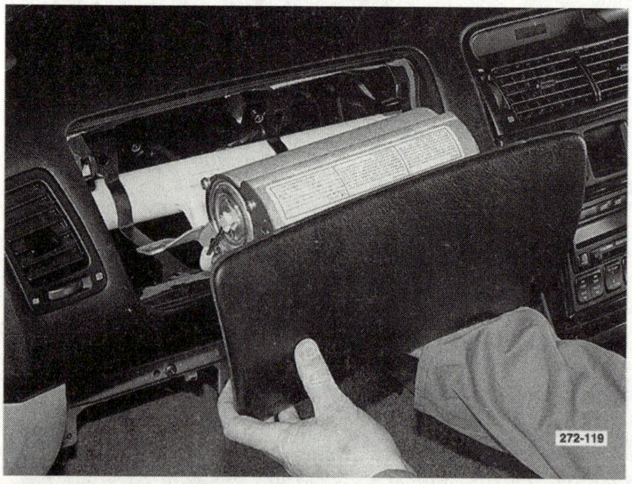

Removing the passenger side airbag.

4 Install the glove compartment and tighten the retaining screws securely.
5 Partially install the lower dash panel and where fitted, install the footwell lamp bulb holder.
6 Install the lower dash panel scrivets ensuring that they engage fully into the holes.
7 Enable the airbag as previously described.

Front Doors

INTERIOR HANDLES AND TRIM PANEL

To Remove and Install

1 Fully lower the window glass.
2 Pull the interior handle to the open position and remove the surround panel retaining screw. Withdraw the interior handle surround panel.
3 On models with manual windows using a suitable wire hook, withdraw the regulator handle retaining clip and remove the handle and washer.

Alternatively, remove the regulator handle using a clip removal tool.

4 On models with power windows, insert a small screwdriver with the blade wrapped in cloth between the door trim and the power window switch assembly. Gently lever the switch assembly upward, remove it from the trim panel and disconnect the wiring.
5 Remove the screw retaining the ventilation supply duct to the door.
6 Using a small screwdriver, remove the cover from the armrest door pull and remove the screw.
7 Grasp one of the bottom corners of the trim panel and carefully pull the panel away from the door, disengaging the retaining clips. Work around the entire door trim until all the clips are disengaged from the door.
8 Lift the door trim straight up to disengage the trim from the top of the door and partially remove the trim.

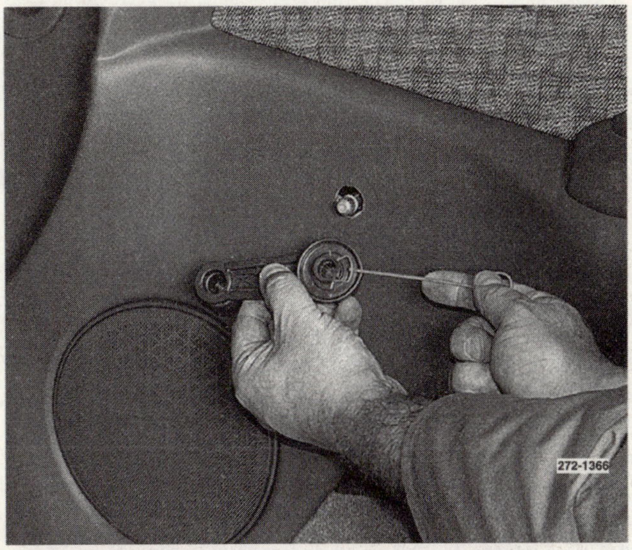

Rear of a regulator handle, showing where to hook a piece of wire to the clip.

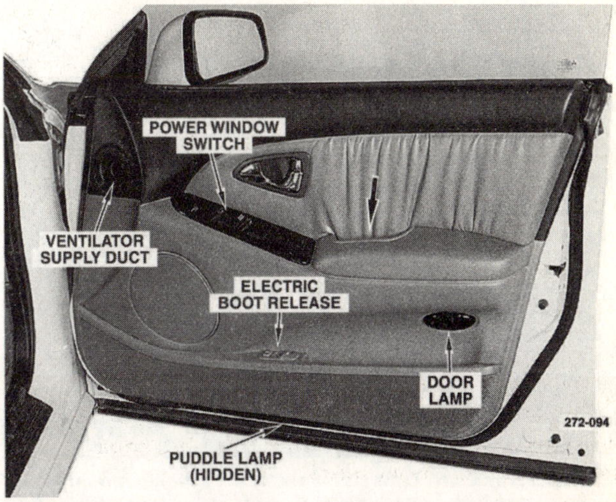

Verada door trim panel. The arrow indicates the location of the pull handle retaining screw and cover.

9 Where fitted, rotate the door lamp bulb holder anticlockwise and remove it from the lamp. Disconnect the lower door lamp wiring connector and the electric boot release and fuel door release wiring connector. If necessary, remove the boot release and fuel door release switch from the trim panel.
10 Remove the screws retaining the interior handle to the door. Partially remove the handle, and disconnect the control rods.
11 Remove the speaker retaining screws and partially remove the speaker. Disconnect the speaker wiring connector and remove the speaker from the door.
12 Unhook the speaker wiring from the speaker mounting bracket. Remove the retaining screws and remove the mounting bracket from the vehicle.
13 Remove the pull handle bracket retaining screws to the door and remove the bracket.
14 Carefully remove the sealing sheet from the inner door panel using a small knife to cut the adhesive, if necessary. Feed the wiring harnesses through the sealing sheet and remove the sheet from the vehicle.

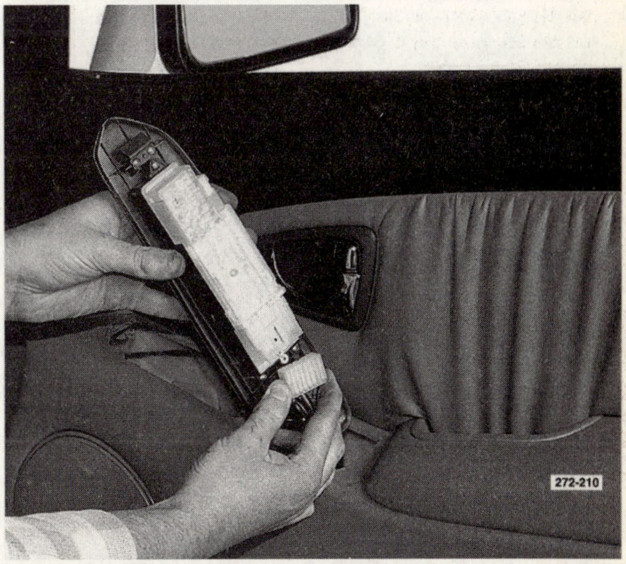

Disconnecting the power window switch.

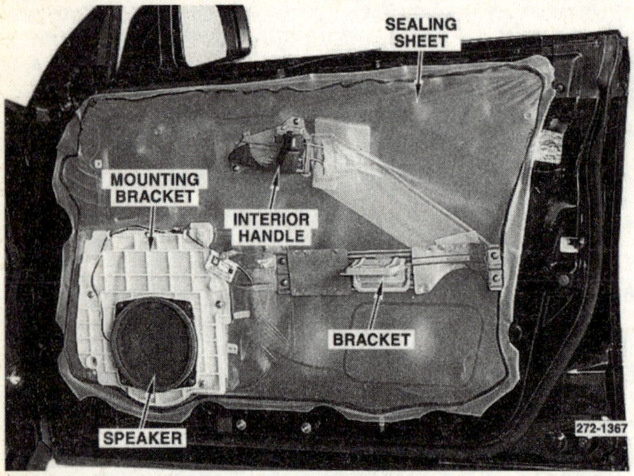

Driver door trim panel removed showing the speaker mounting bracket and plastic sealing sheet.

Installation is a reversal of the removal procedure with attention to the following points:

1. Feed the wiring through the sealing sheet ensuring that it is correctly positioned. Secure the sealing sheet to the door.
2. Connect the interior handle control rods, install the handle and tighten the retaining screws securely.
3. Ensure that the pull handle bracket is installed and the screws are secure.
4. Install the speaker mounting bracket and tighten the retaining screws.
5. Install the speaker wiring to the hooks on the mounting bracket and connect the wiring to the speaker. Position the speaker on the mounting bracket and tighten the retaining screws.
6. Check the upper trim panel retaining clips and renew any damaged clips if necessary.
7. Install the door lamp bulb holder to the door trim. Where fitted connect the puddle lamp, boot release and fuel door release wiring connectors.
8. Install the trim panel to the door and firmly push on the upper edge to locate the trim panel on the door panel.
9. Check that the trim panel retaining clips are aligned with the holes in the door. If not, the trim panel is not fully engaged with the top of the door.
10. On models with power windows, ensure that the power window switch wiring connector is accessible.
11. Press the trim panel retaining clips firmly into the door and check that the trim is sitting flush against the door.
12. On models with manual windows, install the washer and the clip to the regulator handle. Press the handle onto the shaft, ensuring that the handle and window glass are positioned as on the opposite door.
13. On models with power windows, connect the power window wiring switch wiring connector to the switch assembly and install the switch assembly to the door trim.

WINDOW GLASS

To Adjust

TIP *Before adjustment commences, it is advisable to suitably mark the factory settings of all adjusters as a reference.*

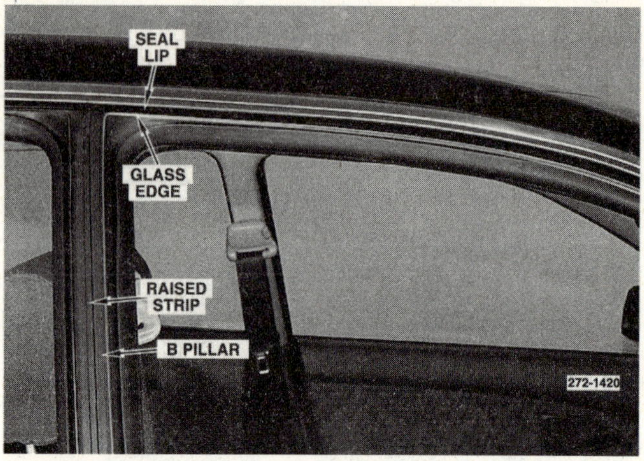

With the door closed to the first stop. Check that the edge of the window glass is parallel with the seal lip.

1. Ensure that there is an even gap between the front of the door and the front guard and also between the rear of the front door and the rear door. If not, loosen the hinge bolts and with the aid of an assistant raise or lower the door assembly until an even gap is obtained between top and bottom of the door. After doing this, it may also be necessary to adjust the lock striker as described later in this chapter.
2. Check that the delta cover and window guide assembly is secure and also that the weather strip along the top outside edge of the door is not worn or damaged. Renew components as necessary.
3. After ensuring that the window is fully up, shut the door to the first stop and observe the position of the window glass.

   The top of the window glass should be parallel with the lip on the upper rubber seal. Refer to the illustration.

TIP *If unsure of determining the window position, compare it to the opposite door after ensuring that the window is fully wound up and the check is made with the door shut only to the first stop.*

4. If the window glass requires adjustment, proceed as follows:
5. Remove the door trim panel and sealing sheet as previously described.

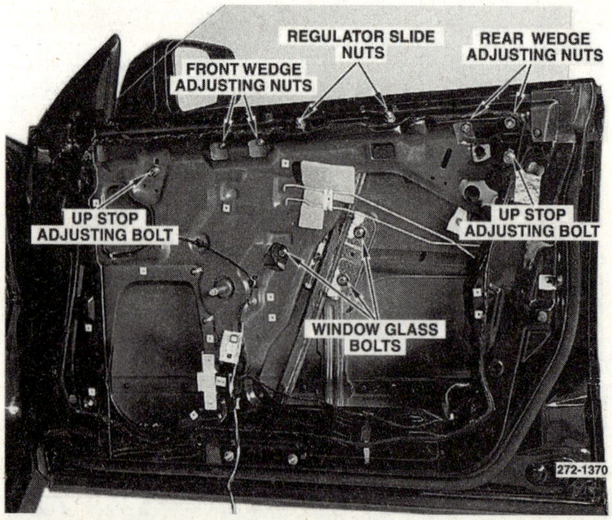

Front window glass adjusting points.

6 Loosen the front and rear up stop bolts and move the stops to achieve the correct height of the window glass. Tighten the bolts and check the window glass position.

> *It will be necessary to lower the window glass slightly to relieve the tension on the stops. Ensure that the window is wound fully up again after each adjustment is made. Also slight adjustment to the wedge adjusters may be necessary to allow sufficient travel. Do not move the wedges excessively or the glass will rattle.*

7 Loosen the bolts retaining the window glass to the regulator slide and tilt the window until the glass is parallel with the lip on the rubber seal.

It may be necessary to repeat both adjustments several times until the correct position of the window is achieved.

8 Wind the window down and up. If it is excessively tight, loosen the nuts retaining the window regulator slide to the top of the door and slide the regulator slide away from the outside of the door. Tighten the nut and recheck the window winding operation.

9 Ensure that the window is wound fully up and close the door completely.

10 As a final check, slowly close the door until the top edge of the window glass contacts the rubber seal. At this point, measure the distance between the upper edge of the door and the vehicle body. The distance should be between 10 – 25 mm.

Also, ensure that the top of the window glass sits approximately 2.5 – 3.5 mm under the lip on the upper rubber seal with the door fully closed. Also, check that the rear of the glass is parallel with the raised edge strip on the B pillar.

> *Due to the contour of the window glass, a difference of approximately 1 mm is allowed between readings taken at the front and rear of the window glass.*
>
> *It will be necessary to make adjustments in small increments, gently shut the door and repeat the adjusting procedure until the specified gaps are achieved.*
>
> *Use caution when closing the door not to shatter the glass when altering the window height adjustment.*

11 Install the trim panel temporarily, shut the door and wind the window up and down. If excessive effort is required to wind the window up and down or, if on models with power windows the motor sounds strained, proceed as follows:

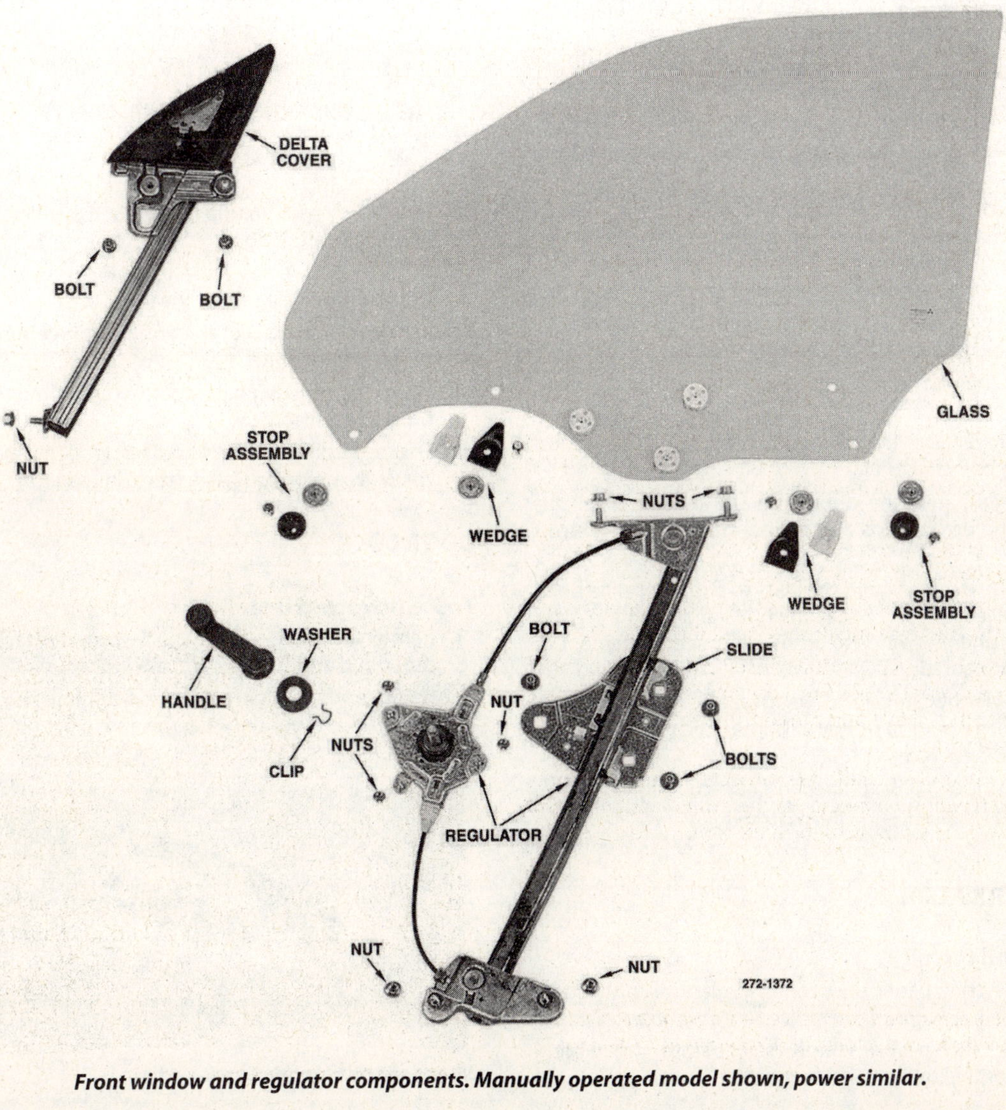

Front window and regulator components. Manually operated model shown, power similar.

- a Remove the trim panel. Loosen the two nuts on the upper window regulator slide and slide the regulator slide away from the outside of the door in small amounts until the window glass is just touching the outer weather strip seal.
- b Tighten the nuts, install the trim panel and check the operation of the window winder mechanism.

12 When the adjustment is satisfactory, install the trim panel as previously described.

To Remove and Install

1 Remove the trim panel and sealing sheet as previously described.

2 Remove the screw retaining the delta cover fascia panel and remove the panel from the door.

3 Remove the clips retaining the ends of the door opening weather seal. Fold down the ends of the seal until the screws retaining the weather strip are visible. Remove the screws and then remove the weather strip from the door.

4 Mark the installed position and then remove the two bolts and nut retaining the delta cover assembly to the front of the door. Tilt the top of the assembly toward the front of the vehicle to release the stud from the door frame and manoeuvre it from the door.

5 Raise the window glass until the regulator slide bolts and stops are visible through the access holes.

On models with manual windows, temporarily install the window regulator handle to raise the glass.

On models with power windows, temporarily connect the power window switch wiring and momentarily turn the ignition On.

6 Mark around the window stops and upper limit wedges using a permanent felt pen or similar. Remove the bolts and remove the stops and wedges from the window glass. Note the position of the retaining nuts and plastic washers.

7 Support the window glass and remove the regulator slide to glass retaining bolts.

8 Manoeuvre the glass through the opening at the top of the door.

Installation is a reversal of the removal procedure with attention to the following points:

1 Insert the glass into the door from the outside with the forward edge downwards and engage the glass in the run channels.

2 Carefully raise the glass until the glass to regulator slide bolts can be installed. Loosely install the bolts and push the glass downwards and rearwards to ensure proper engagement in the rear run channel.

3 Install the window stops and wedges, aligning them with the marks made prior to removal. Tighten the retaining bolts securely.

4 Adjust the window glass as previously described.

5 Install the trim panel as previously described.

WINDOW REGULATOR

To Remove and Install

1 Remove the trim panel and window glass as previously described.

2 On models with power windows, separate the wiring connector for the power window motor.

3 Mark the installed position and then remove the nuts retaining the regulator to the inner door panel. Manoeuvre the regulator assembly from the door.

4 On models with power windows, the electric motor may be removed from the regulator if necessary, after removing the retaining screws.

Installation is a reversal of the removal procedure with attention to the following points:

1 Lubricate the regulator moving components sparingly with multipurpose grease.

2 Install the regulator assembly and tighten the retaining nuts securely.

3 Install and adjust the window glass as previously described.

4 Install the trim panel as previously described.

EXTERIOR HANDLE AND LOCK CYLINDER

To Remove and Install

1 Remove the trim panel and sealing sheet as previously described and wind the window up fully.

2 Where fitted, disconnect the wiring connector from the door lock cylinder.

3 Disconnect the control rods from the rear of the exterior handle and then remove the bolts retaining the exterior door handle.

TIP *By removing the weather strip as described under the Window Glass heading, easier access to the control rod clips can be obtained.*

4 Pull the exterior handle to the open position and remove the handle from the door.

5 If necessary, remove the retaining clip and slide the lock cylinder from the rear of the handle.

Installation is a reversal of the removal procedure with attention to the following points:

1 Lubricate all moving components with multipurpose grease prior to assembly.

2 Ensure that the door handle and lock assembly operates correctly, before installing the trim panel.

DOOR LOCK

To Remove and Install

1 Remove the trim panel and sealing sheet as previously described and then wind the window up fully.

2 Remove the retaining clips and disconnect the control rods from the exterior door handle and lock cylinder assembly.

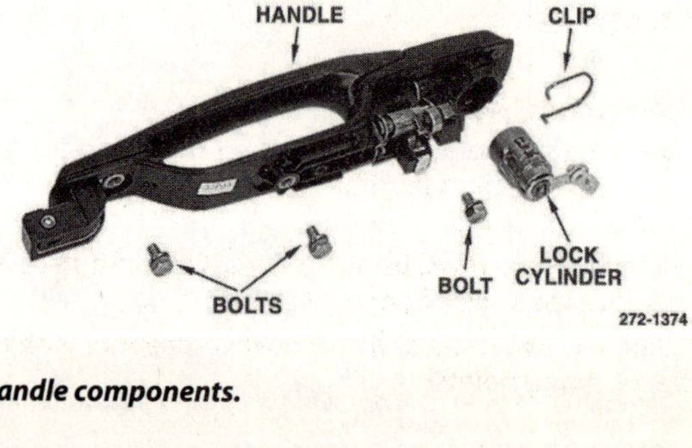

Exterior handle components.

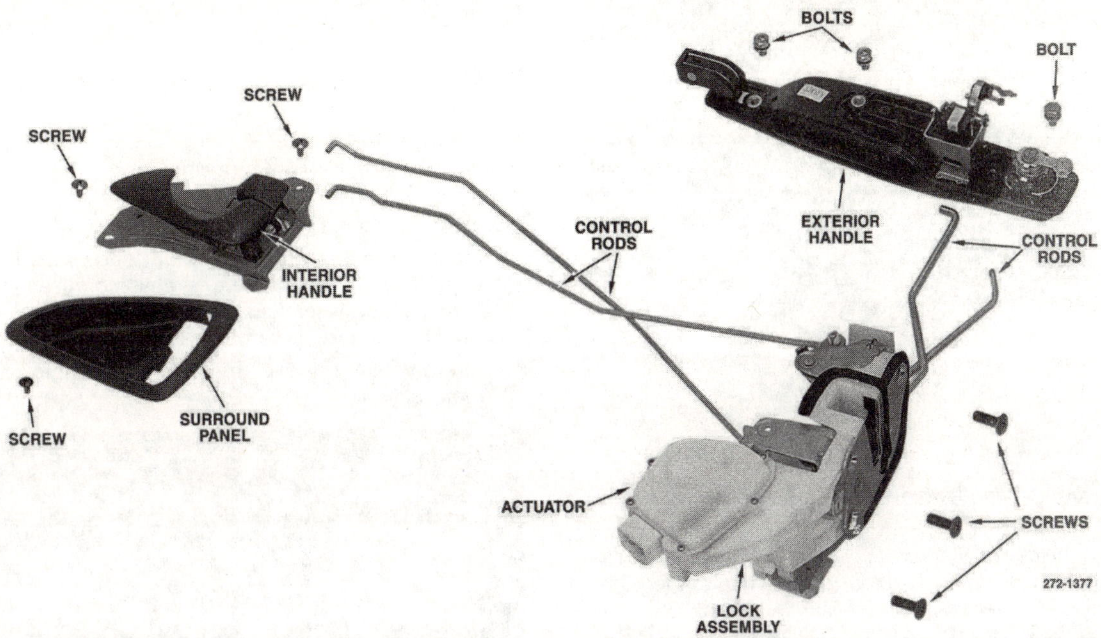

Front door lock components.

> **TIP** *By removing the weather strip as described under the Window Glass heading, easier access to the control rod clips can be obtained.*

3 Remove the retaining clips and disconnect the control rods from the interior door handle assembly.
4 Disconnect the wiring connector from the door lock assembly.
5 Remove the retaining screws, using an impact driver if necessary, and manoeuvre the door lock assembly from the door.

Installation is a reversal of the removal procedure with attention to the following points:

1 Lubricate all moving parts of the lock mechanism with multi-purpose grease before installation.
2 Install the lock assembly into the door cavity and connect the control rods to the exterior door handle and lock cylinder assembly prior to installing the door lock retaining screws.
3 Apply a smear of Permatex Lock Screw to the threads of the door lock retaining screws. Install the retaining screws and tighten securely but not excessively. Connect the control rods to door lock assembly.
4 Check the operation of the lock mechanism before installing the sealing sheet and trim panel.

LOCK STRIKER

To Renew and Adjust

1 Mark the position of the lock striker on the door pillar with a soft lead pencil to facilitate correct replacement of the lock striker.
2 Remove the screws, using an impact driver if necessary, and remove the striker from the vehicle.

Installation is a reversal of the removal procedure with attention to the following points:

1 Align the lock striker with the pencil marks and tighten the screws securely.
2 Close the door and observe the alignment with the rear door. If incorrect, loosen the striker retaining screws and move the striker in or out to achieve correct alignment.

If the door lifts or drops when the lock engages the striker, loosen the screws and move the striker up or down as necessary until the door closes smoothly.

After each adjustment tighten the striker retaining screws and check the adjustment. Readjust as necessary.

EXTERIOR MIRRORS

To Remove and Install

1 Rotate the mirror assembly clockwise, exposing the mirror retaining bolts. Remove the retaining bolts and partially remove the mirror assembly from the door.

Front door lock striker.

Mirror head rotated slightly revealing the mirror retaining bolts.

2 Separate the mirror wiring connector and remove the mirror assembly from the vehicle. Tie a piece of string to the wiring connector to prevent the wiring falling into the door.

Installation is a reversal of the removal procedure.

To Renew Mirror Glass

1 Position the mirror so that it is not tilted. In this position it should be possible to see the tops of the four retaining tangs that hold the mirror glass to the motor plate.

2 Using a piece of string or wire looped around the tangs, disconnect them from the motor plate. When all of the tangs are disconnected, remove the glass from the mirror. Note the position of the spring.

If the mirror glass is cracked or broken, wear gloves to avoid injury.

3 Apply a small amount of petroleum jelly to each retainer of the new mirror glass. Position the spring as noted on removal, ensuring that it is installed into the slide on the backing and push the glass onto the motor plate. Ensure that all the retainers are correctly engaged.

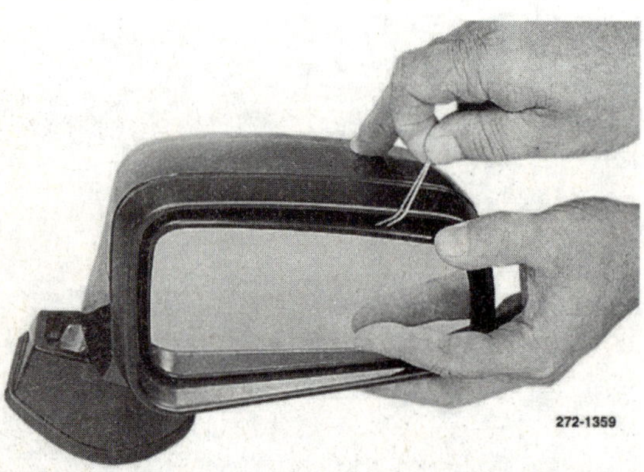

Disconnecting one of the mirror glass tangs using a piece of string.

Rear Doors

INTERIOR HANDLES AND TRIM PANEL

To Remove and Install

1 Pull the interior handle to the open position and remove the surround panel retaining screw. Withdraw the interior handle surround panel.

2 On models with manual windows, using a suitable wire hook withdraw the regulator handle retaining clip and remove the handle and washer. Refer to the illustration under the Front Door heading for more information on the retaining clip.

Alternatively, remove the regulator handle using a clip removal tool.

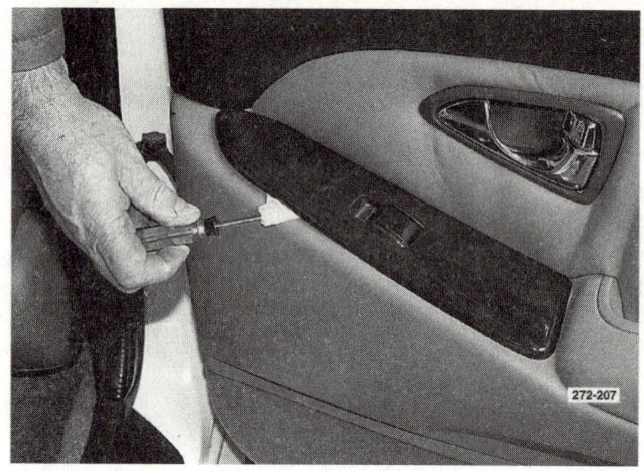

Removing the power window switch from the rear door.

3 On models with power windows, insert a small screwdriver with the blade wrapped in cloth between the door trim and the power window switch assembly. Gently lever the switch assembly upward and remove it from the trim panel.

4 Using a small screwdriver, remove the cover from the inside of the armrest door pull handle and remove the screw.

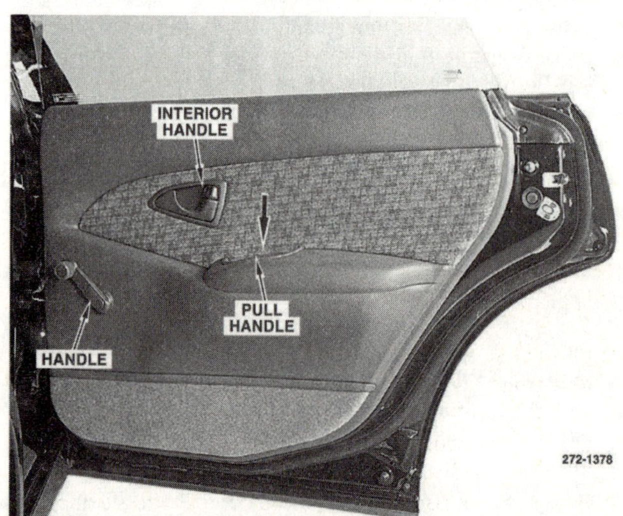

Rear door trim. The arrow indicates the location of the pull handle retaining screw.

5 Grasp one of the bottom corners of the trim panel and carefully pull the panel away from the door, disengaging the retaining clips. Work around the entire door trim until all the clips are disengaged from the door.
6 Lift the door trim straight up to disengage the trim from the top of the door and partially remove the trim.
7 Where fitted, rotate the door lamp bulb holder anticlockwise and remove it from the lamp.
8 Remove the screws retaining the interior handle to the door. Partially remove the handle, and disconnect the control rods.
9 Remove the screws retaining the pull handle bracket to the door and remove the bracket.
10 Carefully remove the sealing sheet from the inner door panel using a small knife to cut the adhesive, if necessary. Feed the wiring harnesses through the sealing sheet and remove the sheet from the vehicle.

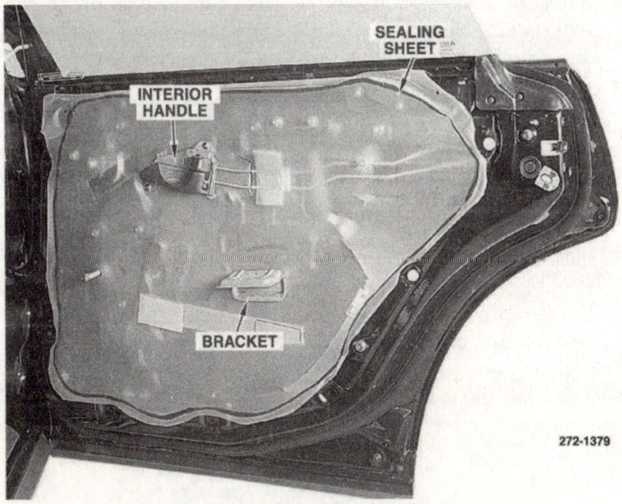

Rear door with the trim panel removed.

Installation is a reversal of the removal procedure with attention to the following points:

1 Feed the wiring through the sealing sheet ensuring that it is correctly positioned. Secure the sealing sheet to the door.
2 Connect the interior handle control rods, install the handle and tighten the retaining screws securely.
3 Ensure that the pull handle bracket is installed and the screws are secure.
4 Check the upper trim panel retaining clips and renew any damaged clips if necessary.
5 Where fitted, install the door lamp bulb holder to the trim panel.
6 Install the trim panel to the door and firmly push on the upper edge to locate the trim panel on the door panel.
7 Check that the trim panel retaining clips are aligned with the holes in the door. If not, the trim panel is not fully engaged with the top of the door.
8 On models with power windows, ensure that the power window switch wiring connector is accessible.
9 Press the trim panel retaining clips firmly into the door and check that the trim is sitting flush against the door.
10 On models with manual windows, install the washer and the clip to the regulator handle. Press the handle onto the shaft, ensuring that the handle and window glass are positioned as on the opposite door.
11 On models with power windows, connect the power window wiring switch wiring connector to the switch assembly and install the switch assembly to the door trim.

WINDOW GLASS

To Adjust

TIP *Before adjustment commences, it is advisable to suitably mark the factory settings of all adjusters as a reference.*

1 Ensure that there is an even gap between the front of the rear door and the front door and also between the rear of the rear door and the rear guard. If not, loosen the hinge bolts and with the aid of an assistant raise or lower the door assembly until an even gap is obtained between top and bottom of the door. After doing this it may also be necessary to adjust the lock striker as described later in this chapter.
2 After ensuring that the window is wound fully up, shut the door to the first stop and observe the position of the window glass.
The top of the window glass should be parallel with the lip on the upper rubber seal. Refer to the illustration.

TIP *If unsure of determining the window position, compare it to the opposite door after ensuring that the window is fully wound up and the check is made with the door shut only to the first stop.*

3 If the window glass requires adjustment, proceed as follows:
4 Remove the door trim panel and sealing sheet as previously described.
5 Loosen the front and rear up stop adjusting bolts and move the stops to achieve the correct height of the window glass. Tighten the bolts and check the window glass position.

i *It will be necessary to lower the window glass slightly to relieve the tension on the stops. Ensure that the window is wound fully up again after each adjustment is made. Also slight adjustment to the wedge adjusters may be necessary to allow sufficient travel. Do not move the wedges excessively or the glass will rattle.*

With the door closed to the first stop, check that the edge of the window glass is parallel with the seal lip.

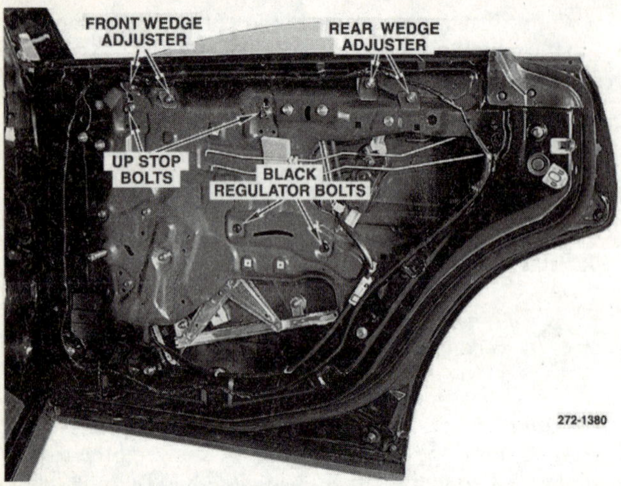

Rear window glass adjusting points.

Loosen the two black window regulator bolts and adjust the position of the bolts along the elongated holes in the door skin until the window glass is parallel with the lip on the rubber seal.

It may be necessary to repeat both adjustments several times until the correct position of the window is achieved.

6 Ensure that the window is wound fully up and close the door completely.

7 As a final check, slowly close the door until the top edge of the window glass contacts the rubber seal. At this point, measure the distance between the upper edge of the door and the vehicle body. The distance should be between 15 – 25 mm.

Close the door fully and ensure that the top of the window glass sits approximately 3 – 4 mm under the lip on the upper rubber seal. Also check that the front of the glass is parallel with the raised edge strip on the B pillar.

Due to the contour of the window glass, a difference of approximately 1 mm is allowed between readings taken at the front and rear of the window glass.

It will be necessary to adjust the window in small increments, gently shut the door and repeat the adjusting procedure until the specified gaps are achieved.

Use caution when closing the door not to shatter the glass when winding the window height adjustment up.

8 Install the trim panel temporarily and wind the window up and down. If excessive effort is required to wind the window up and down or, if on models with power windows the motor sounds very strained, proceed as follows:

9 Remove the trim panel. Loosen the two locknuts on the upper front and rear window guides and rotate the Allen key screws in equal amounts until the window glass is just touching the outer weather strip seal. Tighten the locknuts, install the trim panel and recheck the operation of the window regulator mechanism.

10 When the adjustment is satisfactory, install the trim panel as previously described.

To Remove and Install

1 Remove the trim panel and plastic sealing sheet as previously described.

2 Remove the clips retaining the ends of the door opening weather seal. Fold down the ends of the seal until the screws retaining the weather strip are visible. Remove the screws and then remove the weather strip from the door.

3 Remove the nut retaining the front of the window glass to the window guide.

4 Raise the window glass until the rear wedge is visible. Mark the installed position of the wedge on the window glass, remove the bolt retaining the wedge to the window glass and remove it from the vehicle.

On models with manual windows, temporarily install the window regulator handle to raise the glass.

On models with power windows, connect the power window switch to the rear door for the window to operate.

5 Move the window glass until the lift channel to glass retaining nuts are visible through the access holes. Support the window glass and remove the nuts.

6 Mark the installed position of the bolts in relation to the elongated holes and remove the bolts retaining the glass up stop brackets and remove the brackets from the door.

7 Manoeuvre the window glass from the door.

It may be necessary to lower the regulator assembly slightly to give more room to remove the window glass.

Installation is a reversal of the removal procedure with attention to the following points:

1 Working from the outside, lower the door glass into the lift channel and carefully slide glass into the rear guide.

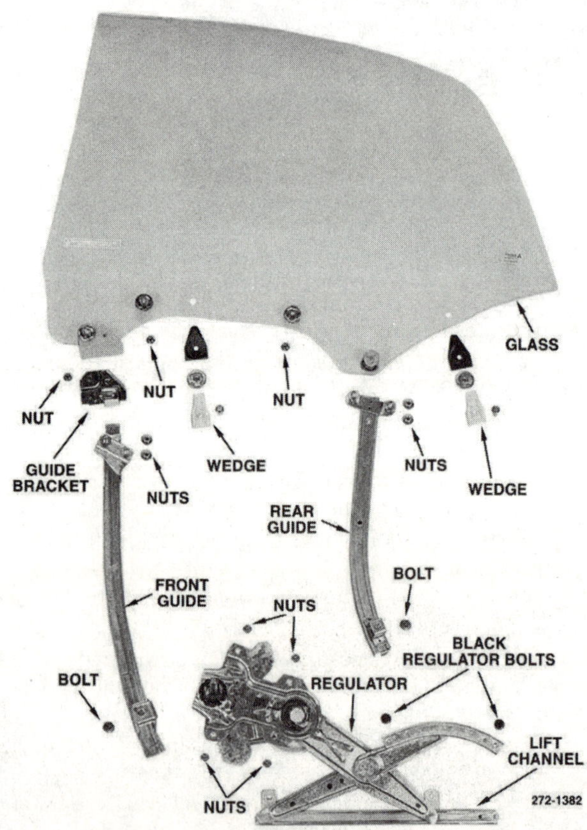

Rear window and regulator components. Manual model shown, power similar.

2 Install the front glass guide bracket to the front guide. Then install the glass on to the guide bracket and tighten the nut.
3 Carefully raise the glass until the glass lift channel nuts can be installed and tighten the nuts.
4 Install the wedge to the position noted on removal and tighten the retaining bolt.
5 Install the up stops to the positions noted on removal and tighten the retaining bolt.
6 Check the window adjustment as previously described.
7 Install the trim panel as previously described.

WINDOW REGULATOR

To Remove and Install

1 Remove the trim panel and window glass as previously described.
2 Loosen the nuts and bolts retaining the glass guides to the door and remove the nuts and bolts retaining the lift channel to the door.
3 Move the glass guides to the side and remove the glass from the top of the door.
4 Mark the installed position of the black bolts and remove the nuts and bolts retaining the regulator to the inner door panel and manoeuvre the regulator assembly from the door.
5 Manoeuvre the regulator from the door.

Installation is a reversal of the removal procedure with attention to the following points:

1 Lubricate the regulator moving components sparingly with multipurpose grease.
2 Install the regulator assembly to the door and tighten the retaining nuts and bolts securely.
3 Install the window glass as previously described.
4 Ensure that the window glass is correctly adjusted.
5 Install the trim panel as previously described.

EXTERIOR HANDLE AND DOOR LOCK

To Remove and Install

1 Remove the trim panel and sealing sheet as previously described and wind the window up fully.
2 Remove the bolts retaining the exterior door handle.
3 Partially remove the handle assembly and disconnect the control rod.
4 Remove the handle from the door.

Installation is a reversal of the removal procedure with attention to the following points:

1 Lubricate all moving components with multipurpose grease prior to assembly.
2 Ensure that the door handle and lock assembly operates correctly, before installing the trim panel.

DOOR LOCK

To Remove and Install

1 Remove the trim panel and sealing sheet as previously described and then wind the window up fully.
2 Remove the retaining clip and disconnect the control rod from the exterior door handle.
3 Mark the control rod on either side of the clip on the door lock actuator. Open the clip, using a small screwdriver and separate the control rod from the clip. Remove the retaining bolts and manoeuvre the actuator from the door.
4 Remove the retaining screws using an impact driver if necessary, and manoeuvre the door lock assembly from the door.

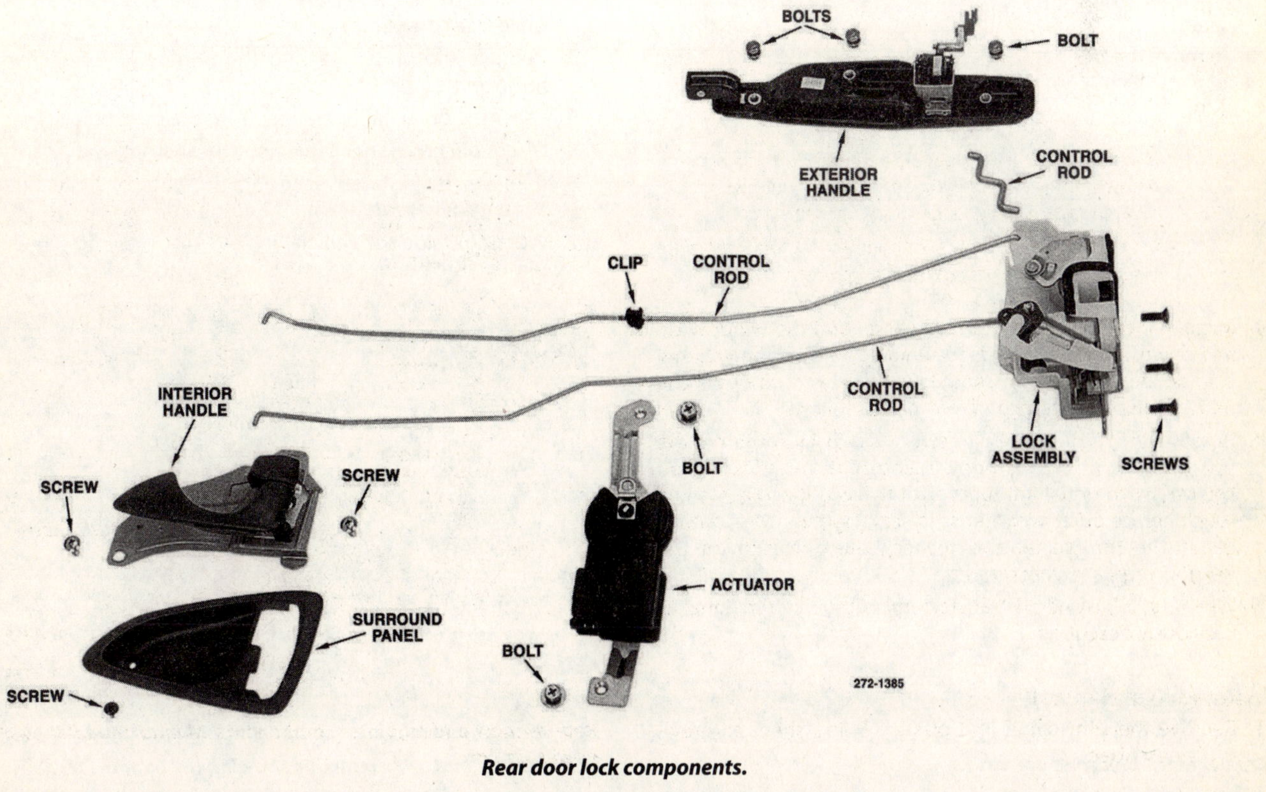

Rear door lock components.

Installation is a reversal of the removal procedure with attention to the following points:

1 Lubricate all moving parts of the lock mechanism with multi-purpose grease before installation.
2 Install the lock assembly into the door cavity and connect the control rods to the door handles prior to installing the door lock retaining screws.
3 Connect the interior handle and door lock control rods to door lock assembly.
4 Install the door lock actuator and clip the control rod to the actuator, ensuring that the control rod is returned to the same position.
5 Check the operation of the lock mechanism before installing the sealing sheet and trim panel.

LOCK STRIKER

To Renew and Adjust

The renewal and adjustment procedure is similar to that of the front door lock striker. Refer to the heading Front Doors in this chapter.

Bonnet

TO REMOVE AND INSTALL

1 Raise the bonnet and place covers on both front mudguards.
2 Disconnect the windscreen washer hose from the bonnet and release the hose from the retaining clips.
3 Mark around the bonnet hinge plates on the bonnet with a soft lead pencil to facilitate installation.
4 With the aid of an assistant supporting the bonnet, remove the bolts retaining the bonnet support struts to the inner guard.
5 Remove the bonnet retaining bolts and lift the bonnet clear of the vehicle.

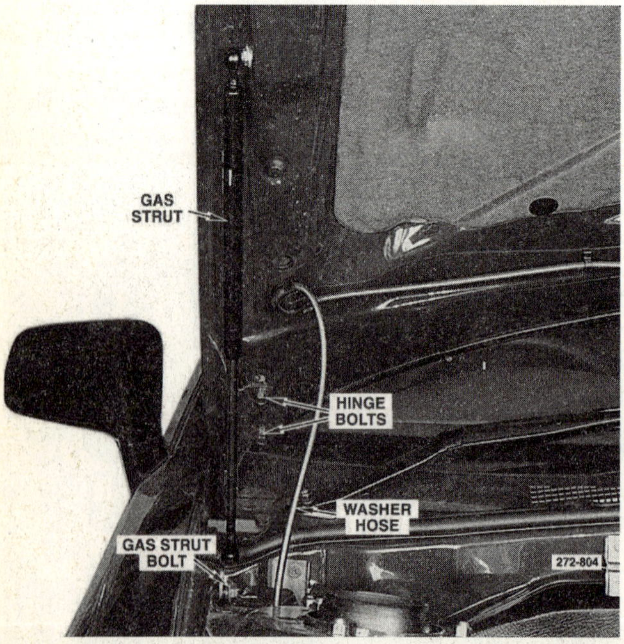

Bonnet hinge bolts and gas strut.

Installation is a reversal of the removal procedure with attention to the following points:

1 Install the bonnet retaining bolts and tighten securely.
2 Install the bonnet support strut retaining bolts and tighten securely.
3 Check the bonnet lock. When the lock is released the bonnet should pop up sufficiently to enable the fingers to release the safety catch and raise the bonnet.
4 Connect the windscreen washer hose to the bonnet.

TO ALIGN

1 Loosen the retaining bolts at the bonnet hinge.
2 Close the bonnet and by pushing with the hand, position the bonnet evenly in its opening.
3 Open the bonnet carefully and tighten the retaining bolts securely.
4 Recheck for correct positioning. If necessary, loosen one hinge at a time to adjust.
5 To align the front of the bonnet to mudguard height, adjust the bonnet bump stop rubbers and bonnet lock as required.

Correct alignment is achieved when the spacing and height of the bonnet and adjacent panels are even.

If adjustment is necessary remove the cover panel, mark the base position of the bonnet lock with a pencil, loosen the retaining bolts and move the lock in the desired direction. Tighten the retaining bolts and recheck.

TO RENEW RELEASE CABLE

1 With the bonnet open, remove the scrivets retaining the cover panel from over the bonnet lock. Withdraw the cover panel.
2 Release the clips retaining the grille and manoeuvre the grille from the vehicle.
3 Remove the bolts retaining the bonnet lock and partially remove the lock.
4 Lever the outer cable retainer from the bonnet lock bracket.
5 Disconnect the inner cable from the lock lever.
6 Release the retainer securing the cable to the radiator support panel and the inner mudguard.
7 Working inside the vehicle, remove the retaining screws and remove the bonnet release lever from the lower dash panel.

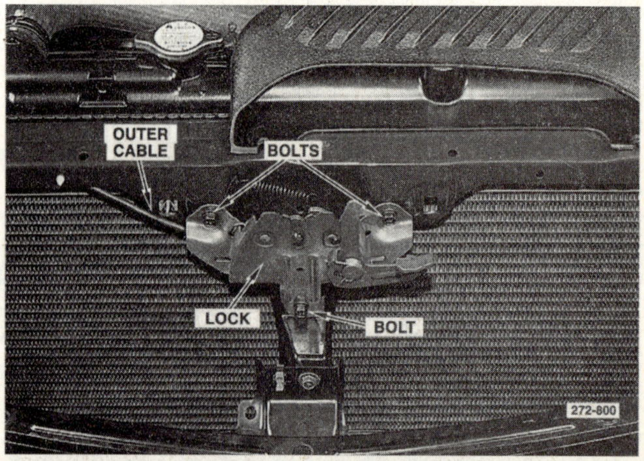

Cover panel and grille removed showing the bonnet lock.

8 Remove the outer cable from the lever and then disconnect the inner cable.
9 Attach a piece of string to the cable end and withdraw the cable through the interior of the vehicle.
10 Tie the string to the new cable and pull the string back into the interior of the vehicle until the cable enters the cabin.
11 Install the cable to the lever and then attach any retainers clips.
12 Connect the cable to the bonnet lock and test the operation before installing the grille and cover panel.

BONNET LOCK

To Remove and Install

1 With the bonnet open, remove the scrivets retaining the cover panel from over the bonnet lock. Withdraw the cover panel.
2 Release the clips retaining the grille and manoeuvre the grille from the vehicle.
3 Mark around the lock assembly with a soft lead pencil to facilitate installation.
4 Remove the retaining bolts and withdraw the lock assembly.
5 Lever the release cable retainer from the bonnet lock bracket.
6 Disconnect the inner cable from the lock lever.

Installation is a reversal of the removal procedure with attention to the following points:

1 If necessary, align the bonnet as previously described.
2 Lubricate the bonnet lock with some dry-lube or similar.

Boot Lid

TO REMOVE, INSTALL AND ADJUST

1 Open the boot lid and with a soft lead pencil, mark around the edges of the hinge mounting plates on both sides of the vehicle.

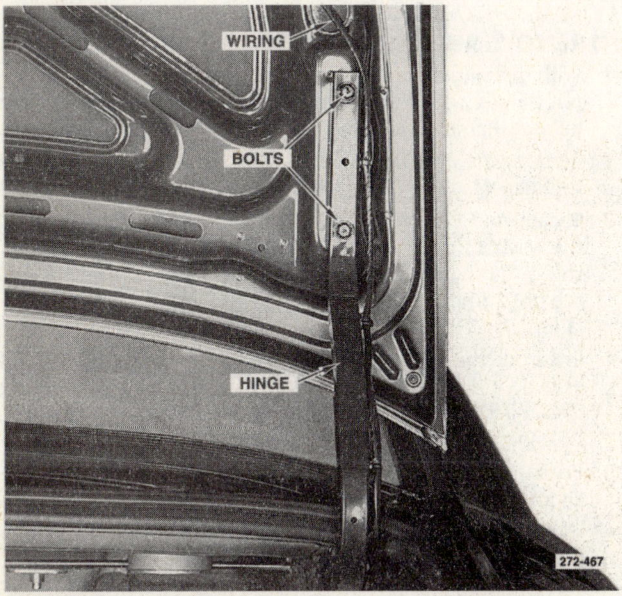

Boot lid hinge and bolts.

2 Disconnect the wiring connector from the mechanism and pull the wiring from inside the boot lid.
3 Remove the release cable from the boot lid as described below.
4 With the aid of an assistant supporting the lid, remove the retaining bolts and remove the lid from the vehicle.

Installation is a reversal of the removal procedure with attention to the following points:

1 Tighten the retaining bolts to just over finger tight and carefully close the lid. Check for an even clearance around the lid and if necessary, push into position by hand.
2 Pass the wiring through the boot lid and then connect the wiring connector.
3 Connect the release cable as described below.
4 When the lid is correctly positioned, carefully open the lid, tighten the retaining bolts securely and recheck the lid alignment.

RELEASE CABLE

To Remove and Install

1 Working in the luggage compartment, remove the plastic retainers from the left hand trim panel. Manoeuvre the panel from the vehicle.
2 Remove the bolts retaining the release actuator to the support bracket. Allow the actuator to hang, supported by the release cables.
3 Remove the nuts and bolts retaining the support bracket. Partially remove the support bracket, disconnect the wiring connector from the rear of the power aerial, where fitted, and manoeuvre the support bracket clear of the work area. If necessary, disconnect the wiring from the boot and fuel door opening relay.
4 Lever the outer cable from the lock actuator and then disconnect the inner cable.
5 Remove the lock mechanism from the boot lid as described below. Separate the outer cable from the retaining clips and feed the cable through the boot lid and from the vehicle.

Installation is a reversal of the removal procedure with attention to the following points:

1 Feed the cable through the boot and connect it to the lock mechanism.

Removing the release cable from the actuator.

2 Connect the inner cable, then the outer cable to the lock actuator.
3 Install the actuator support bracket and then install the actuator to the bracket.

LOCK AND LOCK CYLINDER

To Remove and Install

1 With the boot lid open, disconnect the control rod from the rear of the lock cylinder.

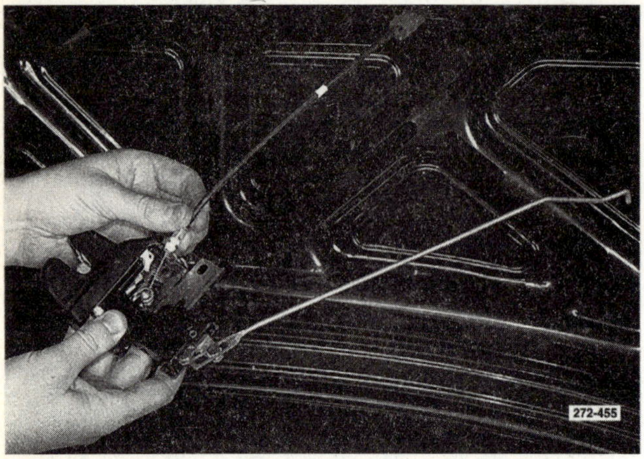

Removing the release cable from the boot lock mechanism.

2 Remove the retaining bolts and remove the lock cylinder from the inside of the boot lid.
3 Disconnect the wiring and remove the bolts retaining the lock to the boot lid. Remove the lock assembly from the boot lid, manoeuvring the control rod from the boot cavity.
4 Lever the outer cable from the lock striker and then disconnect the inner cable.

> *The lock assembly also contains the boot lamp switch.*

Installation is a reversal of the removal procedure.

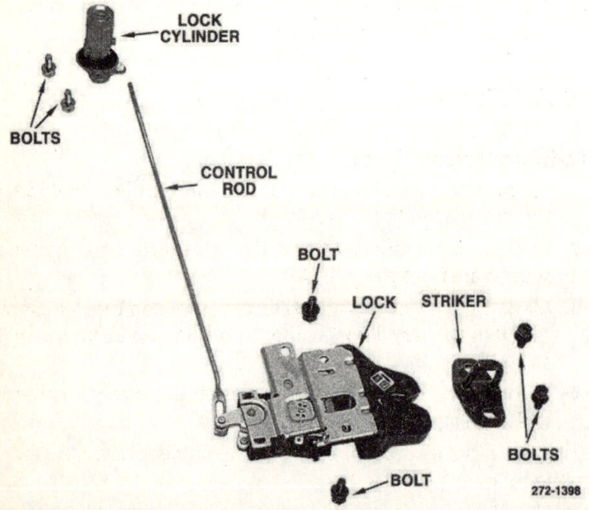

Boot lock components.

LOCK STRIKER PLATE

To Renew and Adjust

1 Pull the scuff panel straight up and remove it from the boot.
2 Mark the installed position of the striker plate using a soft leaded pencil, loosen the retaining bolts and withdraw the striker plate.
3 Check and renew the striker plate if wear or damage is detected.
4 Install the striker plate according to the pencil marks made on removal.
5 Close the boot lid and check the alignment of the lid and for smooth operation of the lock. If necessary loosen the retaining bolts and move the striker plate in the desired direction.

> *After each adjustment tighten the retaining bolts and check the adjustment. Readjust if necessary.*

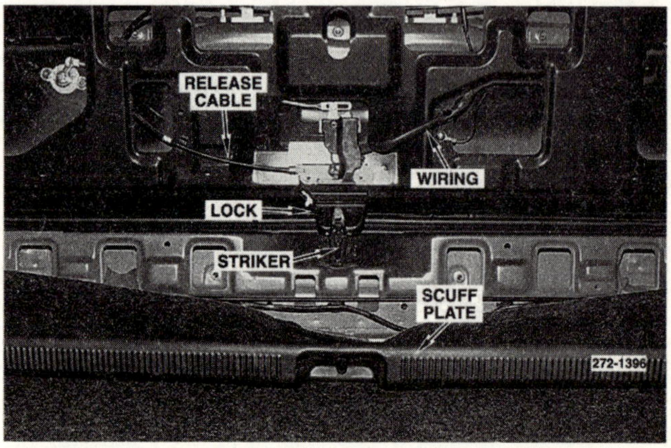

Rear scuff plate removed showing the boot lock and striker.

Tailgate

TO REMOVE AND INSTALL

1 With the tailgate open, prise the covers open on the tailgate pull handle and remove the bolts retaining the handle to the tailgate. Remove the handle.
2 Remove the clips retaining the trim panels and remove the trim panels.
3 Locate and disconnect the tailgate wiring connector and the washer hose from the nozzle and tie a suitable length of string to the hose and wiring. The string will be used to pull the hose and wiring back through the tailgate during installation.
4 Prise out the convoluted tube from the upper right hand side of the tailgate and pull the washer hose and wiring through the panel leaving the string protruding from the hole in the tailgate. Tape the ends of the string in position.
5 With an assistant supporting the tailgate, remove the support strut ball socket clips using a small screwdriver and release the struts from the body.
6 Suitably mark the hinge positions using a soft lead pencil, remove the nuts retaining the tailgate to the hinges and remove withdraw the tailgate from the vehicle.

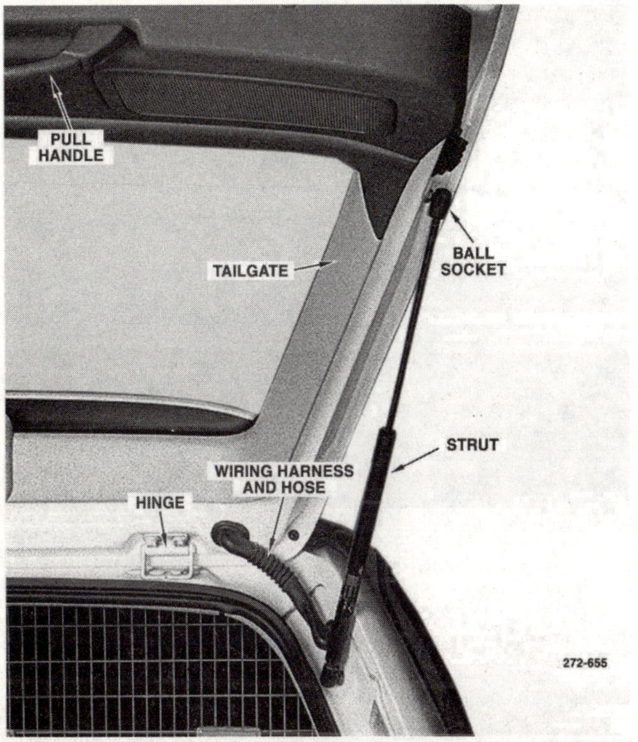

Tailgate hinge and associated components.

Installation is a reversal of the removal procedure with attention to the following points:

1 Ensure that the electrical connector is installed and the relevant components operate correctly.
2 Ensure that the convoluted tube is correctly installed to prevent water and dust entry.

LOCK, HANDLE, DOOR LOCK ACTUATOR AND LOCK CYLINDER

To Remove and Install

1 With the tailgate open, prise the covers open on the tailgate pull handle and remove the bolts retaining the handle to the tailgate. Remove the handle.
2 Remove the clips retaining the trim panels and remove the trim panels.

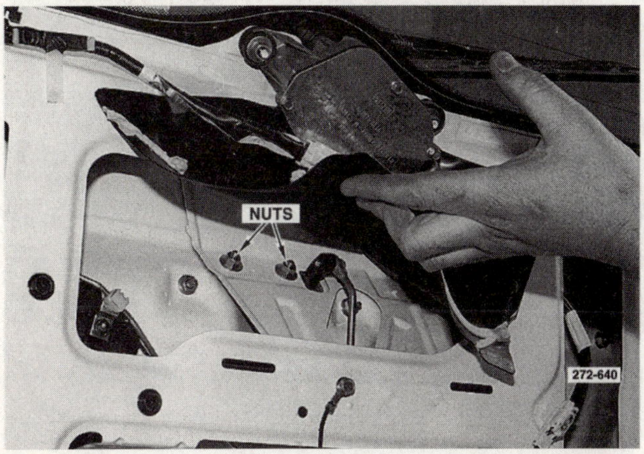

Plastic shield raised showing the tailgate exterior handle nuts.

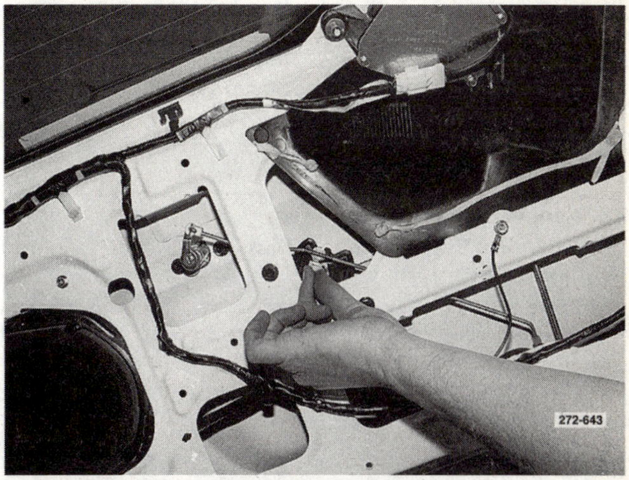

Removing the actuator clip from the tailgate control rod.

3 Disengage the clip and separate the control rod from the lock cylinder. Disconnect the wiring from the lock cylinder, remove the retaining bolts and remove the lock cylinder from inside the tailgate.
4 Mark the control rod on either side of the clip on the door lock actuator. Open the clip, using a small screwdriver and separate the control rod from the clip. Remove the retaining bolts and manoeuvre the actuator from the tailgate.
5 Move the plastic shield around the wiper motor enough to gain access to the handle retaining nuts.
6 Remove the retaining nuts from the handle, unclip the control rod and manoeuvre the handle to a position that will allow the control rod to separate from the handle. Remove the handle from the tailgate.
7 Using a T30 Torx socket, remove the bolts retaining the lock assembly to the lower tailgate and partially withdraw the lock assembly.
8 Remove the lock cylinder, feeding the control rods through the base of the tailgate.

Installation is a reversal of the removal procedure with attention to the following points:

1 Check the operation of the tailgate lock, lock cylinder, actuator and handle prior to installing the trim panels.
2 Install the trim panel, ensuring that the plastic retaining clips at the top of the trim panel engages the holes.

LOCK STRIKER

To Renew and Adjust

1 Fold the carpet away and raise the spare wheel cover. Secure the cover up with the hook.
2 Remove the screws retaining the scuff plate and remove the scuff plate from the vehicle.
3 Mark the position of the striker assembly on the vehicle with a soft lead pencil to facilitate correct replacement of the striker assembly.
4 Remove the screws retaining the striker plate and remove the plate from the vehicle.

Installation is a reversal of the removal procedure with attention to the following points:

1 Install the striker aligning the pencil marks made on removal and tighten the screws securely.

2 Carefully close the tailgate and check the alignment of the tailgate and smooth operation of the lock. If necessary, loosen the screws and move the striker plate as necessary until the alignment and lock operation is correct.

GAS STRUT

To Remove and Install

1 Raise the tailgate and have an assistant take the weight of the tailgate.

2 Using a small screwdriver, remove the clip from the ball sockets on each end of the strut and remove the struts from the vehicle.

Installation is a reversal of the removal procedure, ensuring that the ball sockets are fully engaged on the ball studs.

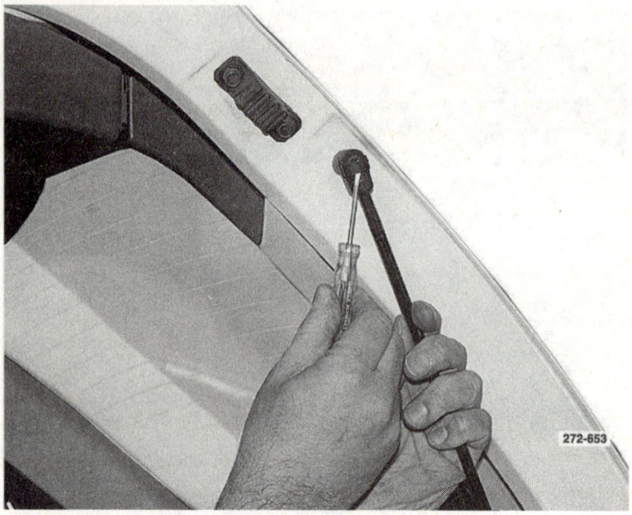

Removing the clip from the gas strut ball socket.

Centre Console

TO REMOVE AND INSTALL

1 On manual transaxle models, unscrew the gear lever knob.

2 On Verada models, proceed as follows:

a Open the coin tray lid and firmly pull the coin tray from the console. Disconnect the wiring connectors from the base of the accessory socket.

b Remove the cup holder assembly from the front of the console by pulling it straight up.

3 Grasp the rear edges of the console fascia panel and pull upward. Working toward the front of the console, raise the fascia panel, releasing the retaining clips. When all of the clips are released, manoeuvre the fascia panel from the console.

> *It will be necessary to open the console lid and move the transaxle selector lever to allow the fascia panel to be removed.*

4 Pull the rear of the console side trim panel away and down from the console to disengage it. Repeat this on the other side trim panel.

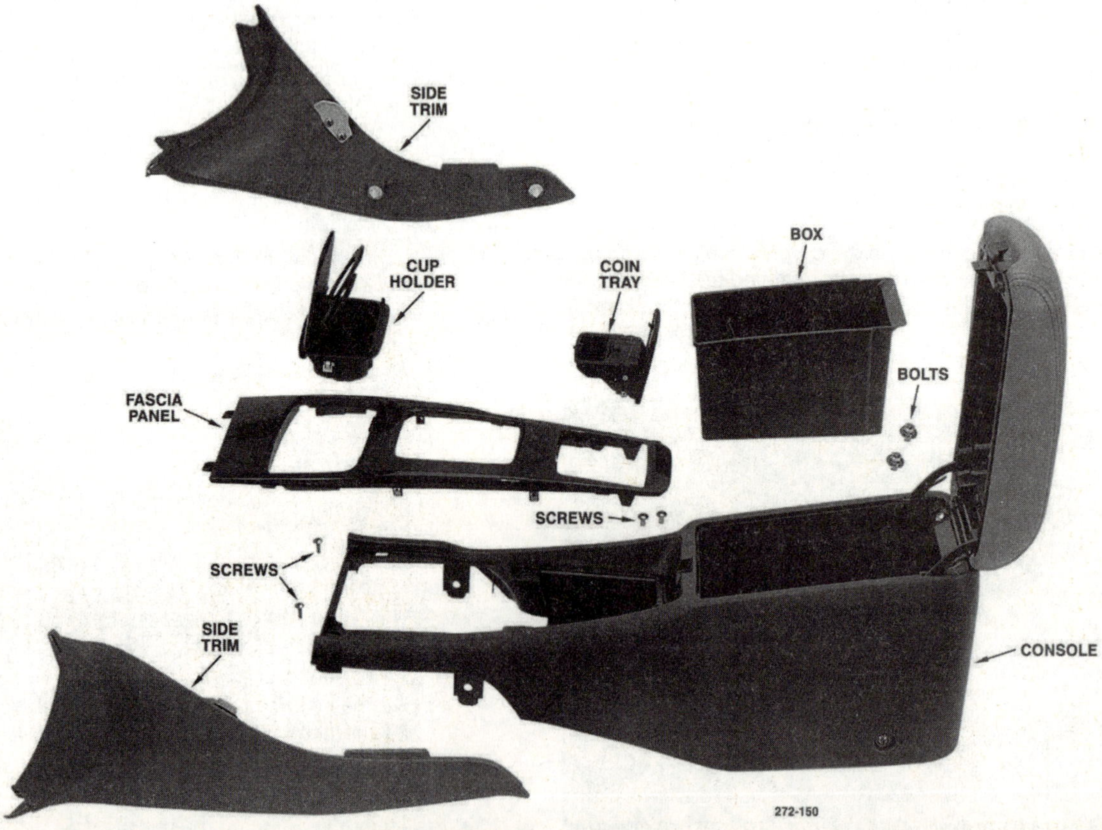

Console components. Verada models.

5 Raise the console lid, remove the console box and remove the bolts retaining the rear of the console bracket to the floor.

6 Remove the screws from the front and the middle of the console. Carefully raise the rear of the console and withdraw it from the vehicle.

It may be necessary hold trim panels away from the rear of the console while removing.

7 If necessary, the console side trim panels can be completely removed as follows:

a Unscrew the plastic scrivets retaining the passenger side lower dash panel. When the screw is removed, pull the clip from the panel. Repeat this on the remaining scrivets and partially remove the panel. Where fitted, rotate the footwell lamp bulb holder anticlockwise and separate it from the panel. Remove the passenger side lower dash panel from the vehicle.

b Remove the screws retaining the bonnet lever to the driver side lower dash panel and position the lever clear of the work area. Remove the screw cover from the top of the panel. Remove the lower dash panel screws and disconnect the panel from each side of the steering column. Partially remove the panel and where fitted, rotate the footwell lamp bulb holder anti clockwise and separate it from the panel. Remove the driver side lower dash panel from the vehicle.

c Remove the screw from the front of the console side trim panel, disengage the front retaining clips and pull the bottom of the panel away from the floor and then down to disengage it from the front of the dashboard. Remove the panel from the vehicle and repeat the procedure on the other side trim panel.

Installation is a reversal of the removal procedure with attention to the following points:

1 Where fitted, ensure that the accessory plug wiring connector is accessible with the console in position.

2 Ensure that the console correctly engages at the front.

3 Tighten all retaining screws securely.

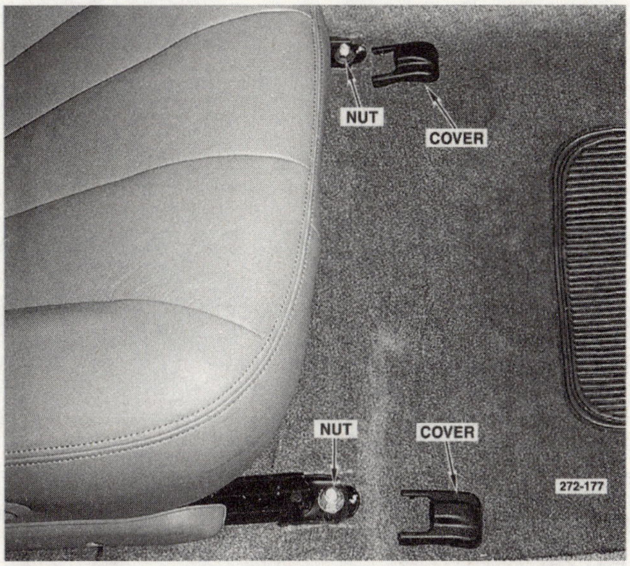

Driver seat slid rearwards showing the front seat track retaining nuts.

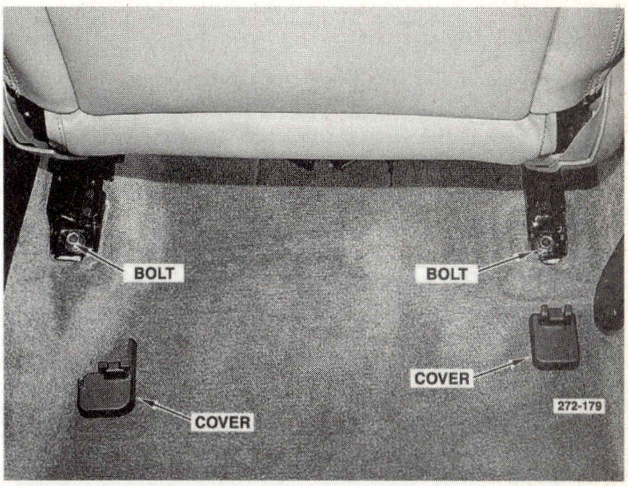

Driver seat slid forward showing the rear seat retaining bolts.

Seats

FRONT SEATS

To Remove and Install

1 Adjust the seat to the most rearward position. Prise and remove the mounting nut covers. Remove the front seat retaining nuts.

2 Adjust the seat to the most forward position. If applicable, prise and remove the clips and covers to the rear of the seat slide.

3 Remove the seat retaining bolts. Lift the seat from the vehicle, disconnecting the wiring for the power seats, where fitted.

Installation is a reversal of the removal procedure with attention to the following points:

1 Lubricate the seat slides with multipurpose grease prior to installation of the seat.

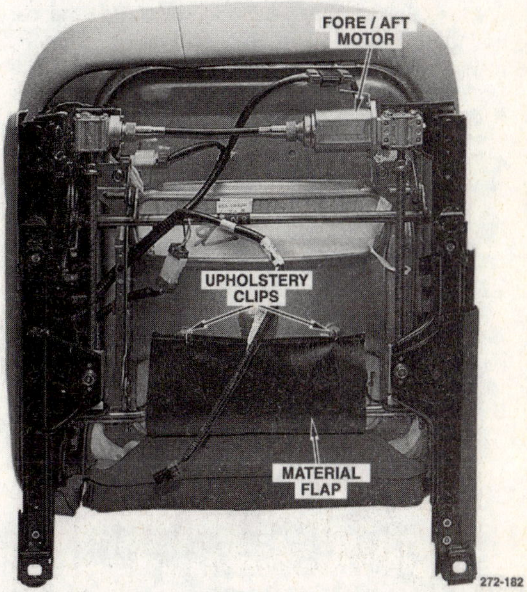

View of underneath the electric seat. Verada Xi models.

2 Check the operation of the recliner mechanism prior to installation of the seat.
3 Install all of the mounting bolts prior to tightening the mounting bolts securely.
4 Ensure that the rear mounting bolt cover is positioned correctly.

To Dismantle Power Seat — Verada Xi

The power seat fitted to the Verada Xi models can be dismantled to renew the switch, seat track or seat cushion and backrest. The seat track is supplied as an assembly consisting of the motors controlling all of the seats functions. The switch is the only other electronic component that is supplied seperately. To dismantle the seat, proceed as follows:

1 Remove the seat from the vehicle as previously described.
2 Pull the switch knob and bezel from the side of the seat cushion.

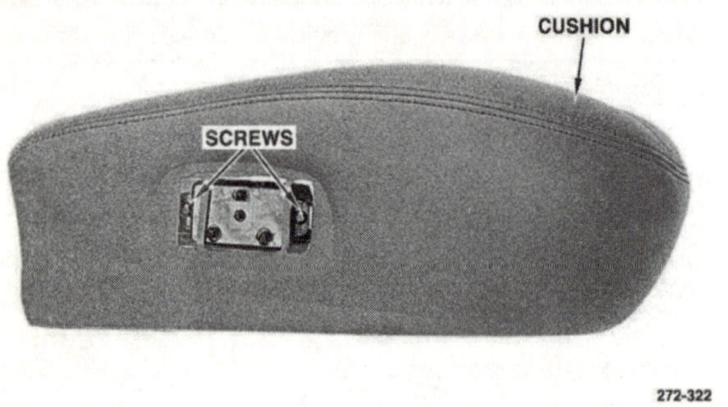

Bezel and knob removed showing the control switch screw locations.

3 Working on the base of the seat, separate the material flap from the seat cushion by releasing the upholstery clips.
4 Remove the bolts and then remove the seat cushion from the base.
5 Remove the screws retaining the control switch to the seat base. Disconnect the wiring and remove the switch from the base.
6 Remove the backrest retaining bolts. Slide the backrest up and then remove it from the seat base.

Installation is a reversal of the removal procedure.

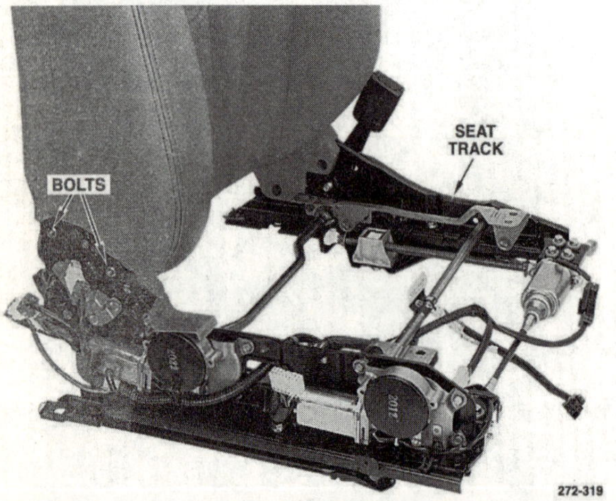

Power seat backrest retaining bolt locations. There are two more bolts in the same position on the opposite side of the seat.

REAR SEAT CUSHION

To Remove and Install

1 Firmly pull the black retaining clip toward the front of the car and lift the seat cushion upwards. Repeat this on the other side of the seat.
2 Lift the seat cushion from the vehicle and place it on a clean surface.

Installation is a reversal of the removal procedure with attention to the following points:

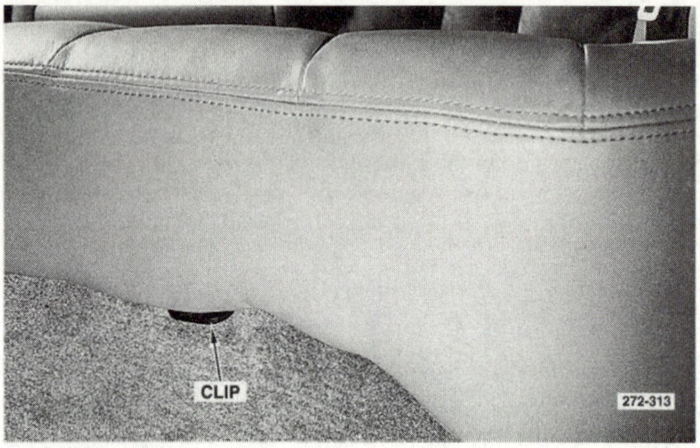

Rear seat cushion retaining clip. A second clip is located on the other side of the seat.

1 Position the centre seat belts between the elastic and the top of the rear seat cushion prior to installation of the cushion.
2 Ensure that the retaining clips are properly engaged when the seat is installed.

REAR SEAT BACKREST — SEDAN

To Remove and Install

1 Remove the rear seat cushion as previously described.
2 Remove the bolts from the lower corners of the backrest.
3 Raise the backrest straight up to disengage it from the hooked brackets and withdraw the seat backrest from the vehicle.

Installation is a reversal of the removal procedure with attention to the following point:

Ensure that the seat backrest is securely installed onto the hooked brackets prior to installing the retaining bolts.

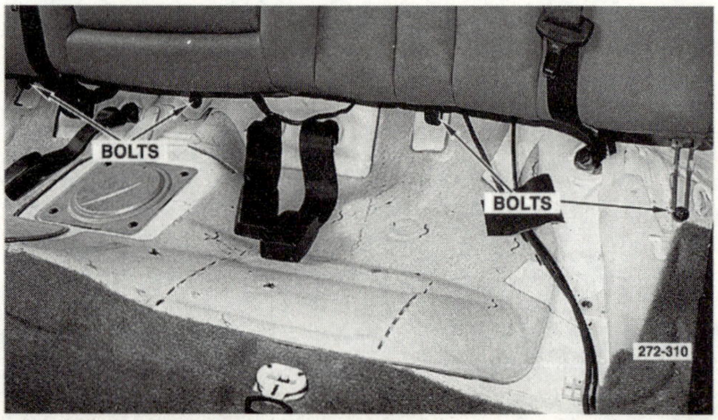

Rear seat backrest bolts. Sedan models.

REAR SEAT BACKREST — STATION WAGON

To Remove and Install

1. Remove the rear seat cushion as previously described.
2. Fold the backrest forward until it locks in the lowered position.

Ensure that the seat is locked in the lowered position before removing the bolts as the left hand side hinge is spring loaded. Do not operate the unlock lever on the top of the seat backrest while the hinge bolts are removed and keep hands clear of the hinge while it is removed.

3. Fold back the carpet and remove the top bolts retaining the seat backrest hinge brackets.
4. With the aid of an assistant, roll the bottom of the seat upward, disengaging the hinge brackets from the lower bolts.

Do not remove the lower hinge bolts from the vehicle.

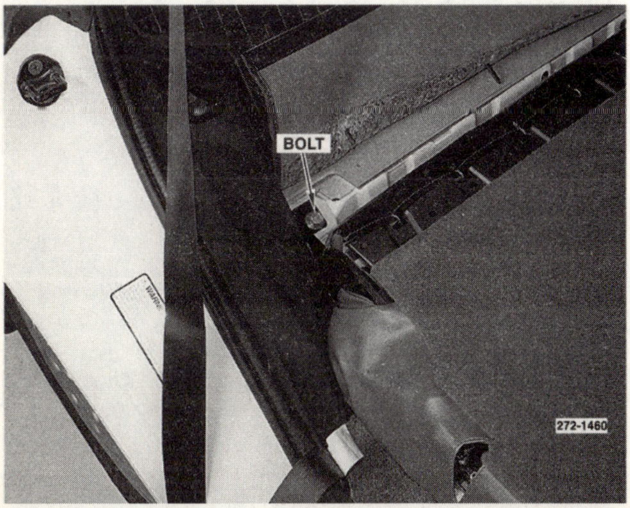

Rear seat backrest lowered showing the backrest retaining bolt. A second bolt is located on the other side of the seat. Station wagon models.

Installation is a reversal of the removal procedure with attention to the following points:

1. Engage the hinge brackets into the lower bolts and then install the top hinge bracket retaining bolts. Tighten the bolts securely.
2. Unlock the seat backrest and lift to the up position, ensuring that the seat locks in that position.

Fuel Filler Door

EMERGENCY RELEASE PROCEDURE

Sedan Models

On models with an electrically operated fuel filler door, an emergency opening cable is provided by the manufacturer and is

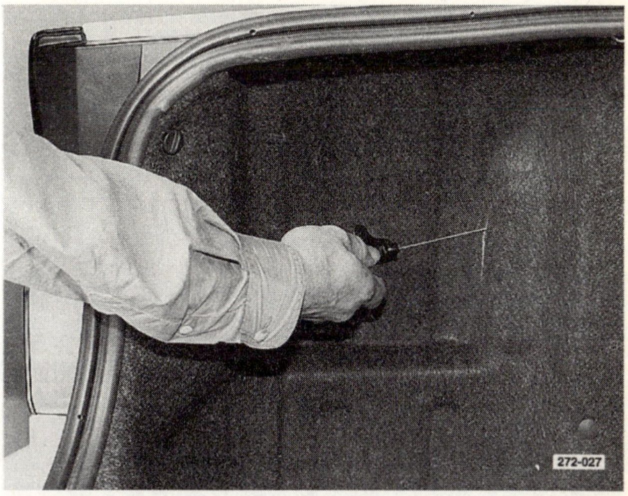

Releasing the fuel door using the emergency release cable. Verada models.

located on the left hand side of the boot. On models with a cable operated fuel filler door, proceed as follows:

1. Remove the clips retaining the trim panel to the left hand side of the boot.
2. Fold the left side trim panel back far enough to reach the fuel filler door cable.
3. Pull the button on the solid part of the cable, near fuel filler recess toward the rear of the vehicle until the fuel filler door opens.
4. Install the trim panel by reversing the removal procedure.

Station Wagon Models

1. Open the tailgate and fold the carpet away from the left hand trim panel.
2. Remove the screws retaining the timber panel to the left of the spare wheel cover and remove the panel from the vehicle.
3. Pull the rear of the lower trim panel from the left side of the luggage compartment. Remove enough panel to allow access to the fuel filler door cable.

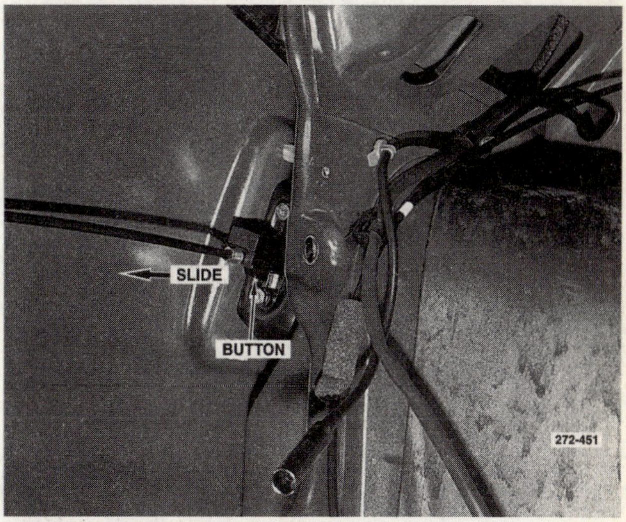

Fuel door emergency release button. Except Verada models. The button is similar on station wagon models.

4 Pull the button on the solid part of the cable, near fuel filler recess toward the rear of the vehicle until the fuel filler door opens.

5 Install the trim panel by reversing the removal procedure.

Seat Belts

The front and rear seats are equipped with lap/sash inertia reel seat belts, which are designed to lock with any sudden change in vehicle motion.

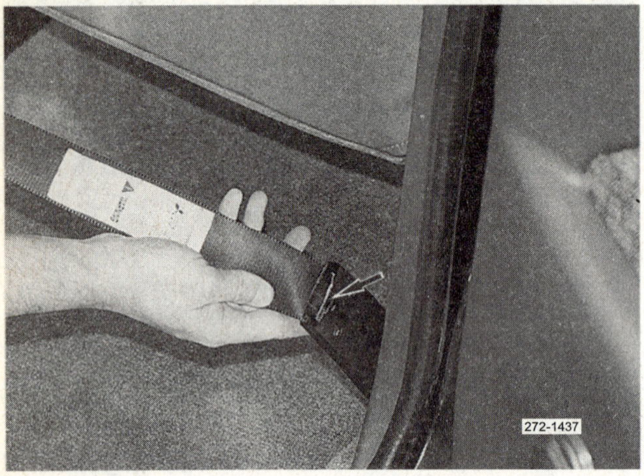

Seat belt rip stitching location, normally concealed by the plastic sleeve. When the red 'replace belt' label becomes visible at the point indicated by the arrow, the belt must be renewed.

A special feature of the seat belts is the rip stitching at the floor anchorage point of each seat belt. In the event of collision, where a large force is placed on the seat belt by the occupant, the small stitching in the belt tears, releasing a looped section of seat belt to provide gradual deceleration of the occupant. When this occurs, a red label will be exposed stating that the seat belt should be renewed.

The seat belts should be periodically inspected as described under the Routine Maintenance heading in this chapter.

Child restraint anchorages are located on the parcel shelf behind the rear seat on sedan models and in the roof panel of the

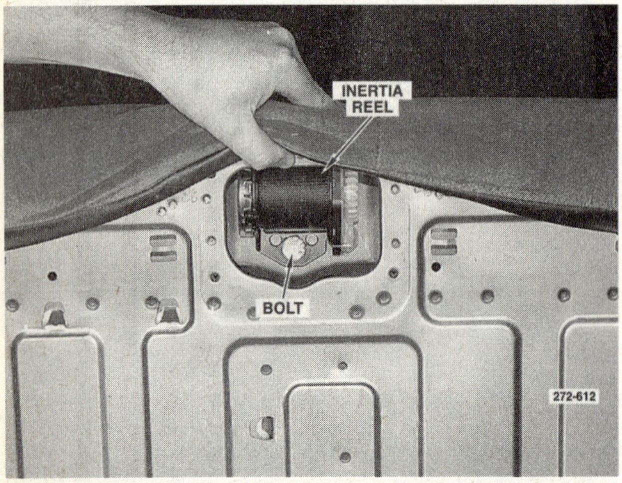

Cover removed showing the centre lap/sash seat belt upper mounting point. Station wagon models.

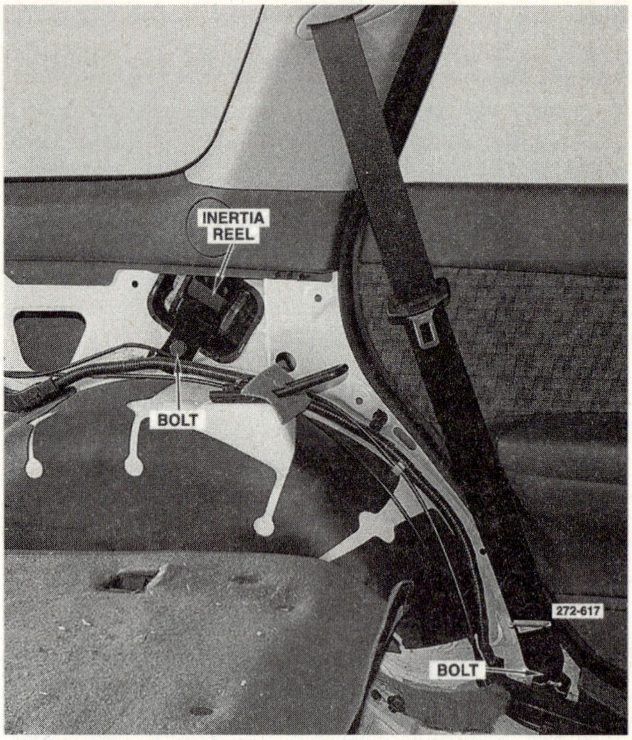

Station wagon side trim panel removed showing the side lap/sash seat belt mounting point.

luggage compartment, below the tailgate hinges, on station wagon models.

These anchorages must be used if a childs seat or capsule is installed to the vehicle. The anchorages should not be used for adult seat belts or harnesses.

On models with a passenger side airbag, never install a rear facing child restraint (baby capsule) to the front seat. In the event of a collision the inflating airbag could cause serious injury to the child.

Dashboard

TO REMOVE AND INSTALL

1 Where fitted, disable and remove the airbag/s as previously described in this chapter.

2 On models without a passenger airbag, proceed as follows:

a Prise out the retaining clips and remove the lower panel from below the glove compartment. Some models may be equipped with a footwell lamp that must be disconnected as the panel is being removed.

b Remove the hinge retaining screws at the bottom edge of the glove compartment and manoeuvre the glove compartment lid from the vehicle.

c Remove the glove compartment latch screws, the bolts at the bottom and the screws around the perimeter and partially withdraw the glove compartment surround panel from the dashboard. Where fitted, disconnect the boot release wiring and the glove compartment lamp.

3 On models without climate control, working in the glove compartment opening, remove the clip retaining the air intake control cable. Mark the installed position of the cable

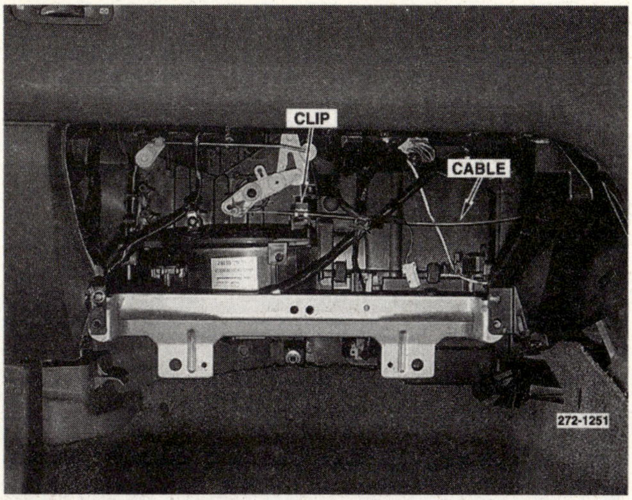

Glove compartment removed showing the air intake cable. Models without climate control.

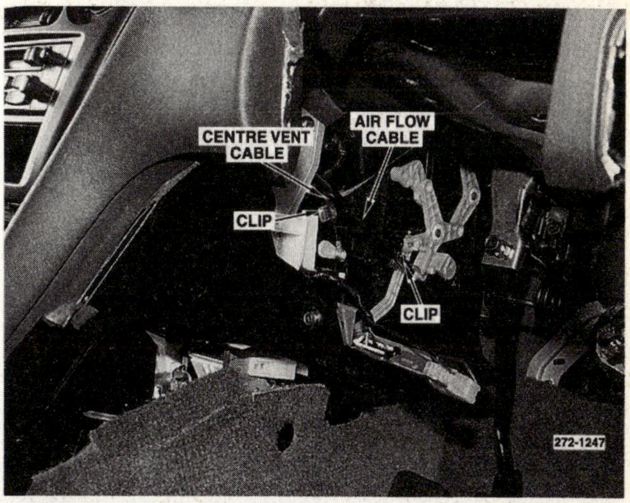

Ventilation control cable locations. Models without climate control.

in relation to the retaining clip and disengage the cable from the lever.

4 On models without a passenger airbag, proceed as follows:

a Remove the screws retaining the bonnet lever to the driver side lower dash panel and position the lever clear of the work area

b Remove the screw cover and the screw from the top right hand side of the panel.

c Remove the remaining dash panel screws and disengage the retaining clip on the top left side of the panel. If the vehicle is fitted with a footwell lamp, partially withdraw the panel, rotate the footwell lamp bulb holder anti clockwise and separate it from the panel. Remove the driver side lower dash panel from the vehicle.

5 Remove the screws retaining the ignition lock surround, remove the shield and carefully prise the surround from the dashboard.

6 Prise the exterior mirror remote control switch from the dashboard, disconnect the wiring connector and withdraw the switch assembly.

7 Remove the centre console and the side trim panels as previously described.

8 Remove the screws and bolts from each side of the vehicle and manoeuvre the metal bracket retaining the centre console.

9 Reach in from the driver side and remove the screw from the air box, located on the bottom of the heater unit. Manoeuvre the air box from the vehicle.

10 Mark the installed position of the temperature control cable in relation to the retaining clip and disengage the cable from the clip and the lever.

11 Remove the steering column as described in Steering (chapter 12).

12 Remove the instrument cluster, as follows:

a Remove the two screws retaining the top of the instrument cluster surround panel. Pull the top of the panel partially out and then pull the bottom of the panel in that order, disengaging the clips from the bottom of the panel.

b Disconnect the switch wiring connectors from the rear of the instrument cluster surround panel and remove the panel from the vehicle.

c Remove the screws retaining the instrument cluster to the dashboard. Partially remove the instrument cluster and disconnect the wiring connector from the rear of the cluster. Remove the instrument cluster from the dashboard.

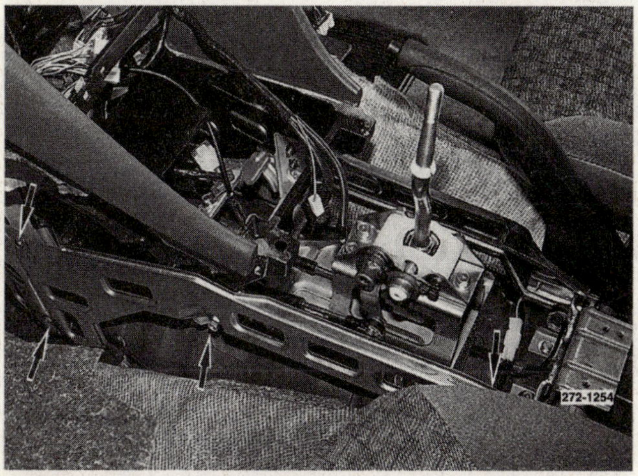

Arrow indicating console bracket retaining screw and bolt locations on the passenger side.

Upper dashboard bolt viewed with the cover or tweeter removed.

13 Working in the driver side footwell, mark the installed position in relation to the retaining clips, and disconnect the ventilation (black) and the airflow (blue) outer control cables. Disengage the inner cables from the linkages on the heater unit.

14 Using a screwdriver wrapped in cloth, carefully prise the sides of the centre ventilation assembly from the dashboard, disengaging the retaining clips. Ensure that the vent control cable does not get caught when removing the vent assembly from the dashboard.

15 Open the ashtray and withdraw it from the dashboard.

16 Place one hand in the ashtray aperture and firmly pull on the astray surround panel until the clips release from the dashboard. Partially withdraw the panel from the dashboard and disconnect the wiring connector and the illumination bulbs.

17 Remove the screws securing the top and bottom of the sound system surround panel to the dashboard.

18 Partially remove the surround panel and disconnect the wiring and aerial lead from the rear of the sound system.

On models with climate control, disconnect the wiring and sniffer motor hose from the rear of the climate control panel, prior to removing the sound system and surround panel assembly from the dashboard.

19 Using a flat bladed screwdriver, prise the speaker covers from the upper corners of the dashboard. If applicable, remove the screws retaining the speakers, partially withdraw the speakers and disconnect the wiring and remove the speakers from the dashboard.

20 Remove the bolts located in the speaker aperture, from each side of the vehicle.

21 Carefully prise the A pillar cover, starting at the top. Disengage the clip at the bottom of the pillar and withdraw the pillar covers from both sides of the vehicle.

22 Disconnect the wiring connectors, located below the power door lock module on the driver side of the vehicle.

23 Remove the remaining retaining bolts and the screw located in the left hand side of the sound system opening, and with an assistant supporting one end of the dashboard, withdraw it from the vehicle.

Installation is a reversal of the removal procedure with attention to the following points:

1 Install the dashboard into position in the vehicle, ensuring that the wiring harnesses on the drivers side are not trapped against the body.

2 Loosely install the mounting bolts and the retaining screw located in the left hand side of the sound system opening. When all of the bolts are installed and the dashboard appears to be centrally located and sitting correctly on the centre dowel, tighten the retaining bolts.

3 Where fitted, install the tweeters to the top of the dashboard, ensuring that the wiring is securely connected and then install the speaker grilles.

4 Install the sound system and the surround panel, ensuring that the wiring is securely connected. On models with climate control, ensure that the sniffer motor hose is correctly installed. Tighten the sound system surround panel retaining screws.

5 Install the ashtray, ensuring that the illumination bulb and cigarette lighter wiring is connected.

6 Install the centre vent assembly, ensuring that the control cable does not foul with the dashboard. Connect the ventilation (black) and the airflow (blue) outer controls cables and then connect the inner cables to the linkages on the heater box. Manoeuvre the air box into position and install the retaining screw.

7 Connect the wiring to the rear of the instrument cluster and then install it to the dashboard. Tighten the retaining screws securely after ensuring that the cluster is sitting correctly on the wiring connectors.

8 Install the instrument cluster surround panel and steering column covers.

9 Install the steering column as described in Steering (chapter 12).

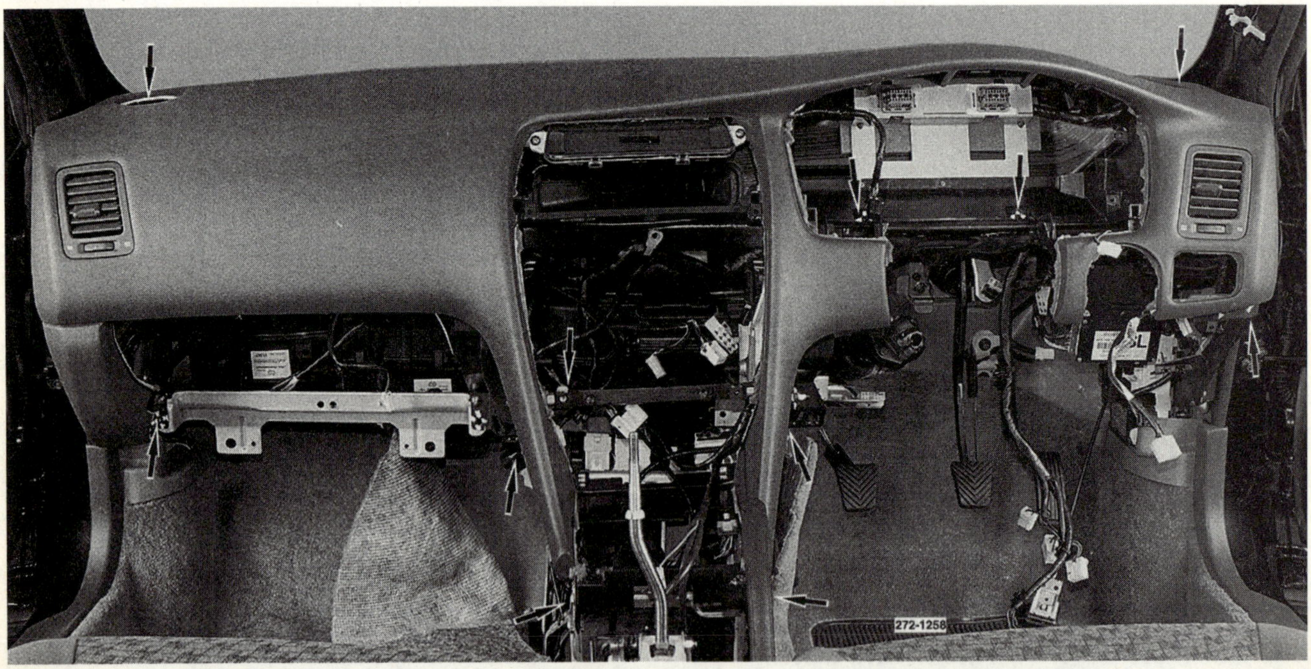

Dashboard ready for removal. Arrows indicate the dashboard retaining screw and bolt locations.

10 On models not fitted with climate control, install the temperature control cable and align the outer cable with the mark made during removal.

11 Install the centre console brackets and tighten the retaining screws and bolts securely. Then install the centre console and the side trim panels as previously described.

12 Connect the wiring to the power mirror switch and install it and the bezel to the dashboard.

13 Install the shield and the ignition lock surround panel, tightening the retaining screws securely.

14 Where fitted, install the passenger airbag as previously described.

15 Where fitted, connect the wiring to the glove compartment lamp and the remote boot release wiring and install the glove compartment surround panel. Tighten the retaining screws securely. Install the glove compartment and tighten the hinge retaining screws securely.

On models with a passenger airbag, ensure that the wiring connector for the passenger airbag is accessible when the surround is installed.

16 On models without an airbag, install the horn pad and tighten the retaining screws securely.

17 On models with an airbag, install the drivers airbag as previously described.

18 Check the operation of all components disturbed during removal and where fitted, enable the airbags as previously described.

Bumpers

FRONT BUMPER

To Remove and Install

1 Raise the front of the vehicle and support on chassis stands.

Front bumper to bracket bolts on the driver side, viewed with the splashguard removed.

2 Remove the bolts and scrivets retaining the centre splashguard to the bumper and vehicle body and withdraw the splashguard.

3 Remove the bolts retaining the front of the wheel housing splashguards to the mudguards on each side of the vehicle. Carefully place the guards clear of the work area.

4 Remove the bolts retaining the corner of the bumper to the body on both sides of the vehicle.

5 Remove the bolts retaining the bumper bracket to the vehicle body.

6 Remove the scrivets retaining the bonnet lock cover panel and withdraw the panel.

7 Release the clips retaining the grille and manoeuvre the grille from the vehicle. Remove the bolts retaining the bumper to the headlamps and to the radiator support brace.

8 Supporting the centre of the bumper, pull the bumper towards the front of the vehicle, and remove. Use an

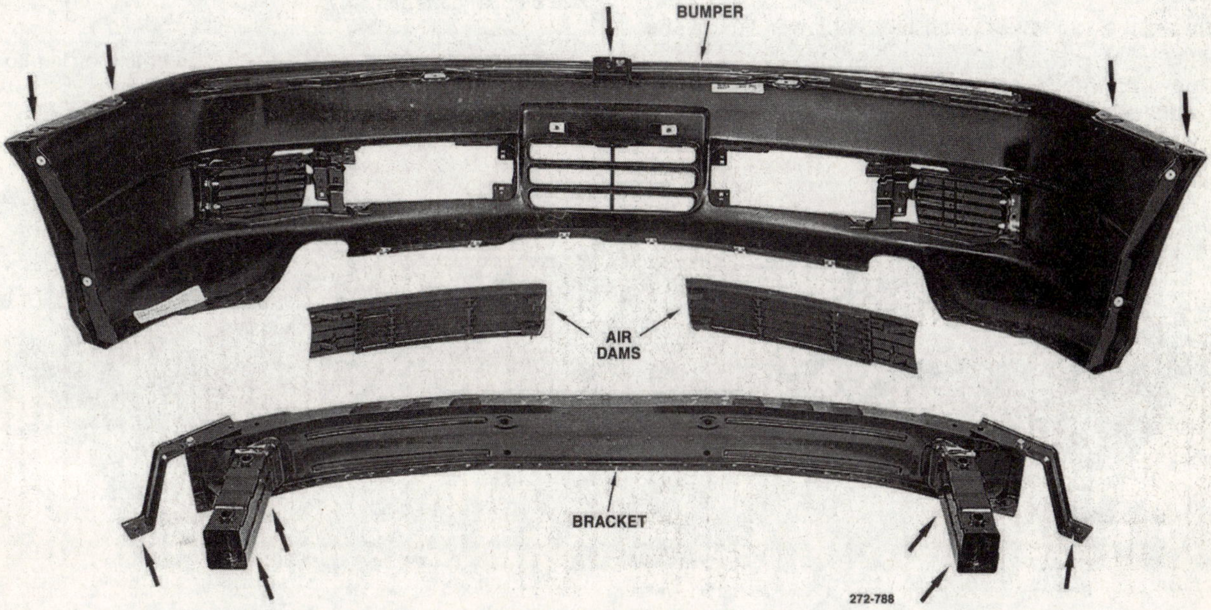

Front bumper dismantled. Arrows indicate the bumper assembly retaining bolt and nut locations.

Front bumper to radiator support bolt.

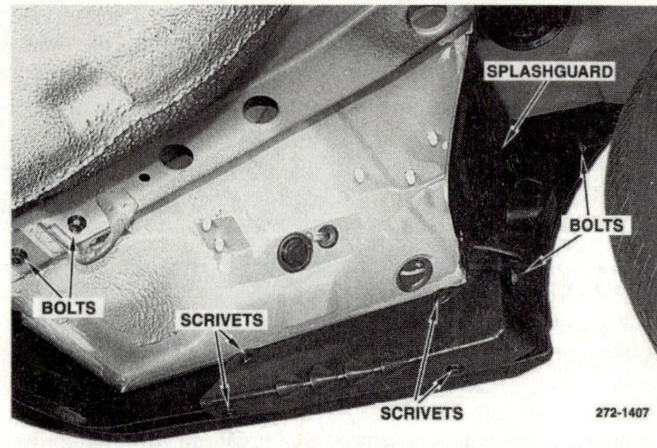

Rear splashguard and bumper bracket retaining bolts. Sedan models.

assistant if necessary. Place the bumper on soft rags to prevent damage.

9 If necessary, remove the screws retaining the air dams and remove the bolts and scrivets retaining the bumper to the bracket. Separate the bracket from the bumper.

Installation is a reversal of the removal procedure with attention to the following point:

1 Install all scrivets and bolts, prior to installing the bolts retaining the bumper to the vehicle body to ensure correct alignment.

REAR BUMPER

To Remove and Install — Sedan

1 Raise the rear of the vehicle and support on chassis stands. Refer to General Information (chapter 1) for the positions of the jacking points.

2 Remove the bolts and scrivets retaining the plastic splash-guard on both sides of the vehicle under the rear of the wheel arch.

3 Remove the scrivets retaining the rear lower side of the bumper.

4 Remove the rubber mountings retaining the rear muffler from the brackets and lower the muffler.

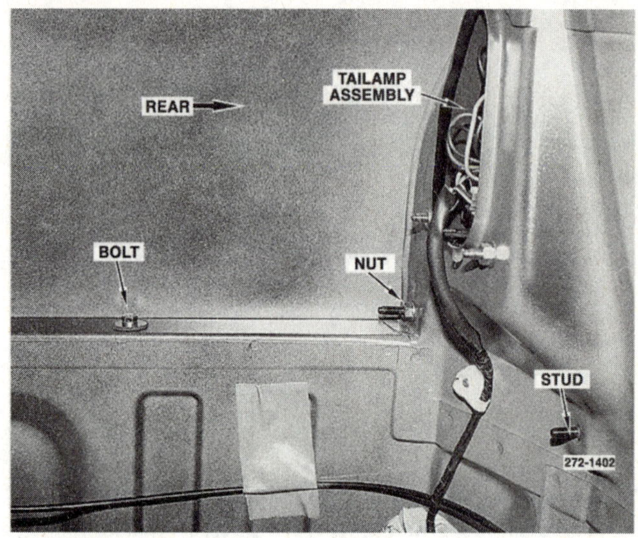

Trim removed from the boot showing the locations of some of the rear bumper bolts and nuts.

5 Remove the bolts retaining the bumper bracket to the vehicle body from both sides of the vehicle.

Sedan model rear bumper removed from the vehicle. Arrows indicate the retaining nuts and bolt locations.

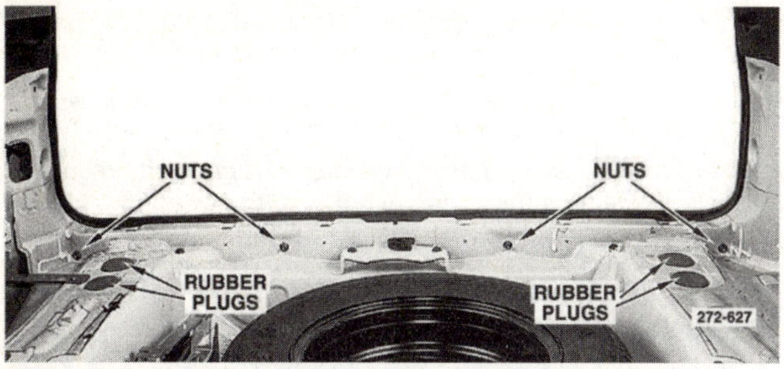

Spare wheel cover and floor cover panels removed showing the rear bumper retaining nuts. Station wagon.

6 Raise the luggage compartment lid. Remove the scuff plate from the inside of the compartment, by firmly gripping and pulling upwards.
7 Remove the scrivets retaining the carpet over the rear lamp assembly and the sides of the compartment from both sides of the vehicle.
8 Remove the nuts and bolts retaining the bumper to the body.
9 With an aid of an assistant, withdraw the bumper from the vehicle. Place the bumper on soft rags to prevent damage.
10 If necessary, remove the scrivets retaining the bumper to the support bracket and separate the bracket from the bumper.

Installation is a reversal of the removal procedure, with attention to the following:

1 Ensure that the rubber pads are all correctly seated on the studs, prior to installing the bumper.
2 Install all retainers prior to tightening the bumper retaining bolts.
3 Apply soapy water to the rubber mountings retaining the rear muffler, prior to installation.

To Remove and Install — Station Wagon

1 Raise the rear of the vehicle and support on chassis stands.
2 Raise the tailgate and raise the spare wheel cover panel. Support the cover up with the hook and string placed onto the gutter.

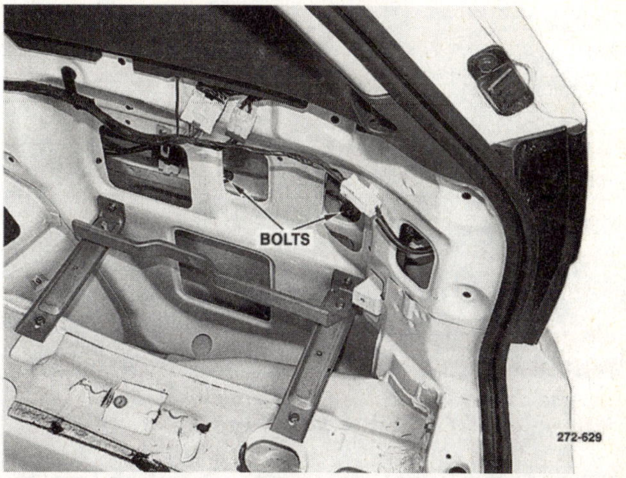

Side trim removed showing the bumper side retaining bolts. Station wagon.

3 Remove the screws retaining the scuff panel, prise the panel straight up and withdraw it from over the striker plate.
4 Remove the screws retaining the timber floor cover panels from either side of the spare wheel cover and withdraw the panels from the vehicle.
5 Remove the nuts retaining the bumper to the vehicle body, located at the rear of the spare tyre cover panel.
6 Prise the rubber plugs covering the bumper retaining bolts and remove the bolts. Lower the spare wheel cover panel.
7 Fold the covers and remove the screws retaining the luggage tiedown strap screws. Remove the tiedown brackets from both sides of the luggage compartment.
8 Carefully prise the retaining clips and partially withdraw the luggage compartment side cover panels from both sides of the vehicle.
9 Working in the openings, remove the bumper retaining bolts from both sides of the vehicle.

TIP *There are two bolts on each side. If difficulty is experienced accessing the left hand side retaining bolts, the aerial may have to be removed.*

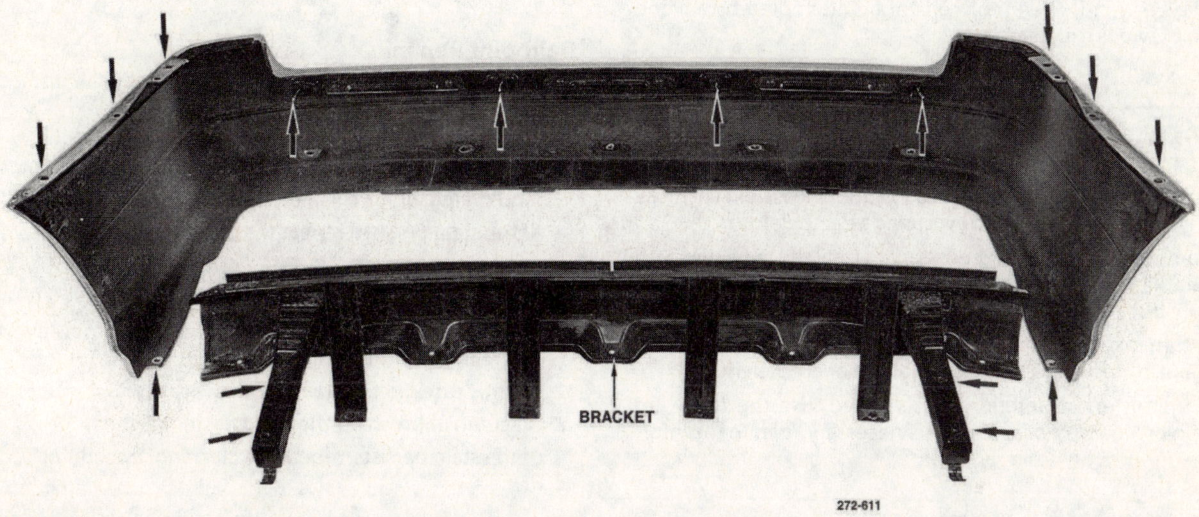

Rear bumper with the support bracket removed. Station wagon models. Arrows indicate the retaining nuts and bolt locations.

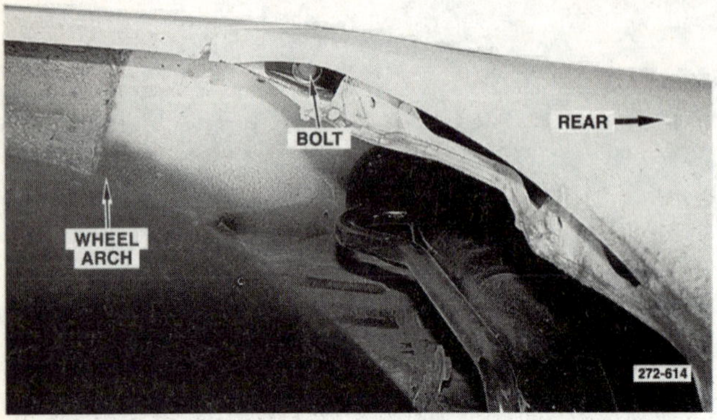

Stone guard removed from the passenger side rear wheel arch, front upper corner bumper bolt. Station wagon.

10 Working in the wheel arches, remove the bolts retaining the stoneguards and withdraw both guards from the vehicle.

11 Remove the screws retaining the upper corner of the bumper to the guard.

12 Remove the screws retaining the lower corner of the bumper to the retaining brackets on both sides of the vehicle.

13 With the aid of an assistant, carefully slide the bumper and bracket towards the rear and withdraw the bumper from the vehicle. Place the bumper on soft rags to prevent damage.

14 If necessary, remove the screws and the clips retaining the bumper to the bracket and separate the components.

Installation is a reversal of the removal procedure with attention to the following points:

1. If removed, install the bumper to the bracket, ensuring that the scrivets are securely installed.
2. Slide the bumper and bracket onto the vehicle and tighten all of the retaining bolts and nuts securely.

Windscreen and Rear Glass

The renewal or replacement of the windscreen and the rear glass follows the same basic procedure.

The successful fitting and sealing of the windscreen or rear glass will depend to a large extent on the technical knowledge and experience of the operator and it is not recommended that rear glass or windscreen replacement be attempted by anyone lacking previous experience.

Vehicle Cleaning

In order to maintain the upholstery and interior trim in good condition, it should be regularly vacuumed to remove loose dirt and dust. The vinyl should then be wiped over using a damp cloth.

From time to time it may be necessary to treat the vinyl trim with a reputable brand of vinyl restorer, ensuring that the manufacturers instructions are carefully followed.

Fabric trim may be given additional protection from liquid spills and stains by treating it with Scotchguard or equivalent.

If the upholstery or interior trim is soiled, use the following cleaning procedure. Following this if necessary, refer to the information on stubborn stain removal.

a Sponge the affected area lightly with warm water containing a little detergent and household ammonia.

b Pat the area using a clean dry cloth. If the soiling is still apparent, repeat the previous operation.

c Alternatively, an upholstery cleaner can be used, ensuring that the manufacturers instructions are followed.

TO REMOVE STUBBORN STAINS ON CLOTH TRIM AND CARPET

Carpet cleaning solution can be made by combining one teaspoon of mild detergent and one teaspoon of white vinegar with one litre of warm water. This solution is useful for cleaning most carpets and fabrics.

Chewing Gum

1. Apply an ice cube to harden the gum.
2. Remove as much of the gum as possible using the edge of a spoon.
3. Dab with a cloth moistened with methylated spirits.
4. Finish by washing the affected area with warm soapy water and drying with a clean dry cloth.

Grease and Other Fatty Stains

1. Remove as much of the grease as possible, taking care not to spread it any further.
2. Carefully dab with a cloth moistened with cigarette lighter fluid or turpentine, working from the edge of the stain.
3. Finish by washing the affected area with warm soapy water and drying with a clean dry cloth.

Chocolate

1. Remove as much of the chocolate as possible.
2. Sponge the stain with carpet cleaner. If the stain persists, treat it with dry-cleaning fluid.
3. For dried stains, sponge with a borax solution. If the stain persists, treat it with a glycerine solution.
4. Finish by washing the affected area with warm soapy water and drying with a clean dry cloth.

Ballpoint Pen Inks

1. Dab stain lightly with a pad soaked in methylated spirits and white vinegar.
2. Sponge with the carpet cleaning solution.
3. Finish by washing the affected area with warm soapy water and drying with a clean dry cloth.
4. If the stain persists, repeat the cleaning procedure.

Blood

1. It is important that blood stains are treated promptly. Make a paste from cornflour and cold water.
2. Lightly rub into the stain and allow to dry.
3. Vacuum or brush off the dried paste.
4. If the stain persists, repeat the cleaning procedure.

Chapter 18
Ventilation System

Contents

Description

The standard ventilation controls consist of rotary temperature and air flow control knobs, a fan speed control switch, an air intake lever and where applicable, an air conditioning push button switch. Cables are used to transfer the movement of the controls to the heater unit and air intake housing.

The air flow control directs air to the vehicle floor, windscreen or dashboard outlets as required. This control also has two bi-level position which directs warm air to the floor outlets and cool air to the dashboard outlets or warm air to the demister and floor outlets at the same time.

A flap in the air intake housing directs fresh air from outside the vehicle or recirculated air from inside the vehicle. The recirculate position is for use in dusty conditions or to allow rapid heating or cooling of the vehicle.

As the temperature control knob is rotated to the warm position, a flap inside the heater case opens and allows air to pass over the heater core. No heater water valve is necessary with this system.

The temperature of the heater core is relative to the temperature of the engine coolant and therefore maximum heating is only available when the engine is at normal operating temperature.

The fan speed control is used to select the desired fan speed. The fan operates at low speed whenever the ignition is in the ACC or On positions.

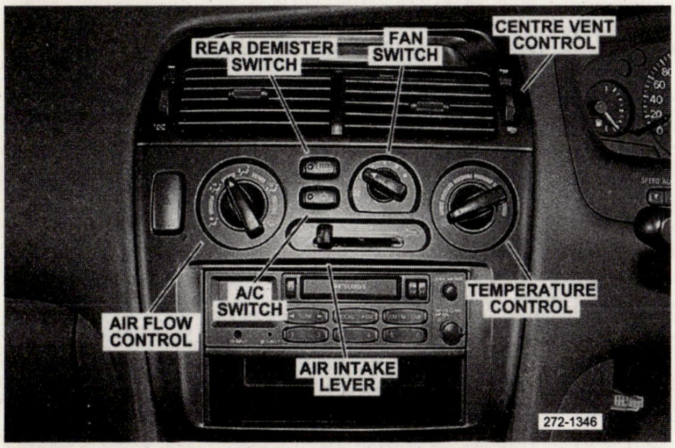

Ventilation controls, models without climate control.

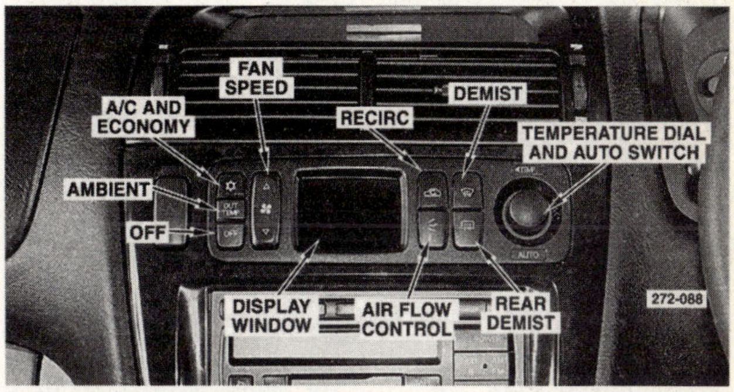

Electronic climate control panel.

Air conditioned models use identical controls with the addition of an air conditioner button.

Verada models are equipped with an electronic climate control system which can automatically control the vehicles interior environment regardless of ambient conditions.

A climate control ECU is used to control the position of the air intake motor and the air flow control motor/position sensor and heater flap control motor/position sensor. The climate control ECU receives information from the sunlight sensor located at the top of the dashboard, the ambient air temperature sensor located behind the front bumper, the fin thermo sensor and the coolant temperature sensor.

The climate control ECU is combined with the controls. The system has a self diagnosis capability, however, due to the complex nature of the system it is recommended that an authorised dealer be consulted should a fault develop that cannot be fixed using the information in this chapter.

It is recommended that any work requiring disconnection of the refrigerant lines be referred to an authorised dealer or an air conditioning specialist.

Routine Maintenance

The ventilation system requires little maintenance other than the following operations, most of which are applicable to the air conditioning system.

1. Regularly check all coolant hoses and connections for leakage including the heater hoses as described in Cooling System (chapter 5).
2. If the system is not cooling sufficiently, have the system recharged by an authorised dealer or an air conditioning specialist. It is considered normal for the system to require periodic regassing.
3. Inspect the air conditioner pipes and hoses for leaks, deterioration and alignment. Evidence of oil leakage is an indication of a refrigerant leak. Repair or renew as necessary.

Check all air conditioning pipes and hoses for oil leakage especially at the points indicated by the arrows, indicating a loss of refrigerant.

4 Inspect the evaporator drain tube for blockage. Clean as necessary.

To prevent problems with the evaporator drain tube blocking, vehicles parked in leafy areas should have the plenum chamber cleaned of leaves and debris from time to time. For the removal procedure of the plenum chamber cover refer to Body Electrical (chapter 16) under the heading Wiper Motor.

5 Inspect the condenser exterior surfaces for blockage. Clean as necessary.

6 Check the drive belt for deterioration. Refer to Engine (chapter 4) for the correct procedure.

7 Operate the air conditioning system once a week for a minimum of five minutes to keep the compressor seals lubricated.

Cleaning the evaporator drain tube using a cable tie.

Trouble Shooting

NO HEATED AIR INSIDE VEHICLE

1 Faulty engine thermostat: Renew the thermostat as described in Cooling System (chapter 5).

2 Temperature selector inoperative: Check that the temperature selector cable (models without climate control) is connected and correctly adjusted.

3 Lack of coolant in the engine: Check the coolant level.

4 Blocked heater hoses or heater core: Remove the blockage or renew the hoses or core as necessary.

5 Air flow doors inoperative: Check for a sticking/binding door flap.

Ensure that the engine is reaching normal operating temperature. If in doubt, check the operation of the thermostat as described under the Thermostat heading in Cooling System (chapter 5). When the engine is at normal operating temperature and the heater is switched on, both heater hoses should feel warm/hot. If the heater core is blocked, one hose will feel hot and the other hose will be noticeably colder.

NO COOLED AIR INSIDE VEHICLE

1 Air conditioning compressor drive belt slipping: Check and if necessary, renew the drive belt as described in Engine (chapter 4).

Splashguard removed from the right hand wheel arch showing the air conditioning compressor and drive belt.

2 Insufficient refrigerant: Refer the to air conditioning heading and if necessary, have an authorised dealer or an air conditioning specialist check the refrigerant charge.

3 Air conditioning (No 17) or ventilation fan relay (No 7) fuse blown: Check the fuses and renew as necessary.

4 Compressor inoperative: With the engine running, switch the air conditioning On and check that the compressor clutch engages with an audible click. If not, check for power at the compressor before having the compressor repaired by a specialist.

If the pressure switch is functioning correctly, the compressor will not engage if the refrigerant charge is low. If the air conditioning system is suspect, it is recommended that the vehicle be taken to an authorised dealer or an air conditioning specialist for testing and repair.

5 Heater system allowing warm air to mix with cooled air: On models without climate control, check that the temperature control cable is connected and correctly adjusted.

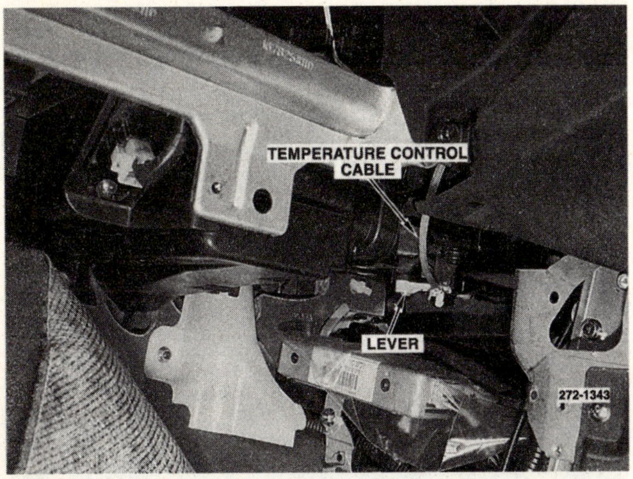

If the air conditioning system is not cooling properly, ensure that the temperature control cable is operating the lever to the full cold position. Adjust the cable as necessary.

VENTILATION CONTROLS INOPERATIVE

Without Climate Control

1 Cable disconnected or split: Gain access to the cable end and ensure that the cable is moving the lever as the relevant control is moved.

2 Controls faulty: Remove the heater controls and renew components as necessary.

With Climate Controls

1 Controls illuminate but do not work: Check that the motors are moving when the relevant control is moved. If not have the system checked by an authorised dealer or air conditioning specialist.

2 Controls do not illuminate: Check circuit power supplies. Refer to the wiring diagrams.

FAN MOTOR INOPERATIVE

1 Fuse blown: Using the wiring diagrams check the relevant fuses and renew as necessary.

2 Faulty ventilation fan relay: Check and renew the relay as necessary.

3 Faulty fan motor: Check and renew the motor as necessary.

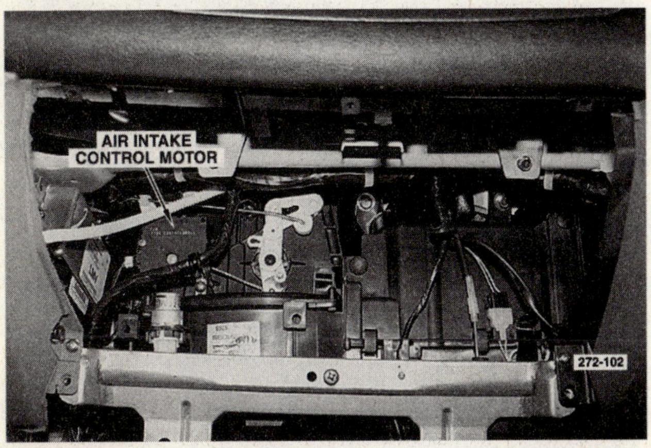

The climate control system uses electric motors to activate the flaps within the ventilation system.

4 Faulty fan motor resistor (models without climate control) or fan speed power transistor (models with climate control): Renew the resistor or controller as necessary.

5 Faulty fan speed switch: Renew the fan speed switch or the heater control assembly as applicable.

FAN MOTOR OPERATES ON HIGH SPEED ONLY

1 Faulty fan motor resistor (models without climate control): Renew the fan motor resistor.

2 Faulty fan speed switch: Renew the fan speed switch or the heater control unit as applicable.

3 Faulty fan speed power transistor (models with climate control): Renew the fan speed power transistor.

WATER LEAKING INTO INTERIOR OF VEHICLE WHEN A/C OPERATING

1 Evaporator drain tube blocked: Raise the front of the vehicle and support it on chassis stands. Carefully clean the drain tube, located on the left hand side of the firewall.

2 Heater unit case cracked or damaged: Check heater unit case.

3 Heater core leaking: Have the vehicle cooling system pressure checked.

To prevent problems with the evaporator drain tube blocking, vehicles parked in leafy areas should have the cover removed and the plenum chamber cleaned of leaves and debris from time to time. For the removal procedure for the plenum chamber cover refer to the Wiper Motor heading in Body Electrical (chapter 17).

Fan Motor and Resistor

TO REMOVE AND INSTALL FAN MOTOR

1 Prise out the retaining clips and remove the lower panel from below the glove compartment. Some models may be equipped with a footwell lamp that must be disconnected as the panel is being removed.

2 Where fitted, disconnect the wiring from the belt lock relay. The relay is situated below the fan motor. Remove the screws and withdraw the relay.

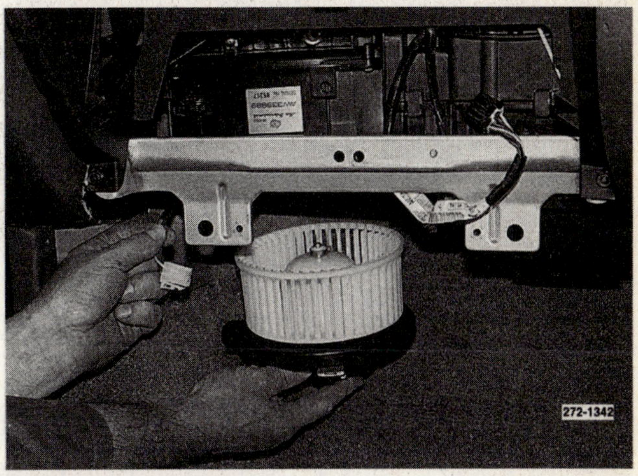

Removing the fan motor.

3 Disconnect the fan wiring connector from the fan.
4 Remove the screws retaining the fan motor and withdraw the fan motor from the air intake housing.
5 Disconnect the wiring at the fan motor, remove the cable tie retaining the wiring to the fan motor and withdraw the motor.

Installation is a reversal of the removal procedure.

TO REMOVE AND INSTALL RESISTOR

1 Prise out the retaining clips and remove the lower panel from below the glove compartment. Some models may be equipped with a footwell lamp that must be disconnected as the panel is being removed.
2 Disconnect the wiring from the resistor.
3 Remove the screws and withdraw the resistor from the vehicle.

Installation is a reversal of the removal procedure, ensuring that the wiring connector is securely connected.

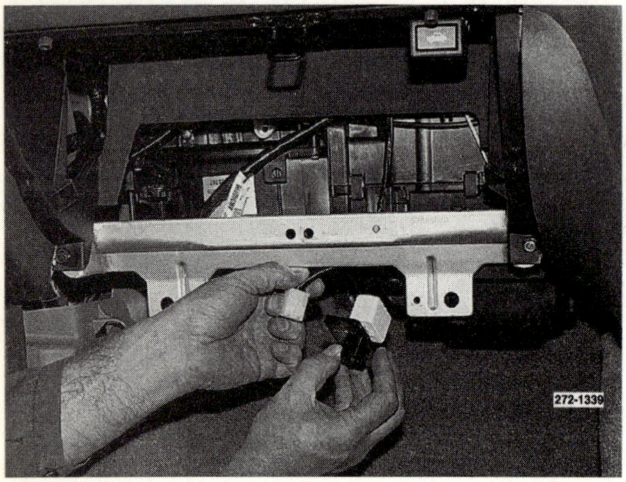

Removing the fan motor resistor.

Ventilation Controls

TO REMOVE AND INSTALL

1 Prise out the retaining clips and remove the lower panel from below the glove compartment. Some models may be equipped with a footwell lamp that must be disconnected as the panel is being removed.
2 Remove the hinge retaining screws at the bottom edge of the glove compartment and manoeuvre the glove compartment lid from the vehicle. On Verada models it may be necessary to disconnect the damper.
3 Remove the glove compartment latch screws, the bolts at the bottom and the screws around the perimeter and partially withdraw the glove compartment surround panel from the dashboard. Where fitted, disconnect the boot release wiring and the glove compartment lamp.
4 Remove the drivers side lower panel from below the steering column, as follows:
a Remove the screws retaining the bonnet lever to the driver side lower dash panel and position the lever clear of the work area. Remove the screw cover from the top of the panel.
b Remove the lower dash panel screws. Partially remove the panel and where fitted, rotate the footwell lamp bulb holder anticlockwise and separate it from the panel. Remove the driver side lower dash panel from the vehicle.

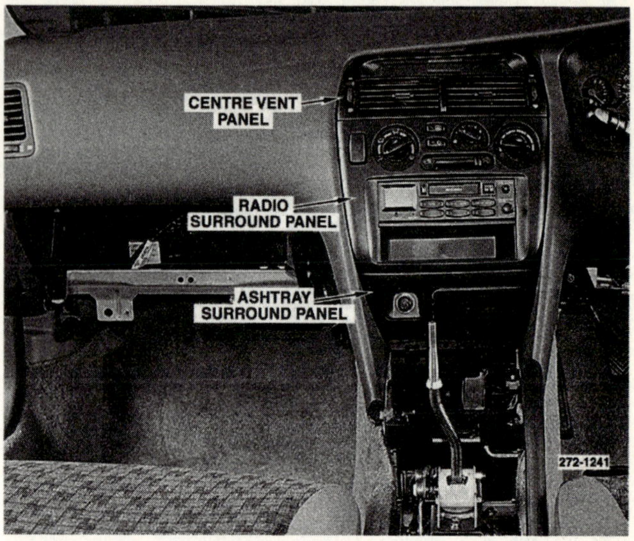

Centre panel and associated components, part way through removing the ventilation controls.

5 Prise out the clip retaining the carpet on each side of the vehicle high up near the centre of the dashboard and fold the carpet down.
6 Remove the fasteners and manoeuvre the metal brackets from each side of the vehicle.
7 Reach in from the driver side and remove the screw from the air box located on the bottom of the heater unit in the centre of the vehicle. Manoeuvre the air box from below the heater unit.
8 On manual transaxle models, unscrew the gear lever knob.
9 On Verada models, open the coin tray lid and firmly pull the coin tray from the console. Disconnect the wiring connectors from the base of the accessory socket. Remove the cup holder from the front of the console by pulling it straight up.
10 Grasp the rear edges of the console fascia panel and pull upward. Working towards the front of the console, raise the fascia panel, releasing the retaining clips. When all of the clips are released, manoeuvre the fascia panel from the console.

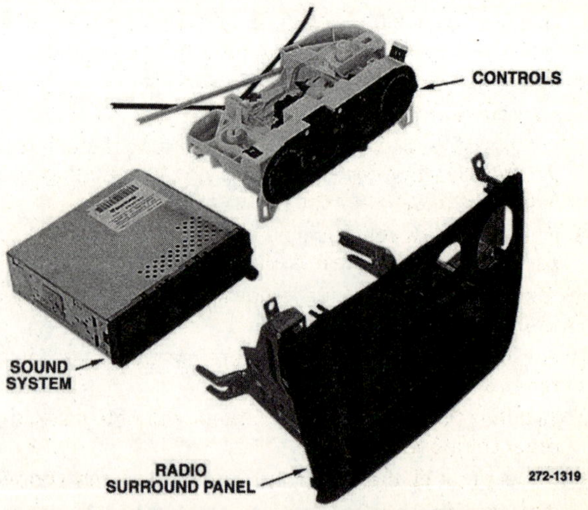

Ventilation controls separated from the surround panel.

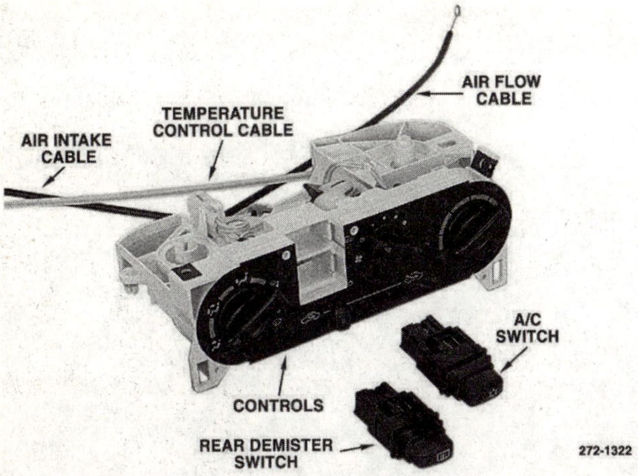

Ventilation controls showing switches removed and cable identification.

> *It will be necessary to open the console lid and move the transaxle selector lever (autos) to allow the fascia panel to be removed.*

11 Open the ashtray and withdraw it from the dashboard.

12 Place one hand in the ashtray aperture and firmly pull on the ashtray surround panel until the clips release from the dashboard. Pull the panel partially out of the dashboard and disconnect the wiring and illumination bulbs.

13 On models without climate control, using a felt pen or quick drying paint suitably mark the position of the heater control cables relative to the retaining clips to preserve the adjustment on installation.

14 Using the same method mark the centre vent cable located on the drivers side adjacent to the air flow control cable.

15 On TE/TF/TH/TJ and KE/KF/KH/KJ models, proceed as follows:

a Using a screwdriver wrapped in cloth, carefully prise the centre vent panel from the dashboard.

b Remove the screws retaining the radio surround panel and partially withdraw it from the dashboard. Disconnect the radio aerial and the wiring connectors. Note the control cable routing as an aid to installation.

17 On TL/TW and KL/KW models, proceed as follows:

a Using a screwdriver wrapped in cloth, carefully prise the centre trim panel and vent assembly from the dashboard.

b Disconnect the electrical wiring connectors from the air conditioning control assembly ECU.

c Pull the knobs and trim rings from the front of the trim panel and remove the control assembly ECU mounting screws. Separate the ECU from the trim panel.

18 If necessary, the controls can be separated from the surround panel by removing the retaining screws.

Installation is a reversal of the removal procedure with attention to the following points:

1 Ensure that the control cables are positioned as noted on removal.

2 Test the operation of the control cables prior to installing the other components.

3 Ensure that all the dashboard wiring is securely connected and correctly routed.

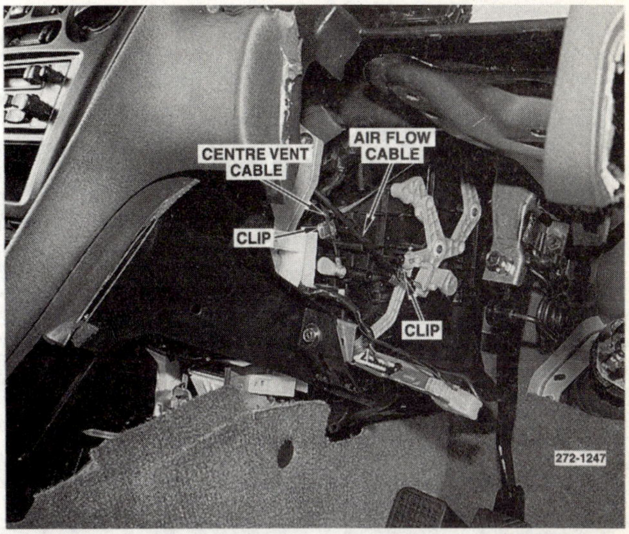

The cables are adjusted by releasing the snap clip, repositioning the outer cable in the required direction and installing the clip.

TO ADJUST CONTROL CABLES

The heater control cables are provided with a limited amount of adjustment. Adjustment is effected by releasing the outer cable retaining clip on the heater unit or air intake housing and moving the outer cable in the required direction. When adjustment is correct, secure the retaining clip and operate the relevant control to ensure satisfactory operation.

To gain access to the cable ends use the relevant information under this heading in Remove and Install.

Heater Unit

TO REMOVE AND INSTALL

1 Remove the dashboard from the vehicle as described in Body (chapter 17).

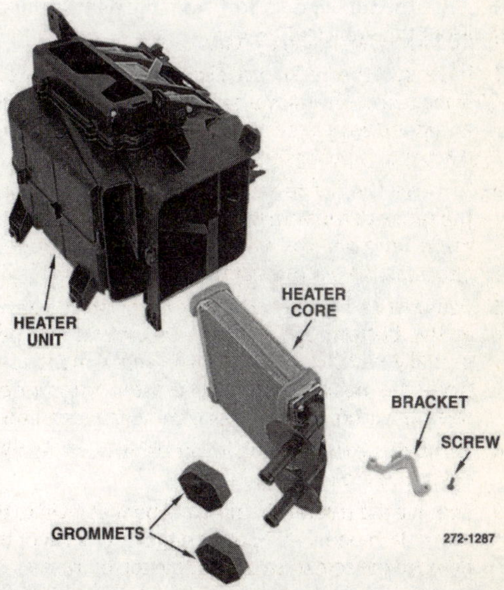

Heater core removed from the heater unit.

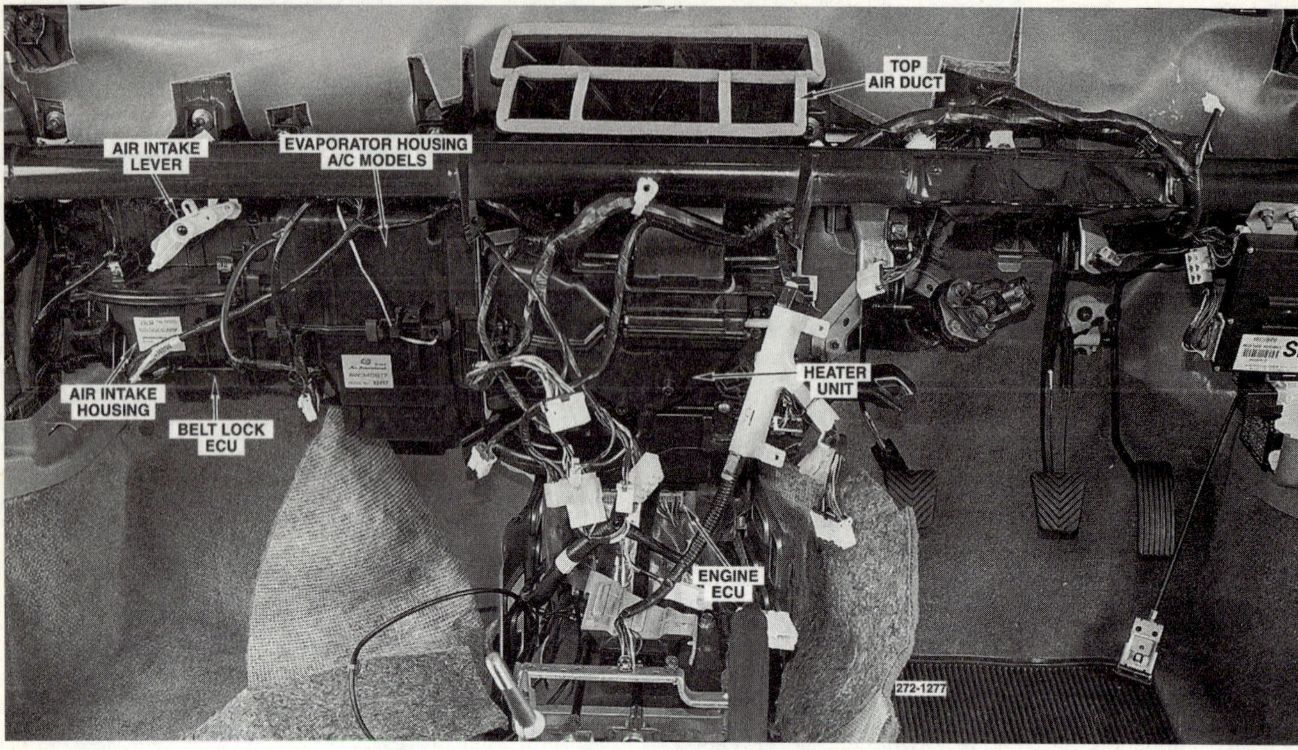

Dashboard and centre frame removed showing the heater unit.

2 Drain the cooling system as described in Cooling System (chapter 5).

3 Release the hose clamps and disconnect the coolant hoses from the heater pipes in the engine compartment.

4 Remove the fasteners retaining the various relays, where applicable the ETACS unit and the engine ECU and place these devices clear of the work area where they will not suffer damage.

5 Mark and disconnect the wiring from the various brackets. Remove the bolts retaining the centre frame to the dashboard crossbar and to the floor brackets. Manoeuvre the frame from the vehicle.

6 Prise the top air duct free from the heater unit and remove it from the vehicle.

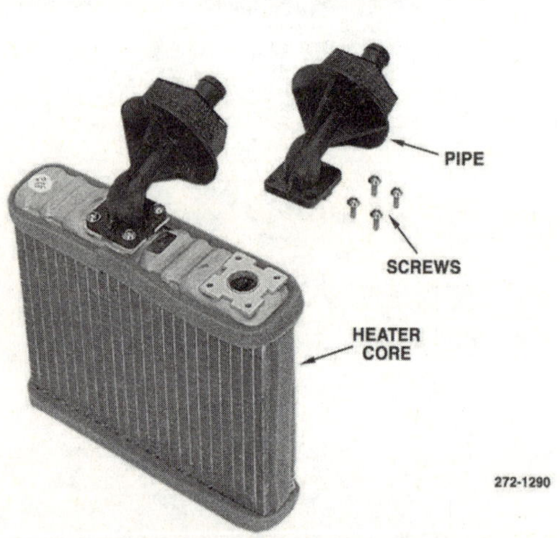

Hose connection pipe removed from the heater core.

7 Prise out the plastic clips retaining the heater unit, the evaporator, on models with air conditioning and the air intake housing (fan assembly).

8 Remove the nuts and bolts retaining the heater unit, the evaporator (A/C) and the air intake housing. Do not disconnect the A/C refrigerant lines.

9 On models with air conditioning, get an assistant to ease the evaporator unit away from the firewall as much as possible without over stressing the refrigerant lines and at the same time manoeuvre the heater unit out from the vehicle. Use care not to spill coolant on the interior of the vehicle or on any electrical device.

10 To remove the heater core, remove the bracket retaining screws and carefully withdraw the core.

TIP *The plastic hose connection pipes are available as a spare part and these are retained with screws. If the pipes are removed they must be renewed. The pipes have a built in seal to seal against the heater core. When renewing, ensure that the mating face of the heater core is free of corrosion.*

Installation is a reversal of the removal procedure with attention to the following points:

1 Renew the sealing grommets if necessary.

2 Ensure that the heater unit is located correctly with the evaporator on air conditioned models.

3 Tighten all fasteners securely but not excessively.

4 Fill the cooling system as described in Cooling System (chapter 5).

5 Run the engine to operating temperature and check the operation of the heater unit. Check for coolant leaks at the heater hose connections and rectify if necessary.

Air Conditioning

DESCRIPTION

The air conditioning system removes heat and moisture from the air within the vehicle.

The system may be operated alone during warm weather or with the heater during colder weather. When used in conjunction with the heater, the system will provide warm, dry air which will rapidly demist the interior of the vehicle.

Basic components of the air conditioning system consist of a belt driven compressor, evaporator, condenser, receiver/dryer, thermal expansion valve (TX valve), pressure switch and pipes/hoses carrying refrigerant through the system. Various other sensors and controls are utilised to monitor temperature, pressures and engine idle speed.

The refrigerant will boil at a very low temperature, therefore it must be contained under pressure. Escaping refrigerant can cause frostbite and eye damage, and will produce a poisonous gas when burnt. If the system suffers a rapid loss of refrigerant, oil from the compressor will be lost with the refrigerant. Any work requiring disconnection of the refrigerant lines must be referred to an authorised dealer or an air conditioning specialist.

In the event of major repairs to the vehicle, such as engine removal, the compressor hoses are of sufficient length to allow the compressor to be removed from the engine and secured to one side of the engine compartment without disturbing any refrigerant connection.

If the operation of the air conditioning system is in doubt, it is recommended that the vehicle be taken to an authorised dealer or air conditioning specialist.

The air conditioning system requires little maintenance and is covered under the Routine Maintenance heading at the beginning of this chapter. It is considered normal for the system to require occasional regassing.

TO CHECK REFRIGERANT

1 With the bonnet open, start the engine and operate the air conditioning. The compressor pulley clutch should be heard to engage and disengage as the air conditioning is turned on and off (cycling).

2 With the air conditioning On and the engine idling, lie down on a clean surface and reach under the front drivers side of the vehicle. The inlet line at the compressor should feel cool the outlet should feel warm to slightly hot.

To avoid burns carry out this procedure with the engine cold and use extreme caution when working near rotating pulleys, belts, fans etc.

3 A system that is low on gas will have less temperature difference between the inlet and outlet lines.

4 No temperature difference indicates a more serious problem and should be referred to authorised dealer or air conditioning specialist.

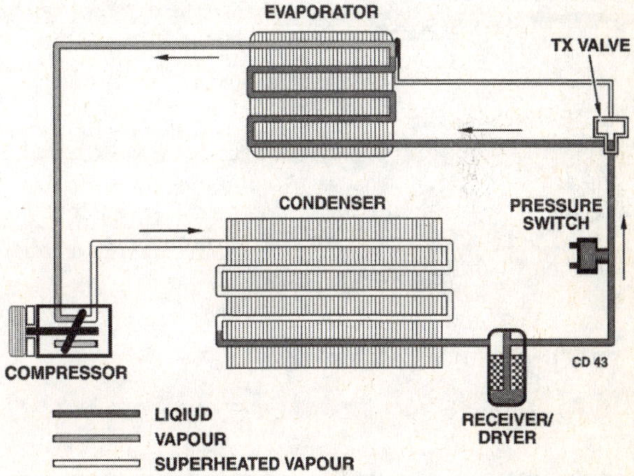

Basic components of the air conditioning system. Various other sensors are utilised to monitor temperature, pressure and engine idle speed.

Chapter 19
Wheels and Tyres

Contents

Specifications

TYRE PRESSURES

Due to the variances in wheel rim sizes, tyre type/construction and load conditions, it is recommended to refer to the tyre placard fitted to the vehicle for the correct inflation pressure pertaining to the wheel/tyre combination and the load being carried. The placard is located on the drivers B pillar.

TORQUE WRENCH SETTINGS

Wheel nut torque .... 90 – 108 Nm

Description

Tyres are available in varying types of diameters, width, aspect ratio (profile), speed ratings and number of plies.

The tyre diameter is measured in inches from bead to bead through the centre line of the tyre. Tyre width is measured in millimetres at the tyres widest point from side wall to side wall. Both figures are included in the tyre size coding.

The aspect ratio, commonly referred to as the profile or series, is the relationship of the width and the cross section height of the

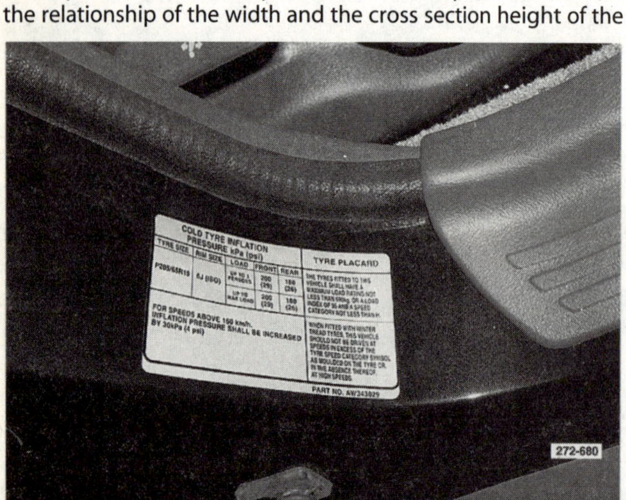

The tyre placard is located on the drivers B pillar.

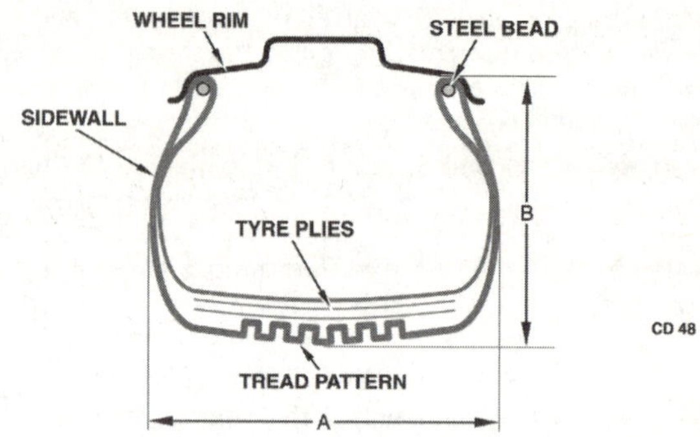

Tubeless tyre and wheel rim showing the various components. Aspect ratio is determined by calculating the height (B) as a percentage of the width (A). For example, if dimension A = 205mm and B = 133.5mm, the tyre aspect ratio is 65 because the height is 65% of the tyre width.

tyre, as measured between the bead and the tread of a new tyre. This relationship is stated as a percentage, for example if a tyre has a width of 205 mm and the height is 133.25 mm the profile would be a 65 series tyre because the height is 65% of the width. This figure is included in the tyre size coding, the higher the number the taller the tyre.

The speed rating gives the maximum speed at which the tyre manufacturer recommends a new tyre to be operated. Speeds above recommendation may cause the tyre to fail due to overheating or separation. Refer to the table below.

SPEED RATING TABLE											
Letter	L	M	N	P	Q	R	S	T	U	H	V
Speed km/h	120	130	140	150	160	170	180	190	200	210	OVER 240

Plies were originally made of cotton, the tyre marked 2 ply, 4 ply, etc. indicated the number of cotton plies used in the construction of that tyre. Development of higher strength plies enabled the actual number of plies to be reduced, while maintaining the same tyre strength. The word rating was added to indicate not the number of plies but the strength of the tyre as one with the equal number of cotton plies.

A variety of materials are used in the construction of a tyre's plies and can consist of one or a combination of the following materials: nylon, polyester, rayon and steel.

The tread pattern of a tyre varies greatly from type and manufacturers. The tread function is to assist the tyre to retain traction under wet road conditions. This is achieved by channelling water away from the tyre contact area.

There are three types of tyre in common use, cross ply, radial ply and retreaded tyres. Each type has specific characteristics,

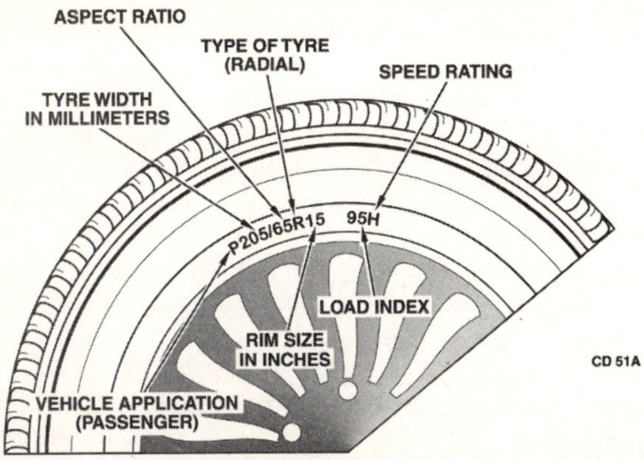

Tyre identification information. In this case the tyre size is a P205/65R15, meaning it is a pasenger vehicle tyre with a normal width of 205mm, a 65% aspect ratio, radial that fits a 15 inch rim.

advantages and disadvantages. Tyres are also available in either tube or tubeless type. A tube type tyre should never be used without a tube fitted.

Two steel beads are part of the construction of all tyres, these beads prevent the tyre from stretching and becoming separated from the wheel rim.

Routine Maintenance

STEEL WHEELS

Any minor damage to the wheel rim can usually be removed using a hammer after the wheel has been removed from the vehicle. However, any major rim damage or buckling will necessitate renewal of the wheel. It is a good policy to occasionally remove the wheels from the vehicle and inspect them for damage, cracks or corrosion.

ALLOY WHEELS

The alloy wheels are coated with a clear protective finish. Do not use abrasive cleaner. Polishing compounds, steel wool etc. when cleaning the wheels. Only mild soap and warm water are recommended. All alloy wheels are susceptible to corrosion damage, particularly if exposed to salt water.

Being relatively soft in comparison to steel wheels, alloy wheels are easily scuffed, however, this will not affect the serviceability of the wheel. Wheels that are buckled or cracked should be repaired or renewed.

Minor cracks and buckles can be repaired by a wheel specialist.

TYRES

Regularly inspect the tread of the tyres for signs of uneven wear. If uneven wear is apparent, refer to the Tyre Wear Trouble Shooting heading in this chapter for possible causes and cures. Also check the tread, grooves and sidewalls for nails, stones, etc. Where detected, these should be removed from the tyre and if necessary, the puncture repaired.

The tyre valves should always have caps installed, be regularly cleaned of dirt or dust and be inspected for wear, leaks or damage every time the tyre pressures are checked.

When the tyre groove depth is less than 1.6mm, or when the tyre is worn to the point where the tyre indicators are level with the tread surface, the tyre should be renewed.

To preserve tyre life, it is a good policy to periodically have the wheels balanced and the wheel alignment checked on a reliable wheel alignment machine.

The wheel and tyre assemblies may be rotated periodically, if desired. Rotation of the wheels and tyres will increase the period between tyre renewal, but it is not essential for safe operation of the vehicle.

The air pressure in the tyre is probably the single most important aspect of tyre care. Too little or too much air pressure in the tyre can cause rapid wear or complete failure through overheating. Where possible, the tyre pressures should be checked and adjusted when the tyres are cold.

As a rule, different tyre types, tread patterns or sizes should never be used on the vehicle at one time. All the tyres on the vehicle, including the spare, should be a matched set to prevent the vehicle behaving erratically under certain conditions. Under no circumstances mix radial ply and conventional ply tyres.

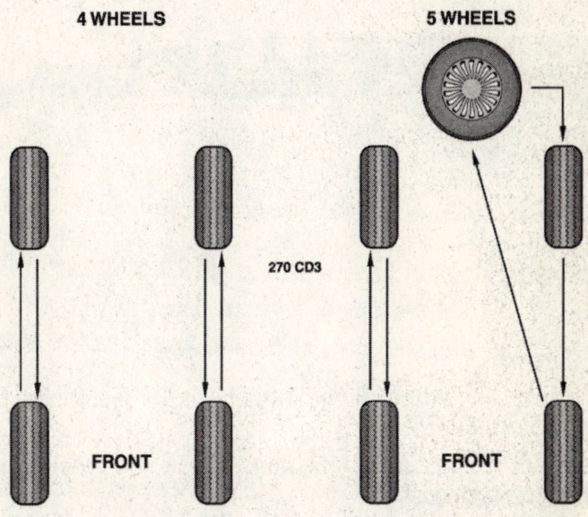

Tyre rotation diagram.

Trouble Shooting

ABNORMAL WEAR ON BOTH SIDES OF TREAD

1 Under inflation of tyres: Check and inflate to the recommended pressures.
2 Overloading: Reduce the maximum loading.

CD11

ABNORMAL WEAR IN CENTRE OF TREAD

1 Over inflation of tyres: Check and reduce to the recommended pressure.

CD12

ABNORMAL WEAR ON INSIDE OF TREAD

Front tyres

1 Insufficient camber angle: Check the front wheel alignment.
2 Sagging front coil springs: Check and renew the faulty springs.
3 Loose or worn front hub bearings: Check and renew the worn hub bearings.
4 Loose or worn suspension components: Check and renew the faulty components. Align the front wheels.
5 Bent steering knuckle: Check and renew the steering knuckle.

CD13

Rear Tyres

1 Incorrect rear wheel alignment: Check the rear wheel toe in and adjust if necessary.
2 Loose or worn rear hub bearings: Check and renew the hub bearings as necessary.
3 Loose or worn suspension components: Check and renew the faulty components as necessary.
4 Bent rear suspension carrier bracket — sedans: Check and renew the carrier bracket as necessary.
5 Bent rear axle assembly — wagons: Have a specialist check the rear axle alignment and renew if necessary.

ABNORMAL WEAR ON OUTSIDE OF TREAD

Front Tyres

1 Excessive camber angle: Check the front wheel alignment.
2 Incorrect coil spring fitted: Check and install the recommended replacement springs.
3 Excessive speed when cornering: Revise driving habits.

Rear Tyres

1 Incorrect rear wheel alignment: Check the rear wheel toe-in and adjust if necessary.
2 Loose or worn suspension components: Check and renew the faulty components as necessary.
3 Bent rear suspension carrier bracket — sedans: Check and renew the carrier bracket as necessary.
4 Bent rear axle assembly — wagons: Have a specialist check the rear axle alignment and renew if necessary.

UNEVEN OR IRREGULAR WEAR

1 Static or dynamic unbalance of the wheel and tyre assembly: Check and balance the wheel and tyre assembly.
2 Lateral runout of wheel: Check and true up or renew the wheel.
3 Faulty suspension unit: Renew the suspension units as necessary.
4 Excessive play in the hub bearing: Check and renew the faulty hub bearings.
5 Excessive play in the front suspension ball joints: Check and renew the ball joints.

LIGHTLY WORN SPOTS AT CENTRE OF TYRE

1 Static or dynamic unbalance of the wheel and tyre assembly: Check and balance the wheel and tyre assembly.
2 Radial runout (eccentricity) of wheel: Check and renew the wheel if necessary.

FLAT SPOTS AT CENTRE OF TREAD

1 Repeated severe brake application: Revise driving habits.
2 Lack of tyre rotation: Periodically rotate the wheel and tyre assembly.

HEEL AND TOE WEAR (SAWTOOTH EFFECT)

1 Overloading: Revise the maximum loading.
2 High speed driving: Avoid as far as possible.
3 Excessive braking: Revise driving habits.

CD14

FEATHERED EDGE ON SIDE OF TREAD PATTERN

Front Tyres

1. Sharp inside edge: Excessive toe in. Check and adjust the wheel alignment.
2. One tyre sharp inside edge, other side sharp outside edge: Check for a bent steering knuckle or tie rod and renew.

Rear Tyres

1. Incorrect rear wheel alignment: Check the rear wheel toe in and adjust if necessary.
2. Loose or worn suspension components: Check and renew the faulty components as necessary.
3. Bent rear suspension carrier bracket — sedans: Check and renew the carrier bracket as necessary.
4. Bent rear axle assembly — wagons: Have a specialist check the rear axle alignment and renew if necessary.

Changing a Wheel

1. Ensure that the vehicle is on firm, level ground and clear from any passing traffic. Switch on the hazard lamps.
2. Firmly apply the handbrake, place the transmission in Reverse on manual transaxle models or Park on automatic transaxle models.
3. Ensure that all passengers are out of the vehicle and chock the wheel diagonally opposite the wheel to be changed.
4. Remove the jack and tools from the storage location applicable to the particular model. Remove the spare wheel from its storage location and place it in close proximity to the wheel being changed.
5. If the wheel nuts are not exposed, remove the wheel cover or centre cap from the wheel being changed, using the tapered end of the wheel spanner.
6. Install the wheel spanner to the wheel nut so that the handle end of the spanner is horizontal and to the left hand side of the wheel. Apply pressure to the end of the spanner handle in an anticlockwise direction and loosen the wheel nut half a turn. Carry out the above procedure on the remaining wheel nuts.

Chock the front and rear of the wheel diagonally opposite the wheel being changed.

To loosen the wheel nuts apply pressure to the end of the spanner handle in an anticlockwise direction.

7. Install the jack to the relevant jacking point and raise the vehicle slightly.
8. Ensure that the jack is correctly positioned and stable.
9. Raise the vehicle so that the tyre clears the ground. Allow enough clearance between the tyre and the ground for the spare wheel to be installed.

Under no circumstances get under the vehicle while the jack is the only means of support.

10. Remove the wheel nuts and withdraw the wheel from the vehicle. Alloy wheels which become seized to the hub may be freed by hitting the tyre with a rubber mallet. Alternatively the spare wheel could be used to bump the seized wheel alternately at the front and rear. Use care not to dislodge the vehicle from the jack.
11. Prior to installing the wheel, ensure that the mounting surfaces are clean and free from corrosion, dirt etc.
12. Install the spare wheel to the vehicle. Align the holes in the wheel with the studs prior to lifting the wheel to facilitate installation. If necessary, use the handle of the wheel spanner as a lever under the tyre and guide the wheel onto the studs.
13. Holding the wheel in position, install the wheel nuts and tighten in a diagonal sequence as much as possible using the wheel spanner.

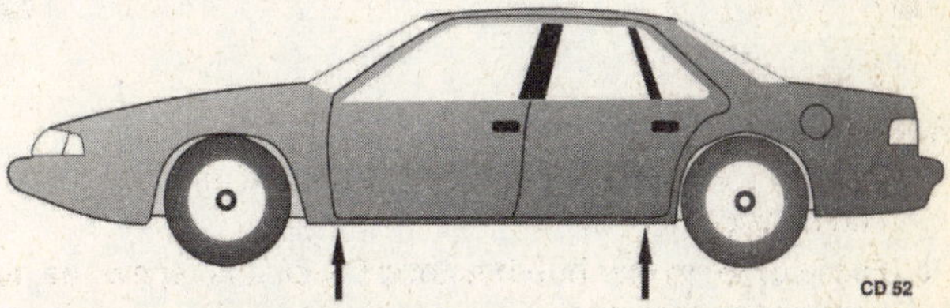

Position jack to the relevant jacking point as illustrated.

Ensure the jack is correctly positioned and stable.

14 Lower the vehicle to the ground and remove the jack from under the vehicle.

15 Using the wheel spanner, tighten the wheel nuts in a diagonal sequence. Repeat the above procedure several times until the wheel nuts are securely tightened.

16 Where removed, install the hub cap or centre cap to the spare wheel. Remove the chocks and install the changed wheel, jack and tools to their respective storage positions.

17 As soon as possible, have the wheel repaired, Check the wheel nut tension again after the vehicle has travelled a short distance. If unsure of the tightness of the wheel nuts, have them checked at the nearest service station.

18 Check the tyre pressure of the spare wheel before driving long distances.

Glossary

A

ABS (Anti-lock brake system) A system, usually electronically controlled, that senses incipient wheel lockup during braking and relieves hydraulic pressure at wheels that are about to skid.

Air bag An inflatable bag hidden in the steering wheel (driver's side) or the dash or glovebox (passenger side). In a head-on collision, the bags inflate, preventing the driver and front passenger from being thrown forward into the steering wheel or windscreen. Side air bags are also used to protect occupants from side impacts.

Air cleaner A metal or plastic housing, containing a filter element, which removes dust and dirt from the air being drawn into the engine.

Air filter element The actual filter in an air cleaner system, usually manufactured from pleated paper and requiring renewal at regular intervals.

Allen key A hexagonal wrench which fits into a recessed hexagonal hole.

Alligator clip A long-nosed spring-loaded metal clip with meshing teeth. Used to make temporary electrical connections.

Alternator A component in the electrical system which converts mechanical energy from a drivebelt into electrical energy to charge the battery and to operate the starting system, ignition system and electrical accessories.

Ampere (amp) A unit of measurement for the flow of electric current. One amp is the mount of current produced by one volt acting through a resistance of one ohm.

Anaerobic sealer A substance used to prevent bolts and screws from loosening. Anaerobic means that it does not require oxygen for activation. The Loctite brand is widely used.

Antifreeze A substance (usually ethylene glycol) mixed with water, and added to a vehicle's cooling system, to prevent freezing of the coolant in winter. Antifreeze also contains chemicals to inhibit corrosion and the formation of rust and other deposits that would tend to clog the radiator and coolant passages and reduce cooling efficiency.

Anti-seize compound A coating that reduces the risk of seizing on fasteners that are subjected to high temperatures, such as exhaust manifold bolts and nuts.

Asbestos A natural fibrous mineral with great heat resistance, commonly used in the composition of brake friction materials. Asbestos is a health hazard and the dust created by brake systems should never be inhaled or ingested.

Axle A shaft on which a wheel revolves, or which revolves with a wheel. Also, a solid beam that connects the two wheels at one end of the vehicle. An axle which also transmits power to the wheels is known as a live axle.

Axleshaft A single rotating shaft, on either side of the differential, which delivers power from the final drive assembly to the drive wheels. Also called a driveshaft or a halfshaft.

B

Ball bearing An anti-friction bearing consisting of hardened inner and outer races with hardened steel balls between the two races.

Bearing The curved surface on a shaft or in a bore, or the part assembled into either, that permits relative motion between them withminimum wear and friction.

Big-end bearing The bearing in the end of the connecting rod that's attached to the crankshaft.

Bleed nipple A valve on a brake wheel cylinder, caliper or other hydraulic component that is opened to purge the hydraulic system of air. Also called a bleed screw.

Brake bleeding Procedure for removing air from lines of a hydraulic brake system.

Brake disc The component of a disc brake that rotates with the wheels.

Brake drum The component of a drum brake that rotates with the wheels.

Brake linings The friction material which contacts the brake disc or drum to retard the vehicle's speed. The linings are bonded or riveted to the brake pads or shoes.

Brake pads The replaceable friction pads that pinch the brake disc when the brakes are applied. Brake pads consist of a friction material bonded or riveted to a rigid backing plate.

Brake shoe The crescent-shaped carrier to which the brake linings are mounted and which forces the lining against the rotating drum during braking.

Braking systems For more information on braking systems, consult the Haynes Automotive Brake Manual.

Breaker bar A long socket wrench handle providing greater leverage.

C

Caliper The non-rotating part of a disc-brake assembly that straddles the disc and carries the brake pads. The caliper also contains the hydraulic components that cause the pads to pinch the disc when the brakes are applied. A caliper is also a measuring tool that can be set to measure inside or outside dimensions of an object.

Camshaft A rotating shaft on which a series of cam lobes operate the valve mechanisms. The camshaft may be driven by gears, by sprockets and chain or by sprockets and a belt.

Canister A container in an evaporative emission control system; contains activated charcoal granules to trap vapours from the fuel system.

Carburettor A device which mixes fuel with air in the proper proportions to provide a desired power output from a spark ignition internal combustion engine.

Castellated Resembling the parapets along the top of a castle wall. For example, a castellated balljoint stud nut.

Castor In wheel alignment, the backward or forward tilt of the steering axis. Castor is positive when the steering axis is inclined rearward at the top.

Catalytic converter A silencer-like device in the exhaust system which converts certain pollutants in the exhaust gases into less harmful substances.

Circlip A ring-shaped clip used to prevent endwise movement of cylindrical parts and shafts. An internal circlip is installed in a groove in a housing; an external circlip fits into a groove on the outside of a cylindrical piece such as a shaft.

Clearance The amount of space between two parts. For example, between a piston and a cylinder, between a bearing and a journal, etc.

Coil spring A spiral of elastic steel found in various sizes throughout a vehicle, for example as a springing medium in the suspension and in the valve train.

Compression Reduction in volume, and increase in pressure and temperature, of a gas, caused by squeezing it into a smaller space.

Compression ratio The relationship between cylinder volume when the piston is at top dead centre and cylinder volume when the piston is at bottom dead centre.

Constant velocity (CV) joint A type of universal joint that allows driving power to be transmitted through an angle without vibration.

Crankcase The lower part of the engine block in which the crankshaft rotates.

Crankshaft The main rotating member, or shaft, running the length of the crankcase, with offset "throws" to which the connecting rods are attached.

Crocodile clip See Alligator clip

D

Diagnostic code Code numbers obtained by accessing the diagnostic mode of an engine management computer. This code can be used to determine the area in the system where a malfunction may be located.

Disc brake A brake design incorporating a rotating disc onto which brake pads are squeezed. The resulting friction converts the energy of a moving vehicle into heat.

Double-overhead cam (DOHC) An engine that uses two overhead camshafts, usually one for the intake valves and one for the exhaust valves.

Drivebelt(s) The belt(s) used to drive accessories such as the alternator, water pump, power steering pump, air conditioning compressor, etc. off the crankshaft pulley.

Driveshaft Any shaft used to transmit motion. Commonly used when referring to the axleshafts on a front wheel drive vehicle.

Drum brake A type of brake using a drumshaped metal cylinder attached to the inner surface of the wheel. When the brake pedal is pressed, curved brake shoes with friction linings press against the inside of the drum to slow or stop the vehicle.

E

EGR valve A valve used to introduce exhaust gases into the intake air stream to reduce the emission of oxides of nitrogen.

Electronic control unit (ECU) A computer which controls (for instance) ignition and fuel injection systems, or an anti-lock braking system. For more information refer to the Haynes Automotive Electrical and Electronic Systems Manual.

Electronic Fuel Injection (EFI) A computer controlled fuel system that distributes fuel through an injector located in each intake port of the engine.

Endfloat The amount of lengthwise movement between two parts. As applied to a crankshaft, the distance that the crankshaft can move forward and back in the cylinder block.

Engine management system (EMS) A computer controlled system which manages the fuel injection and the ignition systems in an integrated fashion.

Exhaust manifold A part with several passages through which exhaust gases leave the engine combustion chambers and enter the exhaust pipe.

F

Fan clutch A viscous (fluid) drive coupling device which permits variable engine fan speeds in relation to engine speeds.

Feeler blade A thin strip or blade of hardened steel, ground to an exact thickness, used to check or measure clearances between parts.

Firewall The insulated partition between the engine and the passenger compartment. Also called the bulkhead

Firing order The order in which the engine cylinders fire, or deliver their power strokes, beginning with the number one cylinder.

Flywheel A heavy spinning wheel in which energy is absorbed and stored by means of momentum. On cars, the flywheel is attached to the crankshaft to smooth out firing impulses.

Free play The amount of travel before any action takes place. The "looseness" in a linkage, or an assembly of parts, between the initial application of force and actual movement. For example, the distance the brake pedal moves before the pistons in the master cylinder are actuated.

Fuse An electrical device which protects a circuit against accidental overload. The typical fuse contains a soft piece of metal which is calibrated to melt at a predetermined current flow (expressed as amps) and break the circuit.

Fusible link A circuit protection device consisting of a conductor surrounded by heat-resistant insulation. The conductor is smaller than the wire it protects, so it acts as the weakest link in the circuit. Unlike a blown fuse, a failed fusible link must frequently be cut from the wire for replacement.

G

Gap The distance the spark must travel in jumping from the centre electrode to the side electrode in a spark plug. Also refers to the spacing between the points in a contact breaker assembly in a conventional pointstype ignition, or to the distance between the reluctor or rotor and the pickup coil in an electronic ignition.

Gasket Any thin, soft material – usually cork, cardboard, asbestos or soft metal – installed between two metal surfaces to ensure a good seal. For instance, the cylinder head gasket seals the joint between the block and the cylinder head.

Gauge An instrument panel display used to monitor engine conditions. A gauge with a movable pointer on a dial or a fixed scale is an analogue gauge. A gauge with a numerical readout is called a digital gauge.

H

Halfshaft A rotating shaft that transmits power from the final drive unit to a drive wheel, usually when referring to a live rear axle.

Handbrake A braking system, independent of the main hydraulic system, that can be used to slow or stop the vehicle if the primary brakes fail, or to hold the vehicle stationary even though the brake pedal isn't depressed. It usually consists of a hand lever that actuates either front or rear brakes mechanically through a series of cables and linkages. Also known as a emergency brake or parking brake.

Harmonic balancer A device designed to reduce torsion or twisting vibration in the crankshaft. May be incorporated in the crankshaft pulley. Also known as a vibration damper.

Hone An abrasive tool for correcting small irregularities or differences in diameter in an engine cylinder, brake cylinder, etc.

Hydraulic tappet A tappet that utilizes hydraulic pressure from the engine's lubrication system to maintain zero clearance (constant contact with both camshaft and valve stem). Automatically adjusts to variation in valve stem length. Hydraulic tappets also reduce valve noise.

I

Ignition timing The moment at which the spark plug fires, usually expressed in the number of crankshaft degrees before the piston reaches the top of its stroke.

Inlet manifold A tube or housing with passages through which flows the air-fuel mixture (carburettor vehicles and vehicles with throttle body injection) or air only (port fuel-injected vehicles) to the port openings in the cylinder head.

J

Jump start Starting the engine of a vehicle with a discharged or weak battery by attaching jump leads from the weak battery to a charged or helper battery.

L

Load Sensing Proportioning Valve (LSPV) A brake hydraulic system control valve that works like a proportioning valve, but also takes into consideration the amount of weight carried by the rear axle.

Locknut A nut used to lock an adjustment nut, or other threaded component, in place. For example, a locknut is employed to keep the adjusting nut on the rocker arm in position.

Lockwasher A form of washer designed to prevent an attaching nut from working loose.

M

MacPherson strut A type of front suspension system devised by Earle MacPherson at Ford of England. In its original form, a simple lateral link with the anti-roll bar creates the lower control arm. A long strut – an integral coil spring and shock absorber – is mounted between the body and the steering knuckle. Many modern so-called MacPherson strut systems use a conventional lower A-arm and don't rely on the anti-roll bar for location.

Multimeter An electrical test instrument with the capability to measure voltage, current and resistance.

N

NOx Oxides of Nitrogen. A common toxic pollutant emitted by petrol and diesel engines at higher temperatures.

O

Ohm The unit of electrical resistance. One volt applied to a resistance of one ohm will produce a current of one amp.

Ohmmeter An instrument for measuring electrical resistance.

O-ring A type of sealing ring made of a special rubber-like material; in use, the O-ring is compressed into a groove to provide the sealing action.

Overhead cam (OHC) engine An engine with the camshaft(s) located on top of the cylinder head(s).

Overhead valve (OHV) engine An engine with the valves located in the cylinder head, but with the camshaft located in the engine block.

Oxygen sensor A device installed in the engine exhaust manifold, which senses the oxygen content in the exhaust and converts this information into an electric current. Also called a Lambda sensor.

P

Phillips head screw A type of screw head having a cross instead of a slot for a corresponding type of screwdriver.

Plastigage A thin strip of plastic thread, available in different sizes, used for measuring clearances. For example, a strip of Plastigage is laid across a bearing journal. The parts are assembled and dismantled; the width of the crushed strip indicates the clearance between journal and bearing.

Propeller shaft The long hollow tube with universal joints at both ends that carries power from the transmission to the differential on front-engined rear wheel drive vehicles. Also called tail shaft.

Proportioning valve A hydraulic control valve which limits the amount of pressure to the rear brakes during emergency stops to prevent wheel lock-up.

R

Rack-and-pinion steering A steering system with a pinion gear on the end of the steering shaft that mates with a rack (think of a geared wheel opened up and laid flat). When the steering wheel is turned, the pinion turns, moving the rack to the left or right. This movement is transmitted through the track rods to the steering arms at the wheels.

Radiator A liquid-to-air heat transfer device designed to reduce the temperature of the coolant in an internal combustion engine cooling system.

Refrigerant Any substance used as a heat transfer agent in an air-conditioning system. R-12, the principle refrigerant for many years, has now been replaced by R-134a, a non-CFC substance that is considered less harmful to the ozone in the upper atmosphere.

Rocker arm A lever arm that rocks on a shaft or pivots on a stud. In an overhead valve engine, the rocker arm converts the upward movement of the pushrod into a downward movement to open a valve.

Rotor In a distributor, the rotating device inside the cap that connects the centre electrode and the outer terminals as it turns, distributing the high voltage from the coil secondary winding to the proper spark plug. Also, that part of an alternator which rotates inside the stator. Also, the rotating assembly of a turbocharger, including the compressor wheel, shaft and turbine wheel.

Runout The amount of wobble (in-and-out movement) of a gear or wheel as it's rotated. The amount a shaft rotates "out-of-true." The out-of-round condition of a rotating part.

S

Sealant A liquid or paste used to prevent leakage at a joint. Sometimes used in conjunction with a gasket.

Sealed beam lamp An older headlight design which integrates the reflector, lens and filaments into a hermetically-sealed one-piece unit. When a filament burns out or the lens cracks, the entire unit is simply replaced.

Serpentine drivebelt A single, long, wide accessory drivebelt commonly used to drive all the accessories, instead of a series of smaller, shorter belts. Serpentine drivebelts are usually tensioned by an automatic tensioner.

Shim Thin spacer, commonly used to adjust the clearance or relative positions between two parts. For example, shims inserted into or under bucket tappets control valve clearances. Clearance is adjusted by changing the thickness of the shim.

Slide hammer A special puller that screws into or hooks onto a component such as a shaft or bearing; a heavy sliding handle on the shaft bottoms against the end of the shaft to knock the component free.

Sprocket A tooth or projection on the periphery of a wheel, shaped to engage with a chain or drivebelt. Commonly used to refer to the sprocket wheel itself.

Starter inhibitor switch On vehicles with an automatic transmission, a switch that prevents starting if the vehicle is not in Neutral or Park.

Strut See MacPherson strut.

T

Tappet A cylindrical component which transmits motion from the cam to the valve stem, either directly or via a pushrod and rocker arm. Also called a cam follower.

Thermostat A heat-controlled valve that regulates the flow of coolant between the cylinder block and the radiator, so maintaining optimum engine operating temperature. A thermostat is also used in some air cleaners in which the temperature is regulated.

Thrust bearing The bearing in the clutch assembly that is moved in to the release levers by clutch pedal action to disengage the clutch. Also referred to as a release bearing or thowout bearing.

Timing belt A toothed belt which drives the camshaft. Serious engine damage may result if it breaks in service.

Timing chain A chain which drives the camshaft.

Toe-in The amount the front wheels are closer together at the front than at the rear. On rear wheel drive vehicles, a slight amount of toe-in is usually specified to keep the front wheels running parallel on the road by offsetting other forces that tend to spread the wheels apart.

Toe-out The amount the front wheels are closer together at the rear than at the front. On front wheel drive vehicles, a slight amount of toe-out is usually specified.

Tools For full information on choosing and using tools, refer to the Haynes Automotive Tools Manual.

Tracer A stripe of a second colour applied to a wire insulator to distinguish that wire from another one with the same colour insulator.

Tune-up A process of accurate and careful adjustments and parts replacement to obtain the best possible engine performance.

Turbocharger A centrifugal device, driven by exhaust gases, that pressurises the intake air.

U

Universal joint or U-joint A double-pivoted connection for transmitting power from a driving to a driven shaft through an angle. A U-joint consists of two Y-shaped yokes and a cross-shaped member called the spider.

V

Valve A device through which the flow of liquid, gas, vacuum, or loose material in bulk may be started, stopped, or regulated by a movable part that opens, shuts, or partially obstructs one or more ports or passageways. A valve is also the movable part of such a device.

Valve clearance The clearance between the valve tip (the end of the valve stem) and the rocker arm or tappet. The valve clearance is measured when the valve is closed.

Vernier caliper A precision measuring instrument that measures inside and outside dimensions. Not quite as accurate as a micrometer, but more convenient.

Viscosity The thickness of a liquid or its resistance to flow.

Volt A unit for expressing electrical "pressure" in a circuit. One volt that will produce a current of one ampere through a resistance of one ohm.

W

Welch plug A disc or cup-shaped metal device inserted in a hole in a casting through which core was removed when the casting was formed. Also known as a freeze plug, core plug or expansion plug.

Welding Various processes used to join metal items by heating the areas to be joined to a molten state and fusing them together. For more information refer to the Haynes Automotive Welding Manual.

Wiring diagram A drawing portraying the components and wires in a vehicle's electrical system, using standardised symbols. For more information refer to the Haynes Automotive Electrical and Electronic Systems Manual.

Index

Conversion Table

METRIC UNIT	CONV FACTOR	IMPERIAL UNIT	CONV FACTOR	METRIC UNIT
TORQUE				
newton metre (Nm)	x 8.9	= lb in	x 0.113	= Nm
newton metre (Nm)	x 0.74	= lb ft	x 1.36	= Nm
newton metre (Nm)	x 0.102	= kgm	x 9.81	= Nm
PRESSURE				
kilopascal (kPa)	x 4.0	= in H_2O	x 0.249	= kPa
kilopascal (kPa)	x 0.30	= in Hg	x 3.38	= kPa
kilopascal (kPa)	x 0.145	= psi	x 6.89	= kPa
megapascal (MPa)	x 145	= psi	x 0.00689	= MPa
Bar	x 14.5	= psi	x 0.0689	= Bar
LENGTH				
millimetre (mm)	x 0.03937	= inch	x 25.4	= mm
(1 inch = 25.4 mm exactly)				
centimetre (cm) 10 mm	x 0.3937	= inch	x 2.54	= cm
metre (m) 1000 mm	x 0.3937	= inch	x 2.54	= mm
metre (m)	x 3.28	= foot	x 0.305	= m
kilometre (km) 1000 m	x 0.62	= mile	x 1.61	= km
AREA				
millimetre^2 (mm^2)	x 0.00155	= inch^2	x 645	= mm^2
centimetre^2 (cm^2)	x 0.155	= inch^2	x 6.45	= cm^2
metre^2 (m^2)	x 10.8	= foot^2	x 0.0929	= m^2
metre^2 (m^2)	x 1.2	= yard^2	x 0.836	= m^2
hectare (ha) 10,000 m^2	x 2.47	= acre	x 0.405	= ha
kilometre^2 (km^2)	x 0.39	= mile^2	x 2.59	= km^2
FUEL CONSUMPTION				

$$L/100\ km = \frac{282.48}{mpg} \qquad mpg = \frac{282.48}{L/100km}$$

METRIC UNIT	CONV FACTOR	IMPERIAL UNIT	CONV FACTOR	METRIC UNIT
TEMPERATURE				
degrees celcius (°C)	= (°F – 32) ÷1.8			
degrees fahrenheit (°F)	= (°C x 1.8) + 32			
VOLUME				
centimetre^3 (cm^3)	x 0.061	= inch^3	x 16.4	= cm^3
litre	x 61	= inch^3	x 0.016	= L
millilitre (mL) (1mL = 1 cm^3)	x 0.034	= oz-liq	x 29.6	= mL
litre (L) 1000 mL	x 1.06	= quart	x 0.946	= L
litre (L)	x 0.26	= gallon	x 4.55	= L
metre^3 (m^3) 1000 L	x 1.3	= yard^3	x 0.76	= m^3
MASS				
gram (g)	x 0.035	= ounce	x 28.3	= g
kilogram (kg) 1000 g	x 2.2	= pound	x 0.454	= kg
metric tonne (t) 1000 kg	x 1.1	= ton (short)	x 0.907	= t
FORCE				
newton (n)	x 0.225	= pound	x 4.45	= N
kilonewton (kN)	x 225	= pound	x 0.00445	= kN
POWER				
kilowatt (kW)	x 1.34	= horsepower	x 0.746	= kW
kilowatt (kW)	x 0.948	= Btu/s	x 1.055	= kW
watt (W)	x 0.74	= ft lb/s	x 1.36	= W
FLOW				
cu. cm/min	x 16.39	= cu.in/min	x 0.061	= cu.cm/min
litres/min	x 0.2642	= gallon/min	x 3.785	= litres/min

DECIMAL AND MILLIMETRE EQUIVALENTS

INCHES	DECIMALS	MILLIMETRES	INCHES	DECIMALS	MILLIMETRES	INCHES	DECIMALS	MILLIMETRES
1/64	0.015625	0.397	23/64	0.359375	9.128	45/64	0.703125	17.859
1/32	0.03125	0.794	3/8	0.3750	9.525	23/32	0.71875	18.256
3/64	0.046875	1.191	25/64	0.390625	9.922	47/64	0.734375	18.653
1/16	0.0625	1.588	13/32	0.40625	10.319	3/4	0.7500	19.050
5/64	0.078125	1.984	27/64	0.421875	10.716	49/64	0.765625	19.447
3/32	0.09375	2.381	7/16	0.4375	11.113	25/32	0.78125	19.844
7/64	0.109375	2.778	29/64	0.453125	11.509	51/64	0.796875	20.241
1/8	0.1250	3.175	15/32	0.46875	11.906	13/16	0.8125	20.638
9/64	0.140625	3.572	31/64	0.484375	12.303	53/64	0.828125	21.034
5/32	0.15625	3.969	1/2	0.5000	12.700	27/32	0.84375	21.431
11/64	0.171875	4.366	33/64	0.515625	13.097	55/64	0.859375	21.828
3/16	0.1875	4.763	17/32	0.53125	13.494	7/8	0.8750	22.225
13/64	0.203125	5.159	35/64	0.546875	13.891	57/64	0.890625	22.622
7/32	0.21875	5.556	9/16	0.5625	14.288	29/32	0.90625	23.019
15/64	0.234375	5.953	37/64	0.578125	14.684	59/64	0.921875	23.416
1/4	0.2500	6.350	19/32	0.59375	15.081	15/16	0.9375	23.813
17/64	0.265625	6.747	39/64	0.609375	15.478	61/64	0.953125	24.209
9/32	0.28125	7.144	5/8	0.6250	15.875	31/32	0.96875	24.606
19/64	0.296875	7.541	41/64	0.640625	16.272	63/64	0.984375	25.003
5/16	0.3125	7.938	21/32	0.65625	16.669	1	1.000	25.400
21/64	0.328125	8.334	43/64	0.671875	17.066		1mm = .03937"	
11/32	0.34375	8.731	11/16	0.6875	17.463		.001" = .0254mm	

Service history

Part No. / Date Odometer	ENGINE OIL	ENGINE OIL FILTER	FAN BELTS	AIR FILTER	FUEL FILTER	RADIATOR HOSE	COOLANT	FRONT BRAKES	REAR BRAKES	BRAKE FLUID	WHEEL ALIGNMENT	WHEEL BALANCE													

Service history

Date Odometer	ENGINE OIL	ENGINE OIL FILTER	FAN BELTS	AIR FILTER	FUEL FILTER	RADIATOR HOSE	COOLANT	FRONT BRAKES	REAR BRAKES	BRAKE FLUID	WHEEL ALIGNMENT	WHEEL BALANCE													
Part No.																									

Service history

	ENGINE OIL	ENGINE OIL FILTER	FAN BELTS	AIR FILTER	FUEL FILTER	RADIATOR HOSE	COOLANT	FRONT BRAKES	REAR BRAKES	BRAKE FLUID	WHEEL ALIGNMENT	WHEEL BALANCE													
Part No. Date Odometer																									

Service history

	ENGINE OIL	ENGINE OIL FILTER	FAN BELTS	AIR FILTER	FUEL FILTER	RADIATOR HOSE	COOLANT	FRONT BRAKES	REAR BRAKES	BRAKE FLUID	WHEEL ALIGNMENT	WHEEL BALANCE													
Part No. / **Date Odometer**																									